AF796380

Handbook of Thin-Film Technology

Hartmut Frey · Hamid R. Khan
Editors

Handbook of Thin-Film Technology

Springer

Editors

Prof. Dr. Hartmut Frey
Esslingen, Germany

Prof. Dr. Hamid R. Khan
Schwäbisch Gmünd, Germany

ISBN 978-3-642-05429-7
ISBN 978-3-642-05430-3 (eBook)
DOI 10.1007/978-3-642-05430-3

Library of Congress Control Number: 2015934012

Springer
© Springer-Verlag Berlin Heidelberg 2015
This work is subject to copyright. All rights are reserved, whether the whole or part of the material is concerned, specifically the rights of translation, reprinting, reuse of illustrations, recitation, broadcasting, reproduction on microfilm or in any other way, and storage in data banks. Duplication of this publication or parts thereof is permitted only under the provisions of the German Copyright Law of September 9, 1965, in its current version, and permission for use must always be obtained from Springer. Violations are liable to prosecution under the German Copyright Law.

The use of general descriptive names, registered names, trademarks, etc. in this publication does not imply, even in the absence of a specific statement, that such names are exempt from the relevant protective laws and regulations and therefore free for general use.

Printed on acid-free paper

Springer-Verlag GmbH Berlin Heidelberg is part of Springer Science+Business Media (www.springer.com)

Acknowledgement

Thin film technology has become an engine for innovation in the computer- and networking industry. The alternative energy and the emission-free drives for vehicles, based also on the thin film technology. The Handbook of Thin-Film Technology presents a collection of current knowledge on coating technologies and their applications. Additionally methods for determining the properties of thin films are also covered. We are very grateful to Mrs. Dipl.-Ing. Carmen Frey for preparing and drawing the figures of this book. We would also like to thank the Management and Staff of Springer Publishing Company, for their support and cooperation.

Contents

Applications and Developments of Thin Film Technology

1

H. Frey

Thin films are generally used to improve the surface properties of solids. Transmission, reflection, absorption, hardness, abrasion resistance, corrosion, permeation and electrical behaviour are only some of the properties of a bulk material surface that can be improved by using a thin film. Nanotechnology also is based on thin film technology.

Thin films are used if no low-priced bulk material that corresponds to the required specifications of the material exists.

Examples from optics are: changes of reflection or fitting of the transmission of glass bodies; micro-, nano- and opto-electronics are based on thin film technology.

Thin film technologies are divided into PVD (physical vapour deposition) and CVD (chemical vapour deposition) processes.

PVD processes include:

- High-vacuum evaporation
- Cathodic sputtering
- Ion plating
- Ion implantation
- Ion beam mixing
- Plasma diffusion methods and pulse implantation
- Plasma spraying.

High-vacuum evaporation is used in five different evaporation sources:

- Resistance-heated sources (heated directly or indirectly)
- Electron beam evaporators with water-cooled Cu crucibles or lined crucibles and different deflection angles of the electron beam
- Anodic arc evaporators
- Cathodic arc evaporators
- Induction evaporators.

Figure 1.1 gives an overview of vacuum evaporation methods.

The coating process with most variants is cathodic sputtering (Fig. 1.2).

A special process is ion beam sputtering with charged or neutral particles. Further versions are techniques, with electrons inserted into the charge space.

Closely related to cathodic sputtering are plasma-supported CVD processes. The important difference between cathode sputtering and the CVD method lies in the basic material. The use of CVD processes is limited because the gaseous basic materials are frequently missing and the available materials are often very toxic.

Plasma treatment methods in principle are not part of thin film coating processes. With these processes the surface of the bulk materials determining their properties are changed. Properties in the range near the solid surface can be completely changed by ion implantation.

Vacuum plasma spraying completes the vacuum coating processes. With this technology very fast thick films with a high affinity to O_2, or N_2, such as Ti, Ta, Zr and Cr can be sprayed onto surfaces.

Further progress in thin film technology depends on efficient surface analysis methods, above all in connection with high resolution electron microscopy.

© Springer-Verlag Berlin Heidelberg 2015

H. Frey, H. R. Khan, *Handbook of Thin-Film Technology*, DOI 10.1007/978-3-642-05430-3_1

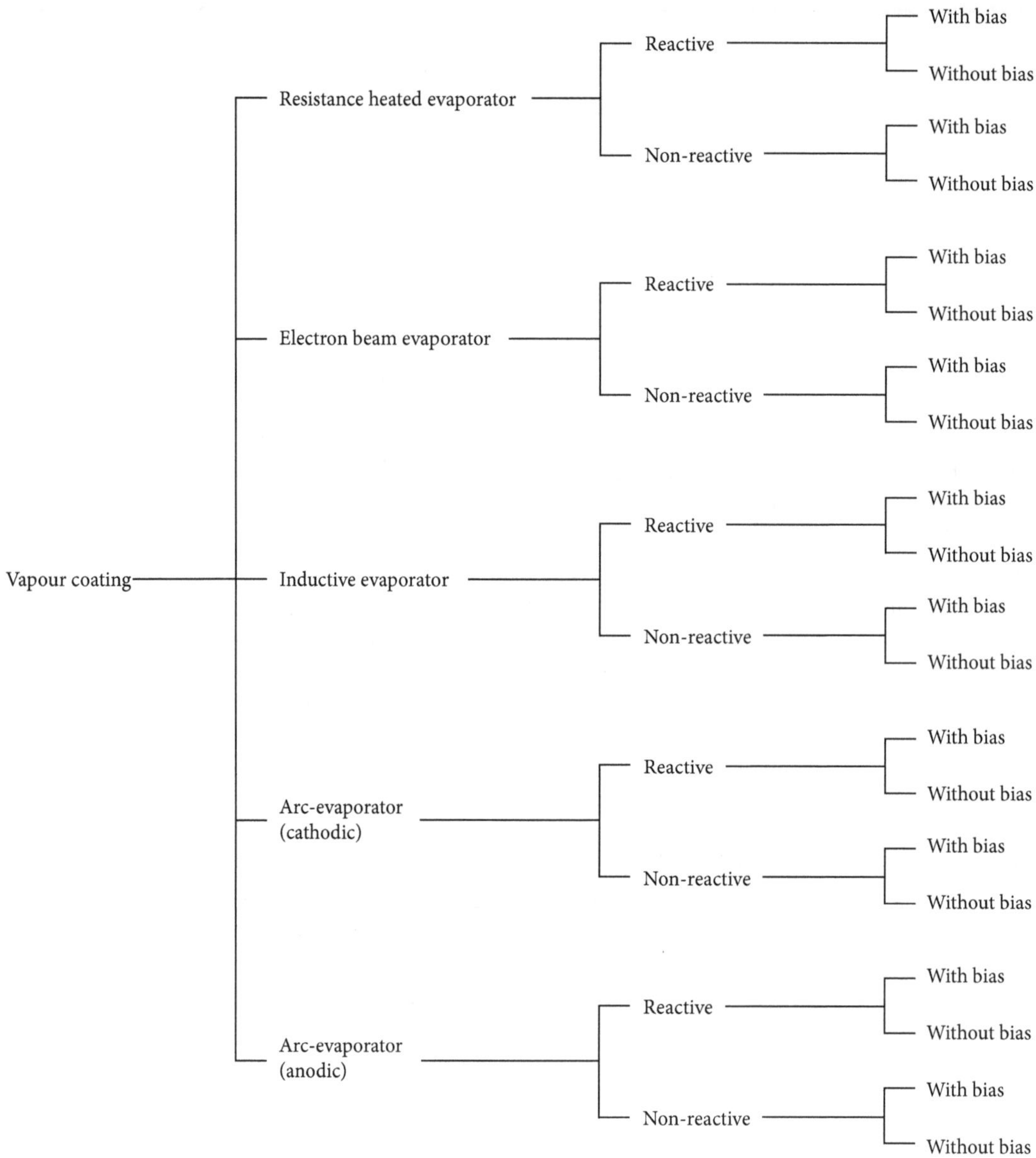

Fig. 1.1 Evaporation in high-vacuum procedure variants

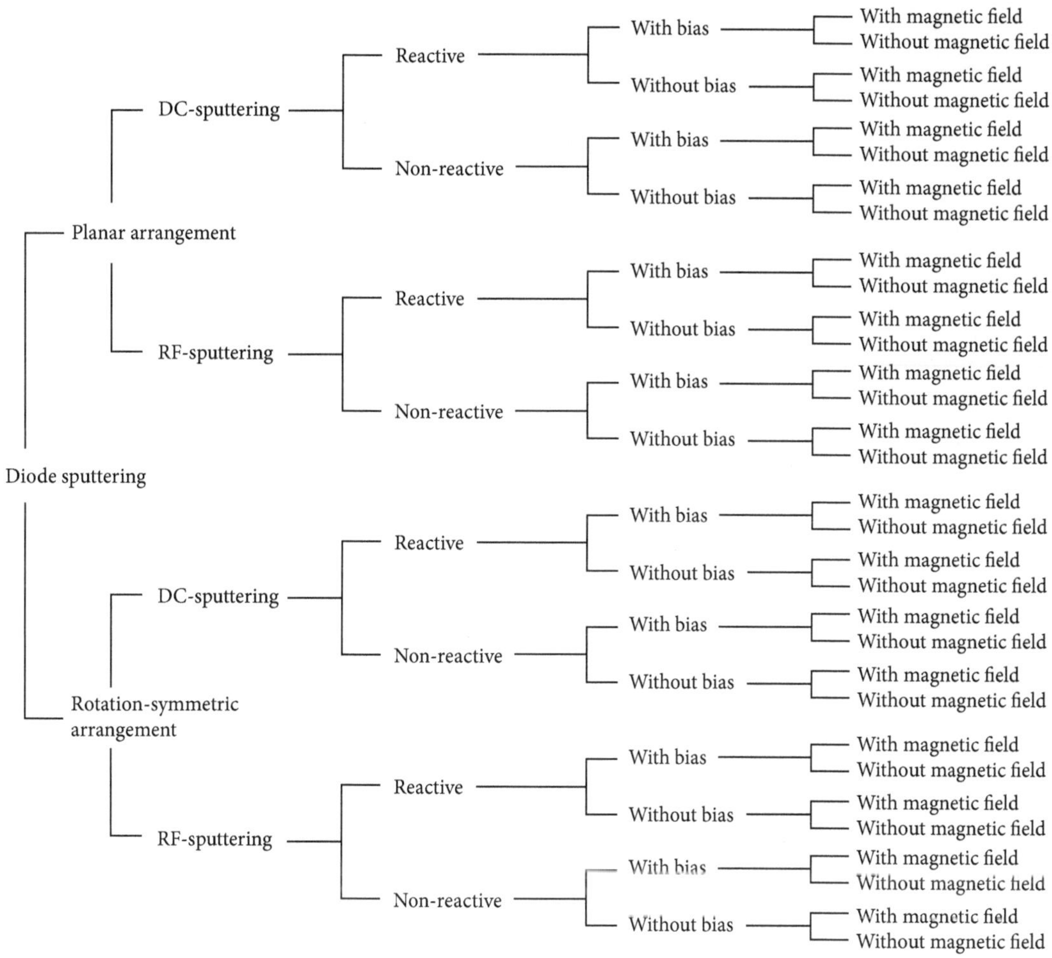

Fig. 1.2 Cathodic sputtering-procedure variants

With magnetic field
Without magnetic field

2 Relevance of the Vacuum Technology for Thin Film Coatings

H. Frey

2.1 Introduction

Vacuum coating processes are characterized by a number of advantages. These include variability of the coating materials, reproducibility of the film properties, and adjustment of the film properties by changing the coating parameters, and the great purity of the coatings. Despite the influence of residual gases in the recipient, and that of the coating material and the condensation rate, the entrapping residual gas molecules into the film can be kept arbitrarily small, if only the residual gas pressure in the recipient is kept accordingly low. Therefore, the conception of the vacuum system during the design and technical execution of vacuum coating plants for Physical Vapor Deposition (PVD) procedures is especially important.

There are many important questions that need to be asked. What is the total pressure necessary, which residual gases are particularly deleterious, and how high can the partial pressure be? Moreover, what is the necessary effective pumping speed of the high vacuum pump in order to achieve within a given time a certain pressure with a still acceptable leakage rate? Further questions are as follows. How large is the gas production during the coating process, how exact must the pressure measurement be, is it sufficient to measure the total pressure, or is it necessary to measure the partial pressure, and which characteristic features of an operating high-vacuum pump are important for a certain coating process? These are only some questions that need to be answered during the dimensioning of a high-vacuum pump system and the selection of the vacuum equipment [1].

2.2 Influence of Residual Gas on Film Quality

Gaseous impurities can occur on the way from the evaporation source to the substrate surface via collisions of vapour particles with the residual gas particles and when residual gas particles strike the substrate surface. The frequency of collisions increase in volume with the number of residual gas molecules and, therefore, with the pressure in the vacuum chamber.

Since the pressure is inversely proportional to the length of the middle free path, we can say that a vapour particle experiences fewer collisions with residual gas particles, the larger the mean free path is. The mean free path is an important feature of vacuum technology. The mean free path describes the distance that a residual gas atom or molecule on average flies without collision. The mean free path is dependent on pressure, gas species and temperature. For air at a temperature of 20 °C the mean free path is with good approximation:

$$\bar{\lambda} = \frac{6.65 \times 10^{-3}}{p} \quad [\text{cm}] \,. \qquad (2.1)$$

$\bar{\lambda}$ Mean free path [cm],
p Residual gas pressure [mbar];

© Springer-Verlag Berlin Heidelberg 2015

H. Frey, H. R. Khan, *Handbook of Thin-Film Technology*, DOI 10.1007/978-3-642-05430-3_2

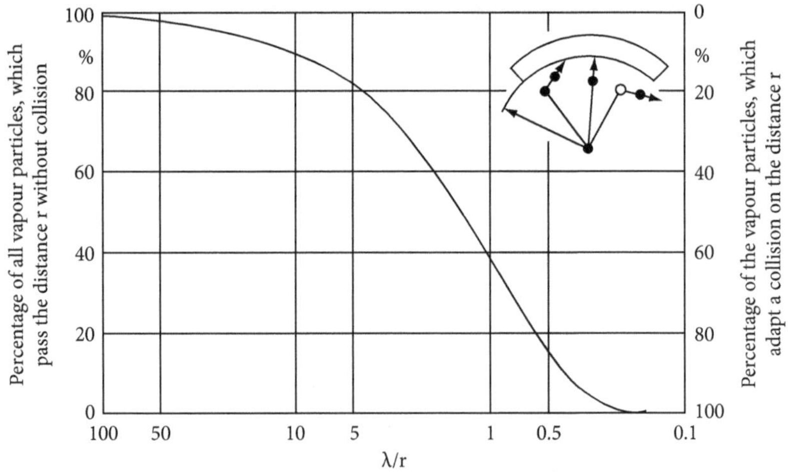

Fig. 2.1 Percentage of vapour particles from an evaporation source colliding with residual gas molecules, as a function of λ/r

at 6.65×10^{-3} mbar the mean free path is about 1 cm, at 6.65×10^{-5} mbar 1 m. For electrons the mean free path is around a factor of $4\sqrt{2}$ greater than for atoms.

Figure 2.1 shows the percentage of evaporated vapour particles colliding with residual gas molecules. If the mean free path is approximately as large as the distance between substrate and evaporation source, then ca. 60 % of the vapour particles sustain a collision with a residual gas particle. Fortunately, a reaction does not take place at each collision, so that the impurities occurring by collisions in the recipient are actually smaller than what would be expected from Fig. 2.1.

Inert gas particles do not react with other particles. Oxygen and nitrogen, in contrast, react with the evaporated particles. The reaction rate increases if the gas particles are ionized. According to the uncertainties in Fig. 2.1, it is insignificant if the diameter of the vapour molecules, the temperature and the real efficiency of the evaporation source are not known exactly.

With reactive evaporation high collision rates and a high reaction capability of the reaction gas and/or gas mixture are desired. The relationship λ/r should be < 0.5.

With reactive coating processes the partial vapour pressure and the reaction gas supply are adjusted exactly. The relationship of the two reaction partners can only be determined experimentally, whereby, however, in order to obtain a high condensation rate the partial vapour pressure and therewith the reactive partial pressure must not be arbitrarily high. Otherwise it comes to numerous collisions between high-energy vapour particles and low-energy gas particles. The energy of the particles that condense on the substrate surface is smaller than that of the particles leaving the evaporation source. The results are smooth films with little adhesiveness, whose properties deviate considerably from the bulk material. Therefore, compound generation by reactive evaporating should take place on the substrate surface. The rule generally is: the mean free path of the gas particles corresponds to the distance of the source substrate. At a distance of 30 cm the partial gas pressure should not be larger than 2×10^{-4} mbar. The mean free path is a statistical size.

The impurities from the collision of residual gas molecules on the substrate surface can be estimated as follows:

- For each time unit and square unit on the substrate the striking residual gas molecules can be calculated with the help of following formula:

$$N_R = \frac{n\bar{c}}{4} = 2.63 \times 10^{22} \frac{p_R}{\sqrt{M_r T_R}} . \quad (2.2)$$

In Eq. (2.2) we have

N_R the number the per second and for each square centimetre on the substrate impacted residual gas molecules in $s^{-1}cm^{-2}$,

p_R the residual gas pressure in mbar,

M_R the relative molecule mass of the residual gas molecules, and

T_R the absolute temperature.

The number of N_V ($s^{-1}cm^{-2}$) applied for each surface unit of the area and time units on the vapour particle impacting on the substrate surface is

$$N_V = 2.63 \times 10^{22} \frac{p_V}{\sqrt{M_V T_V}} ; \tag{2.3}$$

where we have

p_V the saturation vapour pressure of the evaporating material in mbar,

M_V the relative molecule mass of the evaporating material, and

T_V the evaporation temperature in K.

The relationship N_V/N_R

$$\frac{N_V}{N_R} = \frac{p_V \sqrt{M_R T_R}}{p_R \sqrt{M_V T_V}} \tag{2.4}$$

is a value for the impurities resulting from the condensation of residual gases. For aluminium with a relative molecule mass of $M_V = 27$, an evaporation temperature of $T_V = 1440\,K$ and an appropriate saturation vapour pressure of $p_V = 10^{-2}$ mbar, a residual gas pressure of $p_R = 10^{-4}$ mbar and a relative molecule mass of the residual gas of $M_R = 29$, Eq. (2.4), is $N_V/N_R = 47$. In the most unfavourable case, if each residual gas particle striking the substrate surface is adsorbed and reacted with a residual gas particle, a residual gas particle is allotted to 47 aluminium atoms.

With a reduction in the residual gas pressure or an increase in evaporation temperature, drastic improvements of the purity are possible. Since the evaporation temperature is limited, a reduction of the residual gas pressure often remains the only possibility. The probability that a residual gas molecule is received with each impact connection is clearly under 1.

In principle these estimations also apply to for operation of cathodic sputtering equipment. The following must be considered:

a) Conventional sputtering uses total pressures of up to maximum 2×10^{-2} mbar, which is too high for the operation of some kinds of vacuum pumps. The necessary flow control valve decreases the effective pumping speed at the recipient port and equally increases the residual gas partial pressure.
b) At the beginning of the sputtering process there is increased gas pressure for a limited time due to collisions of high-energy or ionized particles, which is particularly important with throttled pumping speed.
c) When noble gas is fed into a sputtering plant a certain portion of other gases always flows into the recipient. When designing a vacuum pump system and with the requirement of a certain degree of purity of the noble gas it should be considered that it has little propose to choose an extremely high degree of purity and to permit a relatively high residual gas and partial pressure at the same time under a too high leakage gas flow or a too small pumping speed.

A noble gas with 0.01 % impurities would lead to a sputtering pressure of 10^{-2} mbar, without consideration of the possible gettering of residual gas at a partial pressure of 10^{-6} mbar. This is irrespective of the pumping speed of the installed vacuum pump system. In the same order of magnitude, also the partial pressure may be appropriate for that at residual gases, due to the equation

$$p = \frac{Q}{S_{eff}} , \tag{2.5}$$

where Q is the accumulation of gas by desorption and leakage and S_{eff} the effective pumping speed at the recipient port.

2.3 Generation of Vacuum

The pressure range in which vacuum coating plants operate extends from approximately 10^{-2} mbar into the ultra-high vacuum range. The initial pressures before the coating process begins are generated, in principle, by pump combinations, since there is no single pump that is able to pump down a recipient between the atmospheric pressure and the ultra-high vacuum. To maintain an operating pressure a pump is sufficient when the pumping speed is based on sorption or condensation (ion getter pumps, cryogenic pumping). The operating mode pressure range and characteristic specifications of vacuum pumps are described in detail in books on vacuum engineering [2, 3].

For many years, the pump combination – oil diffusion pump–oil-sealed rotating-vane pump – was the only possibility to create high vacuum on industrial scale. This combination is still often used today for vacuum coating plants, in particular for evaporating plants.

Advantages of this pumping combination are a constant pumping speed below approximately 10^{-3} mbar and a low dependence of the pumping speed on the type of gas. Pumping speeds from some 10 to 100,000 l/s are available. The unfavourable backstreaming of oil steam in the recipient, which leads among other things to impurities and causes degradation of the adhesive strength of thin films, can be reduced by different methods.

Relatively small backstreaming of oil steam and decomposition products from oil diffusion pumps is reached with high-quality pump fluids at ambient temperature pressures of about 10^{-10} mbar. A cooled top baffle nozzle decreases pump backstreaming by an order of magnitude. The additional installation of a baffle with water cooling, refrigerators or LN_2, results in a further reduction of pump backstreaming.

At pre-evacuation by an oil-sealed rotating-vane pump low-grade oil can get into the diffusion pump and afterwards into the recipient. The quantity of oil diverted in the high vacuum range depends on the contra gas flow and it is the larger, when the residual gas pressure in the volume between diffusion pump and the oil-sealed rotary pump is lower. A pre-evacuation of the recipient over a bypass should, therefore, also not last longer than absolutely necessary.

By gas flushing [4] or installation of a sorption trap oil backstreaming can be decreased drastically, without the pumping speed of the pump used for the pre-evacuation being throttled considerably.

Independently of the type of pre-vacuum pump, because of the extreme loads of the oil in the gasket, oil free pre-vacuum pumps are specially important, e.g. in the semiconductor industry. The oil must possess not only the usual properties like low vapour pressure, small temperature dependence of the viscosity, tight fraction and good lubrication properties, but it also must be stable at high temperatures, inert against chemical attack, e.g. by means of acids, bases and halogens, inert against strong oxidizing agents (oxygen, fluorine), safe when pumping gases with high O_2-concentrations. In addition to lubrication and sealing, the oil also performs the function of filling dead volume, thus increasing the compression ratio. According to on the requirements, different oils are used.

Maintenance of oil-sealed vacuum pumps requires safety arrangements. Cleaning of contaminated oiled parts, removal of dangerous gases and dangerous sludge is not without risk for the maintenance staff. Oil-free pre-vacuum pumps are suitable as pre-pumps for application in the semiconductor industry. They are combined mainly with turbo-molecular pumps. Oil-free operation of vacuum pumps consists of several single pumps with different rotor profiles, arranged one behind the other. On the high pressure side several pairs of rotors are adjusted in series on a common shaft.

The principle of the pump is shown in Fig. 2.2. The two rotors contained in each stage turn, moving in opposite directions in the volumetric displacement. They periodically open and close the inlet and exhaust ports. The rotors separate the pump chamber. On one side of the rotors the gas is sucked in, on the other it is compressed.

In Fig. 2.2a, the suction and compression cycles begin. The final space above the rotors is reduced, that gas is compressed. At the same time, the right rotor begins to open the inlet port, gas is sucked in. In Fig. 2.2b, the left rotor begins to open the exhaust port, and the compressed gas is transported out. In Fig. 2.2c, the compression and sucking procedures are terminated. Inlet and exhaust ports are closed. After the passage of the rotors to the neutral position, both procedures begin again. Generally, oil-free fore-pumps consist

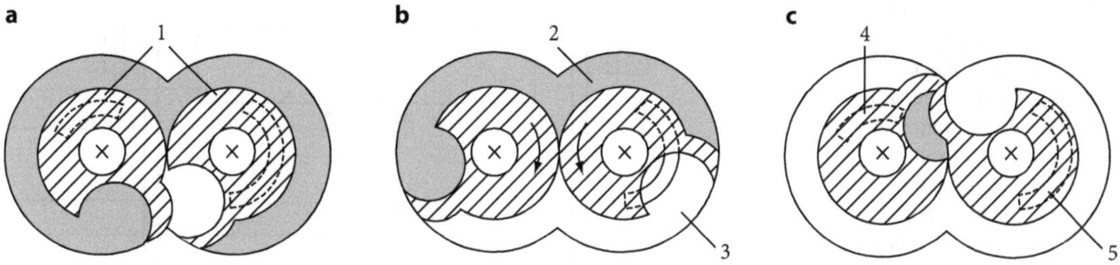

Fig. 2.2 Schematic representation of the pump principle of a claw pump, *1* rotors, *2* compression chamber, *3* suction chamber, *4* exhaust port, *5* inlet port

of four units (claw pumps) that are arranged one behind the other.

A typical pumping speed curve as a function of the intake pressure is shown in Fig. 2.3 [5, 6].

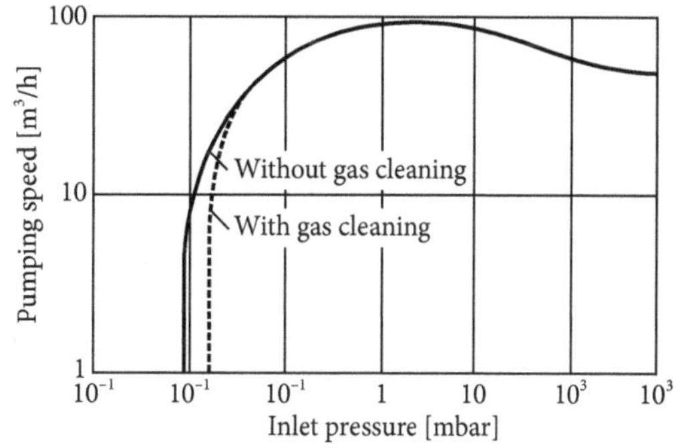

Fig. 2.3 Pumping speed of a four-level claw pump as a function of the intake pressure, ______ without gas flushing, - - - - - - with gas flushing

With the vacuum pump combination – turbo-molecular, oil-sealed rotary pump – there is not danger of contamination of the recipient with professional operation. With a high compression ratio for heavy hydrocarbon molecules the partial pressures keeps the hydrocarbons in the recipient without the use of a baffle under the detection limit. Turbo-molecular pumps have small compression and a smaller pumping speed for lighter gases than for heavier ones. While the pumping speed for hydrogen is only lower than 10 % for air, the compression capacity is distinguished around nearly six orders of magnitude. Due to the dissociation of water vapour, in vacuum coating equipment a considerable portion of hydrogen develops and the pre-vacuum pump must be large enough to hold it.

The advantages of turbo-molecular pumps are especially suitable for cathodic sputtering equipment. The full use of the pumping speed is a significant process-technical advantage of turbo-molecular pumps, because the residual gas partial pressure is substantially smaller than with other equally large pumps with an upstream throttle valve. The correct dimensioning of pre-pumps is particularly important by higher intake pressure.

The importance of the methods that will in the long run take cryogenic pumping into vacuum coating technology is at present not yet foreseeable. High pumping speed properties and thus fast evacuation times and/or low operating pressures are sufficient reasons for the application of evaporating plants operated with low residual gas pressures. A clean vacuum with only light gases in the residual gas atmosphere and with high pumping speeds for hydrogen are a further advantage of cryogenic pumping.

For working processes of cathode sputtering plants cryogenic pumping is being increasingly used [7–10]. The refrigerator principle is mainly used for cryogenic pumping. In the range between 10^{-3} and 10^{-2} mbar without interruption, cryopumps can operate for many hours or days before regeneration is necessary.

With two parallel arranged cryogenic pumps an arbitrarily long operating time of cathode sputtering equipments is possible. Modern cryogenic pumps for sputtering equipment are equipped with a cooled baffle (Fig. 2.4). The angles of inclination of optically opaque baffle surfaces are adjusted in such a way that the pumping speed for the discharge gas, normally argon, is usually only throttled at the absolutely necessary degree.

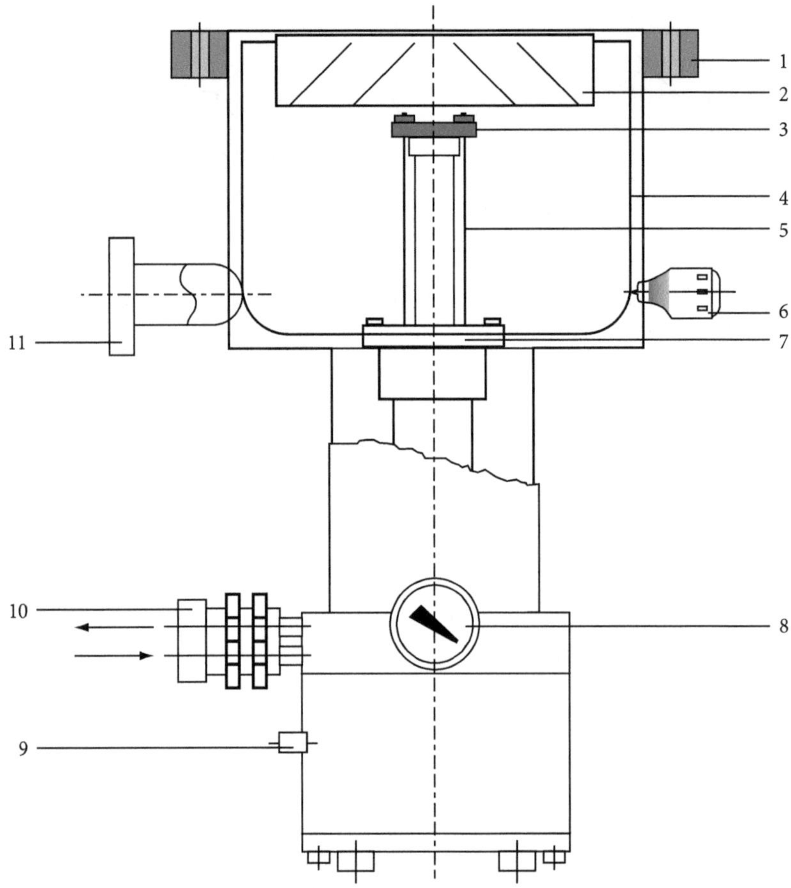

Fig. 2.4 Schema of a refrigerator cryopump. *1* high-vacuum flange, *2* baffle, *3* cold stage, *4* radiation shield, *5* pumping surface, *6* safety valve, *7* cold stage, *8* vapour pressure manometers, *9* electrical supply, *10* gas supply, *11* pre-vacuum flange

The high pumping speed for water vapour remains constant with an arbitrary position of the throttle openings.

2.4 Vacuum Measurement

Currently, the range of thin film technology extends to about 19 orders of magnitude of pressure below atmospheric pressure. Consequently, vacuum measuring techniques have had to be developed to measure low pressures of widely differing magnitudes, from a few mbar to about 10^{-16} mbar. There is no single gauge that is able to cope with such a range, although it is the ideal of scientists and engineers to develop such a gauge. For pressure measurement in partial adjustment ranges, a row of different physical principles exist for vacuum measuring with different characteristics [11].

For the use in vacuum coating equipment essentially four types of gauge are of interest:

a) Thermal conductivity gauge
b) Friction vacuum gauge
c) Ionization vacuum gauge with independent discharge (Penning vacuum gauge)
d) Ionization vacuum gauge with dependent discharge (with hot cathodes).

2.4.1 Thermal Conductivity Gauge

Thermal conductivity gauges are based on a filament mounted in a glass of a metal envelope attached to the vacuum system; the filament being heated by the passage of an electric current. Attainment of the temperature of the filament depends on the rate of supply of electrical energy, heat loss by conductivity through the surrounding gas, heat loss due to radiation (and convection), and heat loss through the support leads to the filament.

The measurable range of thermal conductivity vacuum gauges for technical operation extends from 10^3 to 10^{-3} mbar. Measurement with thermal conductivity gauges is gas species-dependent, and the accuracy of measurement is different in the partial adjustment ranges. In vacuum coating equipment thermal conductivity vacuum gauges serve for pressure monitoring during the pre-evacuation.

2.4.2 Friction Vacuum Gauge

The measuring element of a friction vacuum gauge is a small steel ball, which rotates in a magnetic field. The pressure is determined from the pressure-dependent deceleration of the steel ball. Since the deceleration can be determined physically, an additional calibration of commercial devices is unnecessary.

In the range between 10^{-2} and 10^{-7} mbar there exists a linear dependency in the range between 10^{-2} and 1 mbar, where a correction factor must be used. The unlimited life cycle, good long-term stability and high measuring accuracy within the pressure range, suggest a broad use of friction vacuum gauges in the future.

2.4.3 Cold Cathode Ionization Vacuum Gauge

With many coating processes that are operated in high vacuum, a very exact pressure measurement is unnecessary. In order to reach a continued high quality of the coated films, it is often sufficient to know whether the residual gas pressure before the start of the coating process lies below an experimentally determined limit pressure. Typical examples are plastic foil coating, vapourization of plastic formed components and production of metallic mirrors. For such applications, cold cathode ionization vacuum gauges are suitable.

Measuring errors are attributed mainly to the pressure-dependent discharge current generated partially by secondary electrons, which are released by gas ions striking the cathode. The number of excitation and ionization processes releasing secondary electrons not only depend on the pressure, it also depends on the surface property of the cathode. Contamination, therefore, leads to incorrect measurements. The measuring range of commercial devices extends from 10^{-2} to 10^{-7} mbar, which is possible with specifically created gauges.

2.4.4 Hot-Cathode Ionization Gauge

With some coating processes the exactness of the measurements of cold cathode ionization vacuum gauges is not sufficient to achieve an acceptable reproducibility of the film characteristics. Ionization vacuum gauges with hot cathodes can then be used. Hot cathode ionization gauges use the thermionic emission of a cathode, the emitted electrons being accelerated by the electrostatic field through a grid set at a positive potential relative to the cathode. With exact pressure measurements the physical characteristics of the process gas must be considered.

2.4.5 Total Pressure Measurement in Coating Processes

2.4.5.1 Conventional Cathode Sputtering

Conventional sputtering equipment works with total pressures in the order of 10^{-2} mbar, thus in the limits of the measuring range of cold cathode ionization vacuum gauges and thermal conduction vacuum gauges. Already small pressure fluctuations lead to changes of the sputtering rates and thus to deviations from the desired film thickness. Generally, therefore, actual ionization vacuum gauges with hot cathodes are used with such applications. Friction vacuum gauges or gas-sort-independent measuring diaphragm vacuum gauges with capacitive pressure sensor are also assigned.

2.4.5.2 Reactive Cathode Sputtering

The feed of reaction gas to the noble gas that causes the sputtering effect is very small. With the reactive sputtering process of oxide coatings a too-high feed of oxygen leads to oxidation of the target surface. This, in turn, leads to a drastic decrease of the sputtering and/or condensation rate, film thickness deviations during DC voltage sputtering and changed film properties. With too-low oxygen partial pressure, however, films are part-oxidized and deviate from films with fully-oxidized characteristics. Deviations are also occur if the oxygen partial pressure remains constant and the noble gas partial pressure fluctuates. The gas volume must be kept exactly constant during the coating procedure. This can be achieved production plants by gas flow regulation. Additional precise pressure measurement exercises only control the functioning.

2.4.5.3 Conventional Evaporation

If the residual gas pressure during evaporation in high vacuum is so low that the impurities do not execute a measurable negative influence on the film characteristics, it is completely sufficient not to exceed a given limit pressure during the process cycle. Exact knowledge of the pressure in the recipient is not necessary.

2.4.5.4 Reactive Evaporation

With reactive evaporation, the evaporation speed and/or the condensation rate and the partial pressure of the reactive gas component must be exactly coordinated. Deviations from the standard pressure lead to changes of the film properties. At constant gas flow the pressure is a proportion of the condensation rate and, therefore, it can be used for its regulation.

For precise pressure measurements during reactive coating processes, ionization vacuum gauges with hot cathodes are sufficient. Vacuum gauges with thorium oxide covered cathodes, linearized display and pressure ranges ranging from approximately 1 to 10^{-7} mbar over several decades have a long lifetime and fulfil accuracy requirements.

2.4.6 Partial Pressure Measurement

High purity coatings must be manufactured at a sufficiently low residual gas pressure. Since the effect of the individual residual gas components is very different, we must have a total pressure measurement with the specifications of the still permissible residual gas pressure. This leads to oversizing of vacuum pump systems.

Moreover, a total pressure measurement is not necessary, if the pressure of additional gas feeding in the recipients is to be kept constant, while the partial pressure of the other residual gases vary more or less strongly. This is not always the case with coating equipment due to the volume of gas fluctuating inside the equipment due to heating during evaporation or by bombardment with particles by sputtering. For gas-species-dependent processes, the use of partial pressure gauges for quantitative recording of all gas components is important for the process cycle.

References

1. Einführung in die Hoch- und Ultrahochvakuumerzeugung. (PDF-Datei; 864 kB) Pfeiffer Vacuum, September 2003
2. Wutz M, Adam H, Walcher W (1988) Theorie und Praxis der Vakuumtechnik. Vieweg, Braunschweig, Wiesbaden
3. Wuest W (2002) In: Profos P (ed) Handbuch der industriellen Messtechnik. Oldenbourg
4. Tsutsumi Y, Ueda S, Ikewaga M, Kobyashi J (1990) Prevention of oil vapor backstreaming in vacuum systems by gas purge method. J Vac Sci Tech A 8(3):2764–2767
5. Wycliffe H (1987) Mechanical big vacuum pumps with an oil-free swept volume. J Vac Technol A 5(4):2608–2611
6. Troup AP, Dennis NT (1991) Six years of dry pumping: A review of experience and issues. J Vac Technol A 9(3):2408–2052
7. Kienel G, Frey H (1977) Betriebsverhalten einer Refrigerator-Kryopumpe an einer Aufdampfanlage. Vakuum-Technik 26(2):48–51
8. Baechler WG (1989) Cryopumps for research and industr. Vacuum 37(1/2):21–29
9. Haefer RA (1981) Praktische Ausführung von Kryopumpen. Springer-Verlag, pp 147–217
10. Frey H, Haefer RA, Eder FX (eds) (1981) Tieftemperaturtechnologie. VDI-Verlag, pp 301–310
11. Edelmann C (1998) Vakuumphysik: Grundlagen, Vakuumerzeugung und -messung. Spektrum Akademischer Verlag, Heidelberg

3 Vacuum Evaporation

H. Frey

3.1 Introduction

PVD (physical vacuum deposition) methods are the following: [1]:

- Vacuum evaporation
- Ion plating
- Cathodic sputtering.

These three techniques are also used in reactive processes for coatings with chemical compounds, as well as molecular beam epitaxy [1–5], which is a variant of vacuum evaporation. With ion implantation [6–10], one can change the properties of solid surfaces without coatings. This is not a coating process.

PVD procedures are divided into plasma supported and not plasma supported processes. Vacuum evaporation in high vacuum is not a plasma supported procedure.

The influence of excited or ionized residual gas or vapour particles on the film growth process and, thereby, on the film properties is very favourable for many applications of thin film technology. We therefore, try to produce charged particles also with evaporation.

3.2 Fundamentals

3.2.1 Evaporation Processes

Evaporation in high vacuum uses evaporation sources to heat up the coating material until a sufficiently high vapour pressure is attained, so that a desired evaporation rate is reached. The vapour pressure that develops over a liquid or by sublimation of a solid material, is a function of temperature. If both phases (solid and/or liquid and vapour states) exist side by side in a closed chamber at the same temperature, the equilibrium pressure is called the *vapour pressure* or the *saturation vapour pressure*. In such a state of equilibrium an equal number of atoms of the solid and/or the liquid exchanges into the gaseous phase, as atoms from the gaseous phase condense: the evaporation rate and the condensation rate are identical.

This equilibrium state can realized, e.g. in Knudsen evaporators. With evaporation no ideal conditions are present because the vapour at the lower temperature adjudged substrates, installations and recipient walls condense [11].

The saturation vapour pressure that adjusts over a liquid or a solid as a function of the temperature p_D can be calculated by means of the Clausius Clapeyron equation:

$$\frac{dp_D}{dT} = \frac{\Delta Q_D}{T\left(V_g - V_{fl}\right)}. \qquad (3.1)$$

In Eq. (3.1):

P_D is the saturation vapour pressure,
T the temperature of the evaporation material,
Q_D the heat of vapourization,
V_g the molar volume of vapour,
V_{fl} the molar volume of the evaporating liquid and/or the sublimating solid.

The molar volume in the liquid or solid state is very small in comparison with that of the vapour

© Springer-Verlag Berlin Heidelberg 2015

H. Frey, H. R. Khan, *Handbook of Thin-Film Technology*, DOI 10.1007/978-3-642-05430-3_3

phase. For the saturation vapour pressure at high vacuum evaporation the laws for ideal gases are applicable. This also takes $V_G - V_{fl} \sim V_G - RT/p_D$ from Eq. (3.1),

$$\frac{dp_D}{p_D} = \frac{\Delta Q_D \, dT}{RT^2} \tag{3.2}$$

or

$$\frac{d(\ln p_D)}{d\left(\frac{1}{T}\right)} = \frac{-\Delta Q_D}{R}. \tag{3.3}$$

The reduction of the vapourization heat with increasing temperature is so small in the range interesting for the evaporation process that it can be regarded as constant.

By integration of Eq. (3.3) we obtain

$$\ln p_D = A' - \frac{\Delta Q_D}{RT} \tag{3.4}$$

and

$$p_D = A e^{-\frac{B}{T}}, \tag{3.5}$$

where

A is an integration constant, and

B is a constant depending on the heat of vapourization and, therefore, on the evaporation material.

In Eq. (3.5), which is valued in good approximation to vapour pressures up to about 1 mbar, there is a characteristic of evaporation technology. Between the respective saturation vapour pressure, the condensation rate and the temperature of the material in the evaporation source there exists an exponential correlation, which means that relatively small variations in temperature lead to relatively large changes of the condensation rate. Vapour pressure values of metals as functions of temperature are listed in detailed tables (Fig. 3.4).

To obtain the derivation the relations one proceeds from the equilibrium that the same number of particles leaving the liquid or solid surface return to the surface. Since the number of particles returning to the surface is an explicit function of pressure, temperature and relative molecule mass, the particles can be calculated from gas-kinetic laws:

$$N = \frac{n\overline{c}}{4} = 26.4 \times 10^{21} \times \frac{p_D}{\sqrt{MT}}, \tag{3.6}$$

where

$\overline{c}$ is the average speed of the vapour particles in cm/s,

n the number of particles per cm^3,

p_D the saturation vapour pressure in mbar,

M the relative molecular mass of the evaporating particles, and

T the temperature of the evaporation source in K.

The mass m (in g) of an individual molecule is:

$$m = \frac{M}{L} = 1.66 \times 10^{-24} \times M,$$

$$L = 602.3 \times 10^{21} \text{ molecule/mol}. \tag{3.7}$$

The unit area evaporating quantity of G that evaporates at each time (in $g\,cm^{-2}s^{-1}$) can also be calculated from Eqs. (3.6) and (3.7):

$$G = 0.044 \times p_D \sqrt{\frac{M}{T}}, \tag{3.8}$$

with

p_D in mbar and

T in K.

The evaporation rates as a function of temperature are listed in Fig. 3.1 for some metals. In practice, the margin of option for the evaporation rate is not very large. Slowing evaporation velocity leads to undesirable reactions with residual gases. Undesirable part-oxidic coatings develop, e.g. in connection with oxygen. If we want to avoid impurities or still hold the impurities within permissible limits in the coatings, the residual gas pressure must be kept accordingly low during the coating process, with relatively slow evaporation.

At too fast evaporation, which means that the vapour pressure over the source is too large, vapour particles collide with each other. They do not arrive without collisions at the substrate surface, a part returns to the evaporation source.

Even more serious consequences of too high temperatures of evaporation sources are the spontaneously formed vapour bubbles. The evaporation material is ejected by splashing out of the evaporation source, which arrives partially also the substrate, which leads to film damage.

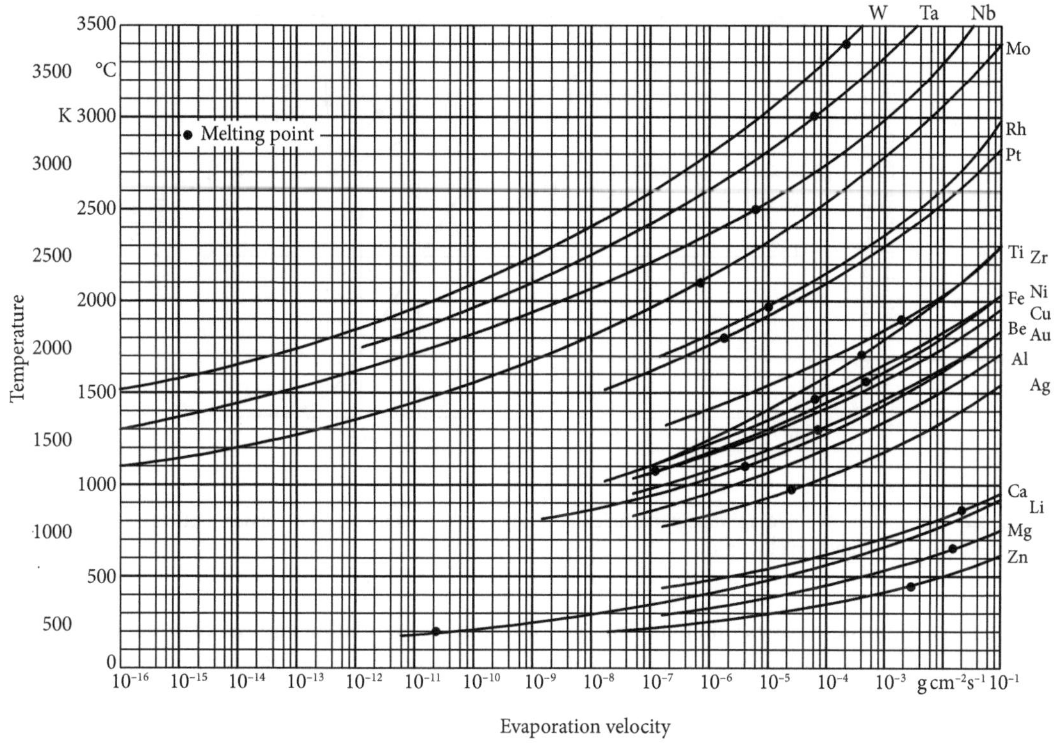

Fig. 3.1 Evaporation rate of metals in the high vacuum range

In practice, vapour pressures are usually in the order of 10^{-2} mbar with evaporation in high vacuum. According to Eq. (3.8) this corresponds, for example, to a relative molecular mass of 100 and a temperature of the evaporation source of 1800 K to an evaporation rate of 10^{-4} (g/(cm^{-}2 s^{-1}).

An increase of the condensation rate is possible with an enlargement of the evaporating surface. Figure 3.1 also shows that the evaporation rates as a function of temperature change strongly. With *zinc* as an example, a material that is relatively simple to vapourize, a rise in temperature of approximately 330 to 400 °C leads to an increase of the evaporation rate from 10^{-4} to 10^{-3} g/(cm^{-2} s^{-1}), i.e. if the temperature in the indicated range changes by only about 7 °C, the evaporation rate changes about 100 %. In order to increase the evaporation rate of titanium from 10^{-4} to 10^{-3} g/(cm^{-2} s^{-1}), the temperature, for example, must be increased from 1560 to approximately 1760 °C.

3.2.2 Transport Phase

The vapour particles emitted from the evaporation source have a mean energy E_D of

$$E_D = \frac{m}{2}v^2 = \frac{3}{2}kT_v = 1.29T_v \quad [\text{eV}], \qquad (3.9)$$

where

m is the mass of a vapour particle in g,
k 8.62×10^{-5} in eV/K,
T_v the temperature of the evaporation source in K, and
v the particle velocity in cm/s.

The energy of the vapour particles depends on the evaporation temperature. The maximum of the Maxwell distribution curve is, for example, at a source temperature of 1500 K at approximately 0.2 eV and at 2000 K source temperature at approximately 0.26 eV, with the assumption that no collisions take place between vapour particles and residual gas particles with lower energy.

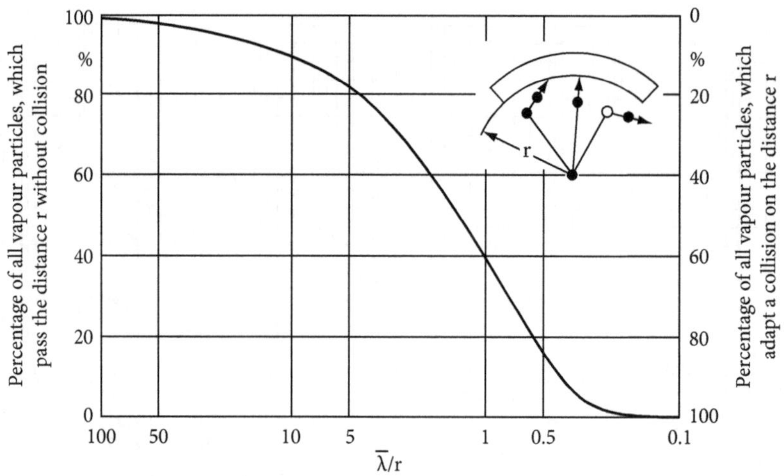

Fig. 3.2 Percentage of outgoing vapour particles at an evaporation source. The vapour particles collide with the remaining gas as a function of λ/r

The two examples show that the energy of vapour particles is relatively low and, therefore, at extremely high evaporation temperatures it also remains low compared to particle energies at sputtering or ion plating. Since for a certain coating material the evaporation temperature may be varied only within close limits, the particle energy is fixed and can be changed only a little.

At higher residual gas pressures the vapour particles transfer energy at each collision to the lower energy residual gas particles, until after sufficient collisions they finally reach energy equilibrium. Then it is valid that

$$E_{\mathrm{D}} = \frac{m}{2}v^2 = \frac{3}{2}kT_{\mathrm{R}}\,, \tag{3.9a}$$

where T_{R} is the temperature of the residual gas and/or that of the recipient wall.

At a recipient temperature of 300 K the vapour particles would only have an energy of less than 0.04 eVafter many collisions with residual gas particles independently of the evaporation temperature. In the transport phase, molecular compounds can develop through collisions between vapour atoms and residual gas, which lessens the purity of the film.

Figure 3.2 shows the percentage of vapour particles emitted from the evaporation source that collide with residual gas molecules, as a function of the relationship between the length of the mean free path and the distance between source and substrate.

3.2.3 Condensation Phase

On collision of a vapour particle with the substrate surface the vapour particle possesses a definitive mobility. The particle moves so long on the surface until it takes a fixed place. Since the binding energy of a vapour atom to the substrate is usually smaller than the cohesion energy of the vapour atoms to each other, a vapour atom diffuses – a sufficient energy is pre-supposed – until it meets another vapour atom and finally forms a nucleus. This nucleus formation at the beginning of the coating process preferentially takes place at defects of the substrate surface and first generates islands. By constant growth, the islands grow into one another, until a continuous film develops.

The growth of the nucleus depends on the coating conditions; it prefers high substrate temperatures, a low melting point of the coating material and a low condensation rate. Of particular importance for the film growth process is the energy of the vapour particles, and with plasma supported processes, the constant bombardment of the surface film by high energy gas molecules or ions.

However, not all molecules that hit the substrate condense. The relationship of the condensing vapour particles to the striking vapour particles is called the *condensation coefficient*. The critical condensation temperature can be shifted to higher temperatures by increasing the vapour beam density. We can also reach a similar effect by applying an intermediate film, the so-called *seed film*. The limit temperatures for progressive condensation are quite different, depending on the material combination.

According to experience, substances with a high boiling point condense better than those with a low boiling point. According to a rule of thumb, for coating materials whose boiling point lies above 1500 °C, one can assume that the condensation coefficient at room temperature is almost 1.

Above the critical condensation temperature at a solid surface, a vapour beam is reflected. One uses this effect, for example, for evaporation from top to bottom, by providing an evaporation source with a heated roof, by which the vapour beams are directed in the desired direction.

3.3 Evaporation of Different Materials

3.3.1 Chemical Elements

With evaporation of a substance heated in high vacuum, the area-specific evaporation rate after Langmuir [12] is

$$\alpha_{v1} = \alpha \times 4.4 \times 10^{-4} \times p_s \sqrt{\frac{M_D}{T}}, \quad (3.10)$$

where

α is the coefficient of evaporation (for ideal evaporation it is $\alpha = 1$),

α_{v1} in $g/(cm^2\,s)$ the surface-related evaporation rate,

p_S in 10^{-2} mbar the saturation vapour pressure at the temperature T,

M_D the mass of the evaporation material, and

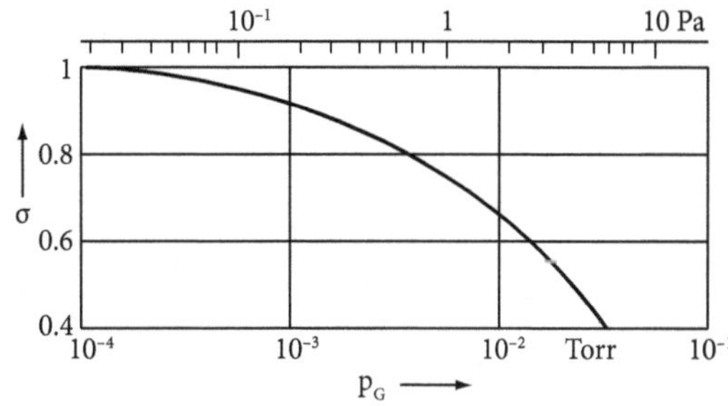

Fig. 3.3 Transmission coefficient τ as function of the gas pressure p_G at the evaporation of copper. Residual gas: air, $p_R < 10^{-4}$ mbar; process gas: argon; vapour pressure over the melt: $p_D = 0.01$–0.03 mbar, $\alpha_{v1} = 3.2 \times 10^{-4}\,g/(cm^2\,s)$ with $p_R < 10^{-4}$ mbar, α_{v1} and α_{v2} determined by weighing by difference

T the absolute temperature of the evaporated material in K.

Equation (3.10) is valid under the condition that an evaporated molecule is not backscattered through gas or a vapour cloud over the evaporator on the vapour-emitted surface. If this assumption is not applied, then the surface-related evaporation rate α_{v2} is reduced by the factor τ:

$$\alpha_{v2} = \alpha_{v1}\tau\,. \quad (3.11)$$

τ is a transmission coefficient and can take values between 0 and 1, depending upon the rate α_{v1} and the gas pressure p_G (Fig. 3.3). A gas pressure of 0.01 mbar already has a substantial influence on the evaporation rate α_{v2}.

To calculate the evaporation rates empirically determined values of the vapour pressure must be used. The saturation vapour pressure depends on the temperature as follows:

$$p_s = K_1 e^{-K_2/T}\,, \quad (3.12)$$

where K_1 and K_2 are material-dependant data from Fig. 3.4. For aluminium, the saturation vapour pressure rises from 0.01 to 0.1 mbar if the temperature increases from approximately 1400 to 1500 K. The saturation vapour pressures for aluminium are reached at temperatures between 500 and 600 K over the melting point. Tungsten has a similar saturation vapour pressure at temperatures between 3600 and 3900 K, i.e. at

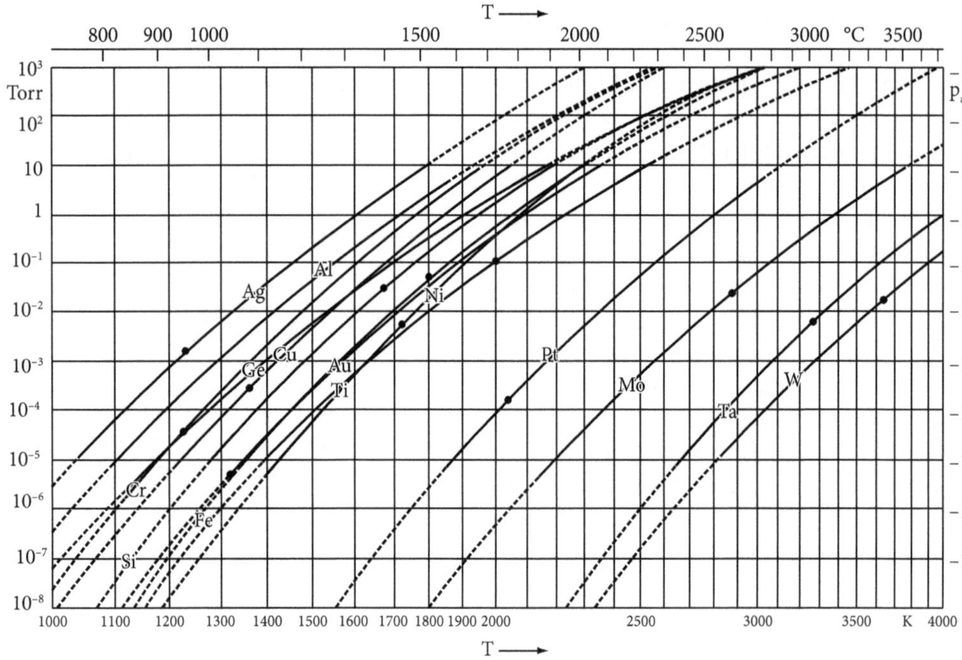

Fig. 3.4 Saturation vapour pressure p_s of some elements in the temperature range between 1000–4000 K

temperatures within the range around the melting point.

The surface-related evaporation rate α_{v1} is with Eq. (3.12)

$$\alpha_{v1} = \alpha \times 4.4 \times 10^{-4} K_1 \sqrt{\frac{M_D}{T}} e^{-K_2/T} . \quad (3.13)$$

Because of the temperature dependence on the saturation vapour pressure, the evaporation rate also grows exponentially with the evaporation temperature (Fig. 3.5). For technical purposes of coating, evaporation rates must be in the range of 10^{-5} to 10^{-2} g/(cm^2 s^1). The curves in Fig. 3.5 show that generally evaporation from the molten phase must take place in order to achieve the required rates. Only in special cases can the evaporation from the solid state, which means via sublimation, take place.

Impurities, like oxides and carbides, whose density is smaller than that of the evaporation material, occur by melting the vapour delivery surface and partly covering it. Thus the evaporation rate is reduced. Such disturbances do not occur if the impurities have high vapour pressure, if the impurities thermally decompose or if the vapour can diffuse through the impurity. The high surface temperature due to direct electron beam heating stimulates the thermal decomposition of disturbing films on the melt. By use of evaporation material of high purity and evaporation from water-cooled crucibles, the influence of impurities on the evaporation rate can be kept small.

3.3.2 Alloys

If alloys are to be manufactured, a constant composition from the components on the entire substrate surface and over the film thickness is necessary. The evaporation of alloy films essentially takes place according to two principles: multi-crucible and single-crucible evaporation (Fig. 3.6).

During multi-crucible evaporation [13–15] the components evaporate separately from several crucibles, to which the number of components corresponds. The components condense together on the substrate surface. During single-crucible

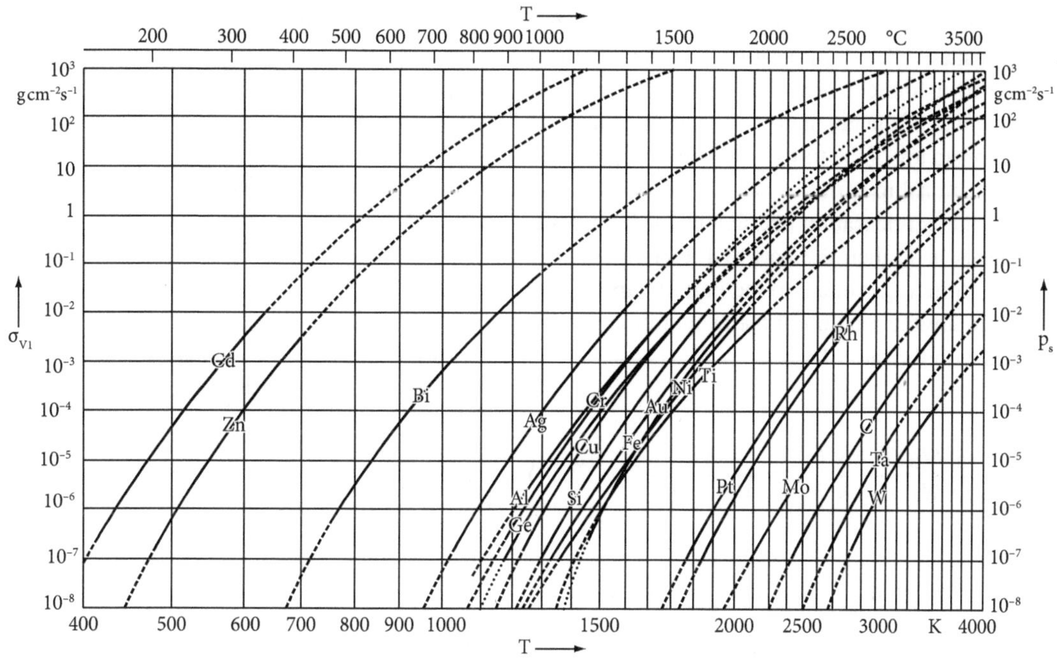

Fig. 3.5 Surface-related evaporation rate α_{v1} of some elements within the range 400–4000 K

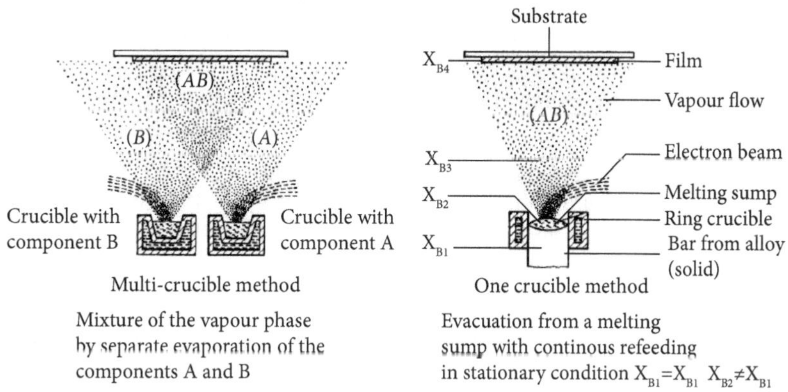

Fig. 3.6 Principles for evaporation in the film of alloys with defined composition (example for a binary alloy A B, X_B... part of component B in percentage by mass)

evaporation a vapour stream with the composition required for the film is produced and condensed on the substrate surface. A variant of single-crucible evaporation comes from a melt that continues after charging the composition with evaporation properties from a required alloy at the film [16].

3.3.2.1 Multi-Crucible Evaporation

Multi-crucible evaporation can be described by the example of the binary alloy AB with the components A and B. Over the two crucibles separate vapour streams develop, with the evaporation rates $\alpha_{vA} = \alpha_{vA}(M_A, T_A)$ and $\alpha_{vB} = \alpha_{vB}(M_B, T_B)$ in accordance with Eqs. (3.10) and/or (3.13).

M_A and M_B are the masses of the components A and B and T_A and T_B the temperatures of the vapour delivery surfaces. If the crucibles are arranged at a distance l from each other, whereby l is small against the distance h_V between the substrate and the crucibles, then an expanded range results, in which the vapour stream contains both components of the alloy.

Due to the directionality of the vapour stream, a constant alloy composition can be obtained only in a limited area of the substrate surface. The effect of the geometry of the arrangement on the constancy of the film composition depends on the relationship l/h_V.

If the condensation coefficient for both components is 1, then the part of the component B in the film x_{B4} is equal to the part in vapour x_{B3}. The following relation of the evaporation parameters with the part of x_{B3} is

$$\frac{\alpha_{vB}(M_B, T_B) \times F_B}{\alpha_{vA}(M_A, T_A) \times F_A} = \frac{X_{B3}}{100 - X_{B3}}. \quad (3.14)$$

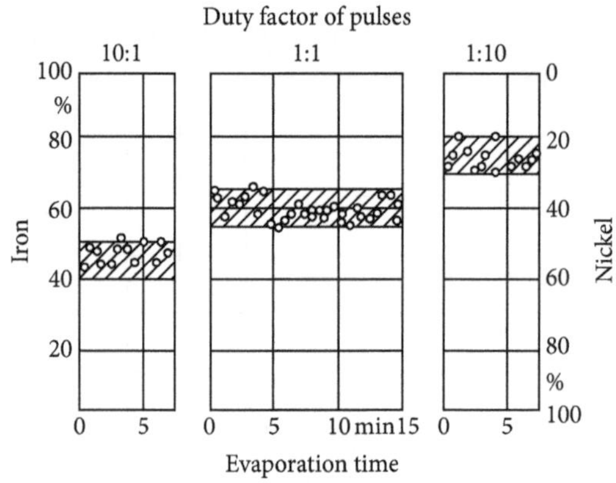

Fig. 3.7 Change of the alloy composition of FeNi films during two-crucible evaporation with a programmed deflected electron beam by variation of the temporal pulse duty factor

The parts of the components are indicated in mass percent and $x_{B3} + x_{A3} = 100$ is valid; F_A and/or F_B are the vapour delivery surfaces. Under the conditions that the surface-related evaporation rates are constant over the evaporator surfaces at which l is $\ll h_V$ and constant for F_B/F_A, a certain portion of the component x_{B4} can be adjusted in the film by the temperatures T_A and T_B in the two crucibles during multi-crucible evaporation.

The high dependence of the evaporation rate on the temperature limits the attainable accuracy of the alloy composition. The fluctuation of a component, e.g. nickel in an iron nickel film, can be determined from Eq. (3.14) with the use of the values from Fig. 3.5 ($T_{Ni} = 1000$ K).

$$\frac{\Delta X_{Ni3}}{X_{Ni3}} \approx 20 \frac{\Delta T_{Ni}}{T_{Ni}}.$$

During a variation in temperature of only 10 K the nickel in the vapour already changes about 10 %. High uniformity of the evaporator temperature as a condition for high constancy of the evaporation rates is the main requirement for generating alloy films with constant composition by multi-crucible evaporation.

Multi-crucible evaporation can take place with a number of electron beam evaporators. Another possibility is to heat several crucibles with an electron beam gun. Electron beam power is divided by programmed deflection on the individual crucibles [17]. The partitioning of the electron beam power to the crucibles takes place via adjustment of a certain temporal pulse control factor for the deflection currents of the beam guide system. In this way, the evaporation rates of the components and thus the alloy composition are adjusted in the film.

Figure 3.7 shows the composition of iron nickel films that were produced by two-crucible evaporation with programmed deflection of an electron beam.

The method of the multi-crucible evaporation to create alloy coatings is used if the evaporation material in the required composition cannot be prepared or is difficult to prepare. In Fe-Ni-NbC coatings produced by multi-crucible evaporation, the NbC part varies from a few tenths to several tens of percent [13]. The mixture in the vapour phase is purposeful, if the vapour pressures of the components differ extremely, i.e. around four or more orders of magnitude. In addition it is used if only very small parts of a component are to be alloyed.

3.3.2.2 Single-Crucible Evaporation

In order to determine the composition of alloy coatings deposited with the method of single-crucible evaporation, the validity of the partial pressure law and the applicability of the Langmuir equation (Eq. (3.9)) with consideration of the thermodynamic properties of the alloy must hold. Thus, the composition in the film from the composition in the melt can be calculated [14–16].

At first, the part of a component in the melt and in the coating is considered. If, e.g. a binary alloy AB evaporated with the mass portion of X_{B2} of the component B in the melt is obtained, then the relation of the mass portion X_{B4} in the film with the portion X_{B2} must be determined. As before, $X_{B4} = X_{B3}$ is to be valid, i.e. the condensation coefficient for both components is 1, and thus the composition of the coating is equal to that of the vapour.

With the evaporation of an alloy, the vapour pressure p_{SAB} over the melt consists of the partial vapour pressures in all, as follows:

$$p_{SAB} = p_{SA^*} + p_{SB^*}\,, \tag{3.15}$$

where p_{SA^*} and p_{SB^*} are the vapour pressures of the components over the melt. For the vapour pressure p_{si^*} of a component and the saturation vapour pressure p_{si}, the following connection is valid:

$$p_{si^*} = \gamma_i N_{i2} \times p_{si}\,, \tag{3.16}$$

where

N_{i2} is the molecular concentration of the component i in the melt,

γ_i the coefficient of activity of this component and

p_{si} the values applied after Eq. (3.12).

The relationship of the evaporation rates over the melt is expressed by:

$$\frac{\alpha_{vB}}{\alpha_{vA}} = \frac{\gamma_B}{\gamma_A}\frac{N_{B2}}{N_{A2}}\frac{p_{SB}}{p_{SA}}\frac{(M_B)^{\frac{1}{2}}}{(M_A)^{\frac{1}{2}}}\,. \tag{3.17}$$

The coefficients of activity γ_A and γ_B demonstrate the interactions between the components of the alloy. The coefficients of activity can be determined from thermodynamic data [18].

For the simple consideration of specific alloys, the magnitudes γ_i can set the relationship γ_B/γ_A equal to 1 [19]. Under consideration of

$$N_{A2} + N_{B2} = 1 \quad \text{and} \quad X_{A2} + X_{B2} = 100\,, \tag{3.18}$$

the relations between the molecular concentrations and the parts in mass percent conform with

$$\frac{N_{B2} \times M_B}{N_{A2} \times M_A} = \frac{X_{B2}}{X_{A2}} \tag{3.19}$$

and

$$\frac{N_{B2}}{N_{A2}} = \frac{X_{B2}}{(100 - X_{B2})} \times \frac{M_A}{M_B}\,. \tag{3.20}$$

Furthermore, the relationship of the evaporation rates is equal to that of the mass fraction in the vapour and/or in the film:

$$\frac{\alpha_{vB}}{\alpha_{vA}} = \frac{X_{B3}}{X_{A3}}\,. \tag{3.21}$$

From Eq. (3.17) the relation between the part of the component B in the film $X_{B4} = X_{B3}$ and the part of the melt X_{B2} results in (3.21)

$$X_{B3} = \frac{100 X_{B2} \times \alpha_{BA}}{100 - X_{2B}(1 - \alpha_{BA})} \tag{3.22}$$

and/or

$$X_{B2} = \frac{100 X_{B3}}{X_{B3} + (100 - X_{B3})\alpha_{BA}}\,. \tag{3.23}$$

In these relations, the evaporation coefficient that describes the evaporation parameters is defined as follows after [20]:

$$\alpha_{BA} = \frac{\gamma_B p_{SB}}{\gamma_A p_{SA}}\left(\frac{M_A}{M_B}\right)^{\frac{1}{2}}\,. \tag{3.24}$$

Equations (3.22) and (3.23) describe the relation between the quantities X_{B3} and X_{B2}. The relation is represented in Fig. 3.8. α_{BA} can be determined with the information of the relationship γ_B/γ_A as a function of the alloy composition and the temperature, or with the simplifying assumption $\gamma_B/\gamma_A \sim 1$ by means of vapour pressure diagrams. Thus, the part of X_{B4} in the film is

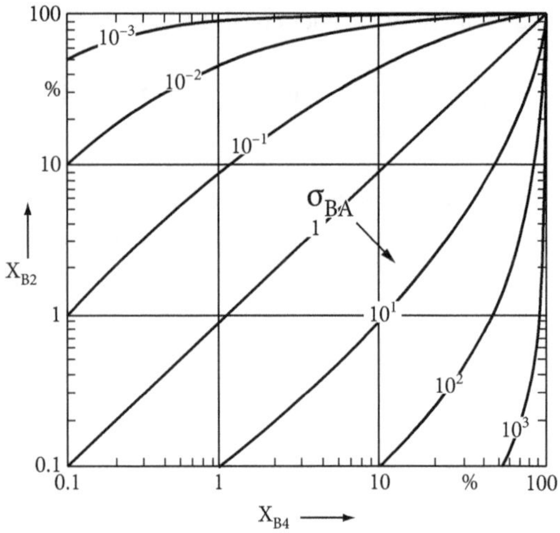

Fig. 3.8 Relation between the part of X_{B4} of an alloy component in the film and the associated portion of X_{B2} in the melt [of 3–20] parameters: evaporator coefficient α_{BA}

determined as function of the part of X_{B2} in the melt. Such a dependence is indicated for FeNi, CrNi and AgCu at assigned temperatures [17].

For $\alpha_{BA} = 1$ the evaporation takes place congruently, i.e. the composition in the film is equal to that in the melt. For $\alpha_{BA} > 1$, the part of component B is larger in the coating than in the melt. This means that in the melt with progressive evaporation component B becomes impoverished. In this case, component B is the more volatile component.

At the evaporation of an alloy from one crucible is $\alpha_{BA} \neq 1$, i.e. the composition of the melt and, therefore, also that of the coating, changes with time. Industrial applications require evaporation with a stationary proportion of components in the coating. If the part of component B in the melt X_{B2} corresponds to the required part in the film X_{B4} in accordance with Eq. (3.23) exactly, then this desired evaporator condition can be kept stationary through reloading of evaporation material into the melt. The conditions for such a stationary evaporation

$$X_{B2} = \text{const}, \qquad X_{B4} = \text{const} \tag{3.25}$$

arise from mass conservation. Additionally, the volume of the melt V_2 must be constant. Thus, it must be valid for each time unit supplied by the mass of the components $^\circ m_{A1}$, $^\circ m_{B1}$ and/or the total mass $^\circ m_1$ is equal to the evaporation rates of the components and/or the total rate:

$$\dot{m}_{A1} = \alpha_{v1A} \times F \tag{3.26}$$

$$\dot{m}_{B1} = \alpha_{v1B} \times F \tag{3.27}$$

and

$$\dot{m}_1 = \dot{m}_{A1} + \dot{m}_{B1} = \alpha_{v1A} F + \alpha_{v1B} F\,. \tag{3.28}$$

The requirement

$$X_{B1} = X_{B4} \tag{3.29}$$

is contained in these conditions, i.e. for the stationary evaporation of an alloy from a crucible the composition of the reloaded material must have exactly the required composition of the alloy in the coating.

For the application of single-crucible evaporation of an alloy with continuous reloading, the transient time is substantial. This is the time necessary until the concentration of component B in the melt, outgoing from the initial concentration of the supplied material X_{B1}, assumes the value X_{B2}, according to Eq. (3.23).

During the transient time the temporal change of the mass of a component in the melt $^\circ m_{B2}$ is equal to the difference between supplied mass at each time unit and evaporated mass at each time unit of this component. Thus,

$$^\circ m_{B2} = {}^\circ m_{B1} - {}^\circ m_{B3} \tag{3.30}$$

applies. With the volume V_2, the density ρ_2 of the vapour delivery surface F and the mass of component B in the melt and in the supplied material results from

$$\frac{1}{100}\frac{\mathrm{d}}{\mathrm{d}t}(V_2 \times \varrho_2 X_{B2}) = \frac{\dot{m}_1 X_{B1}}{100} - \alpha_{VB} \times F\,. \tag{3.31}$$

According to Eqs. (3.9), (3.15) and (3.20) α_{vB} amounts to

$$\alpha_{vB} = 4.4 \times 10^{-4} \times \gamma_B \times p_{SB} \times \left(\frac{M_B}{T}\right)^{\frac{1}{2}} \times \frac{X_{B2}}{M_B}\left(\frac{100 - X_{B2}}{M_A} + \frac{X_{B2}}{M_B}\right)^{-1}. \quad (3.32)$$

The solution of the differential equation (Eq. (3.31)) for the simple case is difficult, since the coefficient of activity is a function of the mass parts, and all parameters are functions of time and additionally partly of temperature. The behaviour is particularly confusing in the heating up phase. For some materials the transient time has been calculated by numeric integration [18].

In closed form Eq. (3.31) was solved on the basis of different simplifications [19]. Such a simplifying assumption exists in the following. During the transient time no reloading takes place, i.e. $^{\circ}m_1 = 0$ applies. The time up to the formation of the melt is negligible in relation to the transient time, and during this time no substantial change of the composition of the melt sump takes place. The change in density during the entire transient time is negligibly small.

With this assumption Eq. (3.31) is given by

$$\frac{\mathrm{d}X_{B2}}{\mathrm{d}t} = -\frac{100\alpha_{vB} \times F}{V_2\varrho_2}. \quad (3.33)$$

Additionally, if $M_A \sim M_B$, then with Eq. (3.32) we obtain

$$\mathrm{d}t = -\frac{V_2\varrho_2\left(\frac{M_B}{M_A}\right)}{4.4 \times 10^{-4}\gamma_B\, p_{SB}\left(\frac{M_B}{T}\right)^{\frac{1}{2}} \times F}\frac{\mathrm{d}X_{B2}}{X_{B2}}. \quad (3.34)$$

During integration of $t = 0$ to $t = t_t$ with $X_{B2}(t = 0) = X_{B1}$ and $X_{B2}(t = t_1) = X_{B2}$ we obtain for the transient time with stationary condition $X_{B1} = X_{b3}$

$$t_t = \tau_t \ln \frac{X_{B3}}{X_{B2}}. \quad (3.35)$$

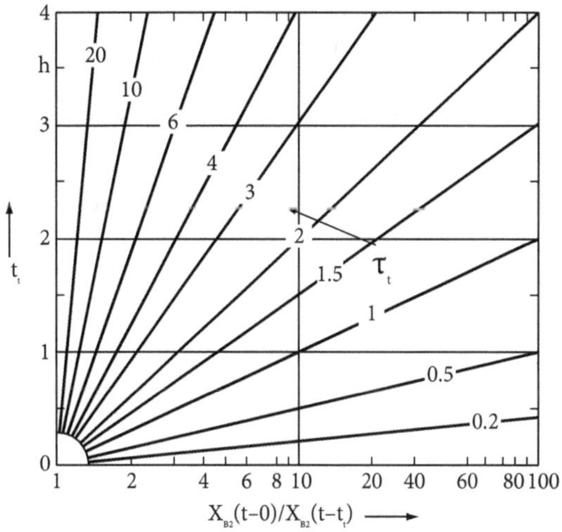

Fig. 3.9 Transient time t_t in dependence on the relationship of the mass component B in the melt at time $t = 0$ and $t = t_t$ (parameters: time constant τ_t in h)

τ_t is the time constant of the transition:

$$\tau_t = \frac{\left(\frac{M_B}{M_A}\right)V_2\varrho_2}{4.4 \times 10^{-4} \times \gamma_B \times p_{SB} \times \left(\frac{M_B}{T}\right)^{\frac{1}{2}} \times F}. \quad (3.36)$$

With consideration of the simplifications we obtain the transient time τ_t (average values of the quantities V_2, ρ_2, F, p_{SB} and γ_B are used for the evaporation temperature T).

Figure 3.9 shows the transient time t_t in dependence on the relationship of the mass component B in the melt at the time $t = 0$ and $t = t_t$.

For the determination of the transient time t_t with the required part of $X_{B4} = X_{B3}$ in the coating, the corresponding concentration in the melt $X_{B2} = X_{B2}$ $(t = t_t)$ is to be calculated according to Eq. (3.23).

Parameters of the evaporator are the melt volume V_2 and the vapour delivery surface F influencing the time constant τ_t and thus the transient time.

For example, during the evaporation of the alloy NiCr, if a composition of 80/20 in the coating is required and if the evaporation takes place at 2000 K, then $\tau_t = 1$ h. For Cr content X_{B2} one obtains a value of 3 % from Eq. (3.23).

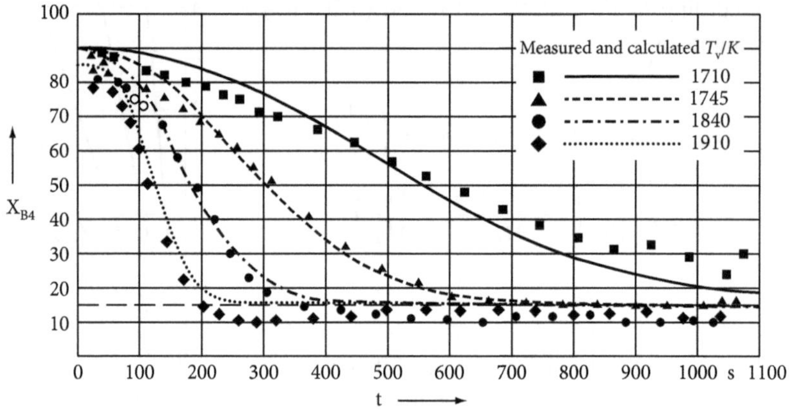

Fig. 3.10 Measured and calculated values of the Ag part in the layer X_{B4} as a function of the time t at the evaporation of an AgCu alloy at different evaporation temperatures T_v until the stationary condition is reached [18]

The crucible is characterized by the relationship $V_2/F = 0.7\,\text{cm}$. With these values a transient time of 48 min occurs, until the stationary condition is reached [19].

The transient time is a non-productive secondary time and must, therefore, be reduced. One obtains a short transition time at a high evaporation temperature with high vapour pressure p_{CB}. With evaporation from crucibles with a large melt volume, i.e. large values of the relationship V_2/F, the transient time is long. In order to obtain a short transient time despite a large relationship V_2/F, a so-called starting insert in the crucible is used. This is a quantity with the volume of the melt sump; its alloy composition corresponds to the melt in the stationary condition. The alloy material composition for the reloading is equal to that of the film to be evaporated; this differs from the composition of the starting insert. On the basis of Eqs. (3.35) and (3.36) also the ability of the so-called flash evaporation [21] for the evaporation of alloys can be seen. By removing volume V_2 of the melt the transient time is reduced; in the limit value $V_2 \rightarrow 0$ the transient time t_t goes to zero. If the material is inserted in small parts in an overheated body, no melt slough is formed. With flash evaporation the concentration of component B of the supplied material X_{B1} is equal to X_{B3} in the vapour, always both at each time, and the easily volatile and less volatile components evaporate completely.

By analysis of the transient time the influence of impurities by evaporation can be estimated. Less volatile impurities, i.e. with small vapour pressure, have a large transient time, which is in most cases large compared to the coating time. During evaporation thereby the part of impurities is smaller in the coating than in the material used. A disadvantage is that the melt with the less volatile impurities is enriched. Thus the evaporation rate is reduced.

During the transient time the alloy composition of the film changes. Figure 3.10 shows calculated and measured curves for the temporal development of the composition during evaporation of an AgCu alloy [15]. One recognizes that the concentration of the more volatile component, silver, is very high (80 to 90 %) at the beginning of the evaporation, decreases with progressive coating time. At the end of the transient time a value of approximately 16 % is reached.

Figure 3.11 shows the increase of the transient time of a silver copper alloy with an increase in the part of X_{B1} of the more volatile component in the supplied material. Further, it follows from this figure that: at low evaporation temperatures the transient time can achieve such large values that for short time intervals, evaporation with practically constant composition of the coating components is possible.

For the AgCu alloy with 29.8 % Ag Ag at an evaporator temperature of 1625 K, within 2 min changes the composition of the film only about approximately 1 %. Depending upon the selected time, interval coatings with a silver part from 85 to 90 % can be evaporated. This behaviour of a melt is used for the production of alloy films by

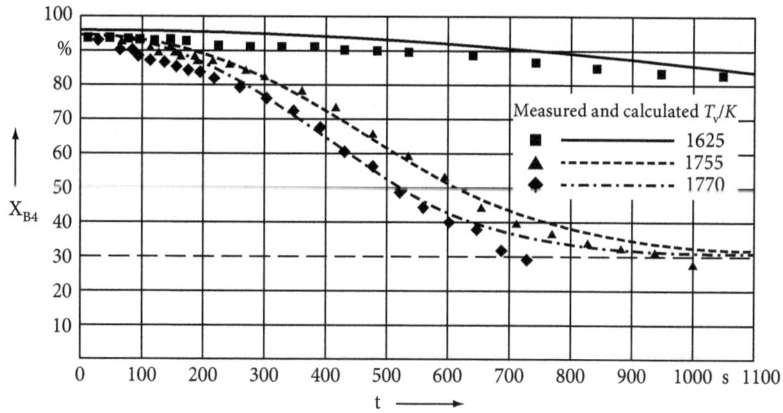

Fig. 3.11 Measured and calculated values of the Ag fraction in the layer X_{B4} as a function of the time t with the evaporation of an AgCu alloy at different evaporation temperatures T_v until the stationary condition composition of the supplied material of 29.8 % Ag is reached [18]

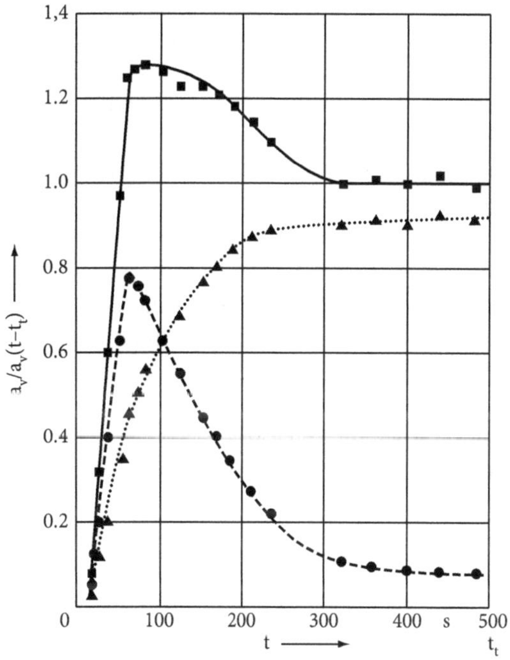

Fig. 3.12 Relative evaporation rate $\alpha_v/\alpha_v(t = t_1)$ of the alloy AgCu (*solid line*) and the components Ag (*dashed line*) and/or Cu (*dotted line*) as a function of the time t_t after [18], final temperature $T_v = 1785$ K, transient time $t_t \sim 500$ s, alloy AgCu with 8.06 % Ag

partial evaporation from a large molten reservoir without reloading. Since the necessary composition of the melt, however, can be maintained only in a limited time interval, with this method of alloy evaporation only a small part of the inserted material can be utilized.

The time before the stationary condition is reached shows the total evaporation rate and the evaporation rates of the components characteristic changes, Fig. 3.12 [18]. As shown at the beginning in the curves, during heating of the evaporation material, the evaporation rates rise steeply. This time is neglected in the preceding explanations for the determination of the transient time. At the beginning of the evaporation process, this very high rate of the easily volatile component in the example Ag, is evident in the high total evaporation rate. The total rate is at first larger than in the stationary case. Up to the end of the transient time the evaporation rate decreases from Ag constantly; the evaporation rate from copper in contrast rises.

Frequently, NiCr is used for the production of resistance coatings and for corrosion protective films, which can be evaporated from a crucible with continuous reloading. The vapour pressures of Ni and Cr differ at an evaporation temperature at $T_v \sim 2000$ K approximately by a factor of 10^2.

The composition of the film after the end of the transient time is shown in Fig. 3.13. The evaporation occurs from a water-cooled ring crucible with reloading in the form of a rod. With this method of alloy evaporation, a constant composition of the coating can be obtained over many hours. The fluctuations of the composition are in the range of the analysis accuracy. Similar results can be reached with other alloys, such as Fe20Cr, other stainless steels of different composition and with Cu10Al [25].

Table 3.1 Comparison of the chemical analysis of evaporation properties and the vapour-deposited layer Ti6Al4V (average values in percent by weight)

Element	Al	V	Fe	C	N	O	H	Other elements	Ti
Evaporation material	6.10	4.0	0.30 max	0.08 max	0.05 max	0.20 max	0.015 max	0.4	Residual
Deposition layer	5.93	3.97	0.144	0.021	0.007	0.12	0.0051	0.4	Residual

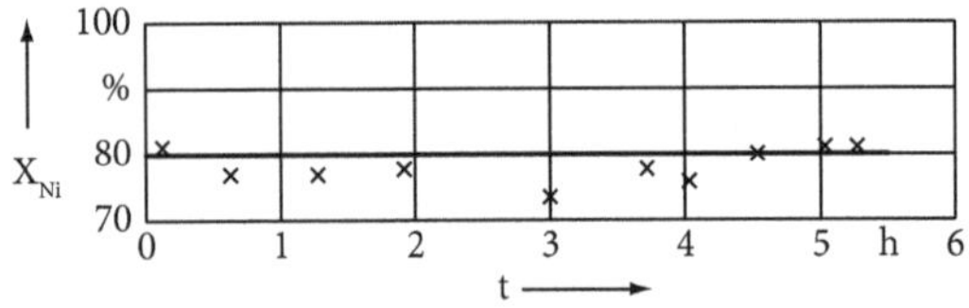

Fig. 3.13 Nickel portion in the layer X_{Ni} with evaporation of NiCr 80/20 from a crucible with continuous charging

Also, for evaporating ternary alloys a constant coating composition can be obtained after the transient time. Foils from Ti6Al4V have been, e.g. manufactured by crucible evaporation with reloading [26].

A certain refining process from impurities is connected with the evaporation. Table 3.1 shows the results of the chemical analysis of the raw material and vapour-deposited films. Due to the higher purity, the values for the tensile strength and bending stress of the vapour-deposited foils are even higher than those of rolled material of the same thickness.

Alloys with strongly differing vapour pressures of components and fluctuations of the composition of the alloys in the melt slough limit the evaporation of alloys from a crucible with continuous reloading. Strong differences in vapour pressures leads to transient times that are too large.

The cause of fluctuations in the composition of the melt slough can be temporal changes by electron beam power or fluctuations in the heat losses in the crucible. Such instabilities lead to changes in the melt slough volume, which can lead to a composition change of the coating. For the dependence of the relative change of the more volatile part in the coating of the relative melt, the slough volume is valid under neglect of the temperature changes, and we obtain approximately:

$$\frac{\Delta X_{B4}}{X_{B4}} = \frac{\Delta V_2}{V_2} \times \frac{X_{B1}}{X_{B2}}, \tag{3.37}$$

whereby $X_{B2} \ll X_{A2}$ and $X_{B2} \ll X_{B4}$ are assumed. The concentration change in the film thus depends on the change of the volume of the melt slough and on the relationship of the concentrations in the feeded material and in the melt slough. Temperature gradients over the evaporator surface $\vec{F}$ have an additional influence on the composition of the film [27].

With a relationship of the concentrations X_{B1}/X_{B2} of 100 for example a change of the proportional part of the component B in the film $\Delta X_{B4}/X_{B4}$ of 10 % results in volume variation of only already 10^{-3}.

The consequence of this influence of the volume variations is that the method of the generation of alloy films has its limits with evaporation from a melt slough with reloading if the vapour pressure of the components differs by more than the factor 10^3.

In special applications, fluctuations in the film composition, which arise in the process of the film deposits, can be balanced by a thermal treatment process [28]. Most compounds dissociate by thermal evaporation, partial or complete, so that thin films with stoichiometric composition cannot be generated by simple evaporation.

3.3.3 Compounds

Not only the thermal dissociation, but also reactions with the crucible material limit the evaporation of compounds. These limits can be partly compensated by use of electron beam evaporation [29].

The use of electron beams can induce difficulties at the evaporation of compounds, for example intensified dissociation by increased surface temperature of the evaporation material and collision dissociation with the evaporated material [30]. Independently, whether a compound is evaporated by a conventional method or by electron beams, the compound A_xB_y of the films, the following applies:

$$\frac{y \times N_A}{x \times N_B} \to 1\,, \tag{3.38}$$

where N_A/N_B is the relationship of the atomic concentrations of the components A and B in the film [31].

An important method of electron beam evaporation is reactive evaporation, i.e. the generation of a chemical compound during the evaporation process with a solid component and a gaseous component. Another possibility to generate compound films is the evaporation of two components, which react by condensation on the substrate. In this way, a compound film Nb_3Sn can be generated by separate electron beam evaporation of niobium and tin. The rate of the condensing molecules α_{KAB} for this type of generation of compounds by evaporation is proportional to the product of the condensation rates of the components α_{KA} and α_{BK} and an average diffusion coefficient D:

$$\alpha_{KAB} = K \times \alpha_{KA} \times \alpha_{KB} \times \overline{D}\,. \tag{3.39}$$

The condensation rates α_{Kt} are proportional to the vapour pressures p_{ei} of the components ($i =$ A, B) at the substrate and to the adhesion coefficient or sticking coefficient β:

$$\alpha_{Ki} \sim \beta(T_S) \times p_{ei}\,. \tag{3.40}$$

From these conditions it can be derived that there is a range for the vapour pressures p_{ei} at certain substrate temperature T_S, the so-called stoichiometric interval in which only the compound A_xB_y condenses.

A variant of this procedure consists of evaporating the components in the necessary relationship, and generating the chemical compound by means of a following temperature treatment. TiC coatings have been generated this way [32]. By reaction of the condensing material with the substrate material or with adsorption films, thin coatings of a certain chemical composition can also be generated. This type of generation of interface coatings plays an important role [33] in the production of adhesive films or adhesive intermediate films.

For reactive evaporation the pressure or partial pressure of the gaseous component necessary for the reaction is fed in the recipient with pressures of 10^{-4} to 10^{-3} mbar.

If the pressure of the reaction gas is chosen to be too high, gas incorporation and losses of energy of the evaporated component cause too frequent collisions of the vapour with the gas, which can have a negative influenced on the film properties. For the generation of oxide and/or nitride films as reaction gas, oxygen and/or nitrogen is usually used. In special cases, the oxidation takes place also with the help of water vapour. For reactive evaporation of carbides, we use easily separable carbon compounds, like C_2H_2 or C_2H_4. The generation of TiC, for example, takes place via the reaction

$$\begin{aligned} &2\,\text{Ti (vapour)} + C_2H_2\ \text{(gas)} \\ &\quad \to 2\,\text{TiC (film)} + H_2\ \text{(gas)}\,. \end{aligned} \tag{3.41}$$

In order to discuss influencing parameters during reactive evaporation, it is assumed that an adsorption of the gaseous component precedes the reaction.

For the relationship of the components in the coating after reaction with an adsorbed atomic reaction gas A is obtained:

$$\frac{N_A}{N_B} \sim \frac{v_A}{v_B}(1-\Theta)^2 \times \frac{Z}{(Z-\Theta)} \times \beta_A \times e^{-\frac{E}{RT_S}}\,. \tag{3.42}$$

Here

- v_A/v_B is the collision number relationship of the components on the substrate,
- θ the degree of coverage,
- Z the number of neighbour adsorption places,
- E the activation energy of chemisorption,

β_A the condensation coefficient,
R the gas constant, and
T_S the substrate temperature.

The result of the reaction depends on the thermodynamic conditions, in particular on the activation energy, the temperature and the condensation coefficient, which decide on the relationship number of collided particles between the reaction gas and the number of condensing vapour particles.

The collision number relationship v_A/v_B is:

$$\frac{v_G}{v_D} \approx 55.5\frac{M_D\, p_G}{\varrho\, \alpha_K}\,,$$

where

- v_G in $cm^{-2}\,s^{-1}$ is the surface-related impingement rate, i.e. the coefficient of restitution of the gas particles, which meet for each time unit the unit area;
- v_D in $cm^{-2}\,s^{-1}$ the surface-related impingement rate, i.e. the coefficient of restitution of the vapour particles, which meet for each time unit the unit area;
- M_D the mass value of the vapour particles (the numerical value corresponds to that of the atomic and/or molecular mass);
- ρ in $g\,cm^{-3}$ the density of the evaporation substance;
- p_G in Pa (10^{-2} mbar) the gas pressure in the recipient;
- α_A in $nm\,s^{-1}$ the condensation rate; here the growth rate of the film.

In order to fulfil the condition of Eq. (3.38), i.e. to attain a stoichiometric composition of the film the relationship v_A/v_B must be much larger than 1, because in Eq. (3.42) only part of the striking gas particles is adsorbed. For reactive evaporation of SiO_2, for example, Eq. (3.38) is only approximately fulfilled for

$$v_A/v_B \geq 10^2\,.$$

The determining influence on the collision frequency relationship of N_A/N_B and thus on the film properties at reactive evaporation of SiO_2 is illustrates in Fig. 3.14.

For high rate reactive evaporation a high partial pressure of the reaction gas must be used. At a pressure of the reaction gas of 10^{-2} mbar the mean free path of a vapour particle amounts to only about 5 mm. The volume collisions can no longer be neglected against the surface collisions.

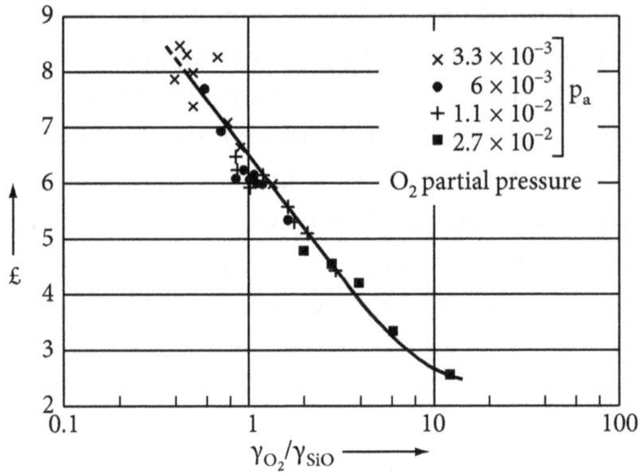

Fig. 3.14 Reactive evaporation of SiO_x; dependence of the dielectric constant ε on the impact number relationship ν_{O_2}/ν_{SiO}

The limit for the increase of the reaction gas pressure depends on the scattering of vapour, and the associated reduction of the evaporation rate and the utilization of material. In addition, losses of energy of the vapour particles arise by scattering.

The method of activated reactive evaporation is the probability of reaction with a certain collision frequency relationship increase, without increase of the reaction gas pressure. The reaction gas, the vapour or both components must be activated. This can be done by ionization methods, for example, by microwaves, by means of an electron beam or by an electrical discharge [31, 34]. An electrical discharge can generated between the evaporation materials and substrate with an electrode with a positive or negative potential applied opposite the evaporation materials [32]. Depending on the polarity, negative or positive charge carriers strike the substrate. In the case of negative polarity on the substrate, the arrangements for activated reactive evaporation are called ion plating.

For example, by activated reactive evaporation of titanium in reaction gas C_2H_2 with a high condensation rate of 0.2 μm/s a common condensation of Ti and C in the stoichiometric relationship was reached [31, 32]. At substrate temperatures of around 1000 °C a reaction of the components was obtained.

3.4 Vapour Distribution

3.4.1 Small Surface Evaporators

The outgoing vapour stream from an evaporator is characterized by the vapour stream density distribution $\theta(\alpha)$. With this function (also called evaporator characteristic) the film thickness distribution on the substrates can be calculated at a given arrangement of the substrates relative to the evaporator. For electron beam evaporation the simple case of a point source [35] with the vapour stream density distribution $\theta(\alpha) = \text{const.}$ has more or less a practical importance. Only the special case of crucibleless evaporation from a molten mass can be characterized by this distribution function.

A better approximation for small surface evaporators is the following approach: the vapour stream that escapes from a plane surface element $\mathrm{d}F$ is given by

$$\theta(\alpha) = \theta_0 \cos\alpha \,. \tag{3.43}$$

Here

$\theta(\alpha)$ is the vapour stream density in a direction with an angle α to the normal of the vapour delivery surface, and

θ_0 the vapour stream density for $\alpha = 0$.

Small surface evaporators are defined by the fact that the dimensions of the vapour delivery surface are small in relation to the distance of the substrate. The vapour stream density distribution of such a small surface evaporator can be described by Eq. (3.43). For a larger evaporator surface or extended source distribution, the vapour stream density is determined by summation of the evaporator characteristic of individual small surface evaporators or surface elements.

Electron beam evaporators with a diameter D_V of the crucible and a distance h_V to the substrate are to be regarded as small surface evaporators if $D_V \ll h_V$.

During evaporation with an electron beam evaporator, the conditions of a small surface evaporator apply; there are a series of influences, by which the vapour stream density distribution is changed. The description by simple cosine distribution is not sufficient for practical application. Influences that cause deviations of the evaporator characteristic of the electron beam evaporator from the simple cosine distribution are shown in Fig. 3.15.

Instead of a plane vapour delivery surface a convex surface can be formed, if the molten evaporation material has a high surface tension and the crucible is not wetted. The expected transition of the evaporator characteristic from the cosine distribution to the isotropic distribution of a point evaporator has not been observed, so it must be assumed that other influences are more substantial.

By using an electron beam a very high vapour pressure over the melt can be developed so that the vapour delivery surface is deformed. A concave vapour delivery surface leads to a stronger decrease of the vapour stream density with the angle α (Eq. (3.43)). In the same way, vapour sinks in the environment have an effect on the vapour delivery surface. If the distribution of the vapour stream is constrained by cold parts of the crucible wall, then the influence on the vapour stream density distribution is called a chimney effect.

At high evaporation rates a vapour cloud forms over the vapour delivery surface. By collisions of the vapour particles between themselves a part of the particles does not strike the substrate surface. If the range of high vapour pressure is limited on the vapour delivery surface, then the expansion of the vapour particles outside of the vapour cloud is linear, and it is calculated with a corrected evaporator characteristic. At small expansion and density of the vapour cloud, (i.e. if it can be assumed that the largest part of the vapour stream without multiple collisions of the particles), the correction can take place via a summand that considers the isotropic behaviour of the colliding particles [36].

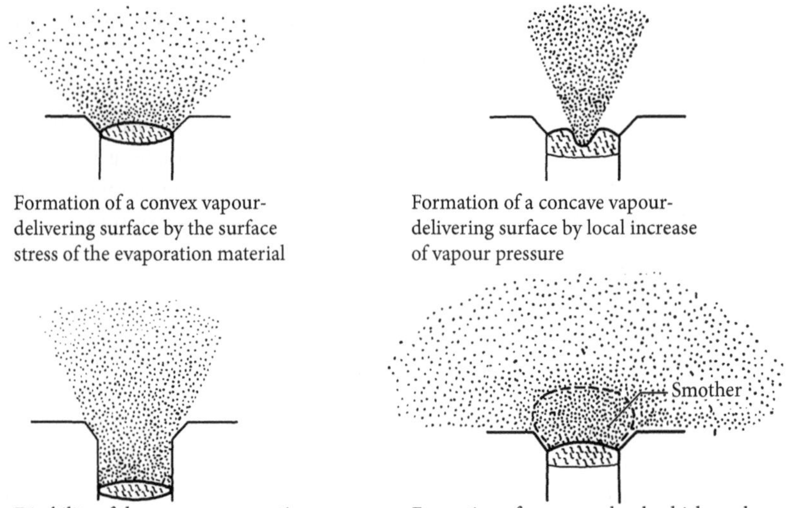

Formation of a convex vapour-delivering surface by the surface stress of the evaporation material

Formation of a concave vapour-delivering surface by local increase of vapour pressure

Disability of the vapour propagation by the crucible wall during insufficient crucible feeding

Formation of a vapour cloud, which works in place of the vapour-delivering surface as source of the vapour stream

Fig. 3.15 Influence of the vapour delivery surface of real small surface evaporators on the vapour stream density distribution

An approach for the description of vapour stream density distribution of real small surface evaporators is a higher-order cosine function in the form

$$\theta(\alpha) = \theta_0 \cos^n \alpha \,, \qquad (3.44)$$

with $n > 1$. A sufficiently exact description of the vapour stream density distribution of electron beam evaporators by Eq. (3.44) in the angle range up to 30° with a not too high evaporation rate s is possible. Figure 3.16 shows the vapour stream density distribution of a 5 kW electron beam evaporator. Eq. (3.44) with $n \sim 2.5$ is valid up to an angle $\alpha = 30°$ for the vapour stream density.

With an increasing evaporation rate, which means an increasing deformation of the melting bath surface, a stronger directionality dependency is to be expected, i.e. the exponent n in Eq. (3.44) depends on the evaporation rate

$$n = f(\alpha_V) \,, \qquad (3.45)$$

whereby n increases with increasing evaporation. For a 12 kW transverse evaporator the correlation $n = K\alpha_V^{1/4}$ was found [33]. For condensation rates in the range around 5 nm s^{-1} for the exponent n between 2 and 3, n can take values up to 6 [37] with rates within the range of 100 nm s^{-1}. For the values mentioned, the distance between the evaporator and substrate is 200–250 mm.

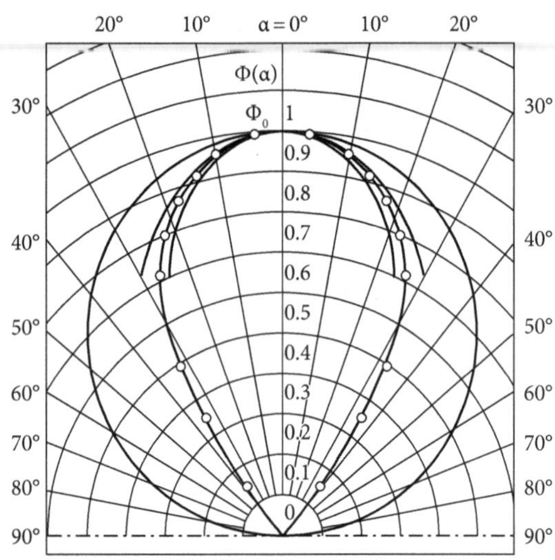

Fig. 3.16 Vapour pressure density distribution of a small area evaporator. 5-kW-electron beam evaporator: evaporation from a water-cooled crucible, evaporation rate 0.4 g/min. *Curve 1* measured characteristic, *Curve 2* $\cos\alpha$-distribution, *Curve 3* $\cos^2\alpha$-distribution, *Curve 4* $\cos^3\alpha$-distribution, θ_0 vapour stream density for $\alpha = 0$, $\theta(\alpha)$ vapour stream density with the angle α

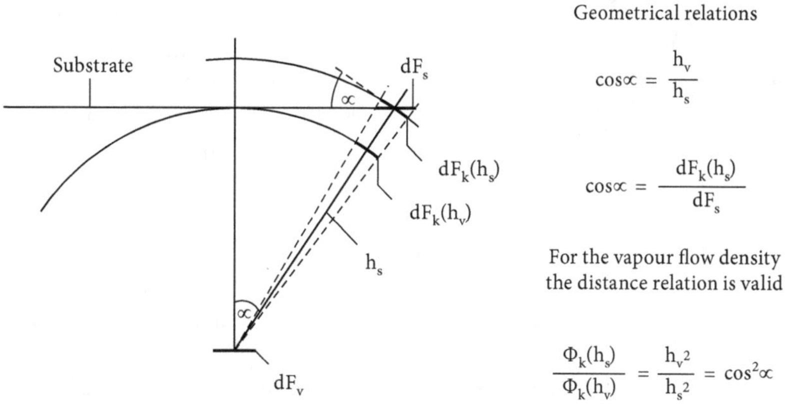

Fig. 3.17 The Determination of the film thickness d_S on a flat substrate from the vapour stream density distribution. dF_v vapour delivery plane element, $dF_K(h_v)$ plane element on the ball around the evaporator with the radius h_V, $dF_K(h_S)$ plane element on the ball with the radius h_S dF_S plane element in the substrate level

For evaporation characteristics according to a cosine function of higher order, the influence of a vapour cloud can be better described by the following approximation [36]:

$$\Phi(\alpha) = \Phi_0 \left(1 - \frac{\Phi_{is}}{\Phi_0}\right) \cos^n \alpha + \Phi_{is} \,. \quad (3.46)$$

θ_{is} is the isotropic part of the vapour stream. At a low evaporation rate it depends on θ_{is} and achieves a constant value with a rising rate. For example, at a 12 kW evaporator, $\theta_{is} = 0.14\,\theta_0$ constitutes this limit value, and a condensation rate of 5 nm/s, measured at a distance of 250 mm, is reached. With Eq. (3.46) and the indicated value for the isotropic part the experimentally determined vapour stream density distribution up to $\alpha = 50°$ with an accuracy of 1 % can be described [36].

The calculation of the film thickness distribution on arbitrary substrate arrangements from the evaporator characteristic is demonstrated by an example. The film thickness distribution is searched for a planar substrate or a planar substrate arrangement parallel to the surface of a small plane evaporator. At a small plane evaporator vapour extinguished on a spherical dish surface is condensed. Then we have

$$\mathrm{d}(\alpha) \sim \theta(\alpha) \,. \quad (3.47)$$

For the film thickness on the planar substrate $d_S(\alpha)$ arises under use of the relations in Fig. 3.17,

$$d_S(\alpha) \sim \cos^2 \alpha \theta(\alpha) \,. \quad (3.48)$$

If for the vapour stream density $\theta(\alpha)$ a simple cosine distribution is used, then with Eq. (3.43) for the film thickness on a planar substrates

$$d_S(\alpha) \sim d_0 \cos^4 \alpha \quad (3.49)$$

is valid.

In the case of the validity of Eq. (3.44), we obtain

$$d_S(\alpha) = d_0 \cos^{n+3} \alpha \,, \quad (3.50)$$

where d_0 is the film thickness for $\alpha = 0$.

With $\tan\alpha = r_S/h_V$ and Eq. (3.50), we obtain

$$\frac{d_S}{d_0} = \frac{1}{\left[1 + \left(\frac{r_S}{h_V}\right)^2\right]^{\frac{n+3}{2}}} \,, \quad (3.51)$$

where

h_V is the distance of the substrate level from the evaporator, and

r_S the distance between the evaporator-normal and the substrate.

In Fig. 3.18 the relative film thickness distribution of d_S/d_0 is illustrated as a function of r_S/h_V for different vapour stream thickness distribution (n = 1–5). The film thickness distribution of d_S/d_0 for $n = 0$ corresponds with point source evaporation [$\theta(\alpha) = \text{const}$].

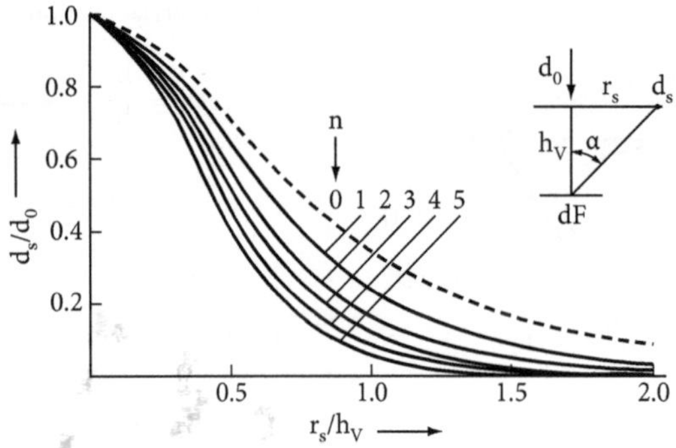

Fig. 3.18 Course of the relative layer thickness d_S/d_O on an flat substrate parallel to the vapour delivery plane element $\mathrm{d}F$ as a function of relationship r_S/h_V for different vapour stream density distributions $\theta(\alpha) = \theta_0 \cos^n \alpha$ [$\theta(\alpha) = 0$ means = const. (point source)]

Instead of the calculation of the film thickness distribution from the vapour stream density distribution after Eq. (3.44) with an exponent $n > 1$ it is also possible to use a simple cosine distribution, i.e. $n = 1$, if a virtual source of vapour is introduced [37]. The virtual vapour source for an evaporator lies with strong directive efficiency around the value Δh over the real vapour delivery surface.

The distance between the virtual vapour source and the substrate surface is h'_V, whereby $h'_V = h_V - \Delta h$. An approximation of a $\cos^n \alpha$ distribution by a virtual $\cos \alpha$ distribution is applicable only for small values of the relationship r_S/h_V. The distance of the virtual source h'_V must correspond to the relationship

$$\left[1 + \left(\frac{r_S}{h'_V}\right)^2\right]^2 = \left[1 + \left(\frac{r_S}{h_V}\right)^2\right]^{\frac{n+3}{2}} . \tag{3.52}$$

The mathematical description by a virtual vapour source simplifies the calculation of the film thickness on any substrate arrangements. The description permits transferring the film thickness distributions calculated for a $\cos \alpha$ characteristic by the Eq. 3.52 to those of the $\cos^n \alpha$ characteristic of real evaporators. With the example of the evaporation of a plane substrate by a small surface evaporator with a $\cos^n \alpha$ characteristic of the vapour stream density, it must calculate exactly on the relative change of film thickness $\Delta d_S/d_S$ on the substrate and on the utilization rate η_m of the evaporated material. On a plane substrate the relative change of film thickness according to Eq. (3.50) is

$$\frac{\Delta d_s}{d_s} = (n + 3) \times \tan \alpha \times \Delta\alpha . \tag{3.53}$$

For substrates that are arranged in the range of small angles ($\alpha \sim \alpha\Delta$), we obtain

$$\frac{\Delta d_s}{d_s} \approx (n + 3) \times \alpha^2 . \tag{3.54}$$

The relative change of film thickness is thus larger when the exponent n is larger. Since by electron beam evaporators the exponent n increases with rising evaporation rate, the uniformity of the film thickness degrades the present substrate arrangement with increasing rate.

The relationship of the mass of the substrate to the total evaporated mass is called the utilization factor η_m of the evaporated material. If the substrate is arranged within the angle α to the normal, then the utilization factor is also equal to relationship of the vapour stream $\theta(\alpha)$ in a right circular cone with a half-cone angle α to the total vapour stream θ_0, which is assumed from a small plane evaporator.

With Eq. (3.44), the calculation of the vapour stream density will be

$$\eta_m = \frac{\overline{\Phi}(\alpha)}{\Phi_0} = \frac{\int_0^{\alpha} \Phi(\alpha)\,\mathrm{d}\alpha}{\int_0^{\pi/2} \Phi(\alpha)\,\mathrm{d}\alpha}$$

$$= \frac{\int_0^{\alpha} \cos^n \alpha \, \sin \alpha \, \mathrm{d}\alpha}{\int_0^{\pi/2} \cos^n \alpha \, \sin \alpha \, \mathrm{d}\alpha} . \tag{3.55}$$

As a result of integration of Eq. (3.55), we obtain

$$\eta_m = 1 - \cos^{n+1} \alpha \tag{3.56}$$

and for small angles α approximately

$$\eta_m \approx \frac{n+1}{2} \alpha^2 . \tag{3.57}$$

One demands a value for evaporation of a plane substrate configuration with a small surface evaporator for a relative change of film thickness $\Delta d_S/d_S$, for example, 5×10^{-2}; thus a permissible angle α results in $\sim 1.1 \times 10^{-1} \sim 6.5°$ for $n = 1$ from Eq. (3.54). Even for $n = 1$ after Eq. (3.57) the utilization factor η_m is only about 1.3×10^{-2}; less than 2 % of the vapour particles arrive at the substrate.

In this evaporation configuration we do not receive a sufficiently extended film thickness on large substrate surfaces. The utilization factor η_m is low. The most important principles to reach a homogeneous film are:

- Distribution of the source strength on an expanded vapour-delivery surface
- Arrangement at the substrates on planes of the same vapour stream density
- Adjustment of the retention time of the substrates in the vapour stream and the vapour stream
- Density distribution through relative motion
- Correction of the film thickness distribution by scattering with the use of increased gas pressure
- Correction of the vapour stream density distribution by means of shutters.

In numerous different substrate arrangements these principles or a combination of several of them are used, see, for example [37, 38].

3.4.2 Evaporation Arrangements with Expanded Source Distribution

A method for realization of an expanded source distribution is to use a number of small surface evaporators. By optimization of the number, one another position and the evaporation rates the required vapour stream density distribution can be obtained with consideration of the characteristics of the individual evaporators.

The resulting film thickness distribution can be determined by accumulation and/or by integration from the film thickness distribution of the individual evaporators. Frequently an arrangement of several small surface evaporators in a line is used. If the substrates are moved perpendicularly to this line, then a high uniformity of film thickness can be achieved (Fig. 3.19). This evaporation principle will be used for tapes and substrates, which on plane carriers or on rotating baskets, are arranged.

Further, a required source distribution can be obtained by using a large-scale crucible and electron beam scanning over the surface of the evaporation material. With the constant power of the electron beam the temperature distribution and thus the site specific evaporation rate are functions of the retention time of the electron beam on each point of the surface.

In the case of sinusoidal deflection of the electron beam in the x-direction of a crucible, different retention times result along the deflection path. At the points of reversal of the electron beam path on the evaporation material the retention time t_w has a maximum. Figure 3.20 shows how the power density distribution of $e(x)$ can be changed along the deflection path by variation of the electron beam diameter d_F and the deflection amplitude. With the limited expansion of the vapour delivery surface to generate a constant film thickness on the complete substrate, the source density within the range of substrate edge must be increased. The power density within the range of the reversal points is increased by sinusoidal deflection of the electron beam. Thus, a correction possibility for the film thickness distribution, which is sufficient for many tasks, is given. A strong rise of a crossfall of the vapour stream density at the crucible edges leads to high film thickness uniformity on the substrate and reduces the utilization factor η_m of the evaporation material.

For this reason it may be necessary for individual applications to select a deflection program with more than two maxima of the power density and herewith also the source density over the deflection path leads (Fig. 3.21). In this arrangement the programmed guidance of the electron beam is generated by a combination of a dynamic deflection field and a static reversal field.

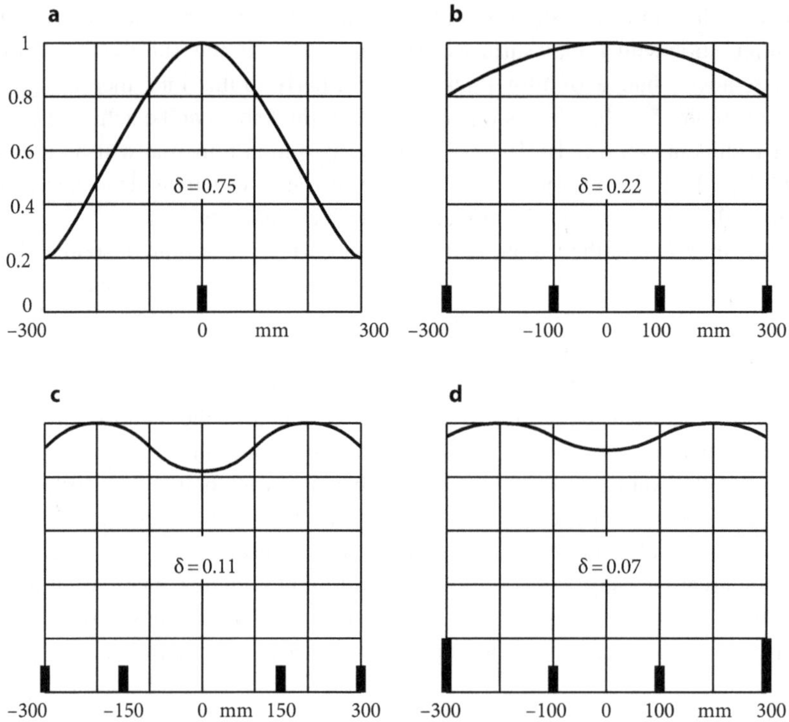

Fig. 3.19 Improvement of the film thickness distribution of d_S/d_0 on vapourization of a plane substrate with the arrangement of several small surface evaporators in a line (calculated). **a** small surface evaporator, **b** four small surface evaporators, same distance between evaporators, same evaporation rate, **c** four small surface evaporators, different distance between evaporators, same evaporation rate, **d** four small surface evaporators, same distance between evaporators, different evaporation rate. Evaporator substrate distance $h_v = 300$ mm, opening angle of the vapour stream $2\alpha = 180°$, vapour stream density distribution $\theta(\alpha) = \theta_0 \cos\alpha$ $\delta = (dS_{max} - dS_{min})/d_0$

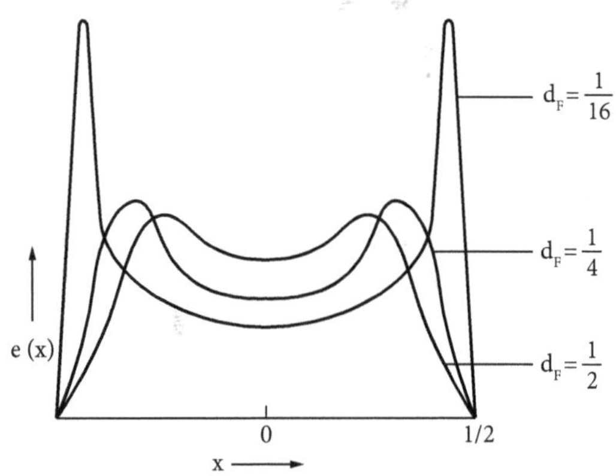

Fig. 3.20 Power density distribution of $e(x)$ on the evaporation material along a deflection course of the length l; sinusoidal deflection of the electron beam, parameter electron beam diameter d_F

3.4.3 Vapour Propagation with Small Free Path Length

In the past considerations of the generation and propagation of a vapour stream, it was considered that essentially a straight line propagation of the vapour particles takes place. For explanation of deviations from vapour stream cosine density distribution, collisions of the vapour particles had to be considered among themselves within the range of the vapour delivery surface.

With increasing vapour pressure or at high gas pressure interaction effects in the entire range between evaporators and substrate take place. In these cases vapour no longer expands in a straight line, and the relations indicated in Sect. 3.4.1 and 3.4.2 lose their relevance.

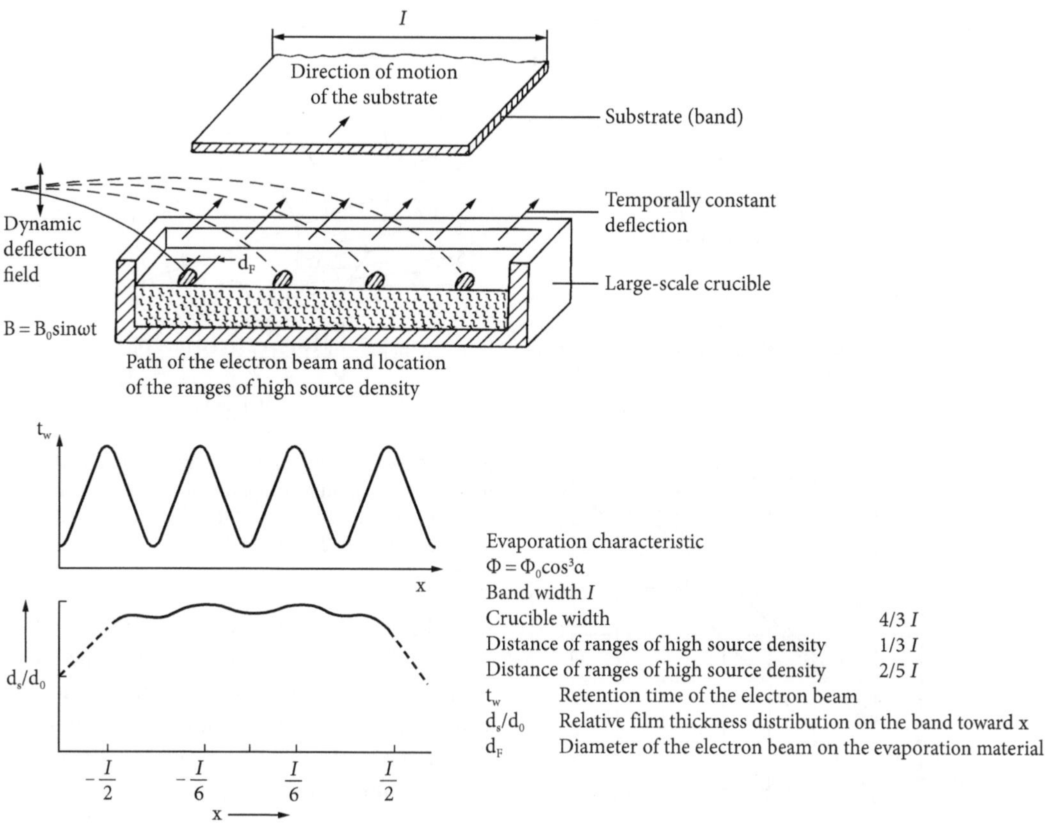

Fig. 3.21 Realization of an expanded source distribution by programmed deflection of an electron beam toward x of a large-plane crucible (schematized)

The propagation of vapour within the range of strong interaction effects of the vapour particles among themselves and the vapour with the gas is characterized by the fact that the mean free path of the vapour particles is small by collisions among themselves l_D, the vapour particles at collisions with gas particles l_{DG} and/or the gas particles with collisions among themselves l_G against the distance h_V between evaporators and substrate.

It thus applies that

$$l_D; l_{DG}; \quad l_G \ll h_V \,. \tag{3.58}$$

These conditions for the expansion of vapour are obtained at a gas pressure of 10^{-2} mbar and with a coating with an extremely high evaporation rate. Since under these circumstances each vapour particle is scattered several times on the way to the substrate surface, the initial orientation and initial energy of the particles are no longer important when the particle strikes the surface. In place of a directed vapour stream area between evaporators and substrate a vapour cloud is formed, which can also surround the substrate.

Figure 3.22 describes the direction modification of the vapour particles by scattering in the vapour cloud. Whereas in the directed vapour stream the vapour particles that strike the substrate only have small direction differences depending on the diameter of the vapour delivery surface and the distance between evaporators and substrate, each emission point P considered within the vapour cloud can be a source of vapour particles, which spread out in all directions.

In extreme cases, frequent interactions of the vapour distribution can be characterized by the pressure of vapour p_D. Under these conditions

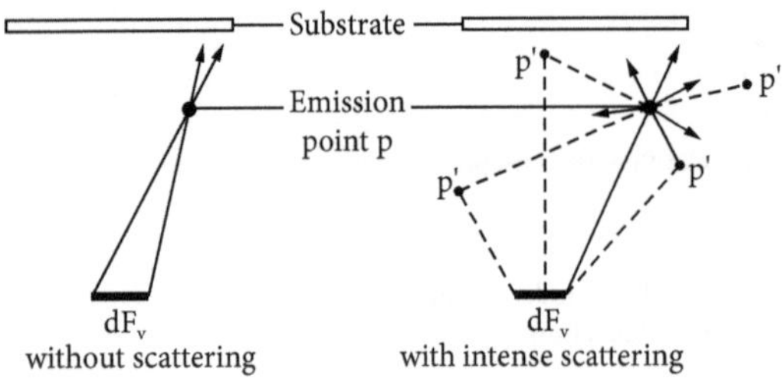

Fig. 3.22 Influence of the scattering on the direction of an up point P in the proximity of the substrate of the outgoing vapour particles. dF_V is a vapour delivery two-dimensional element, P' are scattering centres

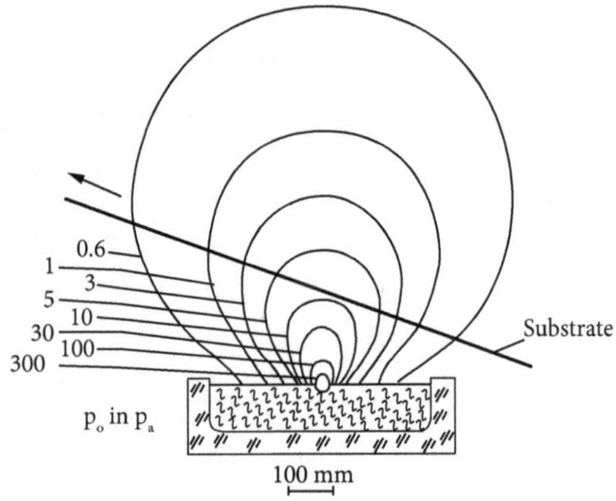

Fig. 3.24 Vapour pressure p_D in the vapour cloud at electron beam aluminium evaporation at a power of $P_0 = 250\,\text{kW}$; power density is within the range of the focal spot, for instance, $2 \times 10^7\,\text{W/m}^2$

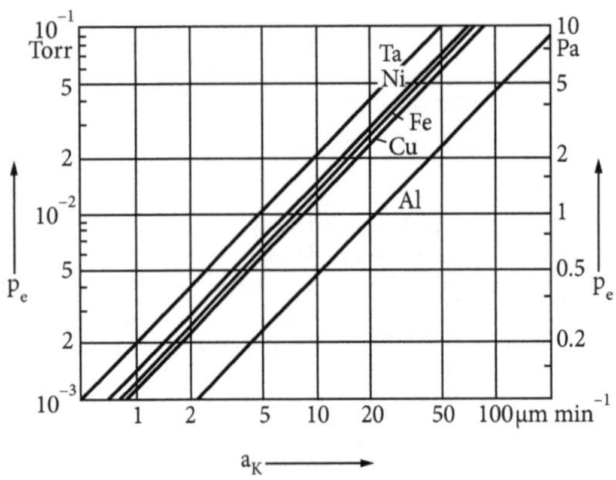

Fig. 3.23 Connection between the vapour pressure p_E within the range of the substrate and the condensation rate α_K (calculated under the assumption of Maxwell distribution of the velocity of the particles in the vapour cloud and the trap coefficient $\beta = 1$)

and on the assumption that the condensation surface does not disturb the pressure ratio in the vapour cloud between the condensation rate α_K and the pressure of the evaporating material p_D in the environment of the substrate p_E we obtain:

$$\alpha_K = K(M, T)\beta p_D \,, \tag{3.59}$$

with the correlation $p_D \sim p_E$.

Figure 3.23 illustrates the vapour pressure p_D in the vapour cloud, which are necessary in order to achieve a certain condensation rate under the conditions of large scattering processes.

If vapour particles are scattered at gas particles, then p_D in Eq. (3.59) is only the vapour partial pressure. The energy of the condensing vapour particles is reduced by repeated collisions with gas particles in relation to their initial value. In the vapour cloud a mean free path can be indicated l_D. For example, at condensation of aluminium with a condensation rate of 5000 nm/s in the environment of the substrate the mean free path adds up to some millimetres. Figure 3.24 illustrates the pressures p_D in the cloud with the evaporated aluminium with a power of 250 kW.

The vapour pressures p_D in certain distances from the evaporator were calculated from the condensation rate in accordance with Eq. (3.59). The values obtained in the environment of the vapour delivery surface were determined by extrapolation.

To which extent the scattering of the vapour particles between themselves changes the film thickness distribution is shown Fig. 3.25. A steel band was led across an evaporator crucible of 500 mm. The electron beam was once on a line in direction of motion of the steel band and another time on a line perpendicular to the steel band. Despite the very different distribution of the vapour source, there were only small differences in the film thickness distributions on the steel band.

At evaporations with small gas pressure and very high vapour pressure no complete isotropy

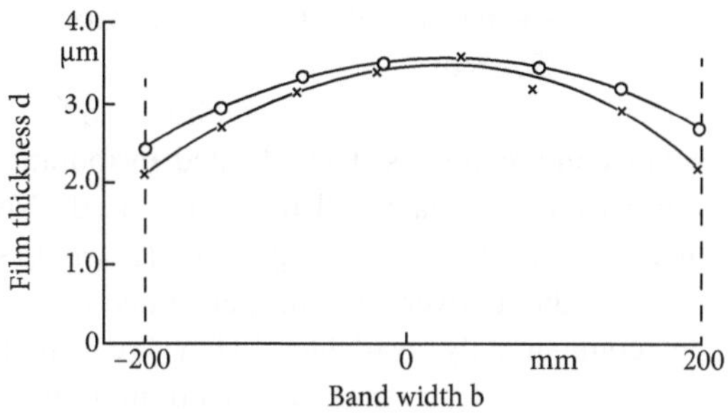

Fig. 3.25 Influence of scattering of vapour particles with high vapour density on the film thickness distribution. Electron beam power 250 kW, programmed deflection of the electron beam in a direction: o transverse to the steel strap movement, × lengthwise to the steel strap movement. Crucible 500 × 500 mm, distance of the strap from the evaporator surface ~ 200 mm

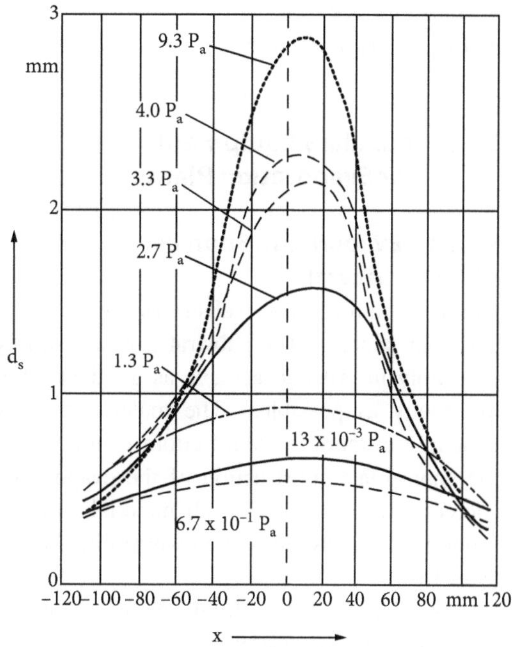

Fig. 3.26 Film thickness distributions on an flat substrate at vapourization with an extremely high evaporation rate for different gas pressures [39]; transverse evaporator, evaporation rate $\alpha_v = 32$ g/min remains constant during pressure increase, compensation of the changed transmission through an increase of the output of 60 kW ($p < 10^{-4}$ mbar) up to 102 kW ($p = 10^{-2}$ mbar); working gas argon, distance evaporator substrate 324 mm

within the range of the substrate prevails, i.e. the condensation is partly intended to occur by a vapour cloud, partly by a directed vapour stream.

With pressure plating [39], reactive vapourization and partially ion plating interaction of the vapour particles with gas particles is the cause of strong scattering of vapour. The transport of the evaporated material through the surrounding gas, takes place via diffusion, if the condition of complete scattering is reached.

Within the range of the vapour delivery surface the interaction with the gas leads to backscattering of vapour. The decrease of the evaporation rate can be described, as in Sect. 3.3.1, by the transmission coefficient τ.

Advantages of scattering in the residual gas are the reduced shadow influence and half-shadow influence at unevenness and edges on the substrate surface. Therefore, with scattering coating the degree of coverage in particular at rough surfaces is improved.

For the expansion of vapour under conditions of extreme high vapour pressure and high gas pressure, a direct influence of the vapour stream with increasing pressure has been observed [39].

Figure 3.26 shows the thickness distribution of films from stainless steel on a plane substrate with an evaporation rate of 32 g/min with different gas pressures.

3.5 Equipment Technology

3.5.1 Prefaces

When designing evaporation equipment the demands in each case, which can be very different, are considered, for example,

- Purity of the film
- Film thickness
- Film thickness tolerances
- Film morphology
- Single film or film system
- Coating process, for example, reactive coating
- Additional ionization, sources of plasma
- Substrate material (glass, plastic, paper, metal, chemical compounds)
- Substrate form (bulk material, shaped parts, foils)

- Substrate size
- Degree of automation.

3.5.2 Auxiliary Equipment for Evaporation Plants

3.5.2.1 Evaporation Sources

3.5.2.1.1 General

During coating, the substrates move over a vapour source. The substrates rotate simply or doubly in order to achieve as good film homogeneity as possible. The material thereby generally evaporates from bottom to top. With the appropriate construction of the evaporation sources, however not with all materials, evaporation is also possible in the reverse direction or sidewise.

Some decisive factors for the choice of the sort of evaporator sources are:

- Film purity
- Condensation rate and film thickness
- Temperature of the substrates during the coating
- Type of coating material.

Only a small part of the energy supplied to the evaporator is needed for the actual evaporation process, for transformation of a solid into a vapour.

The largest part of the energy is lost through thermal conduction over the energy supply cable and/or over the more or less well-cooled wall, for heating the coating material in the crucible and through thermal radiation. Circa half of the supplied energy is exhausted for water cooling. The energy lost through radiation is not considerable and is calculated after Stefan Boltzmann's law:

$$E_S = \sigma \, At\varepsilon T^4 \,. \tag{3.60}$$

Here
E_S is energy in Ws,
A the surface of the radiation source in cm^2,
T the temperature of the radiation source in K,
σ the radiation constant of a black body ($5.67 \times 10^{-2}\,W\,cm^{-2}\,Grad^{-4}$),
ε the correction factor, and
t the duration of the test in seconds.

3.5.2.1.2 Resistance-Heated Evaporation Sources

Because of the simple handling and the small complexity resistance-heated evaporation sources are nowadays still frequently used. Numerous sizes and forms (e.g. spirals, cavities, boxes without covers or with perforated covers) are commercially available. They are usually manufactured from materials with high melting points, such as tungsten, molybdenum or tantalum. For coating materials with low evaporation temperatures (for example, tellurium, selenium) stainless steels are also used as evaporation source materials. Figures 3.27 to 3.39 show the most commonly used forms of evaporation source.

Tungsten spirals are used for highly reflective evaporating aluminium mirrors. Aluminium wires are put, for example, in hairpins form into a spiral coil [40]. Trough-shaped evaporation

Fig. 3.27 Hairpin-shaped evaporator with twisted tungsten wire

Fig. 3.28 Spiral evaporator from twisted tungsten wire

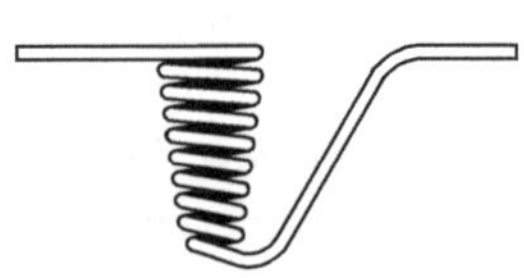

Fig. 3.29 Basket-shaped evaporator

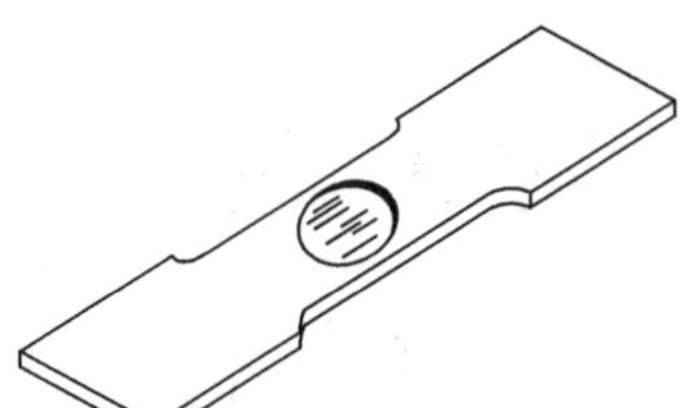

Fig. 3.30 Trough-shaped evaporator

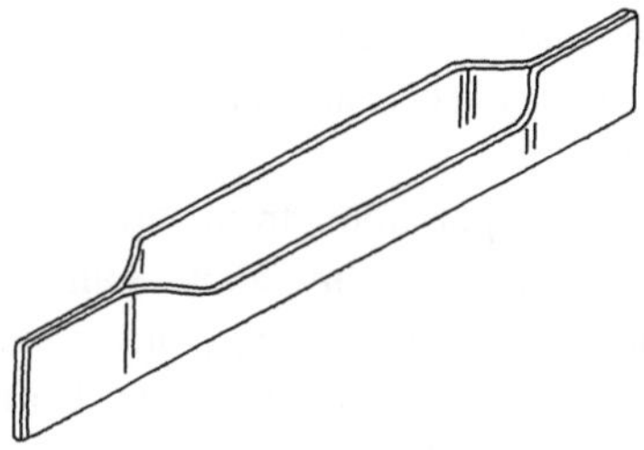

Fig. 3.31 Box-shaped evaporator

Fig. 3.32 Source of evaporation with ceramic crucible

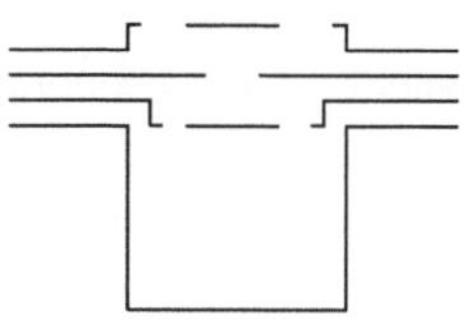

Fig. 3.33 Evaporator with vapour jet turn with round sheets

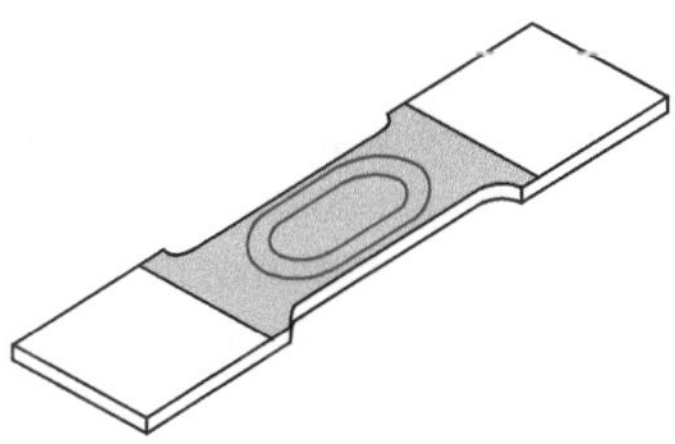

Fig. 3.34 Trough-shaped evaporator with ceramic coating

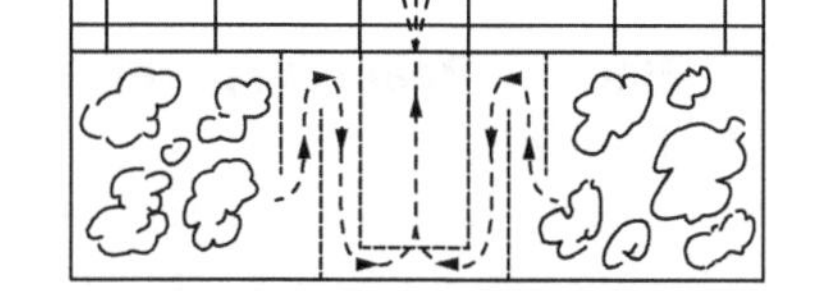

Fig. 3.35 Evaporator with vapour jet turn with round sheets for evaporation from bottom to top, refilling opening

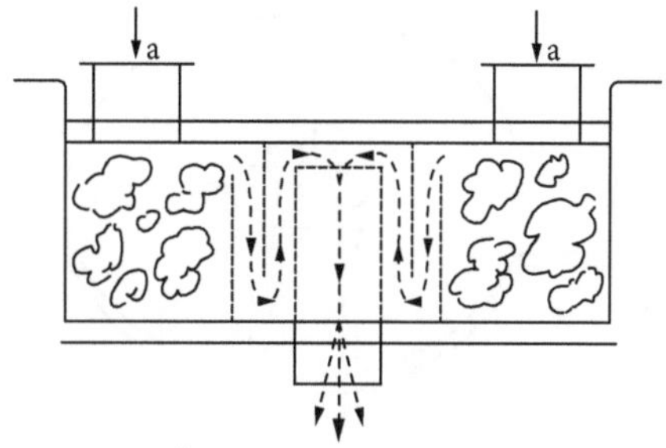

Fig. 3.36 Evaporator with steam jet with round sheets for evaporation from top to bottom refilling opening

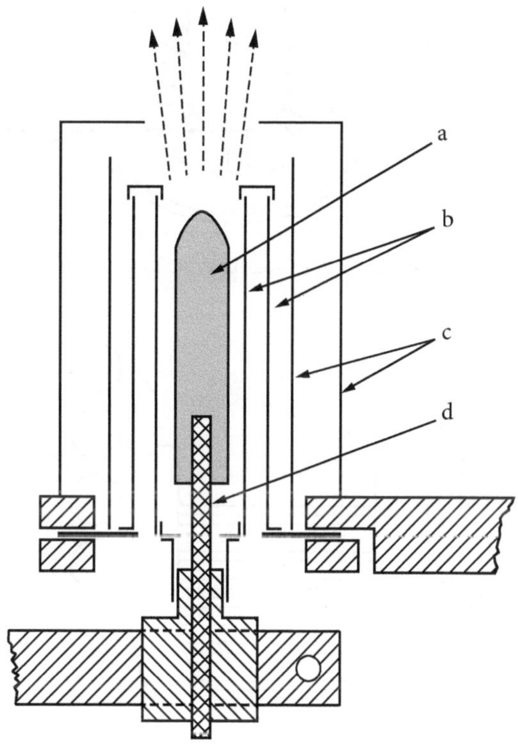

Fig. 3.37 Sublimation of chrome. *a* chrome holder, *b* tantalum heater, *c* radiation protection, *d* tungsten carrier

sources can be regarded in good approximation as point-shaped evaporation sources; they produce a rotationally symmetric vapour lobe.

Box-shaped evaporation sources are used if larger material quantities are to be evaporated. With a perforated cover plate larger losses are avoidable. The cover plate temperature should in no case be lower than the evaporation temperature, since otherwise the danger of condensation exists and thus the openings in the perforated

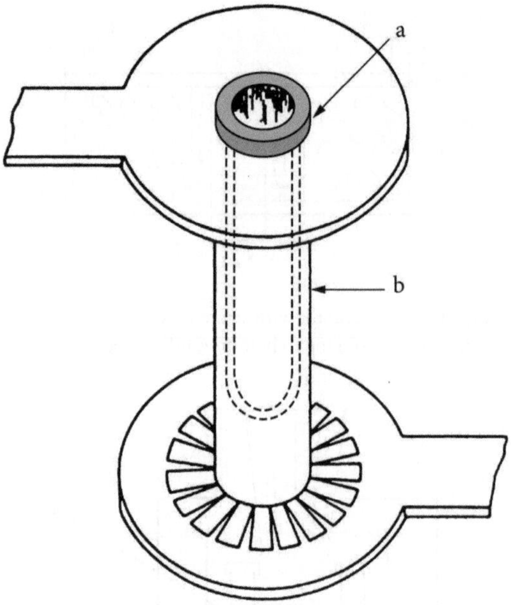

Fig. 3.38 Evaporator using ceramic(s). *a* ceramic crucible, *b* tantalum heater

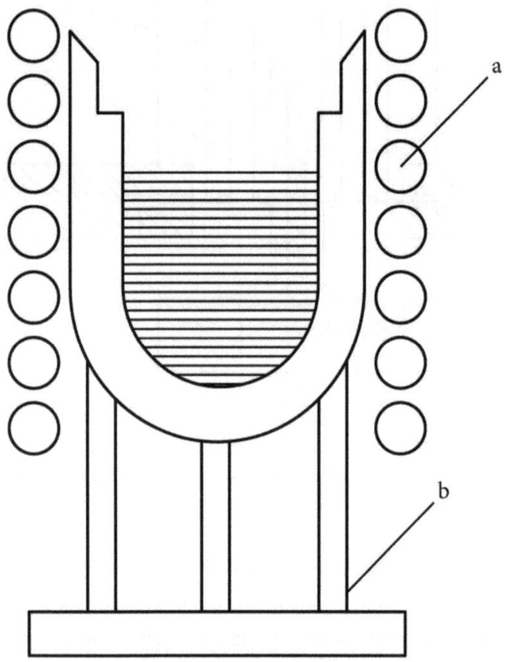

Fig. 3.39 Evaporator with inductive heating. *a* high-frequency coil, *b* ceramic insulators

cover plate can be closed. This can lead to a complete interruption of the coating process. Sublimating materials in pill form or granulates such as silicon monoxides or zinc sulphides are almost exclusively evaporated with covered evaporation sources.

With evaporation from top to bottom resistance-heated evaporation sources receive a heated roof-shaped cover plate, which causes a deflection of the vapour stream. However, the temperature of the deflection plate must be so high that no condensation takes place. Occasionally one provides trough boats with a ceramic coat or with a ceramic cartridge for the extension of the life cycle.

For evaporation of larger material quantities during longer periods, for example, coating foils in coil coating plants, cuboid-formed evaporation sources from semiconducting materials, for example, boride or nitride, are used. They are heated by direct current and held by water-cooled clamping tools, which the thermally caused dimension changes of the evaporation source by counter bearings adjust. The usually wire-shaped coating material is continuously supplied by unspooling from a supply roll.

Typical wire diameters are 1–2 mm. Above the unwinding speed and temperature of the evaporation source, the evaporation speed and in connection with the speed of the foil the desired film thickness can be adjusted. From an evaporation source with an evaporating surface of about 1000 mm^2 about 4.5 g/min aluminium can be evaporated splash-free [41–44].

One method used with certain materials for evaporation is to sublimate wires or metal sheets [45]. A condition is a sufficiently high sublimation rate below the melting point of the coating material. For example, with a coating rate of $10^{-4}\,g\,cm^{-2}\,s^{-1}$, materials for sublimation are Ti, Zr, Fe and Zn.

3.5.2.2 Electron Beam Evaporators

3.5.2.2.1 Principles

An electron beam evaporator consists of an electron beam gun and a crucible. In many cases this configuration is supplemented by a charging device for the evaporation material. With many evaporator constructions the mechanisms for electron emitter generation and electron beam

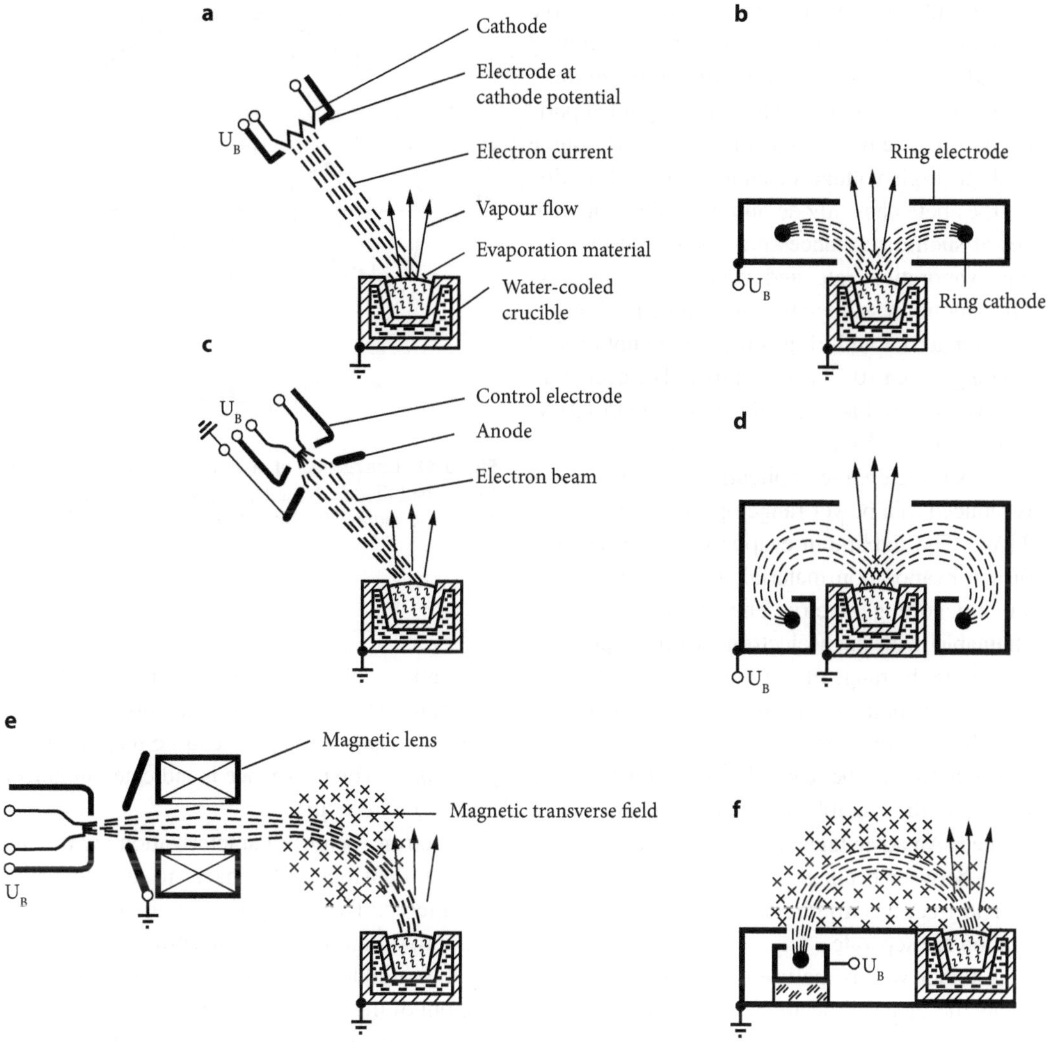

Fig. 3.40 Principles of electron beam evaporators. **a** Evaporator with electron collision heating, linear cathode and electrostatic focusing; **b** evaporator with electron collision heating, ring cathode and electrostatic focusing; **c** evaporator with axial gun and electrostatic long distance focusing; **d** evaporator with electron collision heating, ring cathode, electrostatic focusing and deflection; **e** evaporator with axial gun, magnetic focusing and magnetic 90° deflection; **f** evaporator with transverse gun and magnetic 180° deflection

guidance, as well as the crucible in a unit are combined.

Electron beam evaporators operate according to different principles, which are shown in Fig. 3.40. In simple arrangements with electron collision heating a cathode in the environment of the evaporation material and/or the crucible is used. The electrons flow with an accelerating voltage between cathode and crucible, directed towards the evaporation material.

In order to focus the electron current at a certain diameter on the evaporation material, a control electrode at the cathode potential is used. If ring cathodes are used, then the control electrode also has the form of a ring concentrically surrounding the crucible. Evaporators from this type of ring source were developed in different versions. Because the cathode is arranged in the environment of the evaporation material, a contamination of the evaporated film by evaporation of the cathode material can occur. A suitable form of the gate electrode avoids the evaporating cathode material striking directly the substrate.

The arrangement of high voltage-guided parts in the direct environment of the vapour source leads to high voltage discharges and glow discharges between cathode and crucible. The danger of such disturbances increases with increasing evaporation rate and accelerating voltage. The vapour pressure in the environment of evaporation material with ring sources may not exceed the range from 10^{-3} to 10^{-4} mbar. The accelerating voltage used is generally smaller than 10 kV, usually only 3–5 kV. Because of the limitation of these parameters the application of ring sources is reduced to a power range up to approximately 3 kW. The maximum evaporation rate depends on the evaporation material, the power and the vapourization geometry. The condensation rates attainable with such electron beam evaporators are within the range of 1–10 nm/s with a distance between evaporators and substrate of 250 mm.

When an additional anode is arranged in the range between the cathode, the control electrode and the crucible, the electron beam can be formed. An advantage of this principle is that the range between beam generator and vapour source is field free; electron beam generation and evaporation are separate.

Another way is the use of a magnetic lens to focus the electron beam. The arrangement of a gun over the crucible has the disadvantage that the gun is in the direct vapour stream. Thus, not only is the vapourization of the gun disturbed, but also the usable range, within which the substrates can be arranged is restricted because of shadowing parts of the electron emitter. These disadvantages can be avoided, if the gun is arranged horizontally and the electron beam is deflected electrostatically or magnetically before striking the evaporation material.

The separation of the electron beam generation from the crucible, the magnetic deflection of the electron beam before it strikes the surface of the evaporation material and the use of linear cathodes is realized with transverse electron evaporators. The electron beam is usually deflected at about an angle of 180° or 270°.

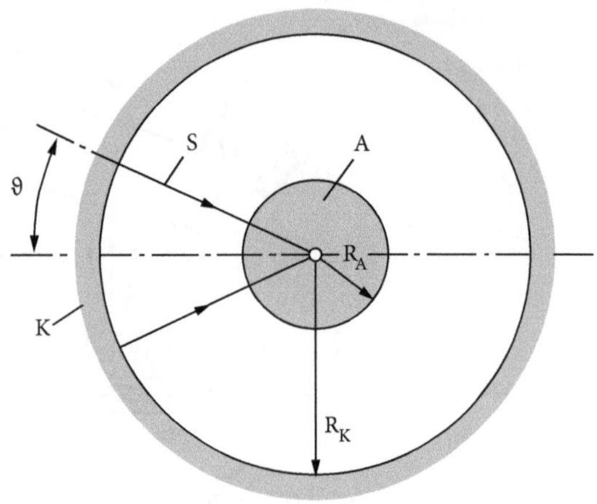

Fig. 3.41 Emergence of a beam producer after Pierce from a ball diode. *A* anode, *K* cathode, *S* edge of beam, R_A anode radius, R_K cathode radius, ∂ half opening angles

High power axial guns, such as those used for electron beam evaporators, are often characterized by means of a space charge magnitude, the *perveance*. The perveance is indicated in units of $10^{-6}\,\mathrm{A/V^{3/2}} = \mu\mathrm{P}$.

Electron beam characterization with high current densities goes back to the theory of J.R. Pierce. The theory is valid up to perveances of approximately 1 μP. Concentrated space cones cut the cathode out of the outside ball and the anode out of the inside ball (see, Fig. 3.41).

With negative space charge, without the focusing electrode the beam would expand itself already before the anode, which would lead to anode loss and/or reduced perveance values. The form of this electrode cannot be calculated analytically and is determined experimentally or by numerical computer calculations.

The potential deformation by the anode opening can be considered by the fact that it is attributed the characteristic of a dispersing lens with the focal length $f = -\sim 4V/V'$ (V = anode potential, V' = gradient on the cathode side). For perveance values larger than 1 μP the theory is no longer valid, since the anode opening comes into the order of magnitude of the distance between cathode and anode, so that the potential deformation becomes inadmissibly large by the

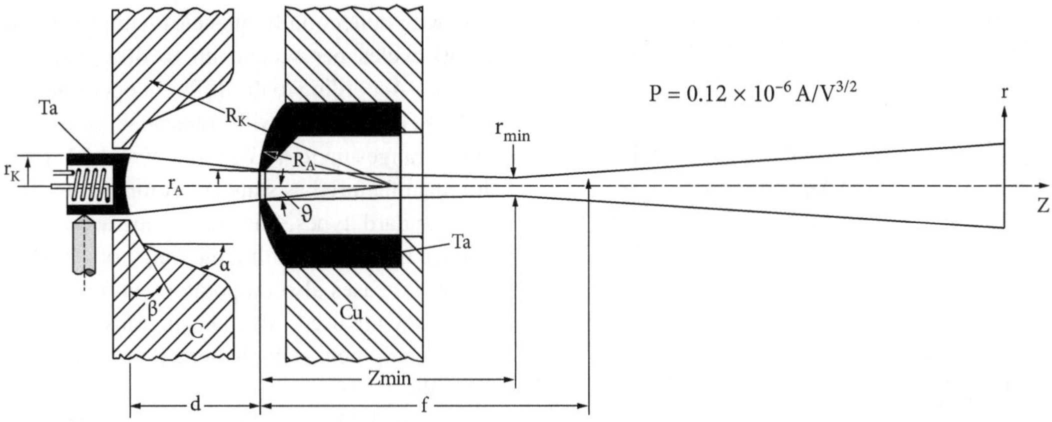

Fig. 3.42 Dimensioning of a Pierce beam generator of 0.12 μP (10 kV, 0.1 A). $r_K = 3$ mm, $r_A = 1.5$ mm, $R_K = 27.08$ mm, $R_A = 13.54$ mm, $r_{min} = 0.87$ mm, $f = 33$ mm, $d = 13.54$ mm, $Z_{min} = 25.7$ mm, $\sin\vartheta = 0.1108$, $\vartheta = 6°\,20°$, $\alpha = 24°$, $\beta = 29°$

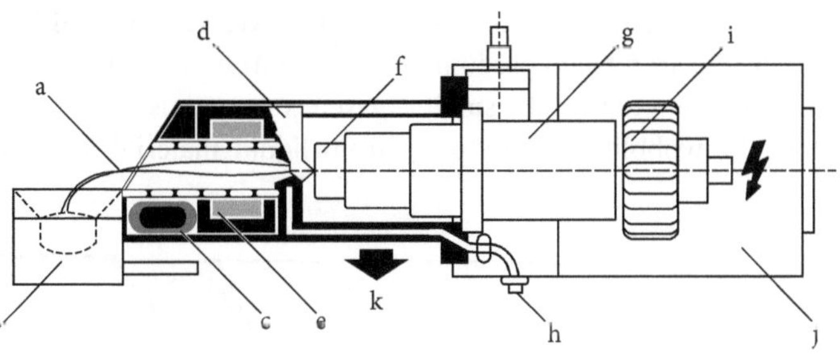

Fig. 3.43 Cut by 25 KW ES evaporator with a Pierce beam generator and 90° beam deflection. *a* electron beam, *b* copper one dish crucible, *c* deflection coil, *d* anode, *e* focusing coil, *f* focusing electrode, *g* high voltage insulator, *h* water cooling, *i* radiator box, *j* protection hood with fan, *k* separate evacuation

anode opening. Large anode current losses and strong beam divergences result.

With special emitting surfaces like concentric toroids, which generate hollow beams, perveance values to 60 μP can be reached. Figure 3.42 shows the result of a calculation for a beam generation system of 0.12 μP (10 kV, 0.1 A) with a Ta cathode of 6 mm in diameter. Industrial evaporators using Pierce beam generators have been used in many different applications.

Figure 3.43 shows the schema of a 90° electron beam evaporator, which is used in the power range of 10–30 kW. The electron beam passes through a crossed deflection field after the focusing lens, which in addition to the 90° deflection permits stirring the beam in the crucible.

3.5.2.2.2 Integrated Guns

For the installation into different vacuum recipients small and compact guns were developed. At both sides of the crucible magnetic transverse fields are arranged, which operate as electron traps (Fig. 3.44). The paths of the scattered electrons drift mirror symmetrically to the incident beam. In this way, the scattered electrons strike in the water-cooled plates in a defined way.

The plane beam of a transverse gun generates an elongated focal spot in the crucible, whose

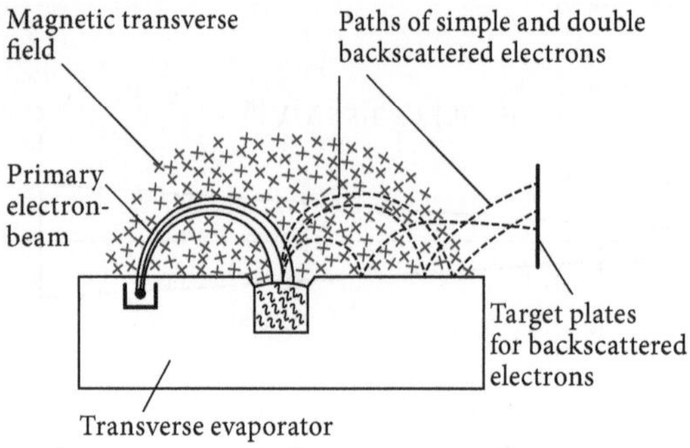

Fig. 3.44 Effect of the magnetic transverse field expanded over the crucible as a trap for backscattered electrons

small axle measures 1–5 mm. The length of the focal spot depends very strongly on the power and at 15 kW it is between 10 and 15 mm. In order to disperse the electron beam power on the evaporation material, the electron beam predominantly wobbles in the direction toward the small axle of the focal spot. The wobbling usually occurs with deflection frequencies within the range of 50 cycles per second up to 400 cycles per second. According to the different operating conditions, evaporators with transverse guns in the power ranges up to 4 kW, 5–7 kW and 12–15 kW are manufactured as standard components. With the standard types, the maximum accelerating voltage has values within the range of 10–15 kV. In special cases also evaporators with outputs of 20–10 kW are used. Much beam transverse evaporator equipment is implemented for an output of 200 kW.

Figure 3.45 shows the technical design of a transverse evaporator.

3.5.3 Electron Emission

3.5.3.1 Generation of Free Electrons

Free electrons are obtained from solid, liquid and gaseous materials. The emission of electrons from liquid material does not play a role with

Fig. 3.45 Intersection by 6 kW an electron beam evaporator with 180° jet deflection. *a* beam generator, *b* *x*-deflection coil, *c* *y*-deflection coil, *d* water-cooled copper crucibles, *e* electron beam

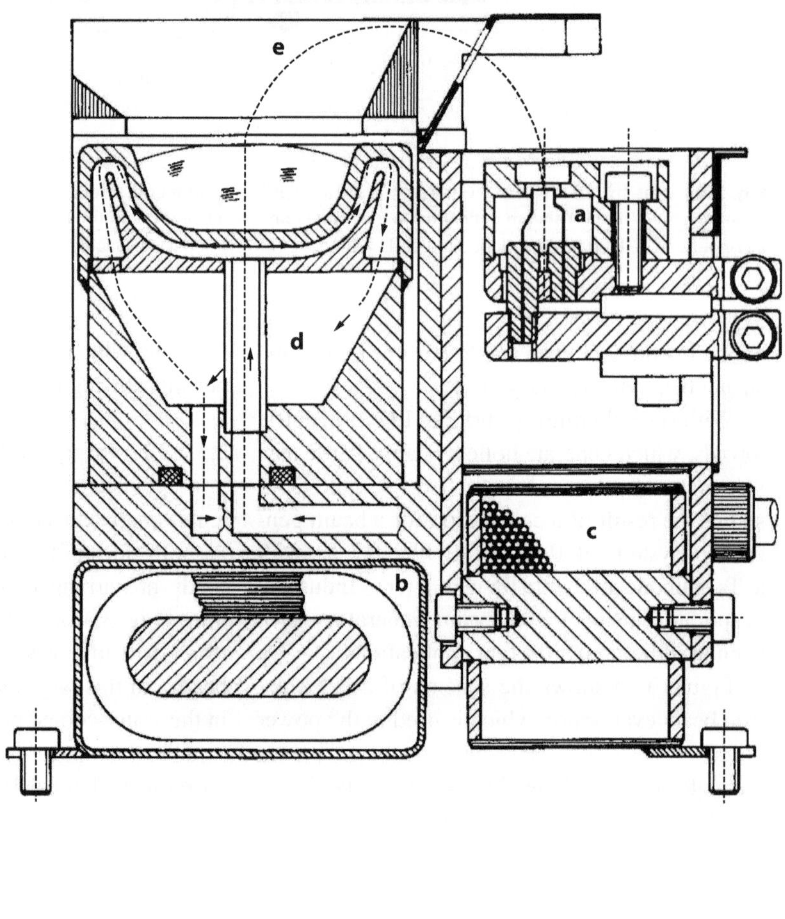

electron guns. For the generation of electrons from gas discharges with cold cathodes and magnetically concentrated low pressure plasmas with low discharge voltage are used [46].

Electron emission from solid materials has gained technical importance. As emitters for electron guns mainly glow cathodes are used. The emission current density is j_{eT} and the Richardson–Dushman law describes the temperature determined saturation current density achievable with these cathodes,

$$j_{eT} = AT^2 e^{-B/T} . \tag{3.61}$$

Here
A is the Richardson constant,
B the characteristic size $B = eU_A/k$ of the electron work function eU_A, and
T the absolute temperature.
The quantities A and B are material-dependent; they are well-known as possible cathode materials.

Tungsten, tantalum and lanthan hexaboride (LaB_6) are the most common materials. Figure 3.46 shows the dependence of the saturation current density of these materials as a function of the cathode temperature. The upper limit for the usable emission current density depends on temperature stability and the material evaporation taking place at high cathode temperature. The current emission density of tungsten occurs at 1–10 A/cm^2. In special cases with reduced lifetime a current density of 15–20 A/cm^2 is reached.

Other cathode materials, such as, for example, LaB_6 or other activated cathodes used to obtain as low an emission temperature as possible at high emission current density. Tungsten is the most important cathode material for electron beam guns. Further, tantalum, LaB_6 and tungsten with emission increasing alloy components are also used. At a given cathode temperature the saturation current density can only be achieved according to Eq. (3.46) if the electrical field intensity before the emitting surface is high enough to suck up all escaped electrons. If this condition is not fulfilled, an electrode cloud is formed before the cathode.

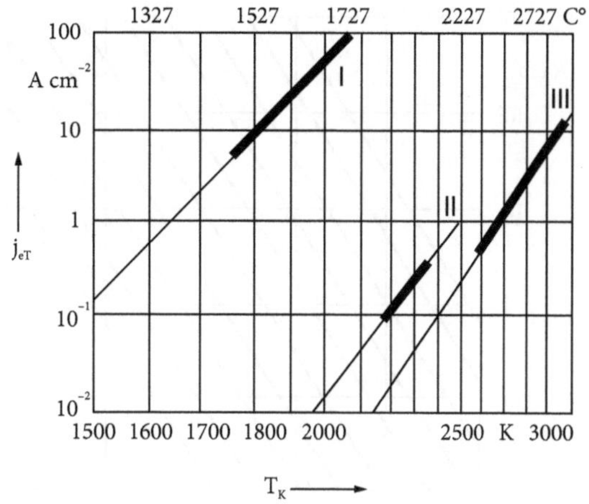

Fig. 3.46 Saturation current density j_{eT} for common cathode materials as a function of the cathode temperature T_K, I LaB_6 on carbonized tantalum, II tantalum, III tungsten, work areas strengths drawn

The space charge of this electron cloud changes the field gradient in such a manner that further emission from the cathode surface is limited. The space charge limited emission current density j_{eQ} arises on the basis of the Child Langmuir law,

$$j_{eQ} = 2.3 \times 10^{-6} K\, U_B^{3/2} ; \tag{3.62}$$

where
j_{eQ} in A/cm^2 (Fig. 3.47),
U_B in V is the accelerating voltage,
K in cm^{-2} a variable dependent on geometry, which in the simplest case of flat parallel electrodes amounts to $K = 1/Z_{KA}^2$, and
Z_{ZA} means the distance between cathode and anode.
The space charge limited current density has been calculated also for other electrode shapes. At a given accelerating voltage and distance between cathode and anode, the space charge limited current density can be changed through the geometry of the field-generated adjustment.

Through the curvature of the potential surfaces the use of small-plane cathodes causes an increase of the field strength at the cathode and thus an increased space charge limited current density, which are described by $K > 1/Z_{KA}^2$ in Eq. (3.62). This effect is used with three-

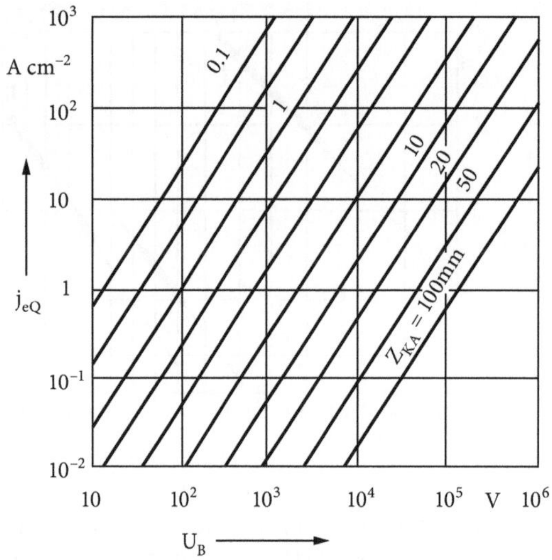

Fig. 3.47 Space charge limited emission current density j_{eQ} as a function of the accelerating voltage U_B with arrangement of cathode and anode in parallel levels with the distance Z_{KA}

electrode beam generators in order to increase the space charge limited emission stream adverse to a plane arrangement by a factor of 2 to 5.

By means of field emission the current density j_{eE} is described by the Fowler–Nordheim relationship. On simplification the assumption of this current density in A/cm^2 is

$$j_{eE} = 1.54 \times 10^{-6} \frac{E^2}{U_A} \times \exp\left(6.83 \times 10^7\, U_A^{3/2}/E\right), \quad (3.63)$$

where

E is the field strength in V/cm, and

eU_A the electron work function in eV.

In the case of an electron work function of 4.5 eV, a field emission current density in the case of a field strength of 3×10^7 V/cm of approximately 10^2 A/cm is reached.

3.5.3.2 Calculation of Beam Generators

For dimensioning of beam generators by means of numerical calculations, the electron beam in the beam generator is characterized by a number of electron trajectories, which are extinguished from the cathode. The envelope of these electron trajectories describes the form of the electron beam and thus the parameters in the smallest beam diameter and/or focal spot.

The electron trajectories must thus be calculated for an emitter configuration that can be described in each case (Fig. 3.48).

The result of such a calculation is the potential distribution in the beam generator, the electron paths and therewith the envelope and the space charge distribution. The current density distribution and/or power density distribution at the cathode and in given cross-sections of the beam generator are also calculated. Additionally, perveance is usually also determined.

For the calculation the equations that describe the electron emission and the electron trajectories in the beam generator are to be used. If one limits the description to electrostatic systems, then this entails the following physical relations:

- The distribution of the electrical potential U in the beam generator is determined by the Poisson differential equation. With the geometrical arrangement of the electrodes and their potentials – the cathode potential U_K, the control electrode potential U_S and the anode potential U_A – the boundary values for the solution of this equation are given.
- The force, which acts to an electron in the electrical field, results in the electrons' equation of motion. The solution of the equation requires the knowledge of the potential U and the electrical field strength E, also the gradients of the potential U in each point of the beam generator.
- The space charge distribution of ρ, in the special case only those of the produced electron beam, changes the potential distribution U and is therefore because of the influence on U also to consider.
- The electron emission of the cathode results from one of the Eqs. (3.61) to (3.63). Is the emission space charge limited, then the electrons are sucked up from the space charge cloud in front of the cathode.

According to Eq. (3.62) the emission current density is a function of the electrical potential U in front of the cathode surface. From

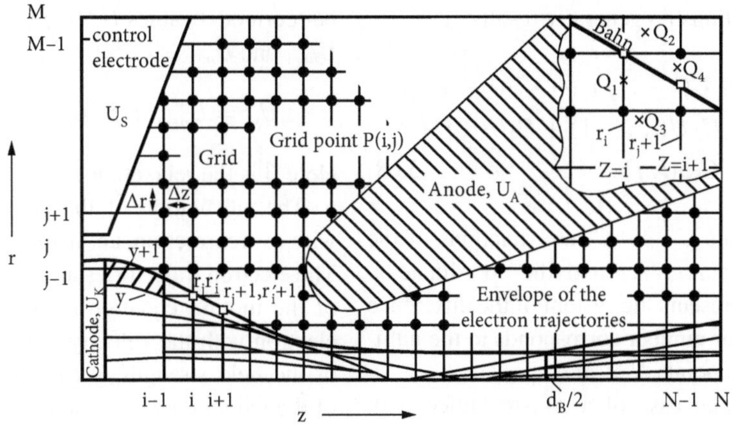

Fig. 3.48 Meridian intersection electrostatic beam generator with orthogonal coordinate lattice of r, z. $\Delta r, \Delta z$ lattice spacing, $N \times M$ number of grid points, electron trajectories 1–$\nu + 1$, elementary current tube between electron trajectories ν and $\nu + 1$ with the current I_ν in a cathode zone, r_i ordinate of the path for $z = i$, $r_i' = dr/dz$ gradient of the path for $z = i$, $Q_1 \dots Q_4$ points of environment of the path section between $z = i$ and $z = i + 1$, d_B of smallest beam diameter

there the regulation of the emission current density requires the knowledge of the potential distribution U.

The solution of the potential equation without space charge takes place by means of a gradual approximation, likewise the solution of the path equation. Because of the influence of the space charge distribution on the potential and the relationship of the potential with the emission current density and/or the electron trajectories, the functional dependence cannot be calculated independently with gradual approximation with the space charge distribution. The solution of the total problem, therefore, requires an iterative approximation procedure. For the determination of the potential distribution U Poisson's equation must solve

$$\Delta U = -\frac{\varrho}{\varepsilon_0}, \tag{3.64}$$

where

Δ is the Laplace differential operator, and

ρ the space charge distribution.

For calculation of the initial potential distribution the Laplace equation is used

$$\Delta U = 0, \tag{3.65}$$

with $\rho = 0$. The electrostatic field in an axial beam is rotationally symmetric; thus,

$$U = U(r, z), \quad \frac{\partial U}{\partial \varphi} = 0. \tag{3.66}$$

In this case, the problem reduces to a two-dimensional problem with the coordinate z, the axle of the beam generator and r the distance from the axle. With these coordinates and Eq. (3.64) the following form is obtained:

$$\frac{\partial^2 U}{\partial r^2} + \frac{1}{r}\frac{\partial U}{\partial r} + \frac{\partial^2 U}{\partial z^2} = -\frac{\varrho}{\varepsilon_0}. \tag{3.67}$$

Numerically this differential equation can be solved by replacing it with a system of different equations.

The transition of the differential equation Eq. (3.67) to a system of difference equations takes place via development of the potential U in the environment of each point $P(i, j)$ into a Taylor array after the lattice spacing $\rho\Delta$ and/or $\zeta\Delta$. This way, the potential in a grid point $P(i, j)$ is represented by its four points of neighbour $P(i-1, j)$, $P(i+1, j)$, $P(i, j-1)$ and $P(i, j+1)$. For the potential in the point $P(i, j)$

the linear relationship generally results in

$$\begin{aligned} U(i,j) &= C_1\,U(i-1,j) + C_2\,U(i+1,j) \\ &\quad + C_3\,U(i,j-1) + C_4 U(i,j+1) \\ &\quad + C_5 \varrho(i,j)\,. \end{aligned} \tag{3.68}$$

The coefficients $C_1 \ldots C_5$ are functions of the lattice spacing Δr, Δz and the coordinate r. Eq. (3.68) represents the system of the difference equations; their number corresponds to the number of the grid points $N \times M$.

In the special case of a square lattice with $\Delta r = \Delta z = h$,

$$\begin{aligned} U(i,j) &= \frac{1}{4}\,[U(i-1,j) + U(i+1,j) \\ &\quad + U(i,j-1) + U(i,j+1)] \\ &\quad + \frac{h}{8\,r_i}\,[U(i,j+1) + U(i,j-1)] \\ &\quad + \frac{h^2}{\varepsilon_0}\varrho(i,j)\,, \end{aligned} \tag{3.69}$$

which is valid for $r \neq 0$. It is to be considered that for points on the axle $P(i,0)$, because of the element with the factor $1/r$ in Eq. (3.67) and/or (3.69), the relationship for $U(i,0)$ must be modified. The solution of the difference equations is achieved via an interactive computing procedure, whereby first a potential distribution $U_0(i,j)$ without space charge is determined, which the Laplace equation (Eq. (3.65)) fulfils. In the next approximation steps, in each case the separately calculated space charge distribution is considered; this way improved potential values $U_n(i,j)$ arise. The calculation must be repeated until the approximations U_{n-1} and U_n apply:

$$|U_n - U_{n-1}| < \varepsilon\,, \tag{3.70}$$

i.e., until for two following each other approximations the difference of the potentials is in all grid points as a given bound ε.

The convergence of the calculation procedure is thereby protected and/or accelerated by a relaxation technology [47]. This technology means that for the next approximation not the values $U_n(i,j)$ are used, calculated with the nth step, but corrected values $U^*(i,j)$, which result from U_{n-1} and U_n:

$$U_n^* = U_{n-1}^* + \beta\left(U_n - U_{n-1}^*\right)\,. \tag{3.71}$$

Here β is the relaxation coefficient.

The calculation of the potential takes place with overcorrection, i.e. it applies for $1 < \beta < 2$. The electron trajectories are calculated by means of the motion equation of the electrons. If an electron with the velocity $\vec{v}$ in an electrical field of the field strength $\vec{E}$ and in a magnetic field of the induction $\vec{B}$ moves, then the Lorenz force acting on the electron is:

$$\vec{F} = -e(\vec{E} + v \times \vec{B})\,. \tag{3.72}$$

Neglecting relativistic effects, thus $v \ll c$, we obtained for the motion equation

$$m\frac{\mathrm{d}v}{\mathrm{d}t} = -e(\vec{E} + v \times \vec{B})\,. \tag{3.73}$$

By means of this equation the electron trajectories can be calculated not only in the beam generator, but also in the beam guiding system. $B = 0$ is limited on electrostatic axial beam generators, and Eq. (3.73) is simplified to

$$\begin{aligned} m\frac{\mathrm{d}^2 r}{\mathrm{d}t^2} &= -e\,E_\mathrm{r} = -e\frac{\partial U}{\partial r} \\ m\frac{\mathrm{d}^2 z}{\mathrm{d}t^2} &= -\varepsilon\,E_z = -e\frac{\partial U}{\partial z}\,, \end{aligned} \tag{3.74}$$

where E_t and E_z are the values of the components of the electrical field strength. The time t can be eliminate by means of the energy theorem

$$\frac{1}{2}m\left[\left(\frac{\mathrm{d}r}{\mathrm{d}t}\right)^2 + \left(\frac{\mathrm{d}z}{\mathrm{d}t}\right)^2\right] = -e\,U\,. \tag{3.75}$$

We obtain the non-linear differential equation of second order for the electron paths in level r, z,

$$\frac{\mathrm{d}^2 r}{\mathrm{d}z^2} = \frac{1}{2U}\left[1 + \left(\frac{\mathrm{d}r}{\mathrm{d}z}\right)^2\right]\left(\frac{\partial U}{\partial r} - \frac{\mathrm{d}r}{\mathrm{d}z}\frac{\partial U}{\partial z}\right)\,. \tag{3.76}$$

The numerical calculation of the paths can take place via procedures of rational extrapolation [47] or via the Runge–Kutta procedure. The procedure supplies the ordinate r and the upward gradient $\mathrm{d}r/\mathrm{d}z$ of the electron trajectories for one point with the abscissa $z = i + 1$ from the values r and $\mathrm{d}r/\mathrm{d}z$ of the abscissa $z = i$.

In addition, r_{i+1} and $\mathrm{d}r_{i+1}/\mathrm{d}z$ are determined by the potential U in each case, and the components of the electrical field at four points $Q(k,l)$ of the surroundings, which is between $z = i$ and $z = i+1$ (see Fig. 3.50). In addition to the matrix with the values $U(i,j)$ for the grid points P, the values $U(k,l)$ are still necessary for the points Q and the values of the derivatives of the potential $\partial U/\partial r\,(k,l)$ and $\partial U/\partial z\,(k,l)$ for the path calculation. For the calculation of the linear equations for the path, we obtained the following form:

$$r_{i+1} = r_i + f\left[\left(U, \frac{\partial U}{\partial r}, \frac{\partial U}{\partial z}\right)_{Q1}; \cdots ; \left(U, \frac{\partial U}{\partial r}, \frac{\partial U}{\partial z}\right)_{Q4}\right]$$

$$\frac{\mathrm{d}r_{i+1}}{\mathrm{d}z} = \frac{\mathrm{d}r_i}{\mathrm{d}z} + g\left[\left(U, \frac{\partial U}{\partial r}, \frac{\partial U}{\partial z}\right)_{Q1}; \cdots ; \left(U, \frac{\partial U}{\partial r}, \frac{\partial U}{\partial z}\right)_{Q4}\right]. \quad (3.77)$$

The starting points of the paths (r_0, z_0) are arranged at a small distance in front of the cathode; the initial directions lie perpendicularly to the cathode surface. The number of paths that can be computed determines the number of ini tial points. In addition, the cathode is divided into zones and a path is assigned to each zone. The values r are then gradually determined by the calculation procedure and $\mathrm{d}r/\mathrm{d}z$ determines the paths toward the z-axis. To calculate the space charge distribution of ρ from the relationship, we use

$$j = \varrho \times v\,. \quad (3.78)$$

A way to numerically calculate the space charge distribution is to determine the current density inside elementary current tubes, which are formed in each case from the two neighbouring paths according to the selected zones of the cathode. The current in a current tube between two paths ν and $\nu+1$ of a cathode zone is the space charge limited current of this zone. In this procedure, the current density $j_\nu(z)$ is reversely proportional to the distance of the neighbouring paths ν and $\nu + 1$.

The part of a current tube for the space charge in the grid point $P(i,j)$ thus comes from the current of the current tube, which calculates the values r and $\mathrm{d}r/\mathrm{d}z$, the paths r and $r + 1$ and the values $U(i,j)$. The space charge $\rho(i,j)$ then results from the sum of the mean space charges of the current tubes, which decline into each grid point when assigned an influence zone.

For the calculation of the values $\rho(i,j)$ a relaxation procedure with a correction factor γ is used

$$\overline{\varrho}_n = \overline{\varrho}_{n-1} + \gamma\left(\varrho_n - \overline{\varrho}_{n-1}\right), \quad (3.79)$$

with $0 < \gamma < 1$.

Another method for the determination of the space charge distribution of each cathode zone is by the use of representative related paths, which come out from the centre of the cathode zone. In this case, the delay time of the electrons on each path in a lattice mesh and the reciprocal distance to the grid point their space charge contribution for this grid point are determined.

The current density at the initial points near the cathode surface j_{eQ}, and thus the current of an elementary tube in and/or a representative path is calculated with the help of the space charge law. The potential of the second or third grid point in front of the cathode is used and inserted for the distance z into Eq. (3.62) according to $2\Delta z$ and/or $3\Delta z$. According to the approximation U_n is improved at the end of each iteration j_{eQ} and/or I_ν. In the first step one frequently assumes estimated values $I_{\nu 0}$.

A requirement of the development of beam generators is the optimization of the electrode configuration with other methods or an empirically determined arrangement. The control electrode geometry is systematically varied and the arrangement with the best beam parameters for the required use is selected. Figure 3.49 shows stages of the optimization of the electrode shape and the gradual improvement of beam parameters, such as the smallest beam diameter, the

a

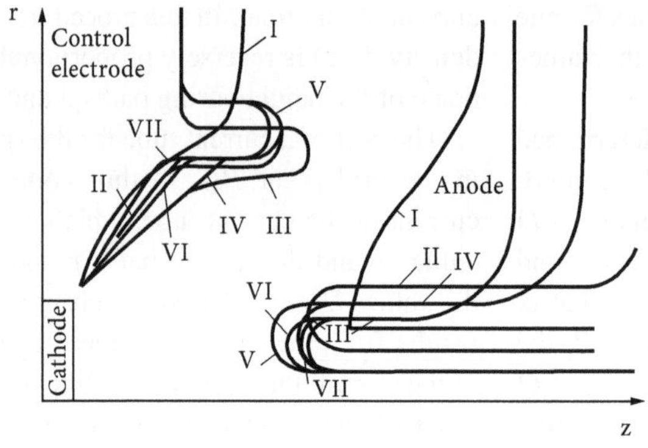

b

Configuration	Beam current I_B mA	Smallest beam diameter d_B mm	Aperture in the smallest beam diameter α_B rad	Leakage current on the anode mA
I	480	3	0,002	200
II	520	2	0,02	10
III	410	0,33	10×10^{-2}	15
IV	560	0,32	8×10^{-2}	5
V	570	0,65	6×10^{-2}	0
VI	520	0,395	$4{,}2 \times 10^{-2}$	0
VII	510	0,38	$3{,}5 \times 10^{-2}$	0

Fig. 3.49 Optimization of an electrostatic jet produced by calculation of characteristic parameters of the electron beam with variation of electrode geometry. **a** variation of the geometry of the electrodes, **b** characteristic parameters of the electron beam. Beam current $I_B = 0.5$ A, acceleration voltage $U_B = 60$ kV, cathode diameter $d_K = 4.5$ mm

aperture in the smallest beam diameter and the leakage current on the anode.

3.5.3.3 Crucibles

The crucible serves as containment for the evaporation material. Material and design of crucibles are determined for electron beam heating, the transfer of energy to the surface of the evaporation material and the power density of the electron beam.

The material of the crucible must have a low vapour pressure at high temperatures and also a small reaction affinity with the evaporation material and the gas in the recipient. Too strong erosion and evaporation of the crucible material lead to contamination of the evaporation material or the evaporated film and also limit the lifetime of the crucible and thus of the coating operating efficiency.

Depending on the different evaporation tasks, water-cooled copper crucibles or so-called hot crucibles, i.e. water-cooled carriers with ceramic crucible cartridges are used, see Fig. 3.50. In special applications, evaporation without a crucible plays a certain role. Water-cooled copper crucibles have the largest application. Because of the thermal conductivity of copper, extremely high temperature differences can develop at the boundary surface between the crucible wall and the evaporation material.

From such crucibles materials that melt at temperatures of 3000–4000 K, such as tungsten and strongly reactive materials, like titanium can be evaporated. The temperature gradient at the boundary layer between copper crucibles and the evaporation material is so large that we can speak of a temperature jump. Figure 3.51 shows the temperature gradient of a water-cooled cop-

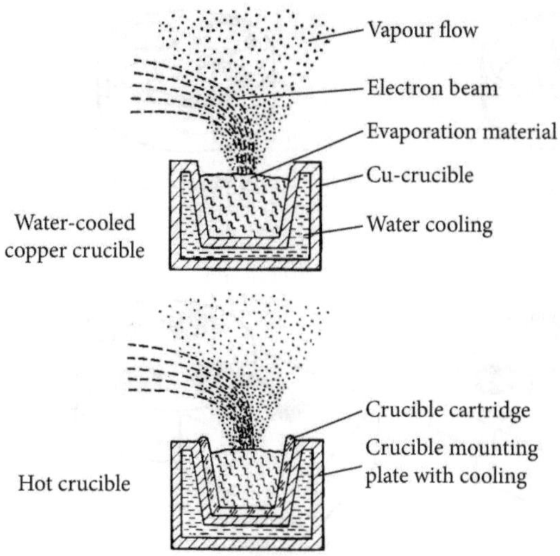

Fig. 3.50 Types of crucible of electron beam evaporators

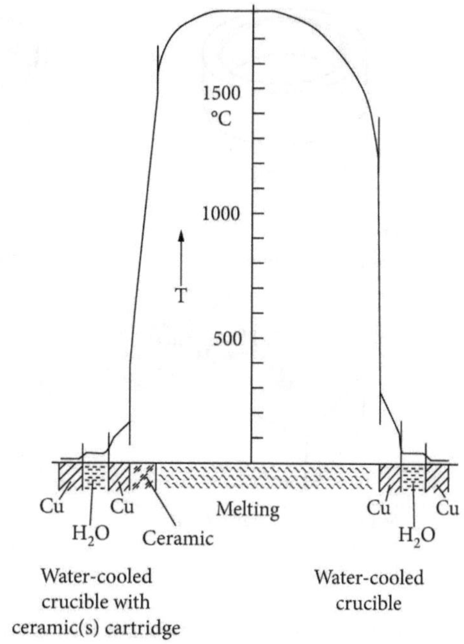

Fig. 3.51 Temperature gradient in the melt and in the crucible (schematized for a water-cooled crucible with and without crucible cartridge)

per crucible. For heat transfer to the crucible, the range within which the liquid melt of the evaporation material contacts the crucible wall is essential. Wetting properties, surface finish of the crucible material and, in particular, the creation of an oxide coating determines the height of the temperature step and the heat transfer.

If high power losses are inadmissible and/or if a higher evaporation rate and thus a better effectiveness is to be obtained, crucible materials that work as a heat barrier must be used.

Figure 3.51 shows the influence of a crucible construction on the temperature gradient in the crucible.

At the boundary layer between evaporation material and crucible wall a smaller temperature gradient takes place, which is induced from the water-cooled crucible. The heat conductivity of the insert material of the crucible determines the heat loss at the wall.

Ceramic crucibles on the basis of Al_2O_3 have attained special importance. For small crucibles, graphite, titanium diboride or boron nitride are frequently used. The lifetime of these materials is about 100 h. The dimensions of the crucibles are determined by parameters of the evaporation process, such as properties of the evaporation material, evaporation rate and the requirement of the energy utilization, as well as the necessary distribution of the vapour stream density. Approximate values for crucible design during given evaporator capacity resulting from the demands that the evaporation material must be kept with liquid melt up to the edge of crucible, on the one hand, and a limited power density, on the other, must not be exceeded in the case of beginning splash creation. The values of the dimensions of the crucible, see Table 3.2, grow proportionally to the power value of the evaporator.

Table 3.2 Dimensions of crucibles for different ranges of evaporator power

Evaporator power kW	Diameter and/or side length of the crucible mm
5–7	15–35
10–15	35–75
40–60	50–150
250	200–500

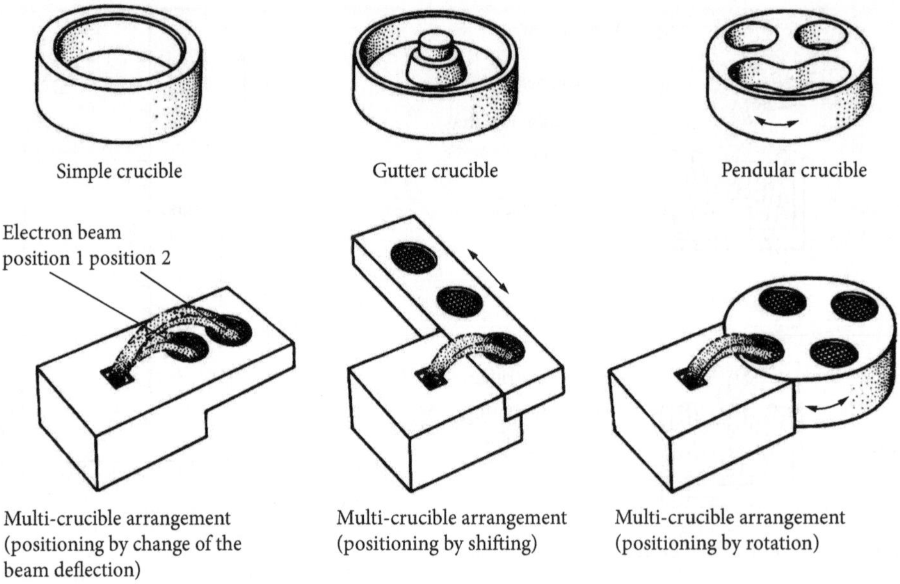

Fig. 3.52 Characteristic crucible designs and/or crucible arrangements of small surface evaporators

When a crucible insert is used a uniformly temperature distribution over the melt sump and a larger molten crater depth results. The choice of materials for the crucible insert not only depends on the heat conductivity, but also on the chemical stability towards the hot evaporation material. Due to repeated heating and cooling, such crucibles are exposed to high mechanical stress; therefore a high temperature change resistance is necessary for the material of the crucible insert.

The predominant crucible designs for small surface evaporators are rotationally symmetric. Only in exceptional cases are square crucibles used for the evaporator for better wobbling adjustment. Crucibles for large surface evaporators with expanded source distribution are adapted against to the size and form of the substrates. The application of such crucibles is usually attached to evaporator capacities of 50 kW and more.

Characteristic crucible forms are shown for small surface evaporators in Fig. 3.52. Simple, pot-shaped crucibles are used for splash-free evaporation of large quantities with electron beam wobbling. In order to improve the uniformity of the beam current density distribution, the crucible is frequently rotated during evaporation.

Fig. 3.53 6 kW electron beam evaporator with different crucibles for replacement

The groove crucible serves for optimal utilization of the evaporation material. If different materials are evaporated in one cycle with only one electron gun, then multi-crucible arrangements must be used. Figs. 3.53 and 3.54 show different crucibles for electron beam evaporation.

To meet demands to avoid contamination of the evaporation material through the material of the crucible, an evaporation method without a crucible has been developed (see Fig. 3.55). The simplest solution is evaporation from a limited melt sump in an expanded block of the evaporation material. For sublimating materials

Fig. 3.54 Indirect electron beam-heated crucible cartridges

the evaporation from rods and plates is applicable; the rods and plates are moved in relation to the electron beam, particularly, if the electron beam is also wobbling. Crater formation in the sublimating material is avoided by a coordinated mechanical shift of the evaporation material and wobbling of the electron beam.

3.5.3.4 Charging

If constant evaporation conditions are to be maintained over long periods of time, then it is necessary to recharge the crucibles during the operation with evaporation material. A recharging of the crucibles and is heated particularly necessary when operating with extremely high evaporation rates and/or if very large quantities must be evaporated in a vapourization cycle. The evaporation of alloys presupposes a recharging of the crucible in order to ensure a defined and constant composition of the vapour stream.

Recharging of the crucibles can take place continuously or intermittently. Simple intermittent charging presupposes that the crucible has a large volume; it makes smaller demands regarding the form and the pre-degassing of the evaporation material. The material supplied must be degassed before being fed into the evaporation equipment. Because the height of the melting bath changes during intermittent feeding, this procedure is applicable only with limited requirements of long-term constancy of the evaporation rate and vapour stream density distribution.

For continuous recharging at each time unit only the quantity that corresponds to the evaporation rate must supplied. Continuous recharging makes higher demands on the charging device; however over long periods of operation constant evaporation parameters must be ensured. The evaporation material used depends on the very different material types, such as, for example, wires, rods or granulates. Figure 3.56 shows different principles of recharging.

A frequently used method of feeding evaporation material is wire charging. The wire is uncoiled with the help of a gearwheel pair by a supply roll and led with the use of a tube to the crucible. For the use of evaporation material in rod form a special crucible shape has been developed, *the ring crucible*. A rod of the evaporation material forms the bottom, a water-cooled copper ring the crucible wall. In the range of the ring crucible the melt sump is formed. Also, hot crucibles with rod refeeding can be made with tubular ceramic inserts.

For pelleted or granulated evaporation material portion charging is used. An appropriate technical solution is the vibration conveyor. A further special method of feeding, in particular for low melting evaporation material, like tin or zinc, is liquid feeding. The material is melted in a pre-crucible and is transferes into the evaporator crucible by the electrode beam at evaporation temperature.

3.5.4 Evaporation Rate

3.5.4.1 Power Density and Evaporation Rate

Due to the exponential dependence of the evaporation rate on the temperature, the evaporation rate increases with the power density of the evaporation material. During given power P_0 a higher power density can be reached by reduction of the electron beam diameter d_F on the evaporation material, by reduction of the electron beam diameter d_F on the evaporation material ($p = P_0/\pi/4\,d_F^2$). Figure 3.57 shows the evaporation

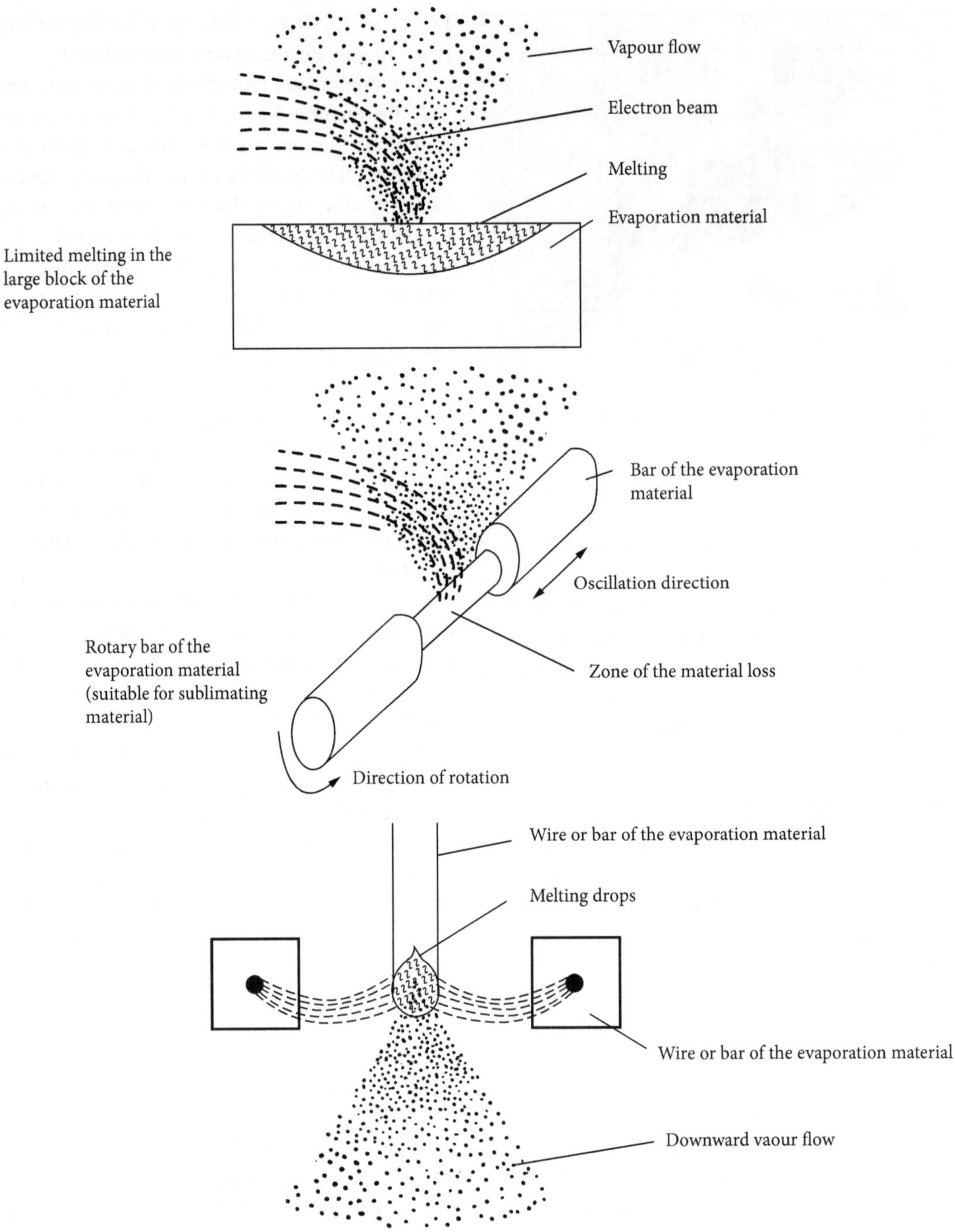

Fig. 3.55 Principles for crucibleless evaporation with electron beams

rate as a function of the current of the magnetic lens for an evaporator with an axial gun. In the case of optimal focusing, i.e. in the case of the smallest beam diameter d_F, a maximum of the evaporation rate results. In Table 3.3 the dimensions of the electron beam are illustrated for the

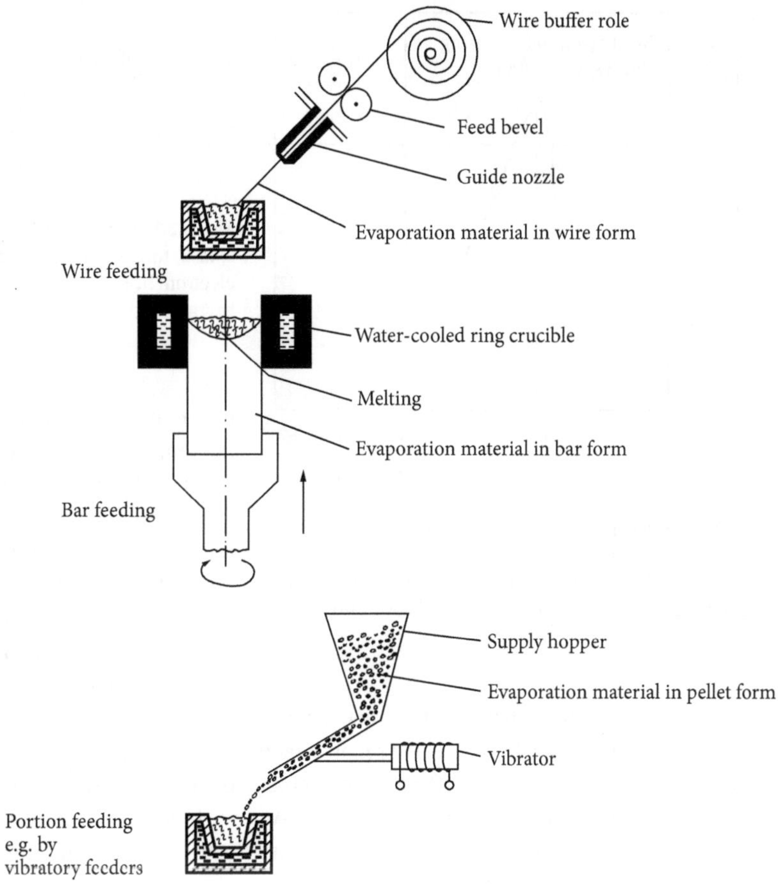

Fig. 3.56 Principles of aftercharging of the crucible with evaporation material

Table 3.3 Influence of the current at the gate electrode U, i.e. focusing of the electron beam on the dimensions of the focal spot F_B and the relative change of the evaporation rate $\Delta a_v/a_v$ during electron beam power P_0 for a transverse evaporator

Evaporation material	P_0 kW	F_B for $U = 0\,V$ mm × mm	for $U = -300\,V$	$\frac{\Delta a_v}{a_v}$ %
Al	8	18 × 16	5 × 8	30
	10	18 × 16	5 × 8	25
	12	18 × 16	5 × 8	15
	14	10 × 12	8 × 14	10
W	8	5 × 12	3 × 8	60
	10	5 × 12	3 × 8	50
	12	5 × 12	3 × 8	40
	14	8 × 16	6 × 10	35

impact of the electron beam on the evaporation material and the evaporation rate as a function of the voltage at the control electrode for certain achievements.

During optimal focusing, the evaporation rate with the power P_0 increases, until the losses in the vapour cloud become active. In many cases, another limit becomes active, resulting in defor-

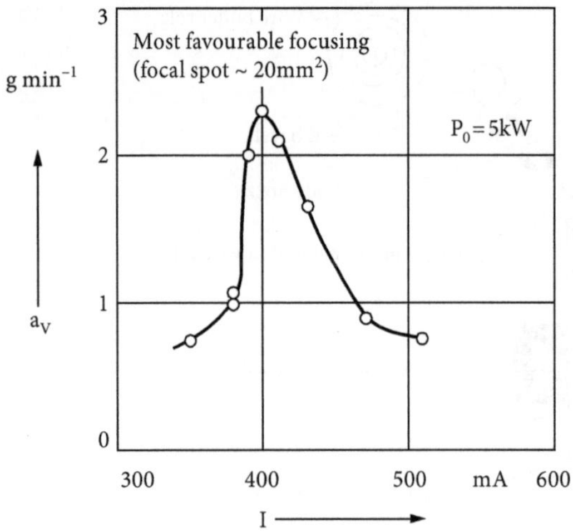

Fig. 3.57 Evaporation rate a_V as function of the power density. Change of the power density with change of the lens current I (water-cooled copper crucible with a diameter of 20 mm, evaporation material CoFe30) [48]

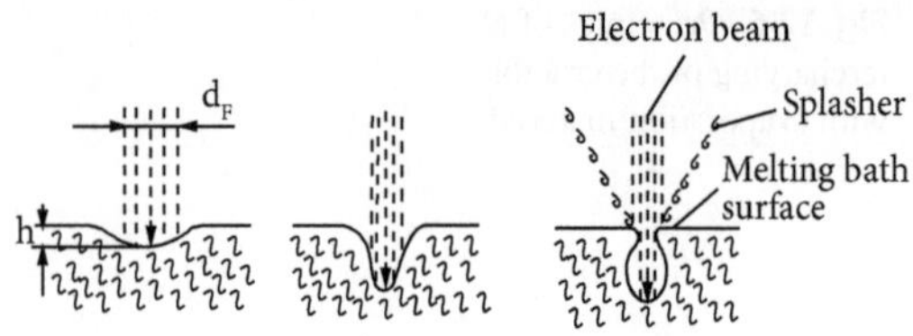

Fig. 3.58 Deformation of the melting bath surface and splash formation with increasing power density of the electron beam (h depth, d_F diameters of the electron beam)

mation of the melting bath surface and splash formations. The reason for this is the recoil effect of the vapour stream, i.e. the effect of the vapour pressure over the melting bath (see Fig. 3.58). At rising power density, i.e. with increasing vapour pressure, the cavity h in the melting bath surface becomes larger. In rough approximation the depth h, measured in mm, also results:

$$h \approx 10^2 \frac{p_D}{\varrho}, \tag{3.80}$$

where

ρ in kg/m^3 is the density of the evaporation material, and

P_D in mbar of $\times 10^{-2}$ pressure of the vapour at the surface of the melting bath.

Among other things, the cavity influences the vapour stream density distribution. With large depth the evaporation rate can be reduced by the so-called chimney effect. If the cavity under inclusion of vapour is intermittently closed, then splashes develop, which are usually inadmissible for the coating.

Local differences of the vapour pressures and bath circulation impact the splash limit unfavourably. The disposition to splash formations not only depends on the power density, but also on the purity of the evaporation material and on the temperature gradient in the crucible.

In order to obtain, with given power and given crucible dimensions, a high evaporation rate without exceeding of the splash limit, a small beam diameter d_F, thus a high power density is usually reduced when the electron beam is led across by programmed deflection over the vapour delivery surface. This operation mode with beam wobbling can still be operated with power densities in the focal spot, by which in stationary operation mode splashes arise.

If it is deflecting only in one direction with saw tooth voltage, it is thus valid for the reduced mean power density $\bar{p}$:

$$\overline{p} = \frac{d_F}{l} p, \tag{3.81}$$

where l is the sweep amplitude. The retention time t_W of the beam in one period is:

$$t_W = \frac{d_F}{l} \times \frac{1}{f_W}, \tag{3.82}$$

where f_W is the wobbling frequency.

Wobbling in the x and y-directions is frequent. Usually, sinusoidal voltages are used, which can be very simply produced. Adjusting a very small wobbling frequency approximates the stationary case and thus does not shift the limit of the splash formations substantially. Operation wobbling leads to higher heat losses because the heated mass is larger than with stationary operation.

Table 3.4 Maximum evaporation rate a_V for crucibles with different diameters of the vapour delivery surface data processing and different crucible depth H during the same electron beam achievement (crucible water-cooled, electron beam achievement 5 kW, beam not wobbling, evaporation material CR 80/20)

D_V	H	a_V
mm	mm	g min^{-1}
10	10	1.45
20	15	1.48

Table 3.5 Time dependence of the condensation rate a_K during the period of operation t_B (evaporator capacity 12 kW, diameters of the water-cooled copper crucibles 50 mm, measurement of the condensation rate a_K in the distance of 250 mm, evaporation material aluminium)

t_B	a_K
min	µm min^{-1}
1	4.1
2	3.6
16	3.2

3.5.4.2 Evaporation Parameters and Evaporation Rate

In the following, the influence of the main parameters of some technical electron beam evaporators on the evaporation rate are discussed. An important parameter is the power density in the focal spot of the electron beam.

The middle power density with programmed deflection is within the range of 2×10^6–2×10^7 W/m^2, with the low values used for operation with crucible inserts. The crucible size and deflection programme are selected according to the necessary evaporation rate and vapour stream density distribution. With small water-cooled crucibles the heat losses have a substantial influence on the evaporation rate. With increasing crucible diameter the temperature gradient between the vapour-delivery surface and the crucible, and thus the heat losses, sinks. Under the same conditions one achieves a higher evaporation rate with a larger crucible, Table 3.4.

Due to increasing heat losses on removal of the filling level of the crucible, the evaporation rate has the highest value with full crucibles and decreases with progressive evaporation. The filling level only has an influence on the time-dependent trend of the evaporation rate, but because of the so-called chimney effect at the crucible wall, the vapour stream density distribution is also time-dependent (see Table 3.5).

Figure 3.59 shows how the heat losses are reduced by a crucible insert and thus can increase in relation to the evaporation rate during evaporation from a water-cooled copper crucible. Despite the smaller electron beam power during evaporation of aluminium with the crucible insert, for instance, a triplication of the evaporation rate is reached.

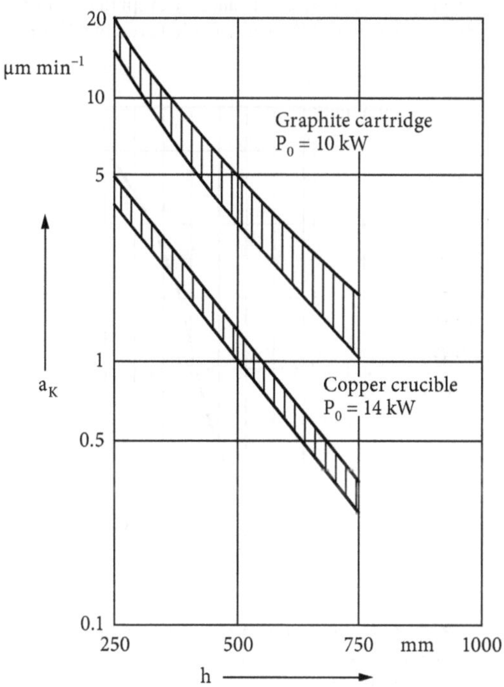

Fig. 3.59 Comparison of the condensation rates a_K of aluminium with evaporation from water-cooled crucibles with and without crucible insert (evaporator capacity P_0, evaporator distance h, crucible diameter 50 mm)

The efficiency of an evaporation mechanism is characterized by the maximum evaporation rate for a certain material as a function of the evaporator power. In place of the evaporation rate the condensation rate perpendicular to the crucible centres in a certain distance h can be used.

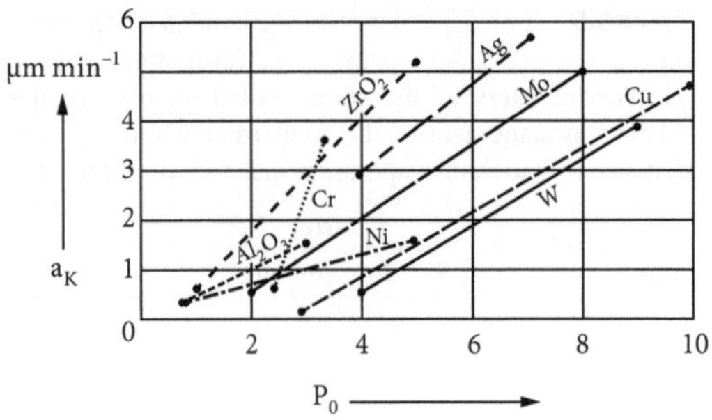

Fig. 3.60 Condensation rates a_K for different materials at evaporation of 10 kW of transverse evaporators as a function of the power (distance of evaporator substrate 400 mm, diameter of the water-cooled crucible 38 mm)

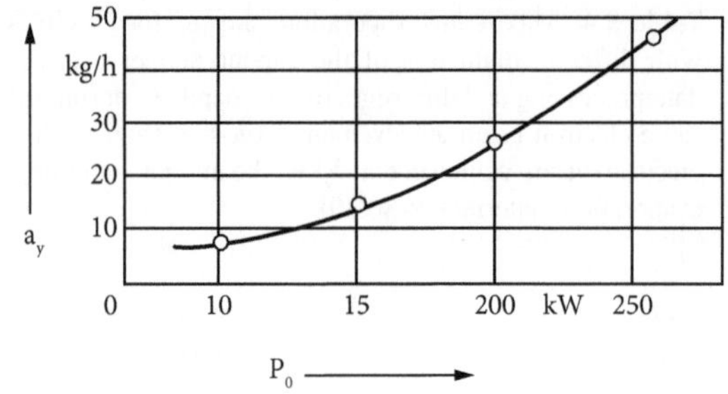

Fig. 3.62 Evaporation rate a_V of aluminium as a function of the evaporation power P_0 (evaporator with axial gun, ceramic crucible 500 × 500 mm) [50]

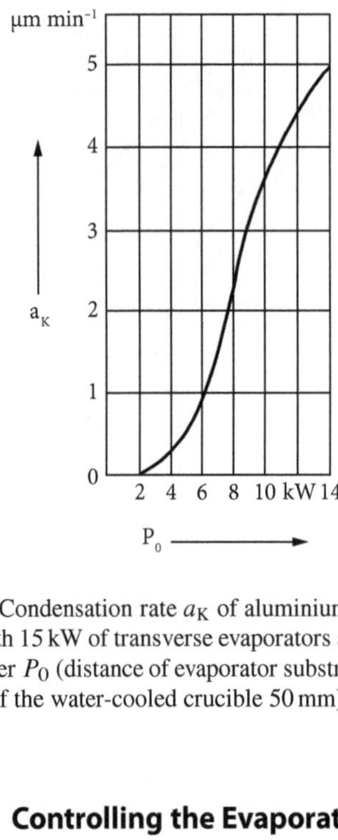

Fig. 3.61 Condensation rate a_K of aluminium with evaporation with 15 kW of transverse evaporators as a function of the power P_0 (distance of evaporator substrate 250 mm, diameter of the water-cooled crucible 50 mm) [49]

3.5.4.3 Controlling the Evaporation Process

The properties of evaporated films depend on many parameters of the coating process or the design of the evaporation equipment, the measurement and regulation of the coating process. The main parameter that determines the evaporation process is the condensation rate, see Fig. 3.60 to 3.62.

An important condition for a stable evaporation rate is the stabilization of the electron beam power. The permissible relative changes of parameters are:

for the accelerating voltage to

$$\Delta U_B / U_B \leq \pm 1\text{–}2\,\% ,$$

and for the beam current to

$$\Delta I_B / I_B \leq \pm 1\,\% .$$

Also the remaining electrical magnitudes, which determine the power density on the surface of the evaporation material, such as the magnetic lens currents, returning currents and the currents of the programmed deflection, must be regulated. As a variable for the evaporation rate, generally the vapour stream and/or directly the power is used. Regulating circuits for the adjustment of a constant evaporation rate of electron beam evaporators up to 15 kW are frequently realized on the basis of condensation rate measurements with swing quartz systems.

When evaporation the rate and the vapour stream density distribution must be kept constant over several hours, then this achieved with evaporators with refeeding systems, unless the beam current in the melting bath level must also be controlled. The most frequently applied method for rate regulation is measure the emission current, Fig. 3.63.

The swing quartz method is limited by the lifetime of the measuring quartz. It must be replaced after an appropriate material mass coating. For thick films and high rates within the range of µm/s its application is, therefore, not reasonable.

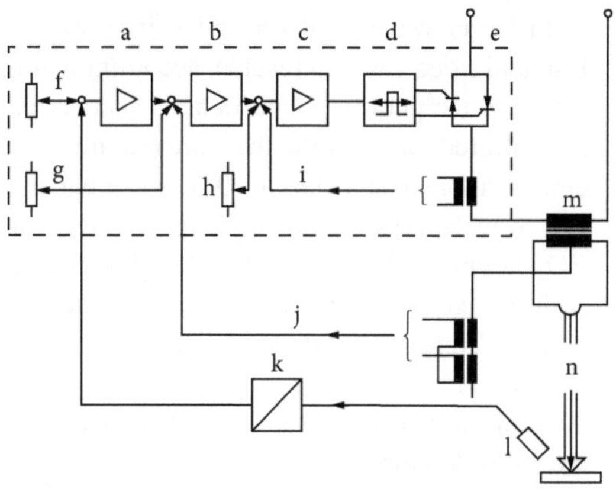

Fig. 3.63 Block control schema for rate regulation. *a* rate automatic controller, *b* emission current automatic controller, *c* filament current automatic controller, *d* impulse equipment, *e* actuator, *f* rate set point value, *g* emission current value, *h* minimum heat current, *i* heat current, *j* emission current, *k* rate actual value, *l* measuring head rate, *m* cathode heat transformer, *n* electron beam

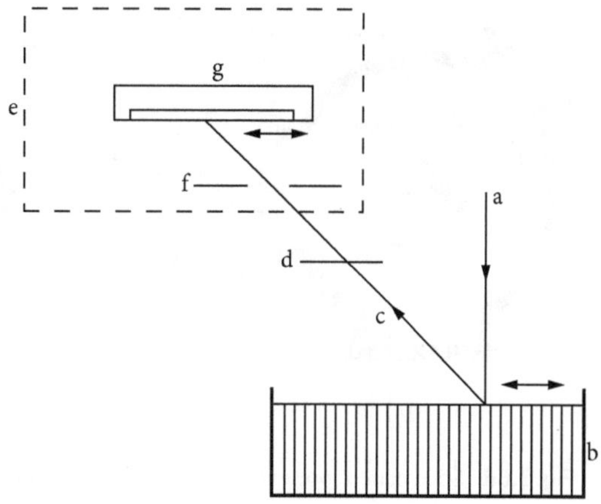

Fig. 3.64 Principle of the X-ray sensor. *a* electron beam, *b* crucible, *c* X-ray, *d* filter foil, *e* camera, *f* hole aperture, *g* photodiode

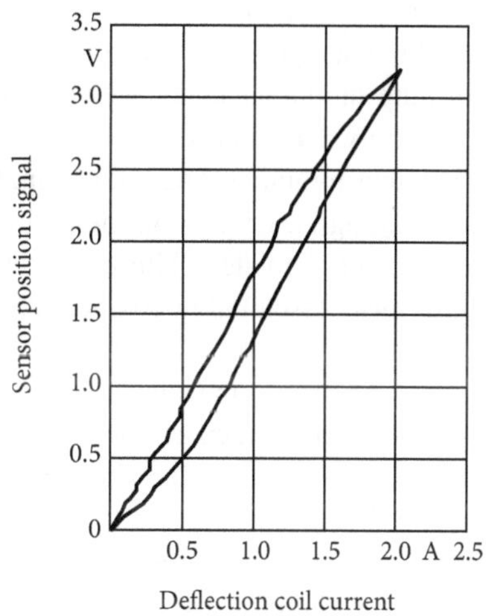

Fig. 3.65 Sensor signal for the position control of an electron beam evaporator as a function of the coil current of the deflection system

3.5.4.4 Focus Control

When using electron beam evaporators the focus in the crucible should be visually adjustable and controllable [51]. The sensor measures the X-rays that develop by collision of the electron beam with the evaporation material. With a pinhole camera the focus is imaged on a photodiode, which converts the signals of the x and y-coordinates into voltage levels to control the gun (see Fig. 3.64).

The measuring accuracy of the beam position is better than 1 mm, and the sensor is used for electron beam evaporators with beam currents over 6 kV and power more than 1 kW. Figure 3.65 illustrates the accuracy of the sensor. The hysteresis of a deflection system is shown by measurement of the beam position as a function of the coil current. With image converters the X-ray focus can be made visible.

3.5.4.5 Other Sources of Evaporation

With the majority of applications of thin film technology resistance-heated evaporation sources or electron beam guns are sufficient. For special tasks other evaporation methods are used.

With the induction evaporator one heats the coating material by induction received in electrically conducting materials in electrical alternating fields. The metal to be evaporated is in a ceramic crucible arranged in the centre of a water-cooled copper spiral and loaded with frequencies of approximately 100 kHz. The energy is transferred without mechanical contact; the utilization of energy is much better than with resistance-heated evaporation sources or evapo-

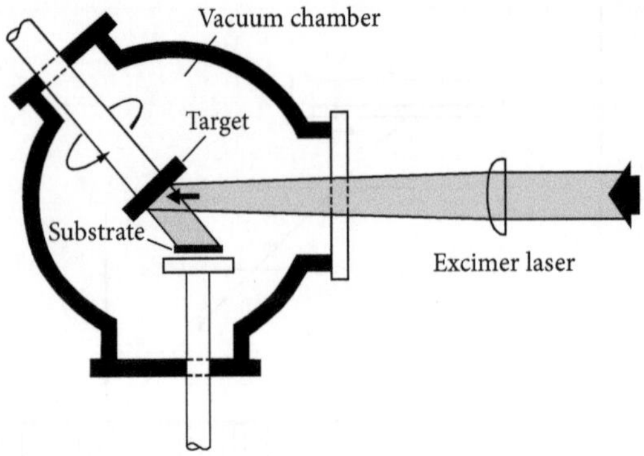

Fig. 3.66 Schematic representation of the laser ablation

ration by electron beams. A further advantage is the high evaporating rates.

Laser evaporation [44] is used particularly for research purposes. Figure 3.66 shows the schema of laser ablation equipment. The target is located on a turning mounting plate, which can be ablated uniformly. Diagonally opposite to the target the substrate is mounted. The laser beam focused by a lens through a window in the recipient and the target material ablated. This evaporating material flies away essentially perpendicularly from the target and strikes the substrate which is to be coated. On the way to the substrate the partially excited atoms deliver their energy in the form of intensive plasma glowing. In principle, any laser with sufficient power can be used for evaporation. The choice of a suitable laser and its operating parameters are, therefore, of great importance.

During evaporation of metals, for example, the following is to be considered. The absorbed part at a metallic surface of a laser beam with perpendicular impact on the relationship of the emission ε depends on:

$$\varepsilon = \frac{4n}{(n+1)^2 + k^2}, \qquad (3.83)$$

where

n is the real part of the complex refractive index,
k the imaginary part of the complex refractive index,
ε a metal parameter, excluded in the spectral region $0.4 < \lambda < 1.0\,\mu\text{m}$, and
n, k and emission ε depend on the wavelength.

With larger wavelengths n and k increase very fast, and ε decreases in reverse, accordingly. Further ε is temperature-dependent; ε increases in the infrared range with the temperature. The emission for some metals at 20 °C with characteristic wavelengths for argon, ruby, ND:YAG and CO_2-lasers are listed in Table 3.6. Experiences with an excimer laser, which has a particularly high efficiency with the evaporation of metals, are reported in [52, 53].

Some important advantages of laser ablation are the following:

- Target and film material also have the same composition as multi-component materials.
- Very high condensation rates, also with very high-melting materials. As consequence, purer films with the same residual gas conditions as in the case of conventional evaporating.
- The target forms at the same time its own crucible, so that also with highly reactive materials at evaporation temperature no reaction occurs between crucible and coating material.
- No glow cathodes or other parts present at high temperature (for example, sources of evaporation) are needed, which is particularly advantageous for laser ablation in the case of reactive coating processes.
- By splitting the laser beam can be evaporated by different targets simultaneously.

There are, however, two disadvantages for which there is still no fully satisfying solution.

- A part of the material condensed on the coupling window weakens the laser beam, decreases the condensation rate and limits the film thickness attainable. An improvement can be achieved by attachment of one or more deflecting mirrors, because the reduction of the reflecting power of the mirrors by condensate reduces the evaporation process less strongly than a transmission reduction with the coupling window. The attachment of a rotary shutter on the vacuum side of the window is also a solution.
- The splashing effect means that microscopically fine particles with sizes between approximately 0.1–10 μm strike the substrate and reduce the film quality. The same problem

Table 3.6 Emission of different metals at 20 °C by some laser wavelengths

Metal	Emission Ar (500 nm)	Rubin (700 nm)	Nd:YAG (1000 nm)	CO_2 (10 µm)
Aluminium	0.09	0.11	0.08	0.019
Cooper	0.56	0.17	0.10	0.015
Gold	0.58	0.07	–	0.017
Iridium	0.36	0.30	0.22	–
Iron	0.68	0.64	–	0.035
Lead	0.38	0.35	0.16	0.045
Molybdenum	0.48	0.48	0.40	0.027
Nickel	0.40	0.32	0.26	0.03
Niobium	0.58	0.50	0.32	0.036
Platinum	0.21	0.15	0.11	0.036
Rhenium	0.47	0.44	0.28	–
Silver	0.05	0.04	0.04	0.014
Tantalum	0.65	0.50	0.18	0.044
Tin	0.20	0.18	0.19	0.034
Titanium	0.48	0.45	0.42	0.08
Tungsten	0.55	0.50	0.41	0.026
Zinc	–	–	0.16	0.27

also arises with arc evaporation. The density and size of these particles also decrease with the energy of the laser beam and with increasing distance from the target substrate.

With arc evaporation one must differentiate between anodic and cathode arc evaporation. Cathode arc evaporation is based on the following principle. [54, 55].

Along with the current voltage characteristic in a gas discharge obtained within a pressure range between approximately 10^{-1}–10^{-3} mbar an abnormal gas discharge adjusts between a cathode and an opposite anode, which is characterized by the fact that with increasing voltage difference also the current increases.

Further increase of the voltage leads to spark breakdown, which is characterized by steeply dropping voltage with increasing current, following from the range of the arc discharge with current densities of 10^7–10^{11} A/m^2. Pressure, gas species and electrode configuration influence the minimum amperage.

The material (cathode) to be evaporated, which can be of different standardized sizes of either circular or rectangular form, situated opposite any formed anode between an electrical arc is ignited (see Fig. 3.67). The arc feeds on the evaporating material, so that such an arc discharge can be operated also under high vacuum conditions.

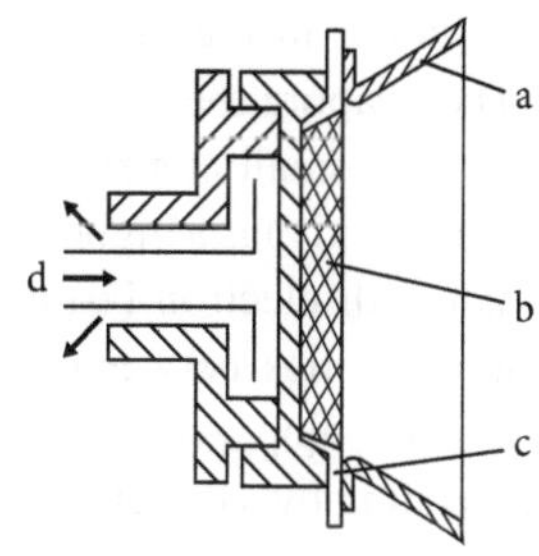

Fig. 3.67 Schema of a vacuum arc evaporator for cathodic evaporation. *a* anode, *b* target, *c* insulator, *d* water cooling

The arc moves in statistic distribution over the cathode surface, whereby also several spots can occur at the same time. Due to the high power density of approximately 10^{11} W/cm^2 within the range of the cathode surface (spot), an arc strikes the cathode material and evaporates it. Within the range of the spots, which have a diameter of approximately 10 µm, temperatures range between 3000 and 6000 K.

Apart from the problem-free evaporation of alloys – target material and film possess the same composition – the high ionization rate of vapour and residual gas for reactive coating processes are also of advantage. With a stabilized magnetic field arc evaporation systems can be manufactured, such as Al_2O_3, AlN, VN, VO_x, TiN, ZrO_2 and ZrN splash free films.

Anodic vacuum arcs [56], which work with spots on cold cathodes and with evaporating hot anodes, have the advantages of thermal evaporation, such as the high evaporation rate, the low thermal load of the substrates, low operating voltage with the advantages of a plasma-supported procedure such as separation of the films from the plasma condition with high adhesion on the substrate, and homogeneous films of compact structure with characteristics close to the respective solids.

A further difference is the different relationship of the anode surface to the cathode surface, which is very small in the anodic process (see Fig. 3.68). The evaporation is done via electron bombardment with small currents; the electrons have a low energy and thus a higher probability of ionization than with electron beam evaporation. The ionization rate lies between 0.5 and 10 %. Additional ions develop, which are accelerated, so substrate cleaning by a dry etching process before the beginning of coating is possible.

Theoretical considerations assume that at the beginning of the process on a small range of the anode, spots form, which lead to heating and evaporation of a small material quantity. The developing higher density inside these ranges causes an increased energy flow, which then again produces more vapour, until finally equilibrium adjusts itself. Discharge is stabilized by the high temperature of the anode.

Typical operational data from anodic arc evaporators are:

Remainder gas pressure 10^{-3} to 10^{-5} mbar
Arc voltage ~ 20 V
Current ~ 20 to 200 A
Energy ~ 0.4 to 4 kW

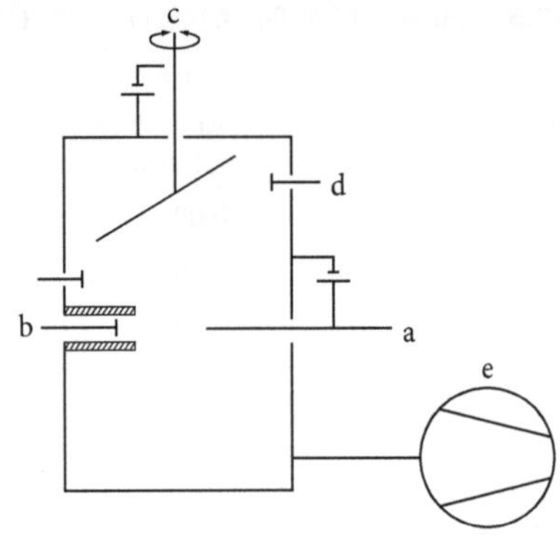

Fig. 3.68 Schema of a coating equipment with anodic deposition by a vacuum arc evaporator. *a* anode, *b* cathode, *c* substrate stops, *d* glowing electrode, *e* evacuation system

Substrate temperature ~ 70 °C
Deposition rate 0.1 to 100 N m/s
Electron temperature 0.4 to 1 eV
Electron density 10^{15} to 10^{18} m^{-3}
Ion temperature 1 eV
Ion energy 5 to 150 eV
Ionization rate of 0.5 to 10 %.

3.6 Ion Plating

3.6.1 Characterization of Ion Plating

Ion plating is a vacuum coating process, by which the condensation of vapour takes place under the influence of ions within a feed gas or of ions within the vapour. The collisions of the ions with the substrate surface and/or the condensing film are connected with a transfer of an energy transmission and an impulse transmission. These are the cause for the change of the film characteristics by the ion effect at the condensation of the film. By the ion plating process the following elementary processes take place [57, 58].

At the collision of high-energy ions sputtering of the substrate surface and/or the adsorb films occurs. If vapour condensation begins, further substrate material sputters, a partial resputtering of the condensing material takes place; with progressive condensation only a part of the condensate is resputtered. In addition, the ions produce a certain defect concentration, whereby the surface is activated.

The influence of the ions on the film properties determines to which extent the energy balance is changed with the condensation. The energy transfer is connected with the condensing vapour stream,

$$E_D = n_D \times e_D \,, \tag{3.84}$$

where

n_D are the number of condensing neutral vapour particles for each time and unit area, and

e_D their mean kinetic energy.

e_D is approximately

$$e_D = \frac{3}{2} k T_V \,, \tag{3.85}$$

where

k is the Boltzmann constant, and

T_v the evaporation temperature.

The energy transfer by the striking ions is

$$E_I = n_I \, e_I \,, \tag{3.86}$$

where

n_I is the number of the ions is for each time and unit area, and

e_I the middle energy of an ion.

To this applies

$$e_I = e \, \overline{U_I} \,, \tag{3.87}$$

where $\overline{U_I}$ is the middle accelerating voltage of the ions. An indicator for the energetic activation of the substrate surface is the energy relationship [58]

$$\varepsilon_I = \frac{E_I + E_D}{E_D} \,. \tag{3.88}$$

The energy relationship denotes the factor by which the condensation energy is increased through the influence of the ions in relation to the condensation energy by evaporation. As long as has not yet led to a rise in temperature of the substrate due to transients, the energetic excitation of the surface atoms corresponds to a higher effective condensation temperature T_K, as it arises from the middle substrate temperature T_S [57]. Among other things, a larger mobility of the condensing particles and thus a higher rate of diffusion velocity result.

Since generally $E_D \ll E_I$, the energy relationship approaches the rate relationship n_I/n_D of ions and vapour particles at the substrate surface. The middle accelerating voltage of the ions is, therefore,

$$\varepsilon_I \approx 6 \times 10^3 \frac{\overline{U_I}}{T_V} \times \frac{n_I}{n_D} \,; \tag{3.89}$$

$$\overline{U_I} \text{ in V}, \quad T_V \text{ in K}.$$

A characteristic value for the middle energy ε_D of a vapour particle at an evaporation temperature of approximately 2000 K is 0.2 eV. Typical values for the middle energy ε_D of the ions at the ion plating are 50–5000 eV.

Table 3.7 shows different values of the relationship n_I/n_D and the size $\overline{U_I}$. With ions of $\overline{U_I} = 500$ V, for example, with a small rate relationship $n_I/n_D \sim 3 \times 10^{-3}$, an energetic activation can be reached, which corresponds to energy conditions at sputtering. By the choice of $\overline{U_I}$, in particular, by change of the relationship n_I/n_D, the energy relationship can be varied by orders of magnitude. Because of the different dependence of the elementary effects on the number of the ions and the ion energy, generally the ion plating effects differ with the same ε_I value. In order to achieve high values for ε_I, ionization facilities must be used for ion plating, with which large values for n_I/n_D can be carried out [59].

Ion plating is limited by an increase of resputtering and by an increase of the temperature of the substrates.

An important condition for the guidance of the ion plating process is that the condensation rate a_K must exceed the resputtering rate a_S,

$$\frac{a_S}{a_K} < 1 \,. \tag{3.90}$$

a_K is the condensation rate, which would result in the case of evaporation without ion effect. For this reason, the relationship of the energy relationship ε_I with the rate relationship is a_S/a_K is of interest. With a simplifying assumption applied in a reduced range, the accelerating voltage for the ions and/or the discharge voltage U_B of the ionization facility,

$$\varepsilon_I = K \sqrt{U_B} \frac{a_S}{a_K} \,. \tag{3.91}$$

Table 3.7 Energy relationship ε_I at the ion plating to the comparison by evaporation and sputtering

Technology of deposition	ε_I	Parameter	
High vacuum evaporation	1	$e_D = 0.2$ eV	
Sputtering	5–10	$e_S = 1$–2 eV	
		$\frac{n_I}{n_D}$	$\overline{U_I}$ in V
Ion plating	1.2	10^{-3}	50
	3.5	10^{-2}	500
		10^{-4}	5000
	25	10^{-1}	50
		10^{-3}	5000
	250	10^{-1}	500
		10^{-2}	5000
	2500	10^{-1}	5000

n_I/n_D relationship of the number of the ions to the number of the vapour particles; $\overline{U_I}$ average accelerating voltage of the ions; e_D average energy of a vapour particle; e_S average energy of a sputtered particle.

3.6.2 Influence on the Film Characteristics and Consequences for the Coating Process

From the mechanism of the ion plating process it follows that between vapour, ions and gas acting elementary processes create more complex condensation conditions than evaporation. Adhesion mechanisms influence the development and growth of nucleation and the morphology of the film in a particular orientation and structure; in addition, stoichiometry and surface topography can be modified with the ion plating.

These physical and partial chemical effects modify the film properties of evaporated films. The sputtering of adsorption films that cover the substrate surface, the creation of mixed interface films from sputtered substrate and film material, the diffusion of both materials, the higher binding energy of activated centres [57] and the changed adhesion coefficient ionized vapour produce ion plated films with good adhesion [49]. Large adhesion is a condition for application of very thick films.

Ion bombardment during the condensation of the film, in situ cleaning, can lead to increased purity of the film. Resputtering and gas adsorption can be a frequent cause for a larger seed film density and thus for finer granular film structure. The energetic excitation, of which resputtering and gas adsorption can cause a change of the nature of the growth, for example change of the orientation of crystals, and a transition of pillared to isotropic or equiaxial growth [53] at substrate temperatures, at which this growth is normally not possible [51]. Under the influence of ions recrystallization at room temperature was observed. An increased binding energy of activated centres and changed adhesion coefficients of ionized gas are causes of increased gas content in the film and selective insertion of alloy components. In this way the implantation of unsolvable components can be achieved, for example helium in gold [52] or phosphorus and lithium in zinc selenide. Scattering in the gas and the differentiated resputtering, particularly at cupids and edges as a consequence of the electrical field gradient, are the cause for the high degree of coverage [53] and small porosity [48]. The use of an ionization system at the start and during the evaporation permits inclusion of sputtering to clean the substrate and in situ cleaning of the film into the coating process.

3.6.3 Equipment for Ion Plating

There are many different possibilities of producing vapour and ions and accordingly diverse systems have been developed [65].

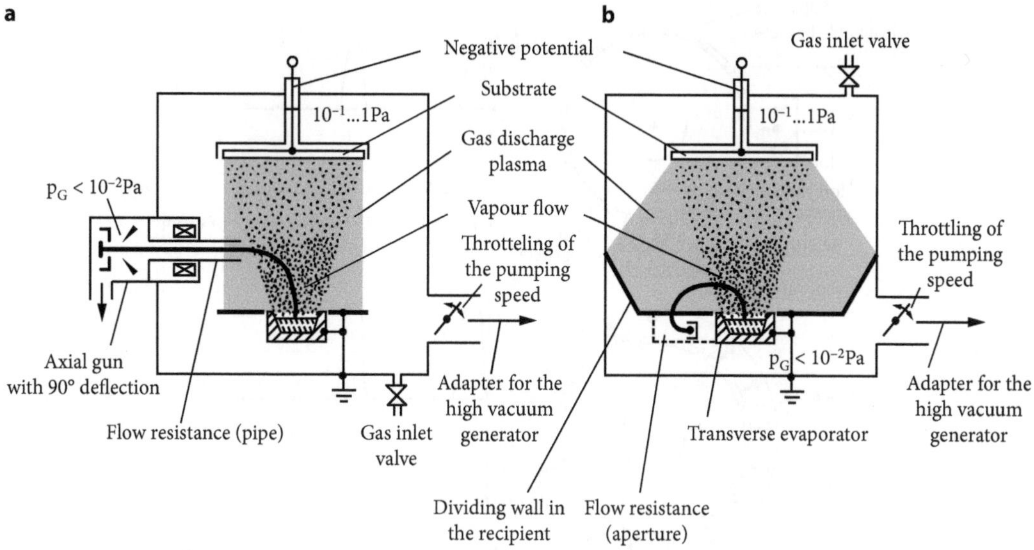

Fig. 3.69 Schematic representation from systems for ion plating with electron beam evaporator and direct current gas discharge between evaporators and substrate, **a** with axial gun, **b** with transverse gun

In order for ion plating to reach high coating rates, electron beam evaporators [66] are used for the evaporation. In the simplest systems the electron beam is used as the energy source for the production of the vapour stream and also for ionization of the vapour stream.

Vapour ions produced by a primary electron beam within the range of the crucible are sucked up by a negative current of some thousand volts and accelerated to the substrate.

With such an arrangement the ion current density reached on the substrate is too small for technical applications. A cause for the low ion current is the low ionization probability of vapour particles by fast ions with energies in the order of 10–30 keV.

An advantage of this solution is that the process can be implemented in high vacuum, at a pressure by 10^{-4} mbar. A possibility for ion plating in high vacuum consists of using the electron beam evaporator only for vapour generation and a separate ion gun for both producing and directing the ion beam toward the substrate. This arrangement permits us control the ion current and the vapour stream separately; however, to control the ion beam is complex.

Usually, a system is used that consists of an electron beam evaporator for vapour generation and a direct current gas discharge between evaporators and substrate for the production of a current of vapour and feed gas ions. Sources of ionization are both the electron beam and the plasma, which are formed between the electrodes for discharge. Figure 3.69 shows the arrangements for electron beam evaporators with a transverse gun and an axial gun.

If the ion plating takes place with a feed gas pressure from $1–7 \times 10^{-2}$ mbar, then a pressure gradient of 100 : 1 is necessary. With the application of self-consistent gas discharges with a current density within the range around 1 mA/cm^2 and high evaporation rates only small values of the relationship are possible for n_I/n_D. An increase of the ion current density, in particular of vapour ions, is reached by arrangement of the crucible in an axial magnetic field. This way slow secondary electrons can be used for additional ionization. Ionization of vapour and/or the feed gas can be also be achieved with high frequency [25]. Such an arrangement makes ion plating possible for non-conductive substrates. The use of a planar plasmatron, a circular

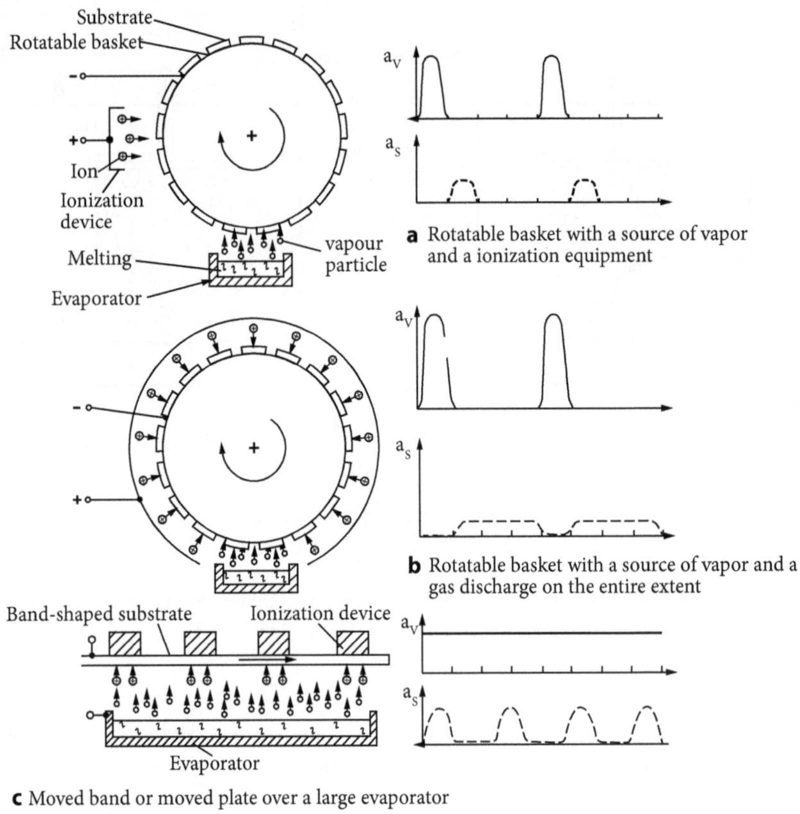

Fig. 3.70 Examples of possible arrangements of the alternating ion plating and timeline of the evaporation rate and the resputtering rate a_S for each substrate place (schematic)

gap discharge system for ion plating [59], permits production from ion current densities over 100 mA/cm^2.

In combination with movement mechanisms for the substrates, however, this process makes the application of ion plating for coating large surfaces with a high rate possible. Moreover, with respect to the discharge pressure and thus the pressure at the coating process, this type of discharge is very flexible.

On the one hand, amperage discharge is still possible in the pressure range of 5×10^{-4} mbar possible, and on the other hand, this disposition qualifies within the pressure range of the gas scattering coating, i.e. up to 10 mbar is suitable. Figure 3.70 shows a plant for alternating ion plating with an electron beam evaporator with flow resistance; an ionization facility of the planar plasmatron type and a rotary disc substrate mounting plate [59].

3.6.4 Structural Constitution of the Ion Plated Films

The structural constitution and the physical characteristics of the future film, the substrate temperatures, the condensation rate, and the pressure of the surrounding gas atmosphere determine, if necessary the relationship of the partial pressure of the gas mixtures and the energy of the ions and neutral particles.

During cathode sputtering with negatively biased substrate the energy of these particles is at some ten eV, with ion plating with a gas discharge energies up to some hundreds eVoccur [67]. Due to these comparatively high energies in an interface film (transition region) it can come to implantation of high energy particles from the gaseous phase and recoil implantation of surface atoms; a procedure that is described as physical mixing or pseudo diffusion.

For film growth, and thus for the quality of the film, the mobility of the condensing particles at the film surface essentially depends on the substrate temperature and the energy of the arriving particles. Aspired is a compact, glass-like or nanocrystalline structure of the film with no inside stress and high adherence.

The correlation between the structural condition of the film and the process parameters is represented in structured models. B.A. Movchan and A.V. Demchishin [68] derived a three-zone model from structure investigations on the basis on thick vapour-deposited nickel, titanium, tungsten-aluminium oxide and zircon oxide. This is shown in Fig. 3.71a. An independent variable is the relationship T/T_M, where T is the substrate temperature during the condensation and T_M is the melting temperature of the coating material. The introduction of the melting temperature here is meaningful, because the binding energy between the condensing particles of this size is proportional to T_M.

Is the relationship smaller with metals than 0.26 (0.3 by chemical compounds), then the adsorbed atoms have less mobility on the surface, shadowing effects can become unbalanced, and a porous dendritically structure from needle-shaped crystals with round end shaped ends (zone 1) results to zone 2, which is characterized from a closely packed stack structure (columnar structure) with a surface of small roughness, which has a developing higher mobility of the adsorbed atoms within the range $0.26 \leq T/T_M \leq 0.45$. Volume diffusion leads to zone 3 with a recrystallized film, high component density and smooth surface film.

The three-zone model for vapour-deposited films was extended by J.A. Thornton [69] for up to 25 µm films of metal (Mo, Co, Ti, Fe, Cu and aluminium alloys) created with cathodic sputtering in argon, (Fig. 3.71b). Impacting particles on the substrate surface have a substantially higher energy (about 4–40 eV) than with by evaporation (0.1–0.2 eV). They also depend on the gas pressure because the particles coming from the cathode lose energy when the particles impact with the gas particles by crossing the discharge area. The higher the gas pressure, the smaller

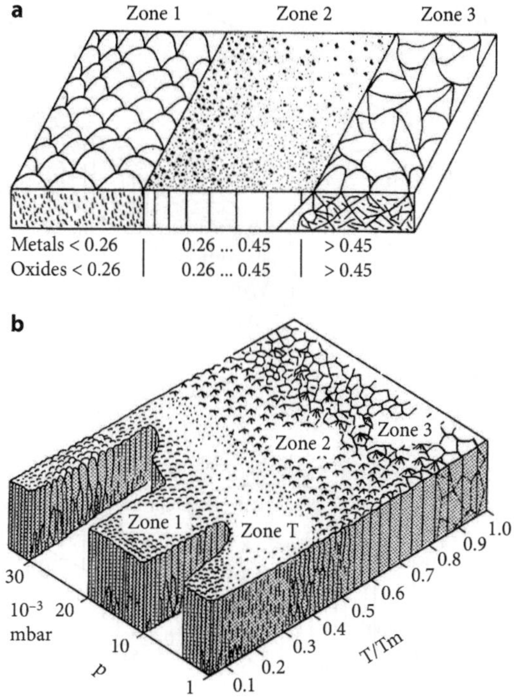

Fig. 3.71 Structured models. **a** according to B.A. Movchan and A.V. Demchishin [68], **b** after J.A. Thornton [69]. T = substrate temperature, T_m = melting temperature

the average kinetic energy of the condensing particles and thus their mobility on the substrate surface. Below $T/T_M \leq 0.5$, with removing pressure we have a more broadly transition region T, which exhibits a close fibrous structure with a smooth surface.

The growing film is exposed to a constant bombardment with particles of high energy, which increases the mobility of the adsorbed atoms. This particularly applies during cathodic sputtering with a negatively biased substrate. Therefore, it is more meaningful to represent the structural constituent as a function of the entire energy reaching the surface [59]. This is shown in Fig. 3.72a for sputtered germanium films. As the schematic representation shows, the limit between zones 1 and T shifts to lower temperatures with increasing energy.

A further structural model [59] is shown in Fig. 3.72 for a higher energy region (200–500 eV) of the impacted ions, as is typical for ion plating

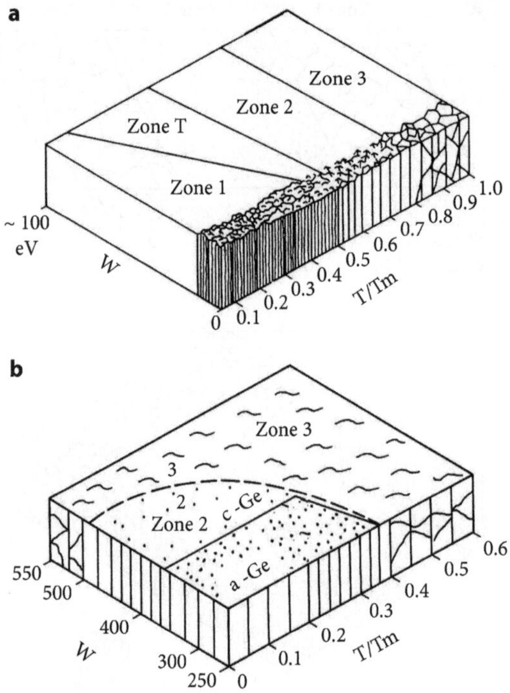

Fig. 3.72 Structure models. **a** after R. Messier et al. [70], **b** after C. Fountzoulas and W.B. Nowak [65]. T = substrate temperature, T_m = melting temperature, W = ion energy

with gas discharge. The structural constituent of germanium films (thickness about 1.5 µm, a condensation rate of about 0.4 µm/min) as a function of the total energy measured in the contrary field of particles striking the substrate is shown in the figure. It is remarkable that there is an absence of zones 1 and T and that there is a tightly packed film structure of zone 3 at room temperature ($T/T_M = 0.25$) and high ion energy (≥ 500 eV).

3.6.5 Adhesive Strength of Ion Plated Films

The basic condition for the usefulness of ion plated films is a high adhesive strength, which depends on the internal stress of the deposited film. At all coating processes stress in the film can occur if the atoms are prevented to take on the state of deepest energy at the growth of the film. Disturbances of the equilibrium are caused by gas incorporation and the formation of defect structures; they are inevitable at typical manufacture conditions of the ion plated films. On the other hand, if by the choice of the process parameters, in particular the ion energy, the film growth can be controlled in such a way that closely packed structures develop, which offer various possibilities such as volume diffusion and recrystallization to stress compensation.

Ion plated films, that have been manufactured under accordingly adjusted conditions, show substantially improved adhesion conditions, also with large film thicknesses as compared to evaporated films (10–100 µm) and material combinations, from which one would not expect good adhesion. Thus, ion plated silver films of 12 to 50 µm thickness were applied on stainless steel with an adhesive strength of 300 N/mm^2 [72].

Ion-plated films can withstand even the extreme demands during cold massive forming. A thick copper film with $MoSi_2$-lubrication lacquer on titanium proved to be an optimal intermediate film to avoid cold welding with cup reverse extrusions with a deformation up to 92 % a 1.3 µm.

Electron microscope pictures show that ion plated copper films cover titanium even after a deformation of 400 % of the coherently covered surface [71].

3.6.6 Reactive Ion Plating

Reactive ion plating designates ion plating processes that operate in the presence of reactive gases (for example, O_2, N_2, C_2H_2).

Oxides, nitrides, carbides, etc. need to develop substrate temperatures around 400 °C; normally temperatures between 800 and 1000 °C are necessary.

Titanium nitride films are frequently used, depending upon process conditions, in different phases (TiN and Ti_2N) and different crystal orientations. Depending on the process parameters, the films have different structural conditions and harnesses between 2000 and 3000 HV [73]. With the production of titanium nitride films the N_2 partial pressure is critical. By variation of the manufacturing parameters titanium nitride films of different forms can be manufactured, as they are used for decorative purposes.

References

1. Mattox DM (2010) Handbook of Physical Vapor Deposition (PVD) Processing, 2nd edn. William Andrew
2. Cho JA (1971) Film Deposition by Molecular Beam Techniques. J Vac Sci Tech 8:S31–S38
3. Cho JA (1975) Molecular Beam Epitaxy. Prog Solid-State Chem 10:157–192
4. Sacks R, Sieg R, Ringel S (1996) Investigation of the Accuracy of Pyrometric Interferometry in Determining $Al_xGa_{1-x}As$ Growth Rates and Compositions. J Vac Sci Tech B 12(3):2157–2162
5. Pinsukanjana P, Jackson A, Tofte J, Maranowski K, Campbell S, English J, Chalmers S, Coldren L, Gossard A (1996) Real-time Simultaneous Optical-Based Flux Monitoring of Al, Ga, and In using Atomic Absorption for Molecular Beam Epitaxy. J Vac Sci Tech B 14(3):2147–2150
6. Williams JS, Poate JM (eds) (1984) Ion Implantation and Beam Processing. Academic Press
7. Ziegler F (2000) Ion Implantation Science and Technology. Edgewater, USA
8. Hilleringmann U (2004) Silizium-Halbleitertechnologie. Teubner
9. Frey H (1992) Ionenstrahlgestützte Halbleitertechnologie. VDI-Verlag, Düsseldorf
10. Marshall A, Natarajan S (2002) SOI Design: Analog, Memory and Digital Techniques. Springer
11. Burrows G (1965) Evaporation in an evacuated container. Vacuum 15(8):289–399
12. Langmuir J (1973) The vapour pressure of metallic tungsten. Phys Rev 2(5):329–342
13. Movcan BA, Domcisin AV, Kulak LD (1974) Structure and mechanical properties of thick Fe, Fe-NbC, Fe-Ni-NbC- condensates. J Vac Sci Technol 11:869–874
14. Schiller S et al (1967) Erfahrungen mit einem 5 kW Elektronenstrahlverdampfer. Vakuum-Technik 16(167/ 9):205–209
15. Taylor RC (1974) Single source evaporation of gadolinium cobalt alloys. J Vac Sci Technol 11(6):1148–1150
16. Oron M, Adams CM (1969) Controlled electron beam co-deposition of copper-nickel films. J Mater Sci 4(169):252–258
17. Swift RA, Noval BA, Merz KM (1968) Fractionation of Ni-Cr-Cu-Al alloys during vacuum evaporation. J Vac Sci Technol 5(3):79–83
18. Santala T, Adams CM (1970) Kinetics and thermodynamics in continuous electron-beam evaporation of binary alloys. J Vac Sci Technol 7(6):22–29
19. Forster JS, Pfeifer WH (1972) Vacuum deposition of alloys – theoretical and practical considerations. J Vac Sci Technol 9(6):1379–1384
20. Andreini RJ, Forster JS (1974) Kinetics of solute removal during electron-beam and vacuum-arc melting. J Vac Sci Technol 11(6):1055–1059
21. Hulgren C et al (1963) Selected Values of Thermodynamic Properties of Metals and Alloys. J. Wiley & Sons Inc., New York, p 732
22. Wagner C (1962) Thermodynamics of Alloys. Addision-Wesley, Reading, Mass., pp 15–17
23. Olette M (1960) Élemination sous vide des oligo-élémentes contenus dans les alliages ferreux. Mem Sci Rev Met 67(6):467–480
24. Richards JL (1966) Flash-evaporation. In: Anderson JC (ed) The Use of Thin Films in Physical Investigations. Academic Press, New York, pp 71–85
25. Harker HR, Hill RJ (1972) The deposition of multicomponent phases by ion plating. J Vac Sci Technol 9(6):1365–1399
26. Smith HR, Kennedy K, Boericke FS (1970) Metallurgical characteristics of titanium-alloy foil prepared by electron beam evaporation. J Vac Sc Technol 7(6):48–51
27. Krutenat RC (1974) Effect of pool temperature gradients on the compositions of electron beam vapour deposited alloys. J Vac Sci Technol 11(6):1123
28. Stowell WR (1973) Single source evaporation of a niobium based alloy containing volatile constituents. J Vac Sci Technol 10(4):489–493
29. Hoffmann D, Leibowitz D (1971) Al_2O_3 Films prepared by electron beam evaporation of hot pressed Al_2O_3 in oxygen ambient. J Vac Sci Technol 8(1):107–111
30. Hoffmann D, Leibowitz D (1972) Effect of substrate potential on Al_2O_3 films prepared by electron beam evaporation. J Vac Sci Technol 9(1):326–329
31. Bunshab RF, Raghuran AC (1972) Activated reactive evaporation process for high rate deposition of compounds. J Vac Sci Technol 9(6):1385–1388
32. Stowell WR (1974) Ion-plated titanium carbide coatings. Thin Solid Films 22(1):111–120
33. Mattox DM (1973) Thin film metallization of oxides in microelectronics. Thin Solid Films 18(2):173–186
34. Heitmann E (1972) Reaktives Aufdampfen in reaktiven Gasen. Vakuum-Technik 21(1):1–11
35. Ramprasad BS, Rhada TS (1974) Optimum geometry for uniform deposits on a rotating substrates from a pint source Vacuum 24, 4, p. 165

36. Graper EB (1973) Distribution and apparent source geometrie of electron-beam-broad evaporation sources. J Vac Sci Technol 10(1):100–103
37. Turner ME (1973) How to specify the right source to the right job. Solid State Technol 16(7):16–18
38. Soa EA, Persch G (1974) Die Dicke der Aufdampfschichten in ihrer Abhängigkeit von technischen Parametern. Exp Techn Phys 22(6):571–583
39. Erikson (1974) E.D.: Thickness distribution of a metall – alloy from a high-rate electron beam source. J Vac Sci Technol 11(1):366–370
40. Mulder BJ (1977) A method for the splash-free evaporation of the aluminium and other metals. Vacuum 28(1):11–12
41. Kirner K (1970) Reaktionen beim Verdampfen von Aluminium in Bornitridtiegeln, CVA, pp. 64/65
42. Parent ED (1974) Power requirements of resistance-heated intermetallic evaporation sources. J Vac Sci Technol 11(4):820–822
43. Dobrowolzki JA (1977) Source configurations for the deposition of uniform thin films onto a moving web. In: Proc. 7th Intern. Vac. Congr. & 3rd Intern. Conf. Solid Surfaces Vienna, pp 1567–1570
44. Reuter W (1976) Optimierung der Verdampferanordnung bei Bandbedampfungsanlagen. Maschinenmarkt, Würzburg 82(81):1479–1482
45. Warren KA, Denison PR, Bills DG (1967) Resistance heated sublimator. The Review of Scientific Instruments 38(8):1019–1022
46. Morley JR, Smith HR (1972) High rate ion production for vacuum deposition. J Vac Sci Technol 9(6):1377–1378
47. Burlirsch R, Stoor J (1966) Numerical treatment of ordinary differential equations by extrapolation methods. Numer Math 8:1–13
48. Schiller S et al (1967) Erfahrungen mit einem 5-kW Elektronenstrahlverdampfer. Vakuum-Technik 16(9):205–209
49. Hill F (1974) Evaporated aluminium films; A review of techniques and recent improvements. Airco Temescal, Berkley, Calif
50. Schiller S et al (1974) Industrial electron beam coating of trip steel. In: Silva RM, Electron Beam Processing Seminar Stratford-upon-Avon, pp. 2d1–2d28
51. Groeschl ME, Benes E, Schmidt E, Siegmund H, Thorn G, Thomas FW (1989) Sensor for the detection of the incident point of an electron beam. Thin Solid Films 174:323–329
52. Scheibe H, Gorbunov AA, Baranova GK, Klassen NV, Konov VI, Kulakov MP, Prokhorov AM, Weis HJ (1990) Thin Film Deposition by Eximer Laser Evaporation. Thin Solid Films 189:283–291
53. Coutal C, Azema A, Roustan JC (1996) Thin Solid Films. Elsevier
54. Barth KL, Lunk A, Ulmer J (1997) Surface and Coatings Technology. Elsevier
55. Rother BJ, Siegel J, Vetter J (1990) Cathodic Arc Evaporation of Graphite with Controlled Cathode Spot Position. Thin Solid Films 188:293–300
56. Ehrlich H, Hasse B, Mausbach M, Müller KG (1990) The anodic vacuum arc and ist application to coating. J Vac Sci Technol A 8(3):2160–2164
57. Mattox DM (1973) Fundamentals of ion plating. J Vac Technol 10(1):47–52
58. Schwan J, Ulrich S, Roth H, Ehrhardt H (1996) Tetrahedral amorphous carbon films prepared by magnetron sputtering and dc ion plating. Journal of Applied Physics 79:1416–1422
59. Schiller S, Heisig U, Goedicke K (1975) Alternating ion plating – a way to high rate ion vapour deposition. J Vac Sci Technol 12(4):858–864
60. McLeod PS, Mah G (1974) The effect of substrate bias voltage on the bonding of vaporated silver coatings. J Vac Sci Technol 11(1):119–121
61. Boone DH, Strangman TE, Wilson LW (1974) Some effects of structure and composition on the properties of electron beam vapour deposited coating for gas turbine superalloys. J Vac Sci Technol 11(4):641–646
62. Bunshah RF (1974) Structure property relationships in evaporated thick films and bulk coatings. J Vac Sci Technol 11(4):633–638
63. Berg RS, Kominiak GJ, Mattox DM (1974) Incorperation and behaviour of helium in ion deposited Films. J Vac Sci Technol 11(1):52–55
64. Jouan PY, Lempérière G (1991) Ion energy distribution at a negatively based electrode in a sputtering discharge. Vacuum 42:927–931
65. McNally JJ (1990) Ion assisted deposition Handbook of Plasma Processing Technology. Noyes Publications, pp 466–482
66. Chambers DL, Cermichael DC (1971) Electron beam techniques for ion plating. Res And Develop 22(5):32–35
67. Ion assisted deposition Handbook of Plasma Processing Technology Noyes Publications (1990) pp. 466–482
68. Movchan BA, Demchishin AV (1969) Study of the structure and properties to thick vacuum condensates of nickel, titanium, tungsten, aluminium oxide and zirconium dioxide. Fiz Metal Metallowed 28:653–660
69. Thornton JA (1974) Influence of apparatus geometry and despositions conditions on the structure and topography of thick sputtered coatings. J Vac Sci Technol 11:666–670
70. Merssier R, Giri AP, Roy RA (1984) Revised structure zone model for thin film physical structure. J Vac Sci Technol A 2:500–503

71. McNally JJ (1990) Ion assisted deposition Handbook of Plasma Processing Technology. Noyes Publications, pp 466–482
72. Riebeling I, Thoma K, Gärtner H (1985) Ion plating of copper layers on titanum: influence of preparation parameters and deformation on the residual stresses. Mater Sci Eng 69:435–444
73. Matthews A (1985) Surface Engineering 6:93–104

4 Basic Principle of Plasma Physics

H. Frey

4.1 Introduction

Thin film technologies are generally based on plasma-supported methods (see Fig. 4.1).

Plasma means:

- When in a liquid or a gas the number of free charge carriers is so large that charge carriers affect the physical properties of the medium substantially.
- When electromagnetic interactions between the charged particles take place.
- When the number of positive and negative charge carriers is for each unit of volume equally large in each case; the total quantity can be arbitrary.

The number of existing neutral particles (atoms, molecules) is not relevant for the definition of plasma. Since plasma contains free charge carriers, it is current conducting. Electric currents produce magnetic fields and so electrically charged particles by electrical and magnetic fields can be influenced, also a plasma can be influenced by external electrical and magnetic fields. Furthermore the plasma produces such fields and, therefore, the plasma can take in interaction with itself. In electrodynamics plasma is generally an anisotropic, non-linear disperser conductor, i.e. the energy for plasma excitation can be supplied by means of electromagnetic fields of different frequency.

4.2 Quasi-Neutrality

A plasma is not electrically charged, because it contains free charge carriers. From the outside it is electrically neutral, since on the average the number of positive and negative charge carriers is equally high for each unit volume. Powerful electrical fields maintain this electrical equilibrium, the quasi-neutrality [1].

If electrons n_e are contained in a cubic centimetre, then the total charge contained in a sphere of the radius r is

$$Q = \frac{-4\pi}{4\pi 3\varepsilon_0} r^3 n_e e\,, \tag{4.1}$$

where e is the elementary charge and ε_0 the electrical field constant.

If we assume a central symmetrical distribution of the electrons, then we can represent the whole charge Q in the centre of the sphere. In the distance r from the average value of the sphere this charge produces an electrical field of the strength

$$E = \frac{Q}{r^2}\,. \tag{4.2}$$

If we select $r = 1$ and $n_e = 10^{15}\,\text{cm}^{-3}$, we obtain $E = 6 \times 10^8\,\text{V/cm}$. In the condition of quasi-neutrality, however, the positive and negative charges are equal for each unit volume. When n_i is the number of ions and $+Ze$ the charge of the Z-fold positively charged ions, the condition of quasi-neutrality is:

$$n_e = Z n_i \tag{4.3}$$

© Springer-Verlag Berlin Heidelberg 2015

H. Frey, H. R. Khan, *Handbook of Thin-Film Technology*, DOI 10.1007/978-3-642-05430-3_4

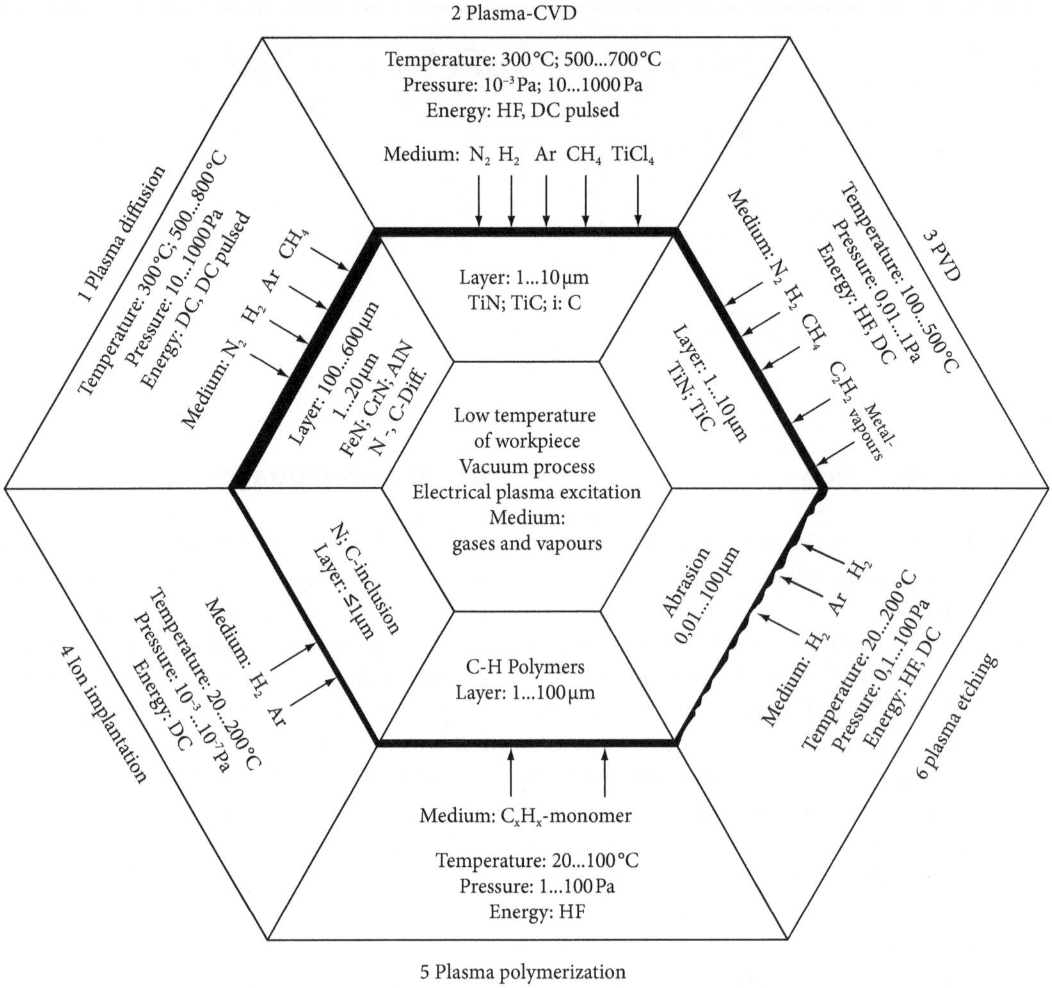

Fig. 4.1 Plasma-supported thin film technologies

or

$$|n_e = Zn_i| \ll n_e\,. \tag{4.4}$$

Since the ions generate a field with $E = +6 \times 10^8\,\mathrm{V/cm}$, the entire electrical field strength is zero in the condition of quasi-neutrality. If, however, – for example by statistic density variations – it comes only in the relationship $1 : 10^{-6}$ to a variation from Eq. (4.3), then the electrical equilibrium is again reconstituted, generating electrical field strength in the amount of $10^{-6} \times 6 \times 10^8 = 600\,\mathrm{V/cm}$.

This strong field reconstitutes the quasi-neutrality immediately. It is only with the occurrence of very high frequency oscillations (for example, of displacement currents) that the ions with their large mass ca no longer follow, and it comes to a spatial separation of the centre of mass of the positive and negative charge clouds.

The oscillating frequency ω_0, up to which the displacement current can be neglected, is determined by the relationship:

$$\frac{\text{displacement current}}{\text{conduction current}} = \frac{\varepsilon_0 \frac{\partial E}{\partial t}}{j} \approx \frac{\varepsilon_0 \omega_0}{\sigma} < 1\,. \tag{4.5}$$

Here, j is the conduction current density, E the electrical field strength and σ the electrical conductivity. With an electrical conductivity σ of the plasma at $100\,\Omega^{-1}\,\mathrm{m}^{-1}$ we obtain $\omega_0 > 10^{13}\,\mathrm{s}^{-1}$ ($\lambda = 3 \times 10^{-5}$ m).

4.3 Characteristic Quantities

4.3.1 Langmuir Plasma Frequency

The force $-eE$, which an electron repulses into the volume (attractive force), is harmonic, i.e. proportionally to the deflection from the rest position [2]; it is

$$-eE = -e\frac{Q}{r^2} = -e^2\frac{1}{3}r^3 n_\mathrm{e}\frac{1}{r^2} = -\frac{1}{3}n_\mathrm{e}e^2 r\,, \tag{4.6}$$

with the linear (radial) equation of motion (usually linear differential equation of second order with constant coefficients), and we obtain

$$m_\mathrm{e}\frac{\mathrm{d}r^2}{\mathrm{d}t} = -\frac{1}{3\varepsilon_0}n_\mathrm{e}e^2 r\,, \tag{4.7}$$

where
m_e is the electron mass.
With the approximated e-potential

$$r(t) = r_0\mathrm{e}^{\mathrm{i}}\omega t\,, \tag{4.8}$$

it follows that the electrons are harmoniously oscillated with the frequency

$$\omega = \left(\frac{n_\mathrm{e}e^2}{3m_\mathrm{e}\varepsilon_0}\right)^{\frac{1}{2}}. \tag{4.9}$$

With the more exact three-dimensional calculation we obtain the equation

$$\omega_\mathrm{PI} = \left(\frac{n_\mathrm{e}e^2}{m_\mathrm{e}\varepsilon_0}\right)^{\frac{1}{2}}. \tag{4.10}$$

All these equations are valid only for cold electrons ($T_\mathrm{e} = 0$ K). If the electrons still have thermal motion, then the conditions are more complicated.

4.3.2 Debye Shielding Length

If the plasma is considered on the whole as quasi-neutral, this does not exclude that within the microscopic range electrical fields result from fluctuations of the charge carrier density. If one regards the direct environment of an ion, then because they are released by the attractive Coulomb force, more electrons remain on the average here, as this corresponds to the medium density of the electrons.

Contrarily, other positively charged particles are kept away due to the repulsive force from this range.

In order to be able to calculate the Coulomb potential in the direct environment of such a test particle, we must use some simplifying assumptions.

- The electrons possess a Maxwell distribution with the temperature T_e.
- Due to the thermal movement of the electrons electrical microfields can develop in the plasma, in which the electrons attain potential energy. This energy $e\varphi$ should be substantially smaller than the thermal energy, i.e. $e\varphi \ll kT_\mathrm{e}$ (φ potential).
- The positively charged ions are to be uniformly distributed over the plasma. Their density is everywhere $n_\mathrm{i} = n_\mathrm{e0}$, whereas n_e0 is the mean density of the electrons.

In a potential field the electron density varies according to the Boltzmann distribution:

$$n_\mathrm{e} = n_\mathrm{e0}\mathrm{e}^{\frac{e\varphi}{kT_\mathrm{e}}}\,. \tag{4.11}$$

Since $e_\varphi \ll kT_\mathrm{e}$ is, we can also write

$$n_\mathrm{e} - n_\mathrm{e0}\left(1 - \frac{e\varphi}{kT_\mathrm{e}}\right). \tag{4.12}$$

Over the considered volume average φ disappears, since it is developed only by fluctuation, i.e. by space charge fluctuations. The potential field φ must be sufficient in the environment of the considered position of the Poisson equation.

With $E = -V_\mathrm{j}$ (see Sect. 4.4.1) and with consideration of Eq. (4.12), we obtain:

$$-\Delta\varphi = \frac{e}{\varepsilon_0}(n_\mathrm{e} - n_\mathrm{i}) = \frac{-e^2}{\varepsilon_0}n_\mathrm{e0}\frac{\varphi}{kT_\mathrm{e}}. \tag{4.13}$$

With the abbreviation (λ_D is the Debye length)

$$\lambda_{\mathrm{D}2} = \varepsilon_0 T_\mathrm{e}/\mathrm{e}^2 n_\mathrm{e0} \tag{4.14}$$

we sustain

$$\Delta\varphi\frac{\varphi}{\lambda_{\mathrm{D}}^2}\,. \tag{4.15}$$

Since the potential in the environment of a test particle is to be considered, this is a spherically symmetric problem. We can therefore write:

$$\frac{1}{r^2}\frac{\mathrm{d}}{\mathrm{d}r}r^2\frac{\mathrm{d}\varphi}{\mathrm{d}r}=\frac{\varphi}{\lambda_{\mathrm{D}}^2}\,. \tag{4.16}$$

The limiting values is $\varphi \to 0$, if $r \to \infty$, and $\varphi \to \varphi_{\mathrm{coulomb}}$, if $r \ll \Delta\lambda$.

Equation (4.16) is solved with the approach

$$\varphi=\frac{c}{r}\mathrm{e}^{\frac{-r}{b}}\,. \tag{4.17}$$

Introducing Eq. (4.17) in Eq. (4.16), we immediately obtain $b = \lambda_{\mathrm{D}}$; the ansatz thereby fulfils the first boundary condition. The constant c is determined by the transition of many small r (as compared to the ion radius). Since λ is usually large against the ion radius, it is possible to develop the exponential function of the Eq. (4.17):

$$\psi=\frac{c}{r}\mathrm{e}^{\frac{-r}{\lambda_{\mathrm{D}}}}=\frac{c}{r}\left(1-\frac{r}{\lambda_{\mathrm{D}}}\right)\cong\frac{c}{r}=\frac{e}{4\pi\varepsilon_0 r}\,. \tag{4.18}$$

From this it follows that:

$$c\cong\frac{e}{4\pi\varepsilon_0} \tag{4.19}$$

or

$$\varphi=\frac{e}{4\pi\varepsilon_0 r}\mathrm{e}^{\frac{-r}{\lambda_{\mathrm{D}}}}\,. \tag{4.20}$$

These considerations show that the potential of a test particle in plasma is correlated by the factor $\mathrm{e}^{r/\lambda_{\mathrm{D}}}$ to the modified Coulomb potential. In the distance $r = \lambda_{\mathrm{D}}$, the undisturbed coulomb potential is already dropped down to an eth of the exponential function part. The electrons, which accumulate around an ion, shield the potential outwards.

The number of electrons in a sphere around the test particle with the plasma length λ_{D} as a radius of the Debye sphere is:

$$n_{\mathrm{eD}}=\frac{4\pi}{3}\lambda_{\mathrm{D}}^3 n_{\mathrm{e0}}=\frac{4\pi}{3}\left(\frac{\varepsilon_0 k T_{\mathrm{e}}}{n_{\mathrm{e0}}e^2}\right)^{\frac{3}{2}}n_{\mathrm{e0}}$$

$$=\frac{4\pi}{3}\left(\frac{\varepsilon_0 k T_{\mathrm{e}}}{e^2}\right)^{\frac{3}{2}}\frac{1}{n_{\mathrm{e0}}^{\frac{1}{2}}}\,, \tag{4.21}$$

or, if we summarize the constant factors:

$$n_{\mathrm{c}}=1.38\times 10^3\frac{T_{\mathrm{e}}^{\frac{3}{2}}}{n_{\mathrm{e0}}^{\frac{1}{2}}}\,. \tag{4.22}$$

The equation becomes invalid, if the number of particles in a Debye sphere becomes $l_{\mathrm{p}} \leq 1$. We must use then quantum-mechanical methods.

4.3.3 Landau Length and Plasma Parameters

The distance of a charge in which the electrostatic energy in the vacuum is equal to the kinetic energy kT is called the ***critical distance*** or the ***Landau length*** [3]:

$$l_{\mathrm{L}}=\frac{e^2}{kT4\pi\varepsilon_0}\,. \tag{4.23}$$

If no recombination (Sect. 4.5.4.2 compares) of electrons and ions takes place, then the mean distance between plasma particles must be larger than the Landau length;

$$\alpha > l_{\mathrm{L}}\,. \tag{4.24}$$

Since the range of influence is given to the electrostatic interaction effect by λ_{D}, we can write:

$$\alpha > \lambda_{\mathrm{D}}\,, \tag{4.25}$$

This means that it does not come to any interaction between the plasma particles; this is a swarm of particles that is independent from each other; they are not interacting particles. ***This is not a plasma***.

The Debye sphere (c.f Sect. 4.3.2) contains $4\pi/3\lambda_{\mathrm{D}}^3 n_{\mathrm{e0}}$ electrons. So we can write:

$$4\pi/3\lambda_{\mathrm{D}}^3 n_{\mathrm{e0}} \geq 1$$

or

$$\Lambda=\frac{1}{n_{\mathrm{e0}}\lambda_{\mathrm{D}}^3}\ll 1 \tag{4.26}$$

or

$$n_{\mathrm{e0}}^{-\frac{1}{3}} \ll \lambda_{\mathrm{D}}\,. \tag{4.27}$$

4.4 Motion of Charged Particles in Electromagnetic Fields

4.4.1 Maxwell Equations

The Maxwell equations describe the behaviour and characteristics of magnetic and electrical fields in fundamental form. They are given – in non-relativistic form – by the following equations [4]:

$$\nabla \times \vec{H} = j + \frac{\partial \vec{D}}{\partial t}\,, \tag{4.28}$$

$$\nabla \times \vec{E} = -\frac{\partial \vec{B}}{\partial t}\,, \tag{4.29}$$

$$\nabla \vec{D} = \varrho_e\,, \tag{4.30}$$

$$\nabla \vec{B} = 0\,, \tag{4.31}$$

where
$\vec{H}$ is the magnetic field intensity,
$\vec{B}$ the magnetic induction,
j the current density,
$\vec{D}$ the electrical displacement, and
$\vec{E}$ the electrical field intensity.
The system of the Maxwell equations is completed by the material equations, which are connected by the vectors $\vec{E}, \vec{D}, \vec{B}$ and $\vec{H}$ with those values that characterize the electrical and magnetic properties of the material:

$$\vec{D} = \varepsilon \vec{E}\,, \tag{4.32}$$

$$\vec{B} = \mu \vec{H}\,, \tag{4.33}$$

where
ε is a dielectric constant, and
μ the permeability.
Since in plasmas ferrous magnetic materials are to be excluded and all currents in the plasma explicitly accumulated, we can exchange μ in all equations by the permeability of the vacuum μ_0. The current density j is defined as the charge passing through a unit area for each time unit.

There exist charge carriers n_i in a plasma volume element of the sort i with the charge q_i, and then the current density is

$$j = \sum_i n_i q_i \vec{v}_i\,, \tag{4.34}$$

where
$\vec{v}_i$ is the mean velocity of the charge carriers i.
For the electrical space charge density we obtain

$$\varrho_{\mathrm{el}} = \sum_i n_i q_i\,. \tag{4.35}$$

A plasma with only unit charged ions and electrons becomes simply

$$j = e(n_{\mathrm{i}} v_{\mathrm{i}} - n_{\mathrm{e}} v_{\mathrm{e}}) \tag{4.36}$$

and

$$\varrho_{\mathrm{el}} = e(n_{\mathrm{i}} - n_{\mathrm{e}})\,, \tag{4.37}$$

where
$n_{\mathrm{i}}, n_{\mathrm{e}}$ is the density of the ions and/or those of the electrons.
With the divergence executed to Eq. (4.28) and with $\nabla(\nabla \times \vec{H}) = 0$, we obtain

$$\nabla j + \frac{\nabla \partial \vec{D}}{\partial t} = 0\,, \tag{4.38}$$

obtaining with Eq. (4.30):

$$\nabla j = -\frac{\partial \varrho_{\mathrm{el}}}{\partial t}\,. \tag{4.39}$$

This equation means that the sources of the current density are determined by the temporal change of the space charge density. At stationary processes ($\partial/\partial t = 0$) this equation predicates that the current density j does not have sources or that the current lines have no beginning and no end, which means that they are closed.

Since $\nabla B = 0$, this means that B is the curl of a vector:

$$\vec{B} = \nabla \times \vec{A}\,, \tag{4.40}$$

where
$\vec{A}$ is the vector potential of $\vec{B}$.
Introducing $\nabla \times \vec{A}$ in Eq. (4.29),

$$\nabla \times \left(\vec{E} + \frac{\partial \vec{A}}{\partial t}\right) = 0\,. \tag{4.41}$$

If the curl of a vector disappears, then we can write:

$$\vec{E} + \frac{\partial \vec{A}}{\partial t} = -\nabla \varphi\,. \tag{4.42}$$

By transformation, we obtain

$$\vec{E} = -\nabla\varphi - \frac{\partial \vec{A}}{\partial t}. \quad (4.43)$$

In a stationary process the electrical field can always be represented by a gradient field of the potential φ.

In order to derive a principle of conservation of energy for the electromagnetic field, we multiply the scalar in Eq. (4.28) with $\vec{E}$ and according Eq. (4.29) with $\vec{H}$:

$$\vec{E}(\nabla \times \vec{H}) = j\vec{E} + \vec{E}\frac{\partial \vec{D}}{\partial t} = j\vec{E} + \frac{\partial}{\partial t}\varepsilon\frac{\vec{E}^2}{2} \quad (4.44)$$

$$\vec{H}(\nabla \times \vec{E}) = -\vec{H}\frac{\partial \vec{B}}{\partial t} \equiv -\frac{\partial}{\partial t}\mu_0\frac{\vec{H}^2}{2}. \quad (4.45)$$

Subtracting Eq. (4.45) from Eq. (4.44) we obtain:

$$\vec{E}(\nabla \times \vec{H}) - \vec{H}(\nabla \times \vec{E}) = \frac{\partial}{\partial t}\left(\varepsilon\frac{\vec{E}^2}{2} + \mu_0\frac{\vec{H}^2}{2}\right) + j\vec{E}, \quad (4.46)$$

or, after a vector transformation:

$$-\nabla(\vec{E} + \vec{H}) = \frac{\partial}{\partial t}\left(\varepsilon\frac{\vec{E}^2}{2} + \mu_0\frac{\vec{H}^2}{2}\right) + j\vec{E}. \quad (4.47)$$

The vector $\vec{E} \times \vec{H}$ is the Poynting vector. If we integrate Eq. (4.47) over any volume, then with application of the Gauss theorem, we obtain:

$$-\int_0\int(\vec{E} \times \vec{H})\,\mathrm{d}\alpha = \int_V\iint\left[\frac{\partial}{\partial t}\left(\varepsilon\frac{\vec{E}^2}{2} + \mu_0\frac{\vec{H}^2}{2}\right)\right]\mathrm{d}V + \int_V\iint j\vec{E}\,\mathrm{d}V. \quad (4.48)$$

The left-hand side of Eq. (4.48) is the expression for the electromagnetic energy flow into the volume. This leads to a temporal change of the electrical and magnetic field energy in the volume ($\varepsilon\vec{E}^2/2 + \mu_0\vec{H}^2/2$ specifying the density of the electrical and/or magnetic field energy), while the last term is the energy dissipated in the volume. However, this may not be identified generally $j\vec{E}$ with the Joule heat.

4.4.2 Equation of Motion and Law of Conservation of Energy

In plasmas, as in ion-supported coating or etching techniques, collective effects predominate; the consideration of the movement of charged single particles is only a very rough approximation. Since, however, many mechanisms lead to instabilities, which can be attributed to the behaviour of the single particles in electrical and magnetic fields, it is necessary to analyze the motion of individually charged particles [5].

It is accepted that the particles move so slowly that the fields produced thereby (essentially proportionally to the velocity of the particle) are so weak that the fields can neglected in relation to the fields generated from the outside (electrodes, magnetic poles).

For an individual particle, we obtain for the equation of motion:

$$m\frac{\mathrm{d}v}{\mathrm{d}t} = q\vec{E} + q(\vec{v} \times \vec{B}) + \vec{F}. \quad (4.49)$$

In addition to the electrical force $q \times \vec{E}$ also the Lorenz force $q(\vec{v} \times \vec{B})$ acts on a charged particle, which is always perpendicular to the particle velocity $\vec{v}$. $\vec{F}$ is the expression for forces other that charge-dependent forces, for example for the force of gravity, and q is the specific charge.

We now multiply scalar Eq. (4.49) by the velocity, obtaining

$$m\frac{\mathrm{d}\frac{\vec{v}^2}{2}}{\mathrm{d}t} = q\vec{v}\vec{E} + \vec{v}\vec{F}. \quad (4.50)$$

The temporal change of the kinetic energy of the test particle is caused by the power of the electrical force and the general forces $\vec{F}$. The magnetic induction $\vec{B}$ is missing in Eq. (4.50) because the product $\vec{v}(\vec{v} \times \vec{B})$ disappears.

This conclusion shows that a magnetic field cannot influence the energy of a particle. This applies also to temporally variable magnetic fields. Only $\partial\vec{B}/\partial t$ connected the $\vec{E}$ field can change the energy of the particles.

Assuming now that stationary conditions are present, the field strength $\vec{E}$ in Eq. (4.42) can be replaced by:

$$\vec{E} = -\nabla\varphi\,, \tag{4.51}$$

where $\varphi = \varphi(x, y, z)$.

Combining Eq. (4.51) with Eq. (4.50) and $\vec{E} = 0$, we obtain:

$$\frac{\mathrm{d}}{\mathrm{d}t} m\frac{\vec{v}^2}{2} = -q\vec{v}\nabla\varphi\,. \tag{4.52}$$

With

$$\vec{v}\nabla\varphi = \frac{\partial\varphi}{\partial x}\frac{\mathrm{d}x}{\mathrm{d}t} + \frac{\partial\varphi}{\partial y}\frac{\mathrm{d}y}{\mathrm{d}t} + \frac{\partial\varphi}{\partial z}\frac{\mathrm{d}z}{\mathrm{d}t} = \frac{\mathrm{d}\varphi}{\mathrm{d}t}$$

we obtain

$$\frac{\mathrm{d}}{\mathrm{d}t}\left(m\frac{v^2}{2} + q\varphi\right) = 0\,, \tag{4.53}$$

which means that the sum of kinetic and potential energy is constant (φ is the potential energy for each unit charge).

4.4.3 Larmor Motion

If the magnetic induction $\vec{B}$ is temporally constant and if no electrostatic field $\vec{E}(x, y, z)$ is present, then the equation of motion is given as follows:

$$m\frac{\mathrm{d}\vec{v}}{\mathrm{d}t} = q(\vec{v} \times \vec{B})\,; \quad \vec{B} = (0, 0, B_z)\,. \tag{4.54}$$

With the assumption of Cartesian coordinates we obtain

$$m\ddot{x} = q\dot{y}B_z, \tag{4.55}$$

$$m\ddot{y} = -q\dot{x}B_z\,, \tag{4.56}$$

$$m\ddot{z} = 0\,. \tag{4.57}$$

For the z-component we immediately find

$$\dot{z} = v = v_{0z} = \text{const}\,; \quad z = v_{0z}t + z_0\,. \tag{4.58}$$

For the solution of the two first component equations (Eqs. (4.55) and (4.56)) we use the complex numbers:

$$u = x + \mathrm{i}y \quad \text{resp.} \quad \dot{u} = \dot{x} + \mathrm{i}\dot{y}\,. \tag{4.59}$$

Now we multiply Eq. (4.56) with i and add it into Eq. (4.55):

$$\ddot{x} + \mathrm{i}\ddot{y} = -\mathrm{i}\frac{q}{m}B_z(\dot{x} + \mathrm{i}\dot{y})\,, \tag{4.60}$$

or with the abbreviation

$$\omega_\mathrm{g} = \frac{qB_z}{m} \tag{4.61}$$

$$\ddot{u} = -\mathrm{i}\omega_\mathrm{g}\dot{u}\,. \tag{4.62}$$

For the solution of this equation we use

$$u = a\mathrm{e}^{\mathrm{i}\omega t} + b\,, \tag{4.63}$$

where a should be a real number;

$$\omega = -\omega_\mathrm{g}\,. \tag{4.64}$$

Thus we have

$$\dot{u} = -\mathrm{i}\omega_\mathrm{g}a\mathrm{e}^{-\mathrm{i}\omega_\mathrm{g}t} = -\mathrm{i}\omega_\mathrm{g}a(\cos\omega_\mathrm{g}t - \mathrm{i}\sin\omega_\mathrm{g}t) \tag{4.65}$$

or

$$v_x = \dot{x} = -a\omega_\mathrm{g}\sin\omega_\mathrm{g}t \tag{4.66}$$

$$v_y = \dot{y} = -a\omega_\mathrm{g}\cos\omega_\mathrm{g}t\,. \tag{4.67}$$

Since the magnetic field cannot change the kinetic energy of the particle, we obtain:

$$v_{0\perp}^2 = v_x^2 + v_y^2 = a^2\omega_\mathrm{g}^2 = \text{const.} \tag{4.68}$$

$V_{0\perp}$ is the velocity component perpendicular to the magnetic field or

$$a = \frac{v_{0\perp}}{\omega_\mathrm{g}}\,. \tag{4.69}$$

Inserting this in Eq. (4.63), whereby $b = x_0 + \mathrm{i}y_0$, we obtain:

$$x - x_0 = \frac{v_{0\perp}}{\omega_\mathrm{g}}\cos\omega_\mathrm{g}t \tag{4.70}$$

$$y - y_0 = -\frac{v_{0\perp}}{\omega_\mathrm{g}}\sin\omega_\mathrm{g}t\,. \tag{4.71}$$

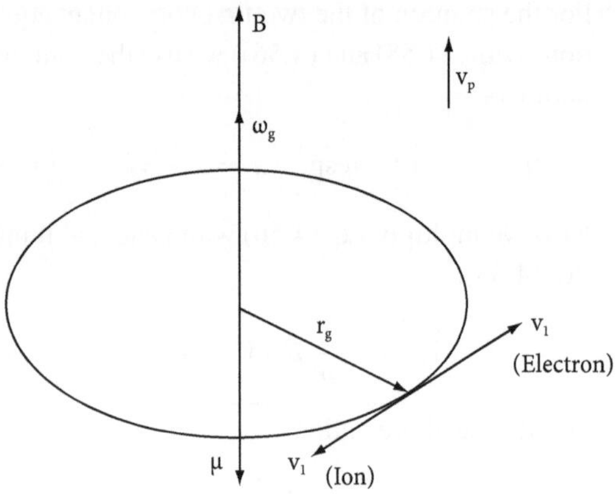

Fig. 4.2 Larmor motion

This is a circular curve, Fig. 4.2, with the radius

$$r_{0g} = \frac{v_{0\perp}}{\omega_g}\frac{v_{0\perp} m}{q\vec{B}}\,. \tag{4.72}$$

The interpretation of Eq. (4.72) reads as: in a temporally constant magnetic field charged particles describe gyration circles with a radius r_g, the Larmor radius, which only depends on the initial velocity of the particles, the specific charge and the induction $\vec{B}$. The gyration frequency ω_g, or Larmor frequency, depends only on the specific charge and the induction.

For electrons and protons for an induction of $|\vec{B}| = 0.1\,\mathrm{T}$ we obtain

$$\omega_{ge} = 1.76 \times 10^{11}\,\mathrm{s}^{-1}$$
$$\omega_{gp} = 9.59 \times 10^{7}\,\mathrm{s}^{-1}\,.$$

Obtaining electrons and protons with the same kinetic energies,

$$\frac{1}{2}m_e v_{e\perp}^2 = \frac{1}{2}m_p v_{p\perp}^2\,; \tag{4.73}$$

are thus applied to the relationship of the gyration radii

$$\frac{r_{ge}}{r_{gp}} = \sqrt{\frac{m_e}{m_p}} \tag{4.74}$$

or $r_{Sp} \cong 40\,r_{Se}$. Therefore, electrons are more closely bound to the magnetic field than ions.

4.5 Collision Determined Plasmas

After collision, the free motion of charge carriers in force fields is treated; the plasma is now to be regarded as a continuum. We assume that in a volume element there are many particles and that due to the thermal motion for each time unit, many collisions between the particles take place and loss of energy and impulse are always joined. Thus, it is to be considered that the collisions can be both purely elastic and inelastic [6]. (Of an elastic collision the sum of the kinetic energies of the collision partners before and after the collision is identical.) Inelastic collisions occur only in the form of excitation and ionization collisions. Since microscopic reactions are not observable, one must always create average values of the particle characteristics in order obtain physical statements.

4.5.1 Distribution of Velocity

Particles have a velocity $\vec{v}'$ between $v_x \le v_x' \le v_x + \mathrm{d}v_x$; $v_y \le v_y' \le v_y + v_y + \mathrm{d}v_y$; $v_z \le v_z' \le v_z + \mathrm{d}v_z$ and a position r between $x \le x' \le x + \mathrm{d}x$, $y \le y' \le y + \mathrm{d}y$ and $z \le z' \le z + \mathrm{d}z$. Then we obtain:

$$\mathrm{d}N = f(r, \vec{v}, t)\,\mathrm{d}x\,\mathrm{d}y\,\mathrm{d}z\,\mathrm{d}v_x\,\mathrm{d}v_y\,\mathrm{d}v_z\,. \tag{4.75}$$

$\mathrm{d}^3r\,\mathrm{d}^3v = \mathrm{d}x\mathrm{d}y\mathrm{d}z\mathrm{d}v_x\mathrm{d}v_y\mathrm{d}v_z$ is a volume element in six-dimensional phase space. The function $f(r, v, t)$ is the distribution function of the considered gas and/or plasma. The distribution function describes how the existing particles distribute themselves in the different volume elements of the phase space at time t. If we integrate Eq. (4.75) over the entire phase space, then we receive the total number of the regarded particles:

$$N = \iiint\!\!\iiint_{-\infty}^{+\infty} f(r, \vec{v}, t)\,\mathrm{d}x\,\mathrm{d}y\,\mathrm{d}z\,\mathrm{d}v_x\,\mathrm{d}v_y\,\mathrm{d}v_z\,. \tag{4.76}$$

The number of particles for each unit volume $n(r,t)$ results from:

$$n(r,t) = \lim \delta V \to 0 \frac{\iiiiiint f(r,\vec{v},t)\,\mathrm{d}^3r\,\mathrm{d}^3\vec{v}}{\iiint \mathrm{d}^3r} = \frac{\iiiiiint f\,\mathrm{d}^3r\,\mathrm{d}^3\vec{v}}{\delta V} = \iiint\!\!\iiint f(r,\vec{v},t)\,\mathrm{d}^3r\,. \tag{4.77}$$

The volume δV is naturally only selected so small that another sufficiently large number of particles are still present in δV.

If we accept an isotropic space in which the distribution function does not depend on the local coordinates, then it follows that:

$$\mathrm{d}n = f(r,t)\,\mathrm{d}v_x\,\mathrm{d}v_y\,\mathrm{d}v_z\,. \tag{4.78}$$

We use the expression

$$\frac{\mathrm{d}n}{n} = \frac{f(r,\vec{v},t)\,\mathrm{d}^3r}{\int\limits_{-\infty}^{+\infty}\int\int f(r,\vec{v},t)\,\mathrm{d}^3r}\,. \tag{4.79}$$

The calculation of the distribution function is the task of the kinetic gas theory. With normal, not degenerate gases Maxwell Boltzmann statistics is valid. Maxwell Boltzmann statistics describes the distribution of gas particles with respect to coordinates and velocities in the presence of any potential field φ. For example, the distribution is frequently d,

$$\mathrm{d}n = \frac{4n_0}{\sqrt{\pi}v_\mathrm{w}^3}\mathrm{e}^{-\frac{1}{v_\mathrm{w}^2}}\left(v^2 + \frac{2\varphi}{m}\right)v^2\,\mathrm{d}\vec{v}\,. \tag{4.80}$$

In the equation

$$v_\mathrm{w} = \sqrt{\frac{2RT}{m}} \tag{4.81}$$

is the most probable velocity of the particles and $\mathrm{d}n$ the number of particles for each volume unit of the plasma around the point (x,y,z). The particle velocity is in the interval between $\vec{v}$ and $\mathrm{d}\vec{v}$; $\varphi(x,y,z)$ is the potential energy of the particles in the considered point of the outer force field, and n_0 is the number of particles for each unit volume in this point, for which $\varphi = 0$ is valid, at which is m the mass, T the temperature and R the gas constant;

$$\mathrm{d}w = \text{const} \times \frac{1}{(2\pi m\,k\,T)^{3/2}} \times \mathrm{e}^{-\frac{p_x^2+p_y^2+p_z^2}{2mkT}}\,\mathrm{d}p_x\,\mathrm{d}p_y\,\mathrm{d}p_z \times \mathrm{e}^{-\frac{\varphi(x,y,z)}{kt}}\,\mathrm{d}x\,\mathrm{d}y\,\mathrm{d}z\,. \tag{4.82}$$

Here

$\mathrm{d}w$ is the probability that the coordinates and the impulse components of the particles determined in the volume element $\delta T = \mathrm{d}x\mathrm{d}y\mathrm{d}z\mathrm{d}p_x\mathrm{d}p_y\mathrm{d}p_z$ around the phase point (x,y,z,p_x,p_y,p_z),

$\varphi(x,y,z)$ the potential energy of the particles in the outside field, and

k the Boltzmann constant.

Written in this form, Maxwell Boltzmann statistics can be regarded as a distribution function, which is observed as a product of the probabilities of two independent events: the probability of a certain impulse value of the particle and an exact position in the space.

The first probability

$$\mathrm{d}w(p) = \frac{1}{(2\pi mkT)^{3/2}} \times \mathrm{e}^{-\left(\frac{p_x^2+p_y^2+p_z^2}{2mkT}\right)}\,\mathrm{d}p_x\,\mathrm{d}p_y\,\mathrm{d}p_z \tag{4.83}$$

is a Maxwell distribution.

The second probability

$$\mathrm{d}w(x,y,z) = \text{const} \times \mathrm{e}^{-\frac{\varphi(x,y,z)}{kT}}\,\mathrm{d}x\,\mathrm{d}y\,\mathrm{d}z \tag{4.84}$$

is a Boltzmann distribution.

We now treat some special cases:

a) no outside forces;

$$\varphi = 0\,;$$

Maxwell distribution

$$f(\vec{v}) = \text{const.} \times \mathrm{e}^{-\frac{mv^2}{2kT}}\,. \tag{4.85}$$

This function describes the Gauss distribution,

b) in the gravity field,

$$\varphi = mgz \quad \text{(barometric level formula)}$$
$$n(z) = n(z=0)\mathrm{e}^{-\frac{mgz}{kT}} . \tag{4.86}$$

c) in the electrostatic field

$$\varphi = -Ze\varphi \quad (Z \text{ number of charges})$$
$$\mathrm{d}n = \frac{4n_0}{\sqrt{\pi} v_\mathrm{w}^3} \mathrm{e}^{-\frac{1}{v_\mathrm{w}^2}\left(v^2 - \frac{2Ze\varphi}{m}\right)} v^2 \,\mathrm{d}\vec{v} . \tag{4.87}$$

d) When a static magnetic field is switched on, the Lorenz force $(\vec{v} \times \vec{B})$ does not execute work on an electrical charge because the force is perpendicular to the particle velocity $\vec{v}$. It develops a macroscopic flow $\vec{u}$ of particle with the charge Ze, then a flow $Ze\vec{u}$ develops, which possesses a magnetic field $\vec{B}$ with the potential $\vec{A}(\vec{B} = \nabla \times \vec{A})$ and the potential energy $eZ(\vec{u}\vec{A})$. We can also derive this from the Hamilton function $\vec{H}$ of a charged particle an electromagnetic field,

$$H = \frac{1}{2m}(p - eZA)^2 + eZ\varphi$$
$$= \frac{1}{2}mv^2 + eZ\varphi . \tag{4.88}$$

In order to obtain the correct equations of motion in Hamiltonian formalism we must exchange the magnetic field $\vec{p} = m\vec{v}$ with $\vec{p} = m\vec{v} + eZ\vec{A}$.

After integration over v, we obtain the equilibrium distribution with the velocity $\vec{u}$:

$$n = n_0\,\mathrm{e}^{-Ze\frac{\varphi-\varphi_0}{kT} + eZ\vec{v}\frac{\vec{A}-\vec{A}_0}{kT}} , \tag{4.89}$$

where n_0 is the density in the position $\vec{A} = \vec{A}_0$ and $\varphi = \varphi_0$.

4.5.2 Mean Free Path and Collision Frequencies

Between two sequential collisions a particle moves in a straight line and on an average travels a certain distance, which is called the ***mean free path*** $\bar{\lambda}$.

The distribution law of the free path determines the probability $\mathrm{d}w(x)$ that a particle without collision travels a path x. In addition it determines whether the particle sustains a collision [7] within the infinitesimal distance $\mathrm{d}x$:

$$\mathrm{d}w(x) = \mathrm{e}^{-n_0 Q x} n_0 Q \,\mathrm{d}x , \tag{4.90}$$

where

n_0 is the number of particles for each cm^3 of the plasma, and

Q the cross-section of the collision.

Figure 4.3 shows the dependence of the collision cross-section Q of electrons against argon ions as a function of electron energy.

The middle distance $\bar{x}$, which a particle can cover without collision, is equal to the mean free path and is defined as

$$\bar{x} = \bar{\lambda} = \int_0^\infty x \,\mathrm{d}w(x) = \int_0^\infty x\mathrm{e}^{-nQx} n_0 Q \,\mathrm{d}x$$
$$= \frac{1}{n_0 Q} . \tag{4.91}$$

If we know the mean free path, the middle free flying time τ, and their reciprocal value, the collision frequency ν is written in the following form:

$$\frac{\bar{\lambda}}{\tau} = \bar{\lambda}\nu = \bar{v} , \tag{4.92}$$

where

$$\bar{v} = \left(\frac{8}{\pi}\frac{kT}{m}\right)^{\frac{1}{2}} \tag{4.93}$$

is the middle thermal velocity.

For a gas mixture with the particle types $1, 2, \ldots n$, we can derive the free distance for the analogous particle types:

$$\lambda_i = \frac{1}{n_1 Q_{1i} + n_2 Q_{2i} + n_3 Q_{3i} + \ldots} . \tag{4.94}$$

For the collision frequencies we find

$$\nu_i = \bar{v}_i\{n_1 Q_{1i} + n_2 Q_{2i} + n_3 Q_{3i} + \ldots + n_{i-1} Q_{i-1,i}\} . \tag{4.95}$$

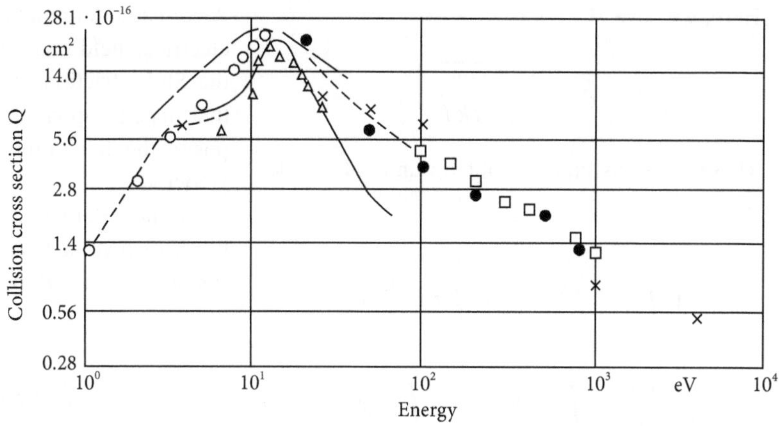

Fig. 4.3 Collision cross-section Q of electrons against argon ions as a function of the electron energy

4.5.3 Drift Motion of Charge Carriers in an Electrical Field with Consideration of Collisions

Before giving an exact explanation of elastic and inelastic collision processes, first let us consider the motion of the charge carrier in an electrical field, in which the particles often suffer collisions. In addition it is accepted that the $\vec{E}$ field is spatially and temporally constant. We can write the equation of motion equation under the assumption that $\vec{E}$ has only one component in the x-direction:

$$m\frac{dv_x}{dt} = q\,E_x\,. \tag{4.96}$$

A double integration supplies:

$$v_x = \frac{q\,E_x}{m}t + v_{0x} \tag{4.97}$$

$$s_x = \frac{q\,E_x\,t'^2}{2m} + v_{0x}t'\,. \tag{4.98}$$

The motion begins after the particle considered has suffered a straight collision. The particle is then accelerated until it suffers a further collision. This takes place after the time t', so that the particle travels back as indicated above in this time s_x from the location of its last collision. Now the middle way $\bar{s}_x$ which a multiplicity of particles travels in this time unit must be calculated. First we must ask how large the probability of the occurrence of an initial velocity between v_{0x} and $v_{0x} + dv_{0x}$ is. According to Eq. (4.83) it is derived as

$$dw_{v_{0x}} = \left(\frac{m}{2\pi kT}\right)^{\frac{1}{2}} e^{-\frac{mv_{0x}^2}{2kT}}\,dv_{0x}\,. \tag{4.99}$$

Let us consider a middle free flying time $\tau = \bar{\lambda}/\bar{v}$. Then the probability for the occurrence of a flying time t' is given through

$$dw_{t'} = \frac{1}{\tau}e^{-\frac{t'}{\tau}}\,dt'\,. \tag{4.100}$$

In order to now obtain the middle travel distance to $\bar{s}$ within the free flying time τ, we must average over all initial velocities and flying times. In addition, Eq. (4.98) is multiplied by $dw_{v_{0x}}$ and $d_{w_{t'}}$ and integrated over all velocities of v_{0x} and all collision times t':

$$\begin{aligned}\bar{s}_x &= \iint s_x\,dw_{v_{0x}}\,dw_{t'}\\ &= \frac{1}{\tau}\sqrt{\frac{m}{2\pi kT}}\int_{-\infty}^{+\infty}\int_{0}^{+\infty}\left(v_{0x}t' + \frac{q\,E_x}{2m}t\right)\\ &\quad\times e^{-\frac{mv_{0x}^2x}{2kT}}e^{-\frac{t'}{\tau}}\,dv_{0x}\,dt\end{aligned} \tag{4.101}$$

$$\begin{aligned}&= A\int_{-\infty}^{+\infty}\int_{0}^{\infty} v_{0x}t'e^{-\frac{mv_{0x}^2x}{2kT}-\frac{t'}{\tau}}\,dv_{0x}\,dt'\\ &\quad+\frac{A\,q\,E_x}{2m}\int_{-\infty}^{+\infty}\int_{0}^{+\infty} t'^2e^{-\frac{mv_{0x}^2x}{2kT}}e^{-\frac{t'}{\tau}}\,dv_{0x}\,dt'\,.\end{aligned} \tag{4.102}$$

Here, we have

$$A = \frac{1}{\tau}\sqrt{\frac{m}{2\tau kT}}\,.$$

These integrals can be split up and solved elementarily:

$$\bar{s}_x = A\int\limits_{-\infty}^{+\infty} v_{0x}\mathrm{e}^{-\beta v_{0x}^2}\,\mathrm{d}v_{0x}\int\limits_0^{\infty} t'\mathrm{e}^{-\frac{t'}{\tau}}\,\mathrm{d}t' + \frac{A\,qE_x}{2m}\int\limits_{-\infty}^{+\infty}\mathrm{e}^{-\beta v_{0x}^2}\,\mathrm{d}v_{0x}\int\limits_0^{\infty} t'^2\mathrm{e}^{-\frac{t'}{\tau}}\,\mathrm{d}t' \qquad (4.103)$$

$$\text{with}\quad \beta = \frac{m}{2kT}\,.$$

The first term of the right-hand side disappears identically; there the first integral $\int_{-\infty}^{+\infty} v_{0x}\mathrm{e}^{-\beta v_{0x}^2}\,\mathrm{d}v_{0x}$ has the value zero. For the second expression we obtain:

$$\bar{s}_x = \frac{A\,q\,E_x}{2m}\sqrt{\frac{\pi}{\beta}}\tau^3\int\limits_0^{\infty}\left(\frac{t'}{\tau}\right)\mathrm{e}^{-\frac{t'}{\tau}}d\left(\frac{t'}{t}\right) = \frac{A\,q\,E_x}{2m}\sqrt{\frac{\pi}{\beta}}2\tau^3\,. \qquad (4.104)$$

If we apply retrogressive the substitution, then it follows that:

$$\bar{s}_x = \frac{1}{\tau}\frac{qE_x}{2m}2\tau^3 = \frac{q\,E_x}{m}\tau^2\,. \qquad (4.105)$$

With the middle free flying time τ we obtain the value for the middle drift velocity

$$\overline{v_{x\mathrm{D}}} = \frac{\overline{s_x}}{\tau} = \frac{q\,E_x}{m}\tau\,. \qquad (4.106)$$

If we again introduce the mean free path, then the following relations result:

$$\overline{v_{x\mathrm{D}}} = \frac{q\,E_x\tau}{m} = \frac{q\bar{\lambda}}{m\bar{v}}E_x = \frac{q\bar{\lambda}E_x}{m\sqrt{\frac{8kT}{\pi m}}} = \frac{q\bar{\lambda}}{\sqrt{\frac{8}{\pi}kTm}}E_x\,. \qquad (4.107)$$

A collision determinate plasma influenced by an electrical field can be described analogously to the Stoke friction with constant drift velocities that depend upon signs of the charge in or opposite the field direction and overlaid thermal velocity.

We now define the proportionality factors between drift velocity and field strength as mobility. For single charged ions and electrons, we obtain

$$b_\mathrm{i} = \frac{e\bar{\lambda}_\mathrm{i}}{\sqrt{\frac{8}{\pi}kT_\mathrm{i}m_\mathrm{i}}} = \frac{e\tau_\mathrm{i}}{m_\mathrm{i}}\,,\qquad b_\mathrm{e} = \frac{e\bar{\lambda}_\mathrm{e}}{\sqrt{\frac{8}{\pi}kT_\mathrm{e}m_\mathrm{e}}} = \frac{e\tau_\mathrm{e}}{m_\mathrm{e}}\,, \qquad (4.108)$$

and for the relationship of the mobility

$$\frac{b_\mathrm{i}}{b_\mathrm{e}} = \frac{\tau_\mathrm{i}\,m_\mathrm{e}}{\tau_\mathrm{e}\,m_\mathrm{i}} = \frac{\bar{\lambda}_\mathrm{i}\,m_\mathrm{e}\,\bar{v}_\mathrm{e}}{\bar{\lambda}_\mathrm{e}\,m_\mathrm{i}\,\bar{v}_\mathrm{i}} = \sqrt{\frac{m_\mathrm{e}}{m_\mathrm{i}}\frac{\tau_\mathrm{e}}{\tau_\mathrm{i}}\frac{\bar{\lambda}_\mathrm{i}}{\bar{\lambda}_\mathrm{e}}}\,. \qquad (4.109)$$

For the middle drift velocities we can write:

$$\vec{v}_\mathrm{iD} = b_\mathrm{i}\vec{E}\,;\quad \vec{v}_\mathrm{eD} = -b_\mathrm{e}\vec{E}\,. \qquad (4.110)$$

By introduction of the electrical current density (j) we obtain:

$$j = j_\mathrm{i} + j_\mathrm{e} = e[n_\mathrm{i}\vec{v}_\mathrm{iD} - n_\mathrm{e}\vec{v}_\mathrm{eD}] = en_\mathrm{i}(b_\mathrm{i} + b_\mathrm{e})\vec{E} = \sigma\vec{E}\,. \qquad (4.111)$$

Quasi-neutrality was presupposed ($n_\mathrm{i} = n_\mathrm{e}$). The size of σ indicates electrical conductivity according to (simple) Ohm's law.

For the relationship of the contributions from ions and electrons to the current density, we can write:

$$\frac{|j_\mathrm{i}|}{|j_\mathrm{e}|} = \frac{b_\mathrm{i}}{b_\mathrm{e}} = \sqrt{\frac{m_\mathrm{e}}{m_\mathrm{i}}\frac{T_\mathrm{e}}{T_\mathrm{i}}\frac{\bar{\lambda}_\mathrm{i}}{\bar{\lambda}_\mathrm{e}}}\,. \qquad (4.112)$$

If the electron and ion temperatures are approximately equal, then we obtain:

$$\frac{|j_\mathrm{i}|}{|j_\mathrm{e}|} \cong \sqrt{\frac{m_\mathrm{e}}{m_\mathrm{i}}}\,. \qquad (4.113)$$

This means in plasma the current is carried predominantly by electrons. Regards electrical conductivity σ, we can write according to Eq. (4.111):

$$\sigma = e n_i b_e \left(1 + \frac{b_i}{b_e}\right) = \frac{e^2 n_e \tau_e}{m_e} \left(1 + \frac{b_i}{b_e}\right). \quad (4.114)$$

Since generally $b_i/b_e \ll 1$, the expression for σ in very good approximation is

$$\sigma = \frac{e^2 n_e \tau_e}{m_e} = \frac{e^2 n_e}{m_e \nu_e}. \quad (4.115)$$

4.5.4 Ionization and Recombination

Ionization [8], i.e. the separation of electrons from atoms or molecules, is an endothermic process. It may be accomplished by a variety of processes (see Sect. 4.5.4.1). We shall describe now in most general terms the sequence of events that might take place in the ionization of a diatomic gas (see Fig. 4.4).

The energy of a diatomic molecule consists of the energy of the translation motion of its centre of gravity $\varepsilon_{\text{transl}}$, the energy of the electron motion in the atoms of the molecule ε_e, the energy of oscillations of the nucleus $\varepsilon_{\text{Schw}}$ and the energy of the rotation ε_{rot}

$$\varepsilon = \varepsilon_{\text{trans}} + \varepsilon_e + \varepsilon_{\text{Schw}} + \varepsilon_{\text{rot}}. \quad (4.116)$$

The translation motion of a diatomic molecule is not quantized and does not differ from an atom. All other forms of the intrinsic energy of a molecule are quantized. The energy $\varepsilon_e, \varepsilon_{\text{Schw}}, \varepsilon_{\text{rot}}$ can take only discrete values.

In a first approximation all three forms of the intrinsic motion in a molecule are independent of each other. If one approaches the absolute zero point, then the molecular energy consists only of the energy of the translation motion; the rotational energy joins with approximately 10 K. At higher energy the nucleus oscillations are considered, which by further energy input finally becomes equal to the molecular binding energy; the molecule is dissociated. With ambient pressure, for example, the dissociation of CO_2 starts at about 3000 K, whereas N_2 starts to dissociate at about 4500 K.

Table 4.1 Dissociation energies of some molecules

Molecule	E_D eV
H_2	4.477
N_2	9.76
O_2	5.08
CO	11.11
NO	6.48
OH	4.37
CO_2	16.56

a) A rigid rotator has three degrees of freedom of the translation movement and two degrees of freedom of the rotation.
b) The energy of a diatomic molecule consists of the translation energy, the rotation energy, the energy of the nucleus oscillations and the energy of the electron movement in the atoms of the molecule.
c) Dissociation of molecule to neutral atoms.
d) Ionization process.

For the dissociation reaction of nitrogen we can write:

$$N_2 + E_D \rightarrow 2\,N. \quad (4.117)$$

The dissociation energy of some diatomic gases is indicated [9] in Table 4.1.

A change of ε_e is the energy of the motion of the electron motion and we do not need to consider this energy. The electrons in the orbits around the atomic nucleus become excited to quantum states above the ground state. This excitation also contributes to the total energy, but the effect is negligible up to about 5000 K. If the temperature continues to rise, the electrons leave their orbit and the atom becomes an ion (see Fig. 4.4). The ionization energy is equal to the absolute size of the binding energy of the electron on the nucleus. The ionization potential ψ is:

$$\psi = \frac{hRZ^2}{en^2}. \quad (4.118)$$

In Eq. (4.120) R is the Rydberg constant [9],

$$R = \frac{m_e e^4}{8\varepsilon_0^2 h^3}, \quad (4.119)$$

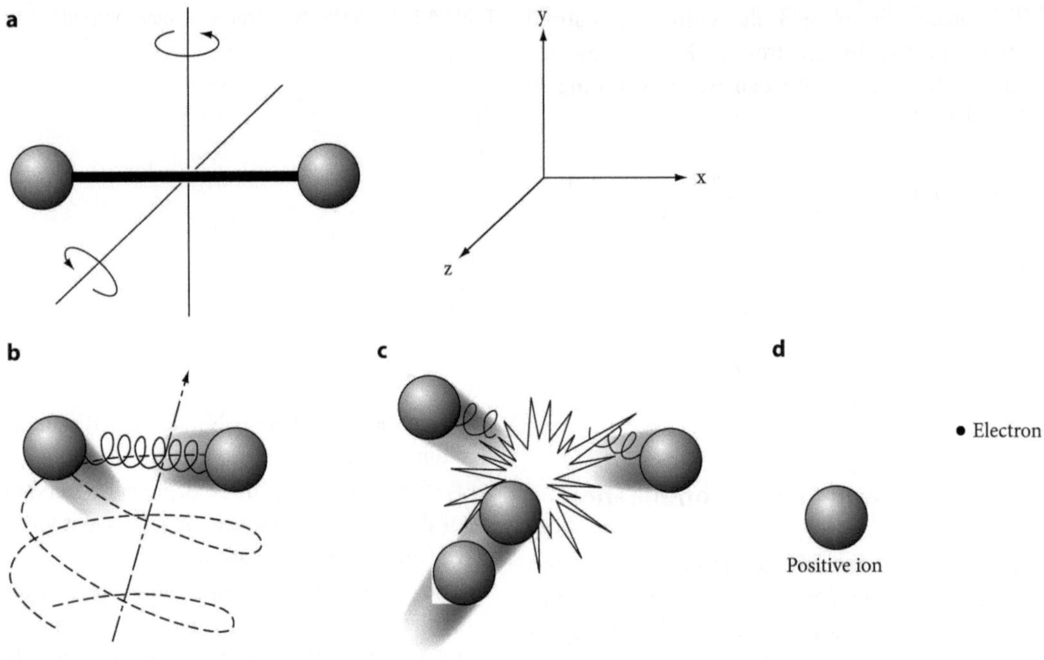

Fig. 4.4 Procedures of the ionization process. **a** A rigid rotator has three degrees of freedom of the translation movement and two degrees of freedom of the rotation. **b** The energy of a diatomic molecule consists of the translation energy, the rotation energy, the energy of the nucleus oscillations and the energy of the electron movement in the atoms of the molecule. **c** Dissociation of molecule to neutral atoms. **d** Ionization process

where
Z is the ordinal number,
n the principal quantum number, and
h the Planck quantum of action.

Figure 4.5 shows the relationship of the ions to the total number of particles in air as a function of the temperature.

The simple ionization reaction of a nitrogen atom is:

$$\mathrm{N} + E_{\mathrm{i}} = \mathrm{N}^{+} + e\,, \qquad (4.120)$$

where
E_{i} is the ionization energy.

Figure 4.6 shows the influence of the temperature on the relationship of the ions to the total number of the particles.

Like dissociation, ionization depends on pressure and temperature. The equilibrium ionization as a function of pressure and temperature is described by the Saha equation (see Sect. 4.5.5.2).

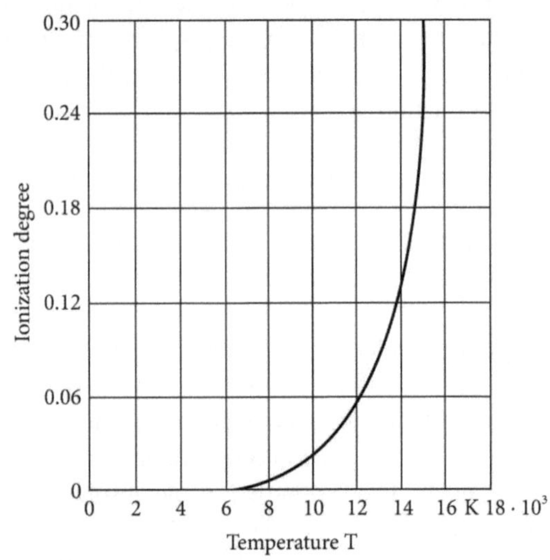

Fig. 4.5 Fraction of molecules thermally ionized in air at 1 bar pressure

The first-level ionization of any neutral atom A and the associated neutralization reaction

Table 4.2 Representative ionization potentials in eV

Element	Ionization level					
	I eV	II eV	III eV	IV eV	V eV	VI eV
Aluminium	5.984	18.823	28.44	119.96	153.77	190.42
Argon	15.755	27.62	40.90	59.79	75.0	91.3
Calcium	6.111	11.87	51.21	67	84.39	
Carbon	11.264	24.376	47.864	64.476	391.986	489.84
Cesium	3.893	25.1				
Helium	24.580	54.400				
Hydrogen	13.595					
Ferrum	7.90	16.18	30.64			
Krypton	13.99	24.56	36.9			
Magnesium	7.644	15.03	80.12	109.29	141.23	186.86
Mercury	10.44	18.8				
Niobium	6.77	14				
Nitrogen	14.54	29.605	47.426	77.450	97.863	551.92
Oxygen	13.614	35.146	54.934	77.394	113.873	138.08
Platinum	(8.9)	18.5				
Potassium	4.339	31.81	46	60.90		90.7
Silicon	8.149	16.34	33.46	45.13	166.73	205.11
Silver	7.574	21.48				
Sodium	5.138	47.29	71.65	98.88	138.60	172.36
Strontium	5.692	11.027				
Titanium	6.83	13.63	28.14	43.24	99.8	120
Tungsten	7.94					
Xenon	12.13	21.2				

may be written as:

$$\mathrm{A} + E_{\mathrm{i}} \underset{K_{\mathrm{n}}}{\overset{K_{\mathrm{i}}}{\rightleftarrows}} \mathrm{A}^{+} + e\,, \tag{4.121}$$

where K_{i} and K_{n} denote, respectively, the ionization rate coefficient and the neutralization rate coefficient. Ionization levels higher than the first may be expected in various applications. Accordingly we can write the general physicochemical ionization reaction as follows (Eq. (4.122)):

$$\mathrm{A}^{j} + E_{\mathrm{i}(j+1)} \underset{K_{j\mathrm{i}}}{\overset{K_{\mathrm{i}j}}{\rightleftarrows}} \mathrm{A}^{j+1} + e\,; \quad j = 0, 1, 2\,, \tag{4.122}$$

where

A^{j} is to be ionized particles,
A^{j+1} is to be ionized particles, and
$E_{\mathrm{i}(j+1)}$ is the ionization energy for $(j+1)$ of the ionization level.

The fist ionization potential is the energy required to remove the first electron from the atom and the second ionization potential is the energy required to remove the second stage, etc. Because the succeeding ionization potentials are correspondingly higher, at low temperatures doubly ionized atoms are less probable than singly ionized atoms.

4.5.4.1 Ionization Processes

Atoms can be ionized by various processes. We might classify the various modes of ionization as follows:

- Ionization by electron collisions
- Ionization by positive ion or neutral atom collisions
- Ionization by collisions of the second kind

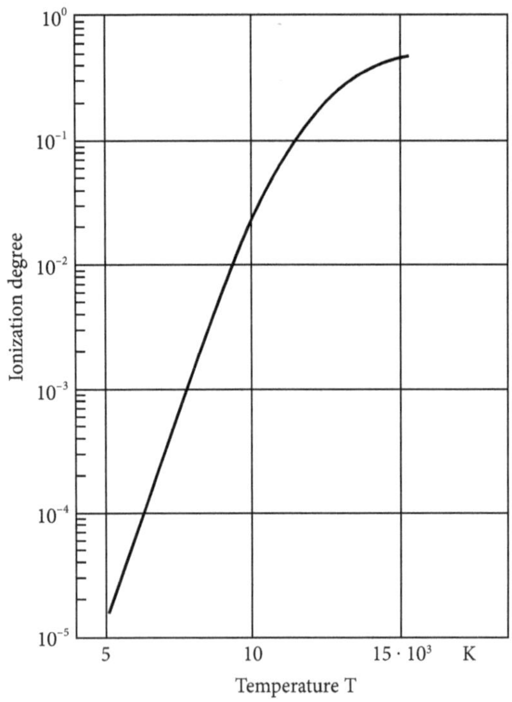

Fig. 4.6 Degree of ionization of argon at 1 bar pressure

- Cumulative ionization
- Ionization by electrical fields
- Surface ionization.

We shall now discuss each of these mechanisms briefly.

4.5.4.1.1 Ionization by Electron Collisions

The probability of ionization of a molecule depends on the ionization cross-section which is a function of the electron energy. If an atom is to be ionized in a single collision, the minimum relative energy of the electron and the atom must equal the ionization energy of the atom [10]. Figure 4.7 shows the ionization cross-sections of noble gases as a function of electron energy.

If an atom is to be ionized by a simple inelastic collision of an electron, then the relative energy between the electron and the atom must be at least the same as the ionization energy E_i of the atom.

If the colliding electron has more than the minimum amount of energy required to ionize the atom, the excess energy may be retained by the electron; it may be transferred to the electron stripped from the atom, or it may be used to excite or further ionize the atom. A combination of these possibilities may also result. Generally, the excess energy transformed into kinetic energy of the ion will be insignificant.

4.5.4.1.2 Ionization by Positive Ion or Neutral Atom Collisions

Compared to electrons, positive ions are generally inefficient in producing ionization by collisions. However, some very energetic positive ions, such as alpha particles, are effective ionizing agents. Because its mass is of the order of the mass of the atoms or molecules of the gas, a positive ion will not attain an energy much larger than that of the neutral gas particles unless it is in a strong electric field or in a very dilute gas. Consequently, except for very-high-temperature gases, most ions will have an energy less than the ionization energy of the gas molecules, and ionization by collision will be improbable.

It is found that positive ions begin to be effective ionizers when their velocities are as great as those of electrons having an energy equal to the ionizing energy of the gas. However, because it is unlikely that the ions will attain such velocities, ionization by ion collision is probably less important than any secondary effects that may occur. These secondary effects can be secondary emission of electrons from walls and electrodes, or they may be inelastic collisions with gas atoms during which the radiation emitted causes photoelectric emission in the gas or at surfaces.

4.5.4.1.3 Radiation-Induced Ionization

A photon may excite or ionize an atom if its energy is greater than the ionization energy of the atom. The energy of radiation is given by ***Planck's law***, namely,

$$E_\mathrm{ph} + h\nu ,$$

where

h is Planck's constant, and

ν the frequency of the radiation.

Ionization by photon collisions is generally of less importance than ionization by electron collisions.

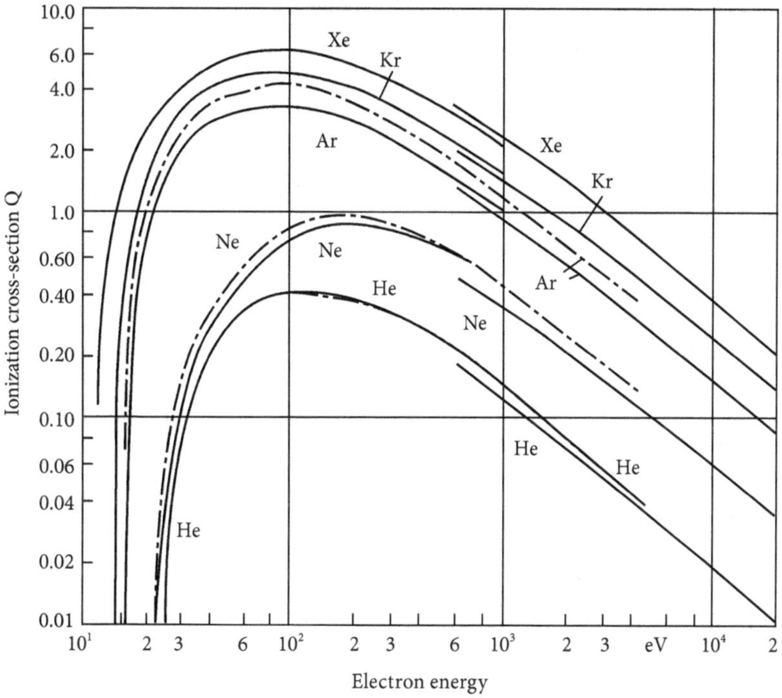

Fig. 4.7 Ionization cross-sections (electron atomic collision) of noble gases (based on different experiments)

4.5.4.1.4 Ionization by Collisions of the Second Kind

An atom can be ionized by energy transfer by an excited or ionized molecule. This process is possible if the available energy is equal to the ionization energy of the atom. The probability of the occurrence of this process at a maximum of the available energy is just sufficient and decreases with an excess of energy. Collisions of the second kind usually occur when gases are mixed, since the energy of radiation emitted when an excited atom of one species returns to its normal state may be greater than the excitation energy of an atom of another species.

4.5.4.1.5 Cumulative Ionization

This process refers to ionization in two or more stages. Thus, an atom excited by one method may have its ionization completed by one or more other processes [11],

$$A + A \rightarrow A^* + A$$
$$A + A^* \rightarrow A^+ + e + A\,.$$

Metastable states provide the most important source of excitation by excited atoms. Metastable states are those from which spontaneous escape by radiation is inhibited according to the principles of quantum theory. Therefore, the lifetime of this state is much longer than that of a normal excited state, lasting as long as 10^{-1} s, compared with the average life of a normal excited state of the order of 10^{-8} s. In a lifetime of this length, the atom can undergo many collisions, and the probability of energy transfer is, consequently, high. Metastable atoms can have a considerable effect on the conductivity of a gas if it contains atoms of a different element with an ionization energy less than the energy corresponding to the metastable excitation level of the gas atoms.

The metastable state can be destroyed by one of the following processes:

(1) The atom may be excited to a higher-energy level by cumulative excitation by any one of the previously mentioned processes.
(2) The atom may undergo a collision of the second kind with an atom of the same element

so that its excitation energy plus its kinetic energy can excite the atom to a higher level.
(3) In mixed gases, the energy of the atoms in the metastable state of one gas may go into the ionization of the atoms of another gas if the available energy is high enough.
(4) Metastable atoms may collide with solid surfaces and cause secondary emission of electrons.

Table 4.3 The most important recombination processes

Ion–ion recombination	
Volume recombination	Radiative recombination
	recombination by charge exchange
Electron–ion recombination	
Volume recombination	Three-body recombination
	Radiative recombination
	Surface recombination

4.5.4.1.6 Ionization by Electrical Fields

In the presence of an oscillating electric field, free electrons in a gas can acquire energies that are sufficient to excite and ionize the gas. The resulting motion of the electrons and ions produces an oscillatory current which increases with the applied field until breakdown occurs. The oscillatory electric field may be produced in the gas of a discharge tube by applying a high-frequency potential to either internal or external electrodes, or by induction from a conductor carrying an oscillating electric current. Under static fields, there is a continual stream of electrons or ions to the electrodes, and the electrons and ions are, therefore, removed from the discharge. This loss is in addition to that occurring at the walls or in the gas, so that the electric field must be high enough to generate electrons an ion to correct this loss. In high-frequency fields, however, this loss to the electrodes can be considerably reduced so that the high-frequency discharge can occur at values of the field that are much lower than those required with the static field. While ionization by a high-frequency field is not a naturally occurring phenomenon, it is of interest because it provides a way by which plasmas can be conveniently produced [12].

4.5.4.1.7 Surface Ionization

When a neutral atom strikes a hot surface, it may lose its outermost electron to the surface and bounce off as an ion. This process is known as surface ionization and is a type of thermal ionization. At the threshold temperature, the atoms begin to boil off the surface, and ionization then begins to take place. The work function of the surface must be greater than the ionization potential of the atom to be ionized for appreciable ionization to take place. Since the highest work functions are of the order of 6 eV, only alkali metals can be ionized by this process with high efficiency. Surfaces can be used for the ionization of alkali metals with methods based on this process.

4.5.4.2 Neutralization Processes

During neutralization or recombination processes, positive and negative particles join to form neutral particles.

For the purpose of a simple analysis we assume that there is one species each of negative and positive particles, and that these are distributed uniformly and isotropically in space. We predicate that the rate of recombination is proportional to the number of encounters between particles of unlike charge, which in turn is proportional to the number of positive particles n_+ and the number of negative particles n_-. The encounters may be of any type. Therefore, the loss in the number of charged particles $\mathrm{d}n$ in a time $\mathrm{d}t$ is:

$$\frac{\mathrm{d}n}{\mathrm{d}t} = -\alpha n_- n_+ \,, \tag{4.123}$$

where α is the coefficient of neutralization or recombination.

If $n_+ = n_-$

$$\frac{\mathrm{d}n}{\mathrm{d}t} = -\alpha n_\mathrm{i}^2 \,. \tag{4.124}$$

The recombination coefficient of neutralization varies with the type of gas, its pressure and temperature, the ions present, etc. Assuming these conditions to be constant, Eq. (4.124) can be integrated directly, yielding the gas species, the pressure, the temperature and the number of charge carriers. If one accepts α as constant, then Eq. (4.124) can be integrated directly:

$$\int_{n_0}^{n} \frac{dn}{n^2} = -\alpha \int_{0}^{t} dt \,. \quad (4.125)$$

Solving for the neutralization coefficient, one obtains

$$\alpha = \frac{1}{t}\left(\frac{1}{n} - \frac{1}{n_0}\right) . \quad (4.126)$$

If n can be measured and $1/n$ is plotted versus t, the slope of the line will be equal to α. It is found in experiments that neutralization rates can greatly exceed the rate of collisions between the gas particles as calculated from kinetic theory. This difference is due to the attractive forces which cause particles of opposite charges to drift together. The intricacies of neutralization phenomena are beyond the scope of this text. However, in Table 4.3 we list the various modes of neutralization based on the detailed discussions in [12, 13].

In cathode sputtering plasmas the neutralization is essentially determined by volume recombination processes in the form of three-body-collisions.

4.5.5 Plasma as Continuum

4.5.5.1 Thermodynamic Basic Equations

In this chapter we present the equations of state of real gases, namely gases which may be dissociated and/or ionized.

Now the plasma is to be regarded as continuum. It consists of three components: electrons, ions and neutral gas; these properties give the state of the system.

For simplification only simply charged ions occur. Thus the quasi-neutrality relationship $n_\mathrm{i} = n_\mathrm{e}$ applies.

The partial densities are defined by the relationship

$$\varrho_j = n_j m_j \,, \quad (4.127)$$

where

m_j is the mass of a particle of the sort j.

Thus the total density of the plasma is defined as follows:

$$\varrho = \sum_{j=1}^{3} n_j m_j = n_\mathrm{e} m_\mathrm{e} + n_\mathrm{i} m_\mathrm{i} + n_\mathrm{n} m_\mathrm{n} = (n_\mathrm{i} + n_\mathrm{n}) m_\mathrm{n} \,. \quad (4.128)$$

The partial pressure is given by the ideal gas equation:

$$p_j = n_j k T_j \,. \quad (4.129)$$

From this follows for the total pressure

$$p = \sum_{j=1}^{3} n_j k T_j \,.$$

If the components have the same temperature, then we obtain

$$p = kT \sum_{j=1}^{3} n_j = kT(n_\mathrm{e} + n_\mathrm{i} + n_\mathrm{n}) = (2n_\mathrm{i} + n_\mathrm{n})kT = nkT \,, \quad (4.130)$$

$$n = n_\mathrm{e} + n_\mathrm{i} + n_\mathrm{n} = 2n_\mathrm{i} + n_\mathrm{e} \,.$$

Dividing Eq. (4.130) with Eq. (4.128), we obtain

$$\frac{p}{\varrho} = \frac{nkT}{\Sigma n_\mathrm{i} m_j} = \frac{kT}{m_\mathrm{n}} \frac{n_\mathrm{e} + n_\mathrm{i} + n_\mathrm{n}}{(n_\mathrm{i} + n_\mathrm{n})} = \frac{kT}{m_\mathrm{n}}\left(1 + \frac{n_\mathrm{i}}{n_\mathrm{i} + n_\mathrm{n}}\right) . \quad (4.131)$$

Let us now define the percent, degree or fraction of ionization by the ratio

$$\alpha = \frac{n_\mathrm{i}}{n_\mathrm{i} + n_\mathrm{n}} \,. \quad (4.132)$$

Thus we can write Eq. (4.131) as follows:

$$\frac{p}{\varrho} = (1 + \alpha)\frac{kT}{m_\mathrm{n}} = (1 + \alpha)\frac{R_0 T}{M} \,, \quad (4.133)$$

where

R_0 is the *general* gas constant, and

M relative molecular mass of the atoms.

If multiple charged ions arise in plasma, then the number of free electrons is

$$n_e = n_1 + 2n_2 + 3n_3 + \ldots + jn_j , \quad (4.134)$$

where
j is the ionization level.
The generalized equation of state for a multiply ionized perfect gas plasma is then from Eq. (4.128)

$$p_j = (n_e + n_n + n_1 + n_2 + \ldots + n_j)kT . \quad (4.135)$$

As a matter of convenience, let us denote by n_s all heavy particles (i.e. ions and atoms) exclusive of electrons. Therefore,

$$n_s = n_n + n_1 + n_2 + \ldots + n_j .$$

Further, we define the ith degree of ionization α_i by

$$\alpha_j = \frac{n_j}{n_s} . \quad (4.136)$$

For the definition of the inner energy ε one proceeds again from an individual particle. The translation energy is

$$\varepsilon_{j\,\mathrm{trans}} = \frac{f}{2} KT , \quad (4.137)$$

where
f is the number of translational degrees of freedom.

For atomic systems $f = 3$ are obtained. If we refer the inner energy to the unit volume, then

$$\varepsilon_{j\,\mathrm{trans}} = \frac{3}{2} n_j kT . \quad (4.138)$$

Related to the masses unit, we obtain

$$\varepsilon_{j\,\mathrm{trans}} = \frac{3}{2} \frac{n_j kT}{\varrho_j} = \frac{3}{2} \frac{kT}{m_j} . \quad (4.139)$$

For the simple case of an ionized monatomic gas, the internal energy is

$$\varepsilon = \frac{1}{\varrho} \Sigma \varrho_j \varepsilon_j = \frac{3}{2} kT \frac{(n_e + n_i + n_n)}{m_n(n_i + n_n)} = \frac{3}{2} \frac{kT}{m_n}(1 + \alpha) . \quad (4.140)$$

Apart from the translation energy also the excitation and the ionization energy must be considered as potential energy. Contrary to molecular gases, for which oscillations and rotations are to be considered, one can neglect the excitation energies at atoms and ions because of the generally short lifetime of exciting particles in relation to the translational energy. This does not apply to the ionization energy, which one attributes to the ions as potential energy.

The inner energy u from an ion of a monatomic gas consists, therefore, of the translation, the ionization energy and the excitation energy of the electrons:

$$u = \frac{3}{2}(1 + \alpha)\frac{kT}{m_n} + \alpha \frac{E_i}{m_n} + \varepsilon_{\mathrm{elec}} , \quad (4.141)$$

where
$\varepsilon_{\mathrm{elec}}$ is the energy of the excited electrons.
In thermodynamics the enthalpy is defined as follows:

$$h = u + \frac{p}{\varrho} . \quad (4.142)$$

If we introduce Eq. (4.141) in Eq. (4.142), then we obtain for monatomic gas plasma the enthalpy h:

$$\begin{aligned} h &= \frac{3}{2}\frac{p}{\varrho} + \frac{\alpha E_i}{m_n} + \frac{p}{\varrho} + u_{\mathrm{elec}} \\ &= \frac{5}{2}\frac{p}{\varrho} + \frac{\alpha E_i}{m_n} + u_{\mathrm{elec}} \\ &= \frac{5}{2}(1 + \alpha)\frac{kT}{m_n} + \frac{\alpha E_i}{m_n} + u_{\mathrm{elec}} . \end{aligned} \quad (4.143)$$

The specific heat may be calculated from the internal energy and the enthalpy because it is defined as follows:

$$c_v = \left(\frac{\partial u}{\partial T}\right)_\varrho \quad (4.144)$$

and

$$c_p = \left(\frac{\partial h}{\partial T}\right)_p . \quad (4.145)$$

Through the differentiation results with neglect of the change of temperature of the excitation energy of the electrons

$$c_v = \frac{3}{2}\frac{k}{m_n}(1+\alpha) + \left(\frac{3}{2}\frac{kT}{m_n} + \frac{E_i}{m_n}\right)\left(\frac{\partial\alpha}{\partial T}\right)_{\varrho} \tag{4.146}$$

and

$$c_p = \frac{5}{2}\frac{k}{m_n}(1+\alpha) + \left(\frac{5}{2}\frac{kT}{m_n} + \frac{E_i}{m_n}\right)\left(\frac{\partial\alpha}{\partial T}\right)_p . \tag{4.147}$$

If $\partial\alpha/\partial T$ goes against zero, i.e. the temperature is too low for a chance of the temperature to effect ionization, or the temperature is so high that the gas plasma is fully ionized – thus $\alpha = 1$ – we obtain very simple expressions:

$$c_{v\alpha=0} = \frac{3}{2}\frac{k}{m_n}, \tag{4.148}$$

$$c_{v\alpha=1} = 3\frac{k}{m_n}, \tag{4.149}$$

$$c_{p\alpha=0} = \frac{5}{2}\frac{k}{m_n} \tag{4.150}$$

and

$$c_{p\alpha=1} = 5\frac{k}{m_n}. \tag{4.151}$$

For the isentropic exponent $\kappa = c_p/c_v$, generally, a complicated expression results,

$$\kappa = \frac{c_p}{c_v} = \frac{5k(1+\alpha) + (5kT + 2E_i)\left(\frac{\partial\alpha}{\partial T}\right)_p}{3k(1+\alpha) + (3kT + 2E_i)\left(\frac{\partial\alpha}{\partial T}\right)_{\varrho}} . \tag{4.152}$$

For $\alpha = 0$ or $\alpha = 1$, $\mathrm{d}\alpha/\mathrm{d}T = 0$ becomes $\kappa = 5/3$, as for each monatomic gas.

While κ in the intermediate range varies marginally, the specific heat shows a large increase, if the ionization begins with rising temperature.

4.5.5.2 Saha's Equation

The ionization degree α in the thermal equilibrium follows from the reaction formula of M.N. Saha [15]:

$$\frac{\alpha^2}{1-\alpha^2} = C\frac{T^{5/2}}{p}\mathrm{e}^{-\frac{E_i}{kT}} . \tag{4.153}$$

Because the constant C in Eq. (4.153) cannot be obtained from classical thermodynamics, we will derive Saha's equation in an alternate manner.

Consider the general ionization reaction

$$\mathrm{A}_j \rightleftarrows \mathrm{A}_{j+1} + e - E_j ; \quad j = 0, 1, 2, \ldots , \tag{4.154}$$

where A_j is the jth level ionized particle and E_j is the ionization energy necessary for ionization from the ith to the $j+1$ level of ionization. The equilibrium constant for this reaction is given in accordance with the law of mass action:

$$K_{j+1} = \frac{n_{j+1}n_e}{n_j} = \frac{Z_{j+1}Z_e}{Z_j}\mathrm{e}^{-\frac{E_i}{kT}} , \tag{4.155}$$

where Z is the total partition function or state sum.

According to the different contributions to the inner energy u the partition function sum Z is

$$Z = Z_{\mathrm{rot}} Z_{\mathrm{trans}} Z_{\mathrm{vib}} Z_{\mathrm{elec}} . \tag{4.156}$$

However, for an atomic gas, only the translational and electronic energies exist. Therefore, for an atomic gas one needs to consider only the translational and the electronic partition functions, and Eq. (4.156) reduces to

$$Z = Z_{\mathrm{trans}} Z_{\mathrm{elec}} . \tag{4.157}$$

Of the partition functions in Eq. (4.157) the translation one is defined by a simple expression, namely,

$$Z_{\mathrm{trans}} = \left(\frac{2\pi mkT}{h^2}\right)^{\frac{3}{2}} , \tag{4.158}$$

where m is the mass of particle and h is Planck's constant.

The partition function for the energy of the electron motion one derives most simply from the probability of state. The probability distribution of different conditions of a final system, i.e. a system with constant energy, which do not interact with the surrounding bodies, is called the micro-canonical Gibb's distribution. Such a system is naturally not realizable, it is an ideal one. Its states are called degenerate. For each energy value there exist different values and to these, appropriate states. One calls the number of states belonging to a given energy the degeneration degree $g(\varepsilon)$.

Each molecule of an ideal gas at not too low temperature, is, for example, a quasi-enclosed subsystem. Its kinetic energy is on average by far larger than its interaction energy with other gas molecules. The probability of state of an atom and/or a molecule, therefore, is

$$w(u_j) = \frac{g_j(\varepsilon_j)\mathrm{e}^{-\frac{\varepsilon_j}{kT}}}{\sum\limits_j g_j(\varepsilon_j)\mathrm{e}^{-\frac{\varepsilon_j}{kT}}} = \frac{g_j(\varepsilon_j)\mathrm{e}^{-\frac{\varepsilon_j}{kT}}}{Z}, \tag{4.159}$$

where $w(\varepsilon_j)$ is the state probability for a quasi-enclosed system remaining in a state with the energy ε_j.

If we develop the partition function Z – the denominator in Eq. (4.159) – for the electrons, then we obtain

$$Z_{\text{elec}} = g_0 + g_1\mathrm{e}^{-\frac{\varepsilon 1}{kT}} + g_2\mathrm{e}^{-\frac{\varepsilon 2}{kT}} + \ldots + g_n\mathrm{e}^{-\frac{\varepsilon n}{kT}} . \tag{4.160}$$

If $\varepsilon_0 \ll \varepsilon_1 \ll \varepsilon_2$, etc., then the partition function is obtained as

$$Z_{\text{elec}} = g_0 . \tag{4.161}$$

The degeneration condition of monatomic gases is calculated through

$$g = 2\vec{S} + 1 , \tag{4.162}$$

where $\vec{S}$ is the vector sum of the spin numbers S

$$\vec{S} = \sum_{j=1}^{N} s_j ,$$

where

N is the number of electrons in the atom.

For an electron $S = 1/2$ and hence

$$Z_{\text{e}} = g_0 = 2\left(\frac{1}{2}\right) + 1 = 2 .$$

Therefore, $Z_{\text{e}} = 2$, and the partition function for an electron reduces to its statistical weight.

In low-temperature problems where calculations of high accuracy are not required, it is possible to approximate the series of Eq. (4.160) with the first two terms. Although, as stated previously, the error introduced in this way increases with increasing temperature, it will be small provided that the energy of the second excited state ε_2 is much greater than kT. For this case, the partition function becomes

$$Z_{\text{elec}} = g_0 + g_1\mathrm{e}^{-\frac{\varepsilon_1}{kT}} . \tag{4.163}$$

To illustrate how the electronic partition function is evaluated, we will consider the argon atom and ion. The fist two energy levels and the ionizational energy of the argon atom are given in [16];

g_j	$e_j/k[K]$
1	0
60	162,500
Ionization energy	183,000

The electronic partition function Z_{elec} of the argon atom is, therefore,

$$Z_{\text{elec}_{\text{Ar}}} = 1 + 60\,\mathrm{e}^{-\frac{162\,000}{T}} + \ldots \tag{4.164}$$

For temperatures below 7500 K the partition function is given to a reasonable degree of accuracy by

$$Z_{\text{elec}_{\text{Ar}}} = g_0 = 1 . \tag{4.165}$$

The first four energy levels and the ionization energy of the singly ionized argon atom are [17]:

g_j	$\varepsilon_j/k[K]$
4	0
2	2062
2	156,560
8	190,550
Ionization energy	320,800

The electronic partition function is

$$Z_{\mathrm{elec}_{\mathrm{Ar}^+}} = 4 + 2\,\mathrm{e}^{-\frac{2062}{T}} + 2\,\mathrm{e}^{-\frac{156\,560}{T}} + 8\,\mathrm{e}^{-\frac{190\,550}{T}} + \ldots \tag{4.166}$$

For $T \ll 12\,000\,\mathrm{K}$, all the terms beyond the second are very small. However, the second term is appreciable except for very low temperatures. The partition function, the most common sputtering gases and their ions were found by O. Laporte [18].

For a further derivation of the Saha equation we now set Eqs. (4.157) and (4.158) in Eq. (4.155), whereby we assume that $m_{j+1} \cong m_j$. We then obtain for the equilibrium constant

$$K_{j+1} = \left(\frac{2\pi m_e kT}{h^2}\right)^{\frac{3}{2}} \frac{(Z_{j+1})_{\mathrm{elec}} Z_e}{Z_{j\,\mathrm{elec}}} . \tag{4.167}$$

We recall that $Z_e = 2$, and hence Eq. (4.167) becomes

$$K_{j+1} = \left(\frac{2\pi m_e kT}{h^2}\right)^{\frac{3}{2}} \frac{(2Z_{j+1})_{\mathrm{elec}}}{Z_{j\,\mathrm{elec}}} . \tag{4.168}$$

Let us consider its final form for the case of single ionization, namely, for the reaction

$$\mathrm{A} \rightleftarrows \mathrm{A}^+ + e - E_i .$$

If we assume that the atom and ion masses are approximately equal, namely $m_n \cong m_i$, and $K_1 = n_e n_i / n_n$, $n_i = n_e$, $\alpha = n_i/(n_i + n_n)$ and $p = (2n_i + n_n)kT$, the Saha equation for single ionization becomes

$$\frac{\alpha^2}{1-\alpha^2} = \left(\frac{2\pi m_e}{h^2}\right)^{\frac{3}{2}} \frac{(kT)^{\frac{5}{2}}}{p} \frac{2Z_{j\,\mathrm{elec}}}{Z_{0\mathrm{elec}}} \mathrm{e}^{-\frac{E_i}{kT}} . \tag{4.169}$$

Modified Saha equations may by derived if both single and double ionization occur. One defines two equilibrium constants K_1 and K_2 from Eq. (4.167). Then proceeding as above, it can be shown that the equilibrium equations are, respectively,

$$\frac{\alpha_1(\alpha_1 + 2\alpha_2)}{(1-\alpha_1-\alpha_2)(1+\alpha_1+2\alpha_2)} = \frac{(2\pi m_e)^{\frac{3}{2}}}{h^2}\left(\frac{kT}{p}\right)^{\frac{5}{2}} \frac{2Z_1}{Z_0} \mathrm{e}^{-\frac{E_{i1}}{kT}}$$
$$\frac{\alpha_2(\alpha_1 + 2\alpha_2)}{\alpha_1(1+\alpha_1+2\alpha_2)} = \left(\frac{2\pi m_e}{h^2}\right)^{\frac{3}{2}}\left(\frac{kT}{p}\right)^{\frac{5}{2}} \frac{2Z_2}{Z_1} \mathrm{e}^{-\frac{E_{i2}}{kT}} . \tag{4.170}$$

As a word of caution, it should be noted that Saha's equation is applicable to homogeneous systems in thermal equilibrium. For actual ionization processes, it serves only as an approximation. For example, the region of ionized gas immediately behind strong shock waves may not be in equilibrium due to a finite relaxation time. In general, equilibrium conditions are enhanced at higher temperatures and densities. On the other hand, as the collision rate increases, it becomes necessary to correct for the effects of particle interactions. The two important corrections to be taken into consideration are the partition function cutoff and the lowering of the ionization potential [19].

4.5.5.3 Law of Conservation of Mass

We now enter into the derivation of the conservation of mass. We consider a volume V and we let the particle density of the component j change only temporally so that through the surface of the volume particles it flows in or out, and thus particles of this sort become newly formed by ionization and recombination processes.

This leads to the relationship

$$\int\limits_V \frac{\partial n_j}{\partial t}\,\mathrm{d}V = -\int\limits_0 n_j \vec{v}_j\,\mathrm{d}f + \int\limits_V \Gamma'_j\,\mathrm{d}v . \tag{4.171}$$

The first integral on the right-hand side of the equation describes the influx of particles. Γ_j describes the number of particles j formed for time and volume units. If we apply the ***Gauss theorem*** to the surface integral, then it continues to follow that:

$$\int_V \left[\frac{\partial n_j}{\partial t} + \nabla(n_j \vec{v}_j) - \Gamma'_j \right] \mathrm{d}V = 0 \quad (4.172)$$

or

$$\frac{\partial n_j}{\partial t} + \nabla(n_j \vec{v}_j) = \Gamma'_j \,. \quad (4.173)$$

If we multiply this equation with m_j and define $m_j \Gamma_j = \Gamma_j$, then we obtain:

$$\frac{\partial \varrho_j}{\partial t} + \nabla(\varrho_j \vec{v}_j) = \Gamma_j \,. \quad (4.174)$$

If we multiply Eq. (4.174) over all three components, then we obtain

$$\frac{\partial \varrho}{\partial t} + \nabla(\varrho \vec{v}) = \sum_{j=1}^{3} \Gamma_j = 0$$

$$\sum_j \Gamma_j = m_\mathrm{e} \Gamma'_\mathrm{e} + m_\mathrm{i} \Gamma'_\mathrm{i} + m_\mathrm{n} \Gamma'_\mathrm{n}$$

$$= (m_\mathrm{e} + m_\mathrm{i}) \Gamma'_\mathrm{e} + m_\mathrm{n} \Gamma'_\mathrm{n} = 0 \,,$$

$$\text{da} \quad \Gamma'_\mathrm{e} = -\Gamma'_0 \,. \quad (4.175)$$

For each electron produced an atom disappears. Equation (4.175) is the differential expression for the law of conservation of mass. Under resolution of $\nabla(\rho \vec{v})$ we can also write:

$$\frac{\partial \varrho}{\partial t} + \nabla(\varrho \vec{v}) = \frac{\partial \varrho_j}{\partial t} + \vec{v} \nabla \varrho + \varrho \nabla \vec{v} = 0 \,. \quad (4.176)$$

For the description of the properties of flowing liquids, we introduce the *substantial derivative* in the hydrodynamics. We regard the properties of the liquid in the moving state by following up a volume element on its flow path and analyze in this moved system the considered property (for example, density or velocity).

For the derivative of the substantial differential quotient of h we obtain:

$$\frac{\mathrm{d}h}{\mathrm{d}h} = \lim_{\Delta t \to 0} \frac{\Delta h}{\Delta t} = \vec{v} \nabla h + \left(\frac{\partial h}{\partial t} \right)_{\vec{v}} \,, \quad (4.177)$$

where
$\vec{v}$ is the flow velocity.

If the considered property is a vector, then we obtain

$$\frac{\partial \vec{h}}{\partial t} + (\vec{v} \nabla) \vec{h} = \frac{\mathrm{d}\vec{h}}{\mathrm{d}t} \,. \quad (4.178)$$

We now combine Eq. (4.178) with Eq. (4.177), obtaining

$$\frac{\mathrm{d}\varrho}{\mathrm{d}t} + \varrho \nabla \vec{v} = 0 \,. \quad (4.179)$$

4.5.5.4 Conservation of Momentum

For each component of the plasma – monatomic and only simply ionized – we can indicate an equation of motion as

$$\varrho_\mathrm{e} \frac{\mathrm{d_e} \vec{v}_\mathrm{e}}{\mathrm{d}t} = -\nabla p_\mathrm{e} - n_\mathrm{e} e (\vec{E} + \vec{v}_\mathrm{e} \times \vec{B}) - n_\mathrm{e} n_\mathrm{i} \mu_\mathrm{ei} (v_\mathrm{e} - v_\mathrm{i}) - n_\mathrm{e} n_\mathrm{n} \mu_\mathrm{en} (\vec{v}_\mathrm{e} - \vec{v}_\mathrm{n}) \quad (4.180)$$

$$\varrho_\mathrm{i} \frac{\mathrm{d_i} \vec{v}_\mathrm{i}}{\mathrm{d}t} = -\nabla p_\mathrm{i} + n_\mathrm{i} e (\vec{E} + \vec{v} \times \vec{B}) - n_\mathrm{i} n_\mathrm{e} \mu_\mathrm{ie} (\vec{v}_\mathrm{i} - \vec{v}_\mathrm{e}) - n_\mathrm{i} n_\mathrm{n} \mu_\mathrm{in} (\vec{v}_\mathrm{i} - \vec{v}_\mathrm{n}) \quad (4.181)$$

$$\varrho_\mathrm{n} \frac{\mathrm{d_n} \vec{v}_\mathrm{n}}{\mathrm{d}t} = -\nabla p_\mathrm{n} - n_\mathrm{n} n_\mathrm{e} \mu_\mathrm{ne} (\vec{v}_\mathrm{n} - \vec{v}_\mathrm{e}) - n_\mathrm{n} n_\mathrm{i} \mu_\mathrm{ni} (\vec{v}_\mathrm{n} - \vec{v}_\mathrm{i}) \,. \quad (4.182)$$

On the left-hand side are the acceleration terms, whereby according to Eq. (4.179) we used $\mathrm{d}_j/\mathrm{d}t = \partial/\partial t + (\vec{v}_j \nabla)$.

On the right-hand side are the forces, which can move the centre of mass of the actual component. Apart from the pressure gradients, however, only the electrical forces are specified. Further forces like, for example, the force of gravity are easy to include. In addition to each component friction terms are introduced, which indicate the momentum exchange between the plasma component considered in each case and the two other components.

These terms are agreed as being proportional to the velocity difference of the components and to the product of the respective particle components. The coefficients of friction μ_{jk} contain only the cross-sections, the temperature and the masses of the actual components. This discussion is also takes place in Sect. 4.5.5.5.1.

This momentum exchange, which penetrates the flows of all components may not be considered with the internal friction of the hydrodynamics, which is not considered here. The internal friction is a momentum exchange within a component if velocities gradients arise.

Addition of Eqs. (4.180), (4.181) and (4.182) gives

$$\varrho_e \frac{\mathrm{d}_e \vec{v}_e}{\mathrm{d}t} + \varrho_i \frac{\mathrm{d}_i \vec{v}_i}{\mathrm{d}t} + \varrho_n \frac{\mathrm{d}_n \vec{v}_n}{\mathrm{d}t} = -\nabla(p_e + p_i + p_n) + (n_i - n_e) e \vec{E} + e(n_i \vec{v}_i - n_e \vec{v}_e) \times \vec{B} \,. \tag{4.183}$$

The friction terms are omitted in the addition, since the momentum loss through friction at a component as momentum gain must arise with the appropriate other component. This follows from $\mu_{jk} = \mu_{kj}$.

First we discuss the right-hand side of Eq. (4.183). Since $p_e + p_i + p_n = p$, the first term is determined by the gradients of the total pressure. The second term is the difference of the particle densities of ions and electrons. If the volume has a space charge, then the electrical field is accelerating.

Since for a plasma the quasi-neutrality relationship applies, $n_i - n_e \ll n_i + n_e$, this term can generally be neglected because this is small in relation to the remaining terms. For the last term, with $n_i = n_e$ we can also write:

$$e(n_i \vec{v}_i - n_e \vec{v}_e) \times \vec{B} = e n_i (\vec{v} - \vec{v}_e) \times \vec{B} = j \times \vec{B} \,. \tag{4.184}$$

The left-hand side of Eq. (4.183) can be transformed with the help of the continuity equations for the single components and for the total plasma in a somewhat more complicated way. With insignificant omissions we have

$$\varrho_e \frac{\mathrm{d}_e \vec{v}_e}{\mathrm{d}t} + \varrho_i \frac{\mathrm{d}_i \vec{v}_i}{\mathrm{d}t} + \varrho_n \frac{\mathrm{d}_n \vec{v}_n}{\mathrm{d}t} = \varrho \frac{\mathrm{d}\vec{v}}{\mathrm{d}t} \,. \tag{4.185}$$

So that we have the following equation of motion for the total plasma:

$$\varrho \frac{\mathrm{d}\vec{v}}{\mathrm{d}t} = -\nabla p + j \times \vec{B} \,. \tag{4.186}$$

The force of inertia holds the equilibrium to the pressure gradient and the Lorenz force.

4.5.5.5 Derivation of Ohm's Law

4.5.5.5.1 Complete Ionized Plasma

On the basis of the two conservation of momentum equations (Eqs. (4.181) and (4.182)), after their subtraction and considering quasi-neutrality $n_e = n_i$ and at the same temperature $T_e = T_i$, we obtain

$$\frac{\mathrm{d}_i \vec{v}_i}{\mathrm{d}t} - \frac{\mathrm{d}_e \vec{v}_e}{\mathrm{d}t} = \frac{m_i - m_e}{m_i m_e} \frac{\nabla p_e}{n} + \frac{m_i - m_e}{m_i m_e} e \vec{E} - n \mu_{ie} \frac{m_e - m_i}{m_i m_e} (\vec{v}_i - \vec{v}_e) + e \left[\frac{\vec{v}_i}{m_i} \times \vec{B} \right] + e \left[\frac{\vec{v}_e}{m_e} \times \vec{B} \right] \,. \tag{4.187}$$

The velocities of $\vec{v}_i$ and $\vec{v}_e$ are to be replaced by macroscopically observable quantities. For this the velocity $\vec{v}_S$ of the centre of mass of the plasma and the current density j are used,

$$\varrho \vec{v}_S = \varrho_e \vec{v}_e + \varrho_i \vec{v}_i \tag{4.188}$$

or

$$(m_e + m_i) \vec{v}_S = m_e \vec{v}_e + m_i \vec{v}_i \tag{4.189}$$

and

$$\vec{j} = en(\vec{v}_i - \vec{v}_e) \,.$$

A simple conversion gives

$$\vec{v}_e = \vec{v}_S - \frac{\vec{j}}{n_e} \frac{m_i}{m_e + m_i} \tag{4.190}$$

and

$$\vec{v}_i = \vec{v}_S + \frac{\vec{j}}{n_e} + \frac{\vec{j}}{n_e} \frac{m_e}{m_e + m_i} \,. \tag{4.191}$$

This is used in Eq. (4.187):

$$\frac{\mathrm{d}_i \vec{v}_i}{\mathrm{d}t} - \frac{\mathrm{d}_e \vec{v}_e}{\mathrm{d}t} = \frac{m_i - m_e}{m_i m_e} \frac{\nabla p_e}{n} + \frac{m_i + m_e}{m_i m_e} e E - \frac{m_i + m_e}{m_i m_e} \mu_{ie} \frac{\vec{j}}{e} + e \left[\left(\frac{\vec{v}_S}{m_i} + \frac{\vec{j}}{n_e} \frac{m_e}{m_i (m_i + m_e)} \right) \times \vec{B} \right] + e \left[\left(\frac{\vec{v}_S}{m_e} - \frac{\vec{j}}{n_e} \frac{m_i}{m_e (m_i + m_e)} \right) \times \vec{B} \right] \,. \tag{4.192}$$

The arrangement of the terms and neglect of m_e against m_i give

$$\frac{\vec{j}}{e}\frac{\mu_{ie}}{m_e}\frac{d_i\vec{v}_i}{dt} - \frac{d_e\vec{v}_e}{dt} = \frac{1}{nm_e}\nabla p_e + \frac{e}{m_e}(\vec{E} + \vec{v}_S \times \vec{B}) - \frac{1}{nm_e}[\vec{j} \times \vec{B}]\,. \tag{4.193}$$

We set for $\sigma = e^2/\mu_{ie}$, and obtain an expression for Ohm's law:

$$\vec{j} = \sigma\left[\vec{E} + \vec{v}_S \times \vec{B} + \frac{1}{n_e}\nabla p_e - \frac{1}{n_e}(\vec{j} \times \vec{B}) + \frac{m_e}{e}\left(\frac{d_e\vec{v}_e}{dt} - \frac{d_i\vec{v}_i}{dt}\right)\right]. \tag{4.194}$$

If we look at the last term on the right-hand side, then we can see that in the bracket, in addition to the electrical field strength $\vec{E}$ there still exist further electromotive forces. One of these is the current due to travelling charges. The other current that must be considered is the current induced due to the presence of the magnetic field. However, the magnetic field will not only induce a current, but due to its interaction with the moving charges, it will rearrange them in the plasma and thuds render the latter anisotropic.

Furthermore, the current generated from the hall term $1/n_e[j \times \vec{B}]$ not must not flow parallel to $\vec{E}$.

In the following we make some further statements about the coefficient of friction μ_{jk}. The friction components introduced to the impulse equations have the impact of a loss of momentum for each time and unit volume,

$$R_{jk} = n_j n_k \mu_{jk}(\vec{v}_j - \vec{v}_k)\,. \tag{4.195}$$

We can express this loss of momentum also with the help of the collision frequency (Sect. 4.5.2),

$$R_{jk} = n_j \nu_{jk}\frac{m_j m_k}{m_j + m_k}(\vec{v}_j - \vec{v}_k)\,, \tag{4.196}$$

where ν_{jk} is the collision frequency of the particles j with k. n_j particles collide for each time unit ν_{jk} times with particles of the sort k and always thereby transfer the momentum $m_j\Delta\vec{v}_j$. By comparison of Eqs. (4.195) and (4.196), then for μ_{jk} we obtain

$$\mu_{jk} = \frac{\nu_{jk}}{n_k}\frac{m_j m_k}{m_j + m_k}\,. \tag{4.197}$$

If we set this value into the expression for σ and consider that it concerns electron ion collisions, then we obtain

$$\sigma = \frac{e^2}{\mu_{ie}} = \frac{e^2 n}{\nu_{ei}}\frac{(m_e + m_i)}{m_e m_i} \cong \frac{ne^2}{m_e\nu_{ei}}\,. \tag{4.198}$$

If we replace the collision frequency in Eq. (4.197) by the mean free path and the mean thermal velocity, then after averaging we find

$$\mu_{jk} = \frac{8}{3}\sqrt{\frac{2}{\pi}kT\frac{m_j m_k}{m_j + m_k}}\,Q_{jk}\,, \tag{4.199}$$

where Q_{jk} is the collision cross-section between the particles j and k (see Sect. 4.5.2).

Neglecting the derivative of the time of $\vec{v}_e$ and $\vec{v}_i$, with Ohm's law we obtain the following:

$$\vec{j} = \sigma\vec{E}^* - \frac{\sigma\vec{B}}{n_e}[\vec{j} \times \vec{b}_0]\,. \tag{4.200}$$

In $\vec{E}^*$ the terms $E^* = \vec{E} + \vec{v} \times \vec{B} + 1/n_e\nabla p_e$ are summarized and $\vec{b}_0$ is the unit vector in the direction of $\vec{B}$.

Also for $\sigma B/n_e$, with Eq. (4.198) we can write

$$\frac{\sigma\vec{B}}{n_e} = \frac{ne^2\vec{B}}{m_e\nu_{ei}n_e} = \frac{e\vec{B}}{m_e}\frac{1}{\nu_{ei}} = \omega_{ge}\tau_{ei} = \beta\,. \tag{4.201}$$

Thus with Eq. (4.198) we have

$$\vec{j} = \sigma\vec{E}^* - \beta(\vec{j} \times \vec{b}_0)\,. \tag{4.202}$$

The hall term $\beta(j \times b_0)$ becomes important, if $\beta \geq 1$. This means that the mean free path becomes larger than the gyration radius or than the circumference of the gyration orbit. If $\beta \leq 1$, then the electrons cannot travel appreciable distances on the gyration orbit, without a collision taking place.

If we now multiply the vectorial in Eq. (4.202) by $\vec{b}_0$, then we obtain

$$\begin{aligned}(\vec{j}\times\vec{b}_0) &= \sigma(\vec{E}^*\times\vec{b}_0) - \beta(\vec{j}\times\vec{b}_0)\times\vec{b}_0 \\ &= \sigma(\vec{E}^*\times\vec{b}_0) + \beta\vec{j}(b_0^2) - \beta\vec{b}_0(\vec{j}\vec{b}_0)\,.\end{aligned} \tag{4.203}$$

Now $\vec{b}_0(\vec{j}\vec{b}_0)$ is $= |j_\parallel|$ parallel to $\vec{B}$. Thereby we can sum up the current density:

$$(\vec{j}\times\vec{b}_0) = \sigma[\vec{E}^*\times\vec{b}_0] + \beta\vec{j} - \beta\vec{j}_\parallel\,. \tag{4.204}$$

If we insert Eq. (4.204) again in Eq. (4.202), then we obtain

$$\vec{j} = \sigma\vec{E}^* - \sigma\beta(\vec{E}^*\times\vec{b}_0)\beta^2\vec{j} + \beta^2\vec{j}_\parallel\,, \tag{4.205}$$

from which follows:

$$\vec{j}(1+\beta^2) - \beta^2\vec{j}_\parallel = \sigma[\vec{E}^* - \beta(\vec{E}^*\times\vec{b}_0)]\,. \tag{4.206}$$

If we divide the vectorial $\vec{j}$ in $\vec{j}_\perp$ and $\vec{j}_\parallel$ in accordance with $\vec{j} = \vec{j}_\perp + \vec{j}_{||}$, then for the two components we obtain:

$$\vec{j}_\parallel = \sigma\vec{E}^*_\parallel = \sigma\left(\vec{E}^*_\parallel + \frac{1}{n_e}\nabla p_{e\parallel}\right) \tag{4.207}$$

$$\vec{j}_\perp - \frac{\sigma}{1+\beta^2}[\vec{E}^*_\perp \quad \beta(\vec{E}^*\times\vec{b}_0)_\perp]\,. \tag{4.208}$$

The current density $\vec{j}_\parallel$ parallel to the $\vec{B}$-field is not affected by $\vec{B}$. Perpendicular to the $\vec{B}$-field the conductivity σ is modified with $(1+\beta^2)^{-1}$. In addition, there is still another term $\beta(\vec{E}^*\times\vec{b}_0)_\perp$ that must be calculated in each individual case.

Now we write the remaining inertial components in the equation of Ohm's law (4.194); thus

$$\begin{aligned}&\frac{\mathrm{d_i}\vec{v}_\mathrm{i}}{\mathrm{d}t} - \frac{\mathrm{d_e}\vec{v}_\mathrm{e}}{\mathrm{d}t} \\ &= \frac{\partial}{\partial t} + (\vec{v}_\mathrm{i} - \vec{v}_\mathrm{e}) + (\vec{v}_\mathrm{i}\nabla)\,\vec{v}_\mathrm{i} - (\vec{v}_\mathrm{e}\nabla)\,\vec{v}_\mathrm{e}\,,\end{aligned}$$

with

$$\vec{\vartheta} = \frac{\vec{j}}{n_\mathrm{e}} = \vec{v}_\mathrm{i} - \vec{v}_\mathrm{e} \quad\text{and}\quad \varrho\vec{v} = \varrho_\mathrm{e}\vec{v}_\mathrm{e} + \varrho_\mathrm{i}\vec{v}_\mathrm{i}\,,$$

we obtain

$$\vec{v}_\mathrm{e} = \vec{v} - \frac{m_\mathrm{i}}{m_\mathrm{i}+m_\mathrm{e}}\vec{\vartheta}\,;\quad \vec{v}_\mathrm{i} = \vec{v} - \frac{m_\mathrm{e}}{m_\mathrm{i}+m_\mathrm{e}}\vec{\vartheta}\,.$$

If we insert this into the equations, we have

$$\begin{aligned}\frac{\mathrm{d_i}\vec{v}_\mathrm{i}}{\mathrm{d}t} - \frac{\mathrm{d_e}\vec{v}_\mathrm{e}}{\mathrm{d}t} &= \frac{\partial\vec{\vartheta}}{\partial t} \\ &\quad + \left[\left(\vec{v} + \frac{\varrho_\mathrm{e}}{\varrho}\vec{\vartheta}\right)\nabla\right]\left(\vec{v} + \frac{\varrho_\mathrm{e}}{\varrho}\vec{\vartheta}\right) \\ &\quad - \left[\left(\vec{v} - \frac{\varrho_\mathrm{i}}{\varrho}\vec{\vartheta}\right)\nabla\right]\left(\vec{v} - \frac{\varrho_\mathrm{i}}{\varrho}\vec{\vartheta}\right) \\ &= \frac{\partial\vec{\vartheta}}{\partial t} + (\vec{v}\nabla)\vec{\vartheta} + (\vec{\vartheta}\nabla)\vec{v} - \left(\frac{\varrho_\mathrm{i}}{\varrho}\vec{\vartheta}\nabla\right)\vec{\vartheta} \\ &\quad + \left(\frac{\varrho_\mathrm{e}}{\varrho}\vartheta\nabla\right)\vartheta \\ &= \frac{\partial\vec{\vartheta}}{\partial t} + (\vec{v}\nabla)\vec{v} + (\vec{\vartheta}\nabla)\vec{v} - (\vec{\vartheta}\nabla)\vec{\vartheta}\,.\end{aligned} \tag{4.209}$$

If $\rho_\mathrm{e}/\rho \ll \rho_\mathrm{i}/\rho$, we finally obtain

$$\frac{\mathrm{d_i}\vec{v}_\mathrm{i}}{\mathrm{d}t} - \frac{\mathrm{d_e}\vec{v}_\mathrm{e}}{\mathrm{d}t} = \frac{\partial\vec{\vartheta}}{\partial t} + (\vec{\vartheta}\nabla)(\vec{v} - \vec{\vartheta})\,. \tag{4.210}$$

We can prove that the term $(\vec{\vartheta}V)\,(\vec{v} - \vec{\vartheta})$ is much smaller than the term $(\vec{v}V)\vec{\vartheta}$. For this purpose, we again express the two terms $\vec{v}$ and $\vec{\vartheta}$ by $\vec{v}_\mathrm{i}$ and $\vec{v}_\mathrm{e}$ and compare them with one another. We can then write:

$$\frac{\mathrm{d_i}\vec{v}_\mathrm{i}}{\mathrm{d}t} - \frac{\mathrm{d_e}\vec{v}_\mathrm{e}}{\mathrm{d}t} = \frac{\mathrm{d}}{\mathrm{d}t}\vec{\vartheta} = \frac{\mathrm{d}}{\mathrm{d}t}\frac{\vec{j}}{en}\,. \tag{4.211}$$

This expression is introduced in Eq. (4.194):

$$\frac{\sigma m_\mathrm{e}}{e^2}\frac{\mathrm{d}\frac{\vec{j}}{n}}{\mathrm{d}t} + \vec{j} = \sigma\left[\vec{E} + (\vec{v}\times\vec{B}) + \frac{1}{n_\mathrm{e}}\nabla p_\mathrm{e} - \frac{1}{n_\mathrm{e}}(\vec{j}\times\vec{B})\right]. \tag{4.212}$$

The first term on the left-hand side is transformed as follows:

$$\begin{aligned}\frac{\sigma m_\mathrm{e}}{e^2}\frac{\mathrm{d}}{\mathrm{d}t}\left(\frac{\vec{j}}{n}\right) &= \frac{\sigma}{\varepsilon_0}\frac{\mathrm{d}}{\mathrm{d}t}\left(\frac{\vec{j}}{\frac{ne^2}{\varepsilon_0 m_\mathrm{e}}}\right) \\ &= \frac{\sigma}{\varepsilon_0}\frac{\mathrm{d}}{\mathrm{d}t}\left(\frac{\vec{j}}{\omega_\mathrm{PI}^2}\right),\end{aligned} \tag{4.213}$$

where

$$\omega_{PI} = \sqrt{\frac{ne^2}{\varepsilon_0 m_e}}$$

is the plasma frequency.

The relation between the plasma frequency and the conductivity is:

$$\sigma = \varepsilon_0 \frac{\omega_{PI}^2}{\nu_{ei}}. \tag{4.214}$$

With Eqs. (4.113) and (4.214) we can finally write Ohm's law for complete ionized plasma in the following form:

$$\frac{\sigma}{\varepsilon_0}\frac{d}{dt}\left(\frac{\vec{j}}{\omega_{PI}^2}\right) + \vec{j} = \sigma\left[\vec{E} + (\vec{v} + \vec{B}) + \frac{1}{n_e}\nabla p_e - \frac{1}{n_e}(\vec{j}\times\vec{B})\right]. \tag{4.215}$$

4.5.5.5.2 Partially Ionized Plasma

During the derivation of Ohm's law the plasma was completely ionized. Now we will consider that neutral particles are also present in the gas plasma. First we will discuss the momentum equations for partially ionized plasma and the terms of the neglect of inertia. With fully ionized plasma these determine the high frequency behaviour and the current density. Quasi-neutrality is also obtained.

By subtraction from Eq. (4.180) of Eq. (4.181) the velocity $\vec{v}_n$ of the neutral particles is eliminated. Furthermore, the current density j is introduced. This way, we obtain the relationship

$$\vec{j} = \sigma\left[\vec{E} + (\vec{v}_{SI}\times\vec{B}) - \left(C + \frac{m_i - m_e}{en_e m_n}\right)(\vec{j}\times\vec{B}) + \left(C + \frac{m_i - m_e}{en_e m_n}\right)\nabla p_e\right]. \tag{4.216}$$

The velocity of the centre of mass $\vec{v}_{st}$ of the charge carrier is introduced as

$$(\varrho_e + \varrho_i)\vec{v}_{SI} = \varrho_e\vec{v}_e + \varrho_i\vec{v}_i = \varrho\vec{v} - \varrho_n\vec{v}_n. \tag{4.217}$$

Now we also have the conductivity σ:

$$\sigma = \frac{e^2 n_e(\mu_{in} + \mu_{en})}{n\mu_{ei}(\mu_{in} + \mu_{en}) + n_n\mu_{in}\mu_{en}}. \tag{4.218}$$

The quantity C in brackets in Eq. (4.216) is

$$C = \frac{\varrho_e\mu_{in} - \varrho_i\mu_{en}}{en_e(\varrho_e + \varrho_i)(\mu_{in} + \mu_{en})} = \frac{m_e\mu_{in} - m_i\mu_{en}}{en_e m_n(\mu_{in} + \mu_{en})}. \tag{4.219}$$

With equation (4.197) we can write the relationship μ_{in}/μ_{en} as follows:

$$\frac{\mu_{in}}{\mu_{en}} = \sqrt{\frac{m_n}{2m_e}\frac{Q_{in}}{Q_{en}}} \tag{4.220}$$

and in all rule the cross-sections Q_{in} and Q_{en} are of the same order of magnitude. From this it follows that $\mu_{in} \gg \mu_{en}$. If we neglect μ_{en} for μ_{in} and m_e for m_i, then for the electrical conductivity we obtain

$$\sigma \approx \frac{e^2 n_e\mu_{in}}{n_e\mu_{in}\mu_{ei} + n_n\mu_{in}\mu_{en}} = \frac{n_e e^2}{n_e\mu_{ei} + n_n\mu_{en}} = \frac{e^2}{\mu_{ei}\left(1 + \frac{n_n\mu_{en}}{n_e\mu_{ei}}\right)}. \tag{4.221}$$

Since now we have the relationship $\mu_{en}/\mu_{ei} = Q_{en}/Q_{ei}$, we obtain

$$\sigma = \frac{e^2}{\mu_{ei}\left(1 + \frac{n_n Q_{en}}{n_e Q_{ei}}\right)} = \frac{\sigma_\infty}{1 + \frac{n_n}{n_e}\frac{Q_{en}}{Q_{ei}}}. \tag{4.222}$$

Here σ_∞ is the conductivity of the completely ionized plasma.

4.5.5.6 Energy Theorem for Plasma

When applied to a control volume the principle of conservation of energy states that the increase of energy within the control volume must be balanced by the sum of the thermal and mechanical energies. On consideration of the kinetic energy we assume from the theorem of momentum, which is multiplied scalar with $\vec{v}$, we obtain

$$\varrho\frac{d\frac{\vec{v}^2}{2}}{dt} = -\vec{v}\nabla p + \vec{v}(j\times\vec{B}). \tag{4.223}$$

We assume that there is no applied electric field.

For the consideration of the thermal energy we use the first law of thermodynamics:

$$dq = du + p\,dv = du + p\,d\frac{1}{\varrho}. \quad (4.224)$$

Multiplication with ρ and differentiation after time result in

$$\varrho\frac{dq}{dt} = \varrho\frac{du}{dt} + \varrho p\frac{d\frac{1}{\varrho}}{dt}. \quad (4.225)$$

If we extend this by addition and subtraction of dp/dt, we obtain

$$\begin{aligned}\varrho\frac{dq}{dt} &= \varrho\frac{du}{dt} + \varrho\left(p\frac{d\frac{1}{\varrho}}{dt} + \frac{1}{\varrho}\frac{dp}{dt}\right) - \frac{dp}{dt}\\ &= \varrho\frac{du}{dt} + \varrho\frac{d\frac{p}{\varrho}}{dt} - \frac{dp}{dt}\\ &= \varrho\frac{d}{dt}\left(u + \frac{p}{\varrho}\right) - \frac{dp}{dt}\\ &= \varrho\frac{dh}{dt} - \frac{dp}{dt} = \varrho\frac{dh}{dt} - \frac{\partial p}{\partial t} - (\vec{v}\nabla)p.\end{aligned} \quad (4.226)$$

The temporal change of the enthalpy is caused by the heat supply and by changes of the pressure. The last term indicates a change of h, if the gas flows into a volume of higher or lower pressure and is thereby compressed or expanded.

We obtain another form if the internal energy u in Eq. (4.225) is used:

$$\varrho\frac{dq}{dt} = \varrho\frac{du}{dt} + p\varrho\frac{d\frac{1}{\varrho}}{dt} = \varrho\frac{du}{dt} - \frac{p}{\varrho}\frac{d\varrho}{dt}. \quad (4.227)$$

If we use the mass conservation law – for Eq. (4.177) –, then it follows that

$$\begin{aligned}\varrho\frac{dq}{dt} &= \varrho\frac{du}{dt} + p\nabla\vec{v} = \varrho\frac{du}{dt} + \nabla p\vec{v} + (\vec{v}\nabla)p\\ \varrho\frac{du}{dt} &= \varrho\frac{dq}{dt} - \nabla p\vec{v} + (\vec{v}\nabla)p.\end{aligned} \quad (4.228)$$

Both equations are useful, but if another notation is used, the heat supply for each unit volume is always given by the Joule effect in the available case. One can, therefore, write:

$$\varrho\frac{dq}{dt} = \frac{j^2}{\varrho}. \quad (4.229)$$

Thus we receive the energy theorem for the thermal energy in the form

$$\varrho\frac{dh}{dt} = \frac{j^2}{\sigma} + \frac{dp}{dt} = \frac{j^2}{\sigma} + \frac{\partial p}{\partial t} + (\vec{v}\nabla)p \quad (4.230)$$

or

$$\varrho\frac{du}{dt} = \frac{j^2}{\sigma} - p\nabla\vec{v} = \frac{j^2}{\sigma} - \nabla p\vec{v} + (v\nabla)p. \quad (4.231)$$

We can continue still another step, which is to replace j^2 by means of the Ohm's law, whereby the time-dependent term in $\vec{j}$ is neglected.

Thus

$$\begin{aligned}\frac{j^2}{\sigma} &= \vec{j}\left[E + (v\times B) + \frac{1}{n_e}\nabla p_e - \frac{1}{n_e}(\vec{j}\times\vec{B})\right]\\ &= \vec{j}\left[\vec{E} + (\vec{v}\times\vec{B}) + \frac{1}{n_e}\nabla p_e\right],\end{aligned}$$

and we obtain the forms:

$$\begin{aligned}\varrho\frac{dh}{dt} &= \vec{j}\vec{E} + \vec{j}(\vec{v}\times\vec{B}) + \frac{j}{n_e}\nabla p_e\\ &\quad + (\vec{v}\nabla)p + \frac{\partial p}{\partial t},\end{aligned} \quad (4.232)$$

$$\begin{aligned}\varrho\frac{du}{dt} &= \vec{j}\vec{E} + \vec{j}(\vec{v}\times\vec{B}) + \frac{\vec{j}}{n_e}\nabla p_e\\ &\quad - \nabla p\vec{v} + (\vec{v}\nabla)p.\end{aligned} \quad (4.233)$$

Here it can be seen that we may not exchange j^2/σ with $\vec{j}\vec{E}$, as it is mostly permitted in elementary electrodynamics. $\vec{j}\vec{E}$ essentially represents the Poynting vector, while j^2/σ is the heat produced by the dissipation.

If we summarize now the energy theorems for the kinetic energy equation (4.223) with Eqs. (4.232) and (4.233), then we obtain the ***energy theorem for the plasma***,

$$\varrho\frac{d}{dt}\left(\frac{\vec{v}^2}{2} + h\right) = \vec{j}\vec{E} + \frac{\vec{j}}{n_e}\nabla p_e + \frac{\partial p}{\partial t} \quad (4.234)$$

or

$$\varrho \frac{\mathrm{d}}{\mathrm{d}t}\left(\frac{v^2}{2}+u\right) = j\vec{E} + \frac{\vec{j}}{n_\mathrm{e}}\nabla p_\mathrm{e} - \nabla(p\vec{v})\,. \tag{4.235}$$

The terms $\vec{j}(\vec{v}\times\vec{B})$ and $(\vec{v}\nabla)p$ are omitted from these equations.

The power of the pressure gradient and the Lorenz force cannot change the total energy of the plasma, but only in each case a form of energy replaced with another. For the pressure gradients this is easily observable: If one lets a gas flow out from nozzle, then the kinetic energy increases, but the temperature degrades itself at the same time. The reversal applies for retain of a gas to.

4.5.6 Transportation Processes

In general, we define transport phenomena as rate processes in which deviations from equilibrium are slight. In such phenomena we are dealing with systems that are not in thermodynamic equilibrium but that nevertheless tend to move toward a state of thermodynamic equilibrium as prescribed by the Boltzmann H-theorem. Indeed, the trend toward equilibrium occurs by virtue of the associated transport mechanism, whereby mass, momentum and energy are transported as a result of the motion and the collisions experienced by particles carrying with them the property being transported.

We now regard the conservation of momentum equations of the individual components of partially ionized plasma, in which at first no $\vec{B}$-field should be present. An acceleration of the individual components takes place only if a partial pressure gradient is present. We now observe a very long, cylindrical discharge column in an isolated tube. On the cylinder wall the temperature is to be so low that only neutral gas is present there, while in the environment of the discharge axis the plasma is to be simply completely ionized.

Then the partial pressure of the electrons and ions in the plasma from the axis must drop to the wall, while the partial pressure of the neutral gas drops in reverse direction. In addition, the pressure over the cross-section must be constant, since the effect of the self-magnetic field of the discharge current must also be negligible.

Due to the partial pressure gradients in radial direction, diffusion currents of the charge carriers (outwards) and of neutral particles (inwards) in the radial direction must be present. The atoms are ionized on the inside; on the other hand, the electrons and ions recombine if they come into colder ranges. Since the walls are isolated, no electric current may flow in the stationary case in the radial direction, and also the centre of mass velocity in this direction must disappear under the same conditions. If we use cylindrical coordinates the following conditions are fulfilled:

$$j_\mathrm{r} = n_\mathrm{e}e(v_\mathrm{ir} - v_\mathrm{er}) = 0 \Rightarrow v_\mathrm{ir} = v_\mathrm{er}\,, \tag{4.236}$$

$$\begin{aligned}\varrho v_\mathrm{r} &= \varrho_\mathrm{e} v_\mathrm{er} + \varrho_\mathrm{i} v_\mathrm{ir} + \varrho_\mathrm{n} v_\mathrm{nr}\\ &= (\varrho_\mathrm{e} + v_\mathrm{i})v_\mathrm{ir} + \varrho_\mathrm{n} v_\mathrm{nr} = 0\end{aligned} \tag{4.237}$$

or

$$n_\mathrm{e} v_\mathrm{ir} = -n_\mathrm{n} v_\mathrm{nr}\,. \tag{4.238}$$

We now describe the radial components of the momentum equations by ions and electrons, whereby the inertia term is again neglected (Stoke friction). Thus, it follows that

$$n_\mathrm{e}n_\mathrm{n}\mu_\mathrm{in}(v_\mathrm{ir} - v_\mathrm{nr}) = -\frac{\mathrm{d}p_\mathrm{i}}{\mathrm{d}r} + n_\mathrm{e}eE_\mathrm{r}\,, \tag{4.239}$$

$$n_\mathrm{e}n_\mathrm{n}\mu_\mathrm{er}(v_\mathrm{er} - v_\mathrm{nr}) = -\frac{\mathrm{d}p_\mathrm{e}}{\mathrm{d}r} - n_\mathrm{e}eE_\mathrm{r}\,. \tag{4.240}$$

An addition of both equations, with $v_\mathrm{ir} = v_\mathrm{er}$ gives

$$n_\mathrm{e}n_\mathrm{n}(\mu_\mathrm{in} + \mu_\mathrm{en})(v_\mathrm{ir} - v_\mathrm{nr}) = -2\frac{\mathrm{d}p_\mathrm{i}}{\mathrm{d}r} = -2\frac{\mathrm{d}p_\mathrm{e}}{\mathrm{d}r}\,. \tag{4.241}$$

From the subtraction it follows that

$$n_\mathrm{e}n_\mathrm{n}(\mu_\mathrm{in} - \mu_\mathrm{en})(v_\mathrm{ir} - v_\mathrm{nr}) = 2n_\mathrm{e}eE_\mathrm{r}\,. \tag{4.242}$$

If we now replace v_nr with Eq. (4.238), then it follows that

$$v_\mathrm{ir} = v_\mathrm{er} = -\frac{2}{n_\mathrm{i}(n_\mathrm{i} + n_\mathrm{n})(\mu_\mathrm{in} + \mu_\mathrm{en})}\frac{\mathrm{d}p_\mathrm{i}}{\mathrm{d}r}\,. \tag{4.243}$$

Accordingly,

$$v_{\mathrm{nr}} = \frac{2}{n_{\mathrm{n}}(n_{\mathrm{n}}+n_{\mathrm{i}})(\mu_{\mathrm{in}}+\mu_{\mathrm{en}})}\frac{\mathrm{d}p_{\mathrm{i}}}{\mathrm{d}r} = -\frac{1}{n_{\mathrm{n}}(n_{\mathrm{n}}+n_{\mathrm{i}})(\mu_{\mathrm{in}}+\mu_{\mathrm{en}})}\frac{\mathrm{d}p_{\mathrm{n}}}{\mathrm{d}r} \quad (4.244)$$

applies. Here, v_{nr} is the diffusions velocity at the ambipolar diffusion.

Because electrons and ions have the same diffusion velocity, a radial electrical field must be formed, which delays the electrons – because of their larger mobility – and accelerates the ions. The field force results from Eq. (4.242), if one appoints for v_{ir} and v_{nr}:

$$E_{\mathrm{r}} = -\frac{1}{n_{\mathrm{e}}e}\frac{\mu_{\mathrm{in}}-\mu_{\mathrm{en}}}{\mu_{\mathrm{in}}+\mu_{\mathrm{en}}}\frac{\mathrm{d}p_{\mathrm{i}}}{\mathrm{d}r}\,. \quad (4.245)$$

Since, generally, due to the mass factor with same cross-sections μ_{in} is much larger than μ_{en},

$$E_{\mathrm{r}} = -\frac{1}{n_{\mathrm{e}}e}\frac{\mathrm{d}p_{\mathrm{e}}}{\mathrm{d}r} \quad (4.246)$$

is always valid with good approximation.

In gas-kinetic theory, we often write the particle flux in the form

$$n_{\mathrm{i}}v_{\mathrm{in}} = -D_{\mathrm{amb}}\nabla n_{\mathrm{i}}\,, \quad (4.247)$$

where D_{amb} represents the ambipolar diffusion coefficient. This iteration is permitted, if the temperature is spatially constant, because then for ∇p_{i}

$$\nabla p_{\mathrm{i}} = \nabla n_{\mathrm{i}}kT = kT\nabla n_{\mathrm{i}} \quad (4.248)$$

is valid.

Equation (4.243) is then permitted to describe

$$n_{\mathrm{i}}v_{\mathrm{ir}} = n_{\mathrm{e}}v_{\mathrm{er}} = -\frac{2kT}{(n_{\mathrm{i}}+n_{\mathrm{n}})(\mu_{\mathrm{in}}+\mu_{\mathrm{en}})}\frac{\mathrm{d}n_{\mathrm{e}}}{\mathrm{d}r} = -D_{\mathrm{amb}}\frac{\mathrm{d}n_{\mathrm{e}}}{\mathrm{d}r}\,. \quad (4.249)$$

D_{amb} is then

$$D_{\mathrm{amb}} = \frac{2kT}{(n_{\mathrm{i}}+n_{\mathrm{n}})(\mu_{\mathrm{in}}+\mu_{\mathrm{en}})} \approx \frac{2kT}{(n_{\mathrm{i}}+n_{\mathrm{n}})\mu_{\mathrm{in}}}(\mu_{\mathrm{in}} \gg \mu_{\mathrm{en}}). \quad (4.250)$$

Finally, if we re-introduce the collision frequency according to Eq. (4.197) for μ_{in}, then it follows that

$$D_{\mathrm{amb}} = \frac{4n_{\mathrm{n}}kT}{(n_{\mathrm{i}}+n_{\mathrm{n}})v_{\mathrm{in}}m_{\mathrm{n}}}\,. \quad (4.251)$$

An energy flow is also connected with the diffusion current, since all diffusing particles carry their energy with themselves.

For the energy flow coupled with the ambipolar diffusion w_{r}, we obtain

$$w_{\mathrm{r}} = \varrho_{\mathrm{er}}v_{\mathrm{er}}h_{\mathrm{e}} + \varrho_{\mathrm{i}}v_{\mathrm{ir}}h_{\mathrm{i}} + \varrho_{\mathrm{n}}v_{\mathrm{nr}}h_{\mathrm{n}} = \sum_j \varrho_j h_j v_j\,. \quad (4.252)$$

Here h_j implies the enthalpies for each unit mass according to the definition equations in Sect. 4.5.5.1 of the excitation energies

$$h_{\mathrm{e}} = \frac{5}{2}\frac{kT}{m_{\mathrm{e}}}\,; \qquad h_{\mathrm{n}} = \frac{5}{2}\frac{kT}{m_{\mathrm{n}}}\,; \qquad h_{\mathrm{i}} = \frac{5}{2}\frac{kT}{m_{\mathrm{i}}} + \frac{E_{\mathrm{i}}}{m_{\mathrm{i}}}\,. \quad (4.253)$$

If we insert this in Eq. (4.252), then we obtain

$$w_{\mathrm{r}} = (n_{\mathrm{e}}v_{\mathrm{er}} + n_{\mathrm{n}}v_{\mathrm{nr}} + n_{\mathrm{i}}v_{\mathrm{ir}})\frac{5}{2}kT + n_{\mathrm{i}}v_{\mathrm{ir}}E_{\mathrm{i}}\,. \quad (4.254)$$

If we consider this with ambipolar diffusion, we obtain

$$n_{\mathrm{e}}v_{\mathrm{er}} = n_{\mathrm{i}}v_{\mathrm{ir}} = -n_{\mathrm{n}}v_{\mathrm{nr}}\,. \quad (4.255)$$

Thus, we immediately receive

$$w_{\mathrm{r}} = n_{\mathrm{i}}v_{\mathrm{ir}}\left(\frac{5}{2}kT + E_{\mathrm{i}}\right)\,. \quad (4.256)$$

If we finally introduce Eq. (4.243) for $n_{\mathrm{i}}v_{\mathrm{ir}}$, then we obtain

$$w_{\mathrm{r}} = -\left(\frac{5}{2}kT + E_{\mathrm{i}}\right)\frac{2}{(n_{\mathrm{i}}+n_{\mathrm{n}})\mu_{\mathrm{in}}}\frac{\mathrm{d}p_{\mathrm{i}}}{\mathrm{d}r}\,. \quad (4.257)$$

Since thermal plasma under constant pressure p_{i} is a function of the temperature, we can further write

$$w_{\mathrm{r}} = -\left(\frac{5}{2}kT + E_{\mathrm{i}}\right)\frac{2}{(n_{\mathrm{i}}+n_{\mathrm{n}})\mu_{\mathrm{in}}}\frac{\mathrm{d}p_{\mathrm{i}}}{\mathrm{d}T}\frac{\mathrm{d}T}{\mathrm{d}r} = -\varkappa_{\mathrm{R}}\frac{\mathrm{d}T}{\mathrm{d}r} \quad (4.258)$$

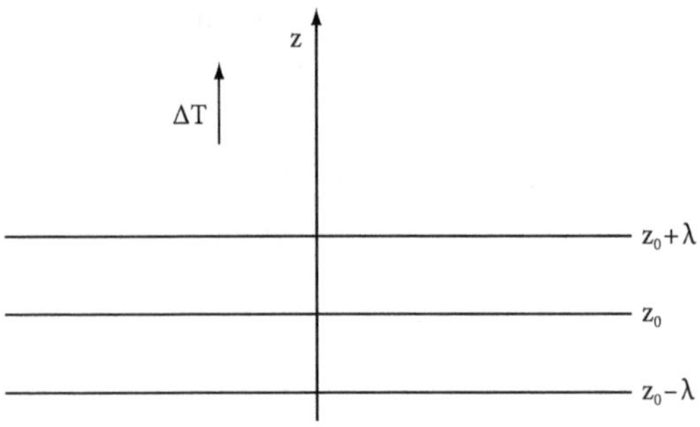

Fig. 4.8 The derivative of the contact heat conductivity κ_K

and

$$\varkappa_R = \left(\frac{5}{2}kT + E_i\right)\frac{2}{(n_i + n_n)\mu_{in}}\frac{dp_i}{dT}\,. \tag{4.259}$$

$p_i = p_i(T, p)$ is well known as a thermal plasma; therefore we can calculate the so-called heat of the reaction conductivity κ_r as a function of the temperature. From the heat of the reaction conductivity introduced here, which is based on the transport of the reaction energy (ionization energy), the normal contact heat conductivity is to be differentiated as κ_K, which is determined by energy exchange due collisions by particles of different energies.

A very simple, however quite well describing derivative for the contact heat conductivity κ_K one receives in the following way: In a uniform medium with only one particle species is present a temperature gradient in z-direction. We stretch in the place $z = z_0$. Up, then we can accept in Fig. 4.8 a xy-plane that the temperature in this level is equal T_0 and/or the energy of the particles is E_0.

Above and below z_0 we stretch two further levels in the distance of the free distance λ. All particles that fly in $\pm$ direction , have sustained collisions in the distance λ of their last collision, before they reach the level z_0 and thus, on average, the dominant temperature in this area must be presumed.

We now obtain the heat flow through z_0, if one of the two particles fluxes, which penetrate in each case the layer $(z_0 + \lambda)$ and $(z_0 - \lambda)$ from the surface z_0, with the respective energies multiples. This leads to the heat flows

$$w_1 = -\frac{1}{6}\left(n_0 + \frac{\partial n}{\partial z}\lambda\right) \times \left(\bar{v}_0 + \frac{\partial \bar{v}}{\partial z}\lambda\right)\left(E_0 + \frac{\partial E}{\partial z}\lambda\right) \tag{4.260}$$

and

$$w_2 = -\frac{1}{6}\left(n_0 - \frac{\partial n}{\partial z}\lambda\right) \times \left(\bar{v}_0 - \frac{\partial \bar{v}}{\partial z}\lambda\right)\left(E_0 - \frac{\partial E}{\partial z}\lambda\right)\,, \tag{4.261}$$

where $\bar{v}$ is the middle thermal velocity. If we sum up both heat flows, then it follows under neglect of terms with λ^2 that

$$w = -\frac{\lambda}{3}\left(n_0\bar{v}_0\frac{\partial E}{\partial z} + n_0E_0\frac{\partial \bar{v}}{\partial z} + \bar{v}_0E_0\frac{\partial n}{\partial z}\right) = -\frac{\lambda}{3}\frac{d}{dz}n_0\bar{v}_0E_0 \tag{4.262}$$

$$w = -\frac{\lambda}{3}E_0\frac{d(n\bar{v})}{dz} - \frac{\lambda}{3}n_0\bar{v}_0\frac{dE}{dz}\,. \tag{4.263}$$

As a secondary condition it must be still demanded that no net stream flows through the control area, which can be only achieved we one introduce a balance current with the velocity v_a. For the particle fluxes we then obtain:

$$0 = -\frac{1}{6}\left(n_0 + \frac{\partial n}{\partial z}\lambda\right)\left(\bar{v}_0 + \frac{\partial \bar{v}}{\partial z}\lambda\right) + \frac{1}{6}\left(n_0 - \frac{\partial n}{\partial z}\lambda\right)\left(\bar{v} - \frac{\partial \bar{v}}{\partial z}\lambda\right) + nv_a\,, \tag{4.264}$$

$$0 = -\frac{\lambda}{3}\frac{d(n\bar{v})}{dz} + nv_a\,. \tag{4.265}$$

Since this equalizing current also contributes to the energy flow, we must supplement the heat flow according to Eq. (4.263)

$$w = -\frac{\lambda}{3}E\frac{d(n\bar{v})}{dz} - \frac{\lambda}{3}n\bar{v}\frac{dE}{dz} + nv_0E\,. \tag{4.266}$$

With Eq. (4.265)

$$w = -\frac{\lambda}{3}n\bar{v}\frac{dE}{dz} \tag{4.267}$$

develops. Since the energy of the particles depends only on the temperature, we can further write:

$$w = -\frac{\lambda}{3} n \bar{v} \frac{\partial E}{\partial t} \frac{dT}{dz} = \varkappa_K \frac{dT}{dz} . \quad (4.268)$$

Since E is the middle thermal energy of the single particle

$$\frac{\partial E}{\partial t} = c'_v , \quad (4.269)$$

where c'_v is the specific heat of the single particle. With $E = u = 3/2\,kT$ it follows that

$$c'_v = \frac{3}{2} k . \quad (4.270)$$

Thus we can write for the heat conductivity κ_K

$$\varkappa_K = \frac{n \bar{v}}{3} c'_v \lambda = \frac{n \bar{v}}{3} \frac{3}{2} k \lambda . \quad (4.271)$$

During a gas mixture the heat conductivities of the individual components add together

$$\varkappa_K = \sum_j \varkappa_{Kj} . \quad (4.272)$$

If we now observe complete ionized plasma, then we can express the middle thermal velocity and the free length of path by the temperature. For the electrons, we then obtain

$$\varkappa_{Ke} = \frac{n_e}{e} \sqrt{\frac{8kT}{\pi m_e} \frac{3}{2} k \frac{1}{n_e Q_{ie}}} , \quad (4.273)$$

and for the ions

$$\varkappa_{Ki} = \frac{n_i}{e} \sqrt{\frac{8kT}{\pi m_i} \frac{3}{2} k \frac{1}{n_i Q_{ei}}} \quad (4.274)$$

is valid accordingly. Only functional dependence is shown by these simple views; for more exact calculations the kinetic theory of gases must be used.

It should be considered that the electrons contribute substantially more to thermal conduction. The heat losses must naturally be included into the energy equation, i.e. the energy equation (4.235) must be extended by the term $(\nabla_k)\nabla T$, where $\varkappa$ now means the total conductivity:

$$\varkappa = \varkappa_K + \varkappa_R . \quad (4.275)$$

The energy equation is now:

$$\varrho \frac{d}{dt} \left(\frac{\vec{v}^2}{2} + h \right) = \vec{j} \vec{E} + \frac{1}{n_e e} \nabla p_e + \frac{\partial p}{\partial t} - \nabla \varkappa \nabla T . \quad (4.276)$$

4.6 Discharge Modes

4.6.1 Direct Current Discharges

4.6.1.1 Autonomous Gas Discharges

Figure 4.9 shows the characteristic curve of the voltage during gas discharge between electrodes [20].

If the gas becomes conductive by the use of outer ionization sources, then we speak of electrical passage through this gas of an *autonomous discharge*. At small voltages the current density j in a discharge of the field strength $\vec{E}$ is proportional:

$$\vec{j} = e n_0 (b_+ + b_-) \vec{E} , \quad (4.277)$$

where

- b_+ and b_- are the mobility of the positive and negative ions,
- n_0 are the number of pairs from electrons and simply positive ions for each unit volume, and
- e the electronic charge.

In the range 10^{-4}–10^2 bar b_+ and b_- are proportional to the gas pressure. With a further increase of the field strength $\vec{E}$ the ion concentration in the discharge decreases, and the intensity of current is no longer a linear function of the voltage.

The maximum intensity of the current, which is possible by the given ionization degree α, is called the saturation current I_s. Here all ions created in the gas reach the electrodes: $I_s = e \times N_0$, with N_0 as maximum number of singly ionized ion pairs, which are produced by the outer source of ionization per second in the gas volume.

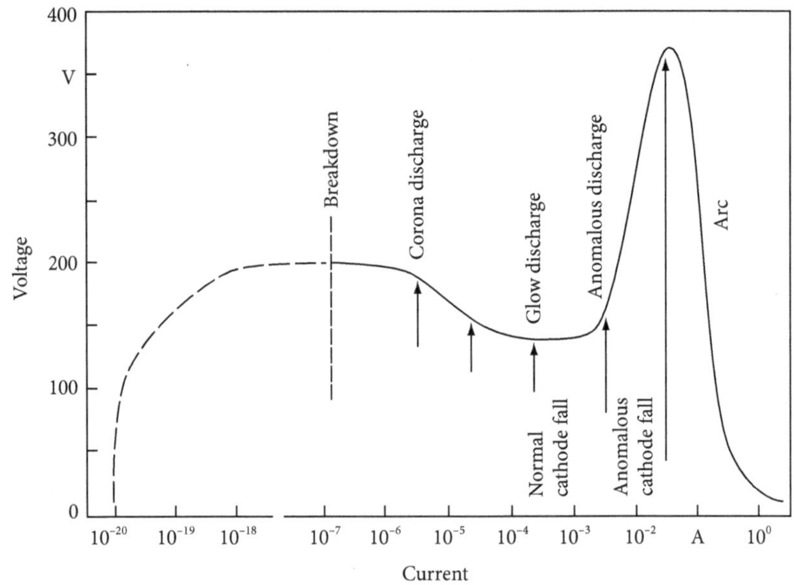

Fig. 4.9 Current voltage characteristic of a gas discharge distance between electrodes

4.6.1.2 Independent Gas Discharge

An electrical gas discharge that continues after the source of ionization has been switched off, is called *independent discharge*. The free electrical charges, which are necessary for such a discharge, are essentially generated by two procedures. On the one hand, free electrical charges are generated via collision ionization of electrons to the gas molecules, on the other hand, by striking out electrons from the cathode surface by bombardment with positive ions.

By momentum transfer of the gas ions to the cathode, the cathode is sputtered. Cathode sputtering equipment works within the range of abnormal glow discharge. The transition of a dependent gas discharge to an independent one is called *breakdown*; it occurs when the breakdown voltage U_d is reached. According to Townsend's approximation theory the burning condition for an independent gas discharge between two electrode plates can be indicated through

$$\gamma(\mathrm{e}^{\alpha d} - 1) = 1\,, \tag{4.278}$$

where

d is the electrode gap,

α the coefficient of the volume ionization of the gas by electron collisions, and

γ the coefficient of surface ionization.

The coefficient is equal to the middle number of electrons that a positive ion strikes out from the cathode. For α as function of the gas pressure and the field strength we obtain:

$$\frac{\alpha}{p} = A\mathrm{e}^{-\frac{Bp}{E}} = A\mathrm{e}^{-\frac{Bpd}{U}}\,, \tag{4.279}$$

where A and B are gas species-dependent constants.

The current density is obtained as

$$j = j_0 \frac{\mathrm{e}^{\alpha d}}{1 - \gamma(\mathrm{e}^{\alpha d} - 1)}\,, \tag{4.280}$$

where d is the distance travelled by the electrons and j_0 the initial current density for $d = 0$. For $j = 0$ the discharge stops. The breakdown voltage U_d depends on pd (Paschen's law [21]):

$$U_\mathrm{d} = B \frac{pd}{\ln pd + \ln \frac{A}{\ln\left(1+\frac{1}{\gamma}\right)}}\,. \tag{4.281}$$

The dependence is graphically illustrated in Fig. 4.10. With small pd, U_d has a minimum.

Further decreases of the pressure or the electrode gap again leads to a rise of U_d [22]. The cause for this is the decreasing probability of ionizing collisions. In Fig. 4.11 the electric

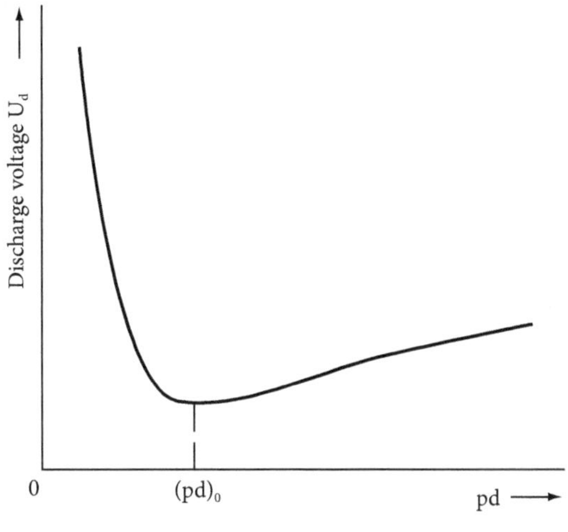

Fig. 4.10 Diagram of Paschen's law

breakthrough voltage for different gases at 1 bar and 20 °C as a function of the sparking difference is shown.

Figure 4.12 shows the dependence of the breakdown voltage and the middle breakdown

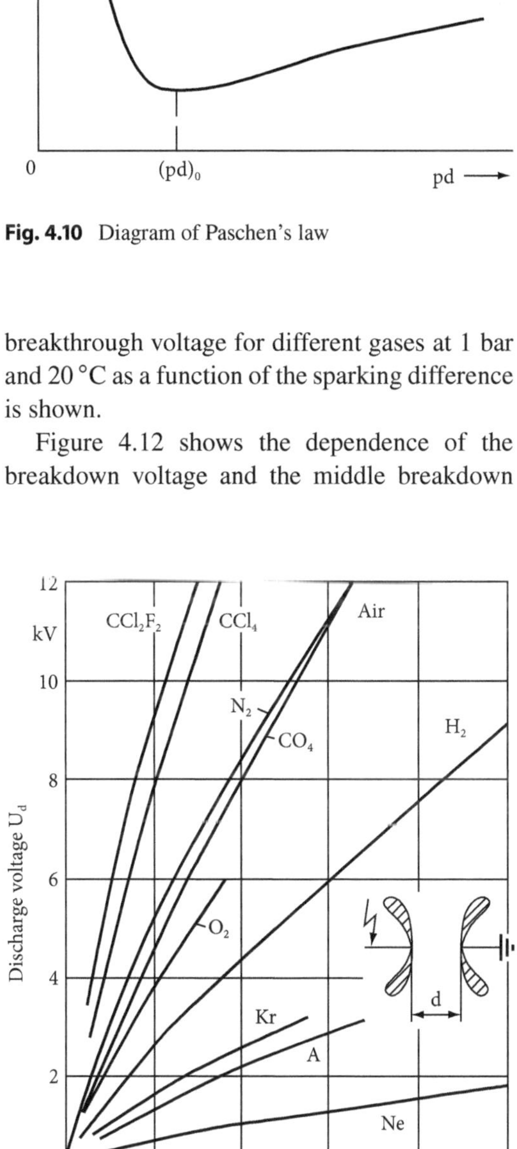

Fig. 4.11 Breakthrough strength of different gases at 1 bar and 20 °C

electric field strength of the sparking distance with an air pressure of 10^{-2} mbar. The independent glow discharge consists of four main ranges, Fig. 4.13.

In the environment of the cathode a strong decrease of potential can be observed, which is connected with a high concentration of positive ions at the boundary between ranges I and II. The glow discharge is essentially a consequence of the recombination of electrons and ions in neutral atoms or molecules. The glow light has a continuous spectrum. In the positive column we observe a high and remaining constant concentration of electrons and positive ions due to the ionization of the gas molecules caused by electron collisions. The decrease of potential within the range of the positive column is relatively small and increases under conditions that are otherwise the same with changing of the diameter of the gas discharge column. The glow of the positive column, which determines the optical characteristics of the glow light, is connected with the emission of excited atoms (molecules) of the gas.

Electrons and positive ions essentially recombine at the walls of the gas discharge tube and cause it to heat up. The positive column often shows a laminated structure, i.e. it alternates bright layers (volumes, strips) and dark gaps. In its form the positive column follows the form of the discharge tube, independently of form and arrangement of the electrodes. This is connected with the transversal (radial) electrical field, which the electrons that attached the tube walls, produce. At glow discharges in short tubes or broad containers no bright positive column is observed. The dark area in front of the cathode is the most important range for the glow discharge. In this range the volume ionization of the gas takes place. The length l_k of the cathodic dark range is equal to the distance between the cathode and the point in the discharge, at which the curve $U = U(x)$ (Fig. 4.13) has a maximum or a turning point. A glow discharge takes place only if the electrode gap is $d \geq l_k$. The change of potential ΔU_k along the cathodic dark space is the cathode drop range. Due to the high potential gradient the electrons and ions reach a high

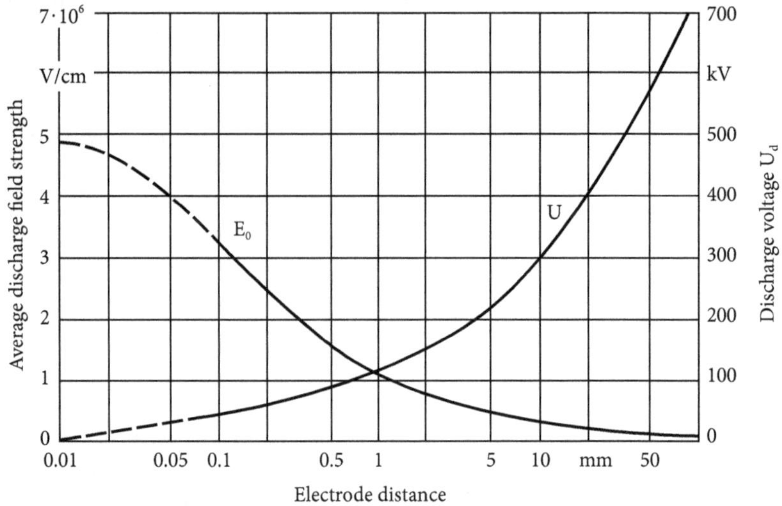

Fig. 4.12 Dependence of the breakdown voltage and the middle break through field strength on the sparking distance *d* with an air pressure of 10^{-2} mbar

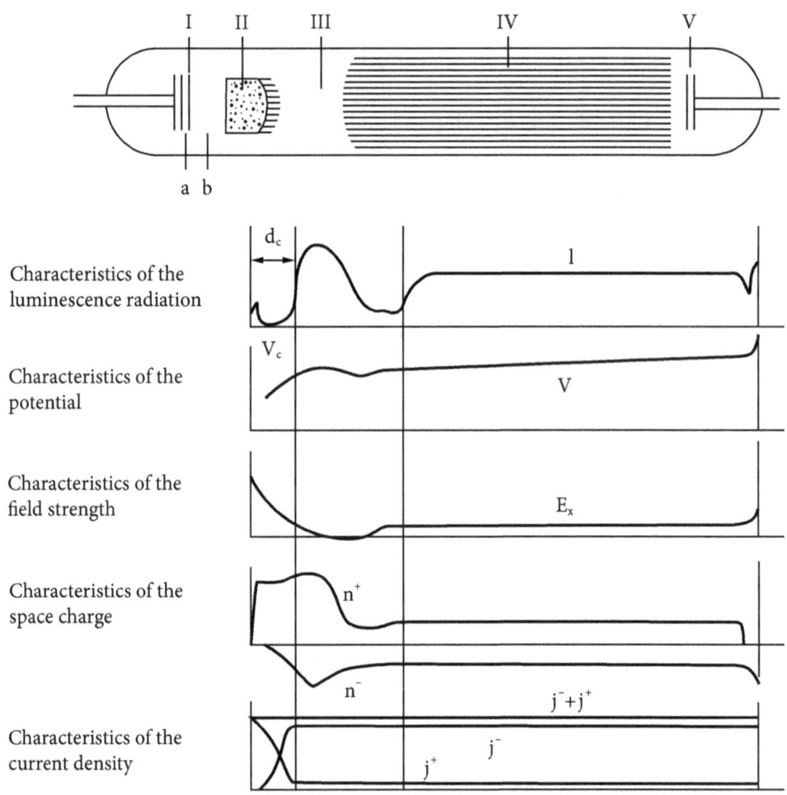

Fig. 4.13 Glow discharge of neon in 50 cm tubes at a pressure of 1 mbar. *I* Cathode glow light, *II* negative glow light, *III* Faraday's dark area, *IV* positive column, *V* anode glow light, *a* Aston's dark area, *b* Crook's dark area

velocity. The velocity of the electrons shortly before the glow edge (Fig. 4.13) corresponds almost to the cathode drop voltage eU_k. One differentiates two sorts of glow discharges (Fig. 4.9): on the one hand, the normal glow discharge, by which the discharge current density does not depend on the intensity of the current, which can be changed by means of an outer resistance, and on the other hand, the abnormal glow discharge, by which the current density increases.

In the first case, the cathode is not completely covered with negative glow light and in the second case it is completely covered. During a normal glow discharge process the cathodic dark range l_k is reversely proportional to the gas pressure. $\Delta U(x)$ depends then on the nature of the gas, the material and the condition of the cathode and increases with the electron affinity to the electrons from the cathode. With increasing discharge current the decrease of potential in the positive column becomes smaller. Therefore, the voltage characteristic of a normal glow discharge falls, i.e. the electrode voltage is reduced with increasing current. With an abnormal discharge I_k increases ΔU_k with increasing discharge current. The voltage characteristic of an abnormal glow discharge is increasing.

4.6.1.2.1 Energy Distribution of Gas Ions in Abnormal Glow Discharges

Since in abnormal glow discharges l_k for a given electrode voltage to a large extent exceeds the mean free path of the gas ions, the gas ions in the cathode-fall sustain energy losses. According to the model of W. D. Davies and T. Vanderslice [23] develop these through elastic ion-gas atom collisions with simultaneous charge exchange, i.e. the collision between a high energy ion and a neutral particle generates a high energy neutral particle and an ion with thermal energy. An ion spluttered this way through the cathode fall range, loses through acceleration the energy gathered in the field with nearly each collision. Thereby only a few ions with energy corresponding to the electrode voltage arrive at the cathode.

The following conditions are necessary:

- Only loss of energy by elastic collisions with charge exchange exists.
- The effective cross-section of the loss of energy is independent of the kinetic energy of the ions.
- Ions outside of l_k are to be generated (the power consumption of the ions begins at the dark space limit, which is identical to the beginning of l_k)

We then obtain for the distribution function [24]

$$\frac{U_E}{N_0}\frac{dN}{dU} = \frac{l_k}{2\lambda}\left[1-\frac{U}{U_E}\right]^{-\frac{1}{2}} \times e^{-\frac{l_k}{\lambda}}\left[1+\left(1-\frac{U}{U_E}\right)^{-\frac{1}{2}}\right]. \tag{4.282}$$

Here l_k is the length of the cathode fall area, λ the mean free path of the ions and U_E the electrode voltage.

From Eq. (4.282) we obtain,

$$N = N_0 e^{-\frac{l_k}{\lambda}}, \tag{4.283}$$

that ions reach the cathode with an energy that corresponds to U_E.

W.D. Davis and T.A. Vanderslice [23] indicate for the average velocity of the ions:

$$\bar{v} = \frac{2eU_E}{M}\left[\frac{2\lambda_1}{pl_K} - \left(\frac{\lambda_1}{pl_K}\right)^2\right]^{\frac{1}{2}}, \tag{4.284}$$

where e is the electronic charge, M the molecular mass and λ_1 the mean free path for the charge exchange interaction effect at 1 mbar.

Since with the effective voltage, which also considers ions with higher energies in the distribution function, Eq. (4.282), the sputter rate of the cathode is a complex situation, no quantitative correlation between the sputter rate S and U_E is to be found in the literature. For an argon glow discharge with 2.2×10^{-2} mbar W.D. Westwood and P.C. Wilcox [24] estimate that the measured sputter rate amounts to 50 % of the sputter rate, which corresponds to the energy of the voltage applied on U_E.

Under the condition of the same or similar discharge properties, S lets itself increase with,

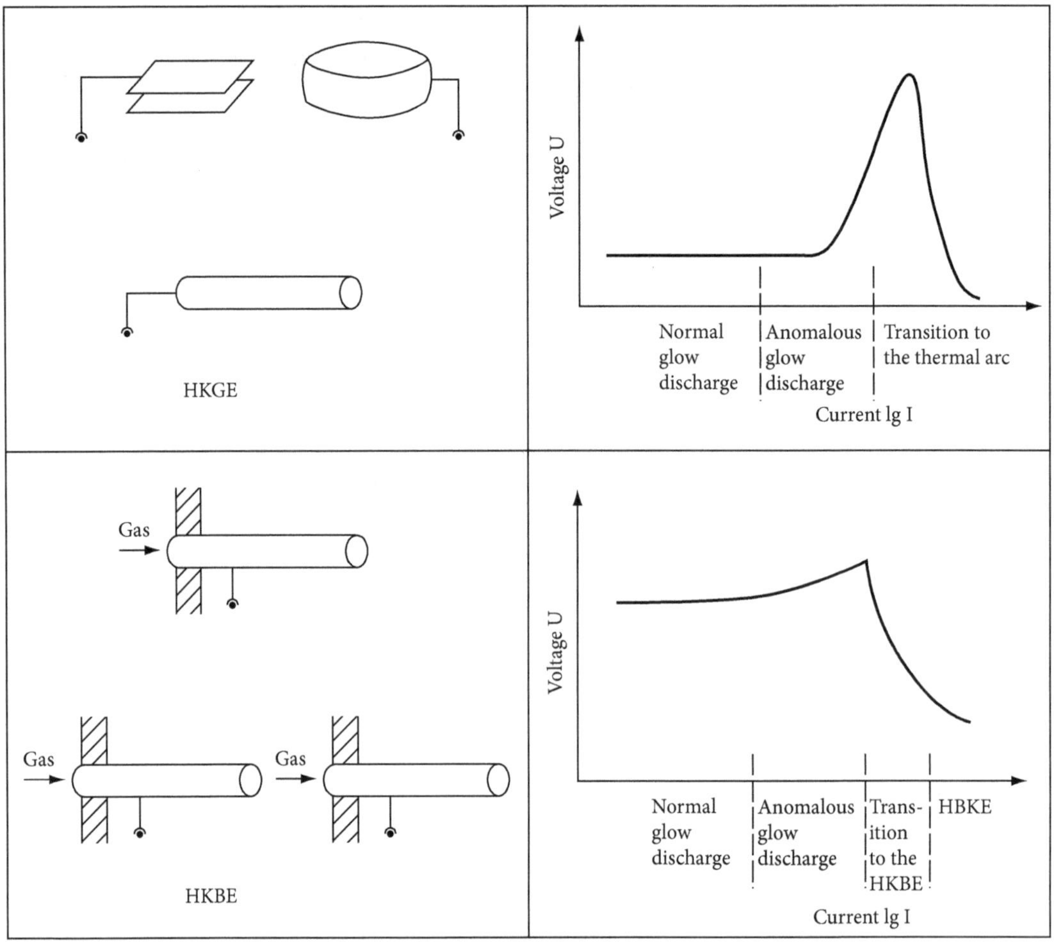

Fig. 4.14 Geometry and voltage characteristic of hollow cathode glow discharge (HKGE) and hollow cathode arc discharge (HKBE)

among others, higher U_E values. This leads to a distribution with a higher number of ions with higher energies. A further increase is conceivable with a decrease of the discharge pressure, since thereby the influence of backscattering is reduced. Since, however, in an unsupported discharge, current and pressure are two parameters that dependent on each other, they cannot be varied arbitrarily. As is desirable from the energetic point of view, decreasing pressure leads to smaller currents and concomitantly to a smaller sputter rate.

4.6.1.2.2 Hollow Cathode Discharge

Hollow cathode discharges [25–28] differentiate between hollow cathode glow discharge and hollow cathode arc discharge. In Fig. 4.14 some possible cathode geometries for a hollow cathode discharge are shown, at the top the hollow cathode glow discharge, at the bottom for hollow cathode arc discharge on the left-hand side.

For hollow cathode glow discharge (HKGE), different cathode geometries can be used:

- double cathodes, circular or tubular cathodes.

For hollow cathode arc discharge (HKBE) tubes are used. These are either individual tubes, or compact tubes, in which holes are drilled, or a tube, in which smaller tubes are integrated. For generation of hollow cathode arc discharge a gas supply is not necessary.

In the context of thin section technology, hollow cathode arc discharge attained relevance as

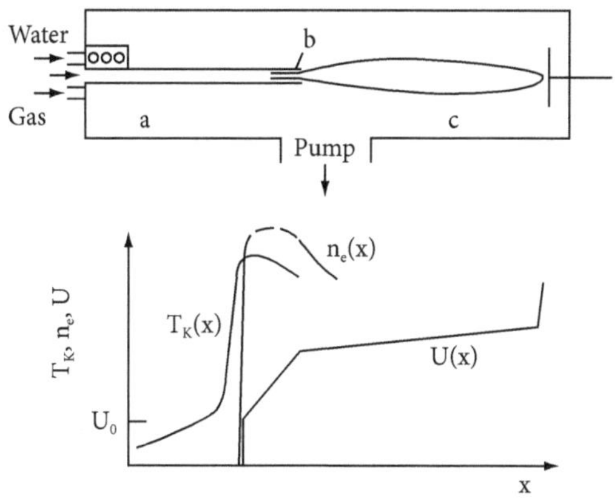

Fig. 4.15 Schema of hollow cathode arc discharge with characteristics of the cathode temperature R_K, the electron concentration n_e and the potential U. *a* active zone, *b* internal positive zone, *d* external zone

source of vapour for high rate coating in ion-supported processes. Figure 4.15 shows this type of discharge schematically. The plasma inside the cathode tube (a) is separated by a thin cathode fall range of the cathode. The fall space voltage is appropriate for vacuum chambers between 12 and 15 V. The active zone (b) follows towards the internal positive column. It is characterized by a high field strength (approx. 8 V/cm). The area between the cathode end and the anode is filled out by the outer positive column (c). The field strength of the outside positive column amounts to approximately 0.2 V/cm.

The cathode mechanism of hollow cathode arc discharge can be described as follows [29]. Due to high temperature thermal emission of electrons occurs. These electrons are accelerated in the cathode fall area and enter with the cathode fall voltage into the adjacent plasma. In the plasma these beam electrons lose their energy by flexible collisions, excitation and ionization of neutral particles.

During hollow cathode arc discharge acting in the normal range, the outside plasma has the high ionization degree of $\alpha > 0.5$, ion temperatures within the range $2\,\text{eV} < T_i < 10\,\text{eV}$ and electron temperatures $2\,\text{eV} < T_c < 10\,\text{eV}$. The electrical field force lies between 0.1 and 0.2 V/cm.

4.6.2 High Frequency Discharges

High frequency AC discharges show, on the one hand, nearly similar behaviour as direct current DC discharges; on the other hand, they deviate very strongly from direct current discharges. At AC discharge no real cathode or anode exist, since contrary to DC discharges the net current from charge carriers to each electrode is zero. Also, AC discharges have no clearly defined potential distribution.

If we use a high frequency voltage $\hat{u}\cos\omega t$ between two even electrodes with the distance d, then the electrical field $\vec{E}$ appears on an electron present between the plates. The electron then obeys – if it does not collide with the walls – the equation of motion (Eq. (4.49)):

$$\frac{d\vec{v}_e}{dt} + \nu_{ej}\vec{v}_e = \frac{e}{m_e}\vec{E}\,, \tag{4.285}$$

where

ν_{ej} is the middle collision frequency of the electron with the atoms and/or ions.

Without collisions the electron would move with a phase shift of 90° in relation to the electrical field, i.e. it would store only reactive energy. With collisions the electron takes on an average the power

$$N = e\vec{E}\vec{v} = \frac{e^2\hat{u}^2}{2m_e\nu_{ej}d^2}\left(\frac{\nu_{ej}^2}{\nu_{ej}^2 + \omega^2}\right), \tag{4.286}$$

as in [30, 31]. In relation to the effect of a low frequency or a direct current discharge, with which $\omega \ll \nu_{ej}$, the effect of the electrical field is reduced by the factor

$$\gamma = \sqrt{\frac{\nu_{ej}^2}{\nu_{ej}^2 + \omega^2}}\,. \tag{4.287}$$

One can thereby define an effective current $u_e = \hat{u}\gamma$, for which Paschen's law is valid. An exact specification of the breakdown voltage is usually only possible with difficulty. In practice, the distance d, the pressure p and the frequency ω are more important to develop a breakthrough. Igniting is possible whenever the carrier losses are

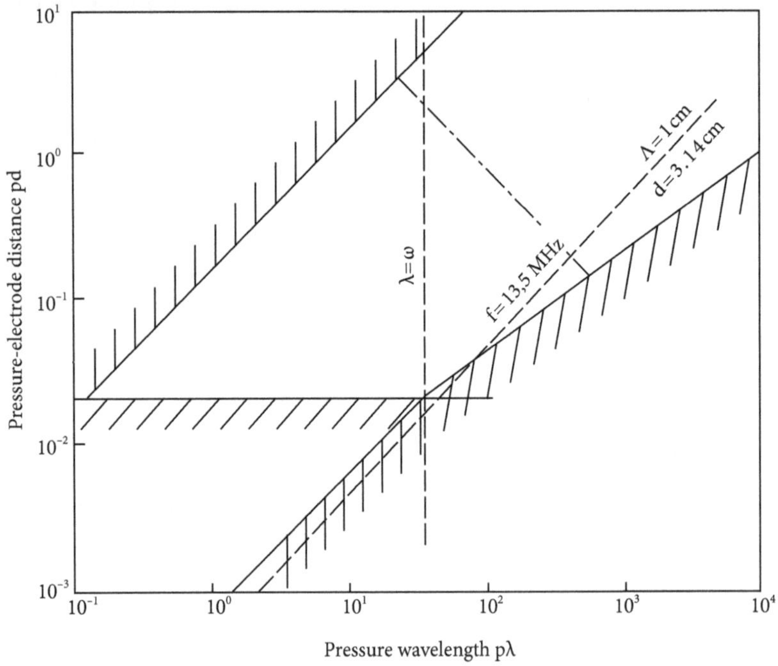

Fig. 4.16 Boundaries for the high frequency range dominated by diffusion losses

only due to diffusion. In high frequency AC discharge limits in the application of this theory are applied for the following reasons:

- At low pressures the mean free path becomes larger than the chamber dimension.
- Is the amplitude of the motion of the electron calculated from Eq. (4.285) larger than the distance of the electrodes, then the electrons push to the walls with each half wave, whereby additional losses develop.
- If the wavelength of the high-frequency field is not large compared to the electrode gap (very high frequency), then the field distribution must be considered.

These limits, which depend somewhat on the nature of the gas and the electrode material, are shown in Fig. 4.16. The product of the pressure and electrode gap ($\Lambda \cong d/\pi$) is plotted vertically, the product of pressure and wavelength λ is plotted horizontally. On the straight line with the denotation „optimal breakthrough" is the appropriate value of the breakdown voltage minimum. The straight line electron resonance boundary is plotted in the area of the mean free path. Above this curve a breakthrough caused by releasing secondary electrons is possible. In the hachured area no discharge can be normally ignited [32].

The following considerations refer to plasma with Maxwell velocity distribution of the electrons. The high frequency AC discharge can be described in good approximation.

If we bring an electrically charged electrode into contact with gas plasma, then a layer forms around the electrode, in which the electrical field decreases rapidly. We assume that the space potential electrode surface is φ^*, and thus the potential difference

$$U = \varphi - \varphi^* . \tag{4.288}$$

If $e|U| < kT$, for the distribution of the carriers in the field of the electrode we can use the Boltzmann distribution:

$$n_\mathrm{i} = n_{\mathrm{i}0} \mathrm{e}^{-\frac{eU}{kT_\mathrm{i}}} , \tag{4.289}$$

$$n_\mathrm{e} = n_{\mathrm{e}0} \mathrm{e}^{-\frac{eU}{kT_\mathrm{e}}} . \tag{4.290}$$

When we also use quasi-neutrality, at some distance of the plasma affecting electrode $n_{\mathrm{i}0} =$

$n_{e0} = n_e$. For the Poisson equation, which generates the relationship between the potential in the environment of the electrode and the space charge, we obtain

$$\Delta U = -\frac{e}{\varepsilon_0}\left(n_e e^{-\frac{eU}{kT_i}} - n_e e^{-\frac{eU}{kT_e}}\right). \quad (4.291)$$

If the electrode has the plasma potential, then the current to the electrode – independent of the sign of the charge – is

$$i_j = \frac{e n_j \bar{v}_j}{4} A_0 , \quad (4.292)$$

where
$\bar{v}$ means the middle velocity of the particles, and
A_0 the surface of the electrode.
If a potential difference between plasma and electrode exists, then the current taken up by the electrode is regulated by the available fields. The quantity A_0 of the current taking up area will thereby change. As an expression for the charged particle flux in first approximation we obtain

$$j_j = \frac{e n_j \bar{v}_j}{4} A e^{-\frac{eU}{kT}}). \quad (4.293)$$

This equation is based on the perception that the Boltzmann distribution is valid for the charge carriers. The general current/voltage characteristics for the general process are shown in Fig. 4.17; at $U = 0$ the electrode collects the charge carriers (Eq. (4.292). Since with $T_i = T_e = \bar{v}_i = \bar{v}_e m_e/m_i$, the obtained $i_e \ggg i_i$ for potential values $\varphi < \varphi^*$ is valid for the electron flow equation (4.293). In the point φ' the saturation current of the ions is reached. Thus the junction region must correspond, $\varphi' <$ to $\varphi < v^*$, to the sum of electron flux and ion flux.

Of special interest is the electron's running current i_e. This is appropriate in the voltage range $\varphi' < \varphi < \varphi^*$. If we plot i_e as a function of U on a semi-logarithmic scale, then we obtain:

$$\frac{kT_e}{e} = \frac{1}{\frac{\mathrm{d}\ln_e}{\mathrm{d}U}} . \quad (4.294)$$

For the analysis of the ion saturation current with strongly negative potential in relation to the plasma we use the expression:

$$\Delta U = \frac{e}{\varepsilon_0} n_e e^{-\frac{eU}{kT_e}} - \frac{e}{\varepsilon_0} n_i(U) . \quad (4.295)$$

For the range of values $U \gg kT_e/e$, which corresponds to the saturation current of the ions, this means that in the vicinity of l_0 no electrons are present (see Fig. 4.18).

An appreciable number of electrons occur only by $U \sim kT_e/e$, since the electrons cannot reach the electrode with higher inverse voltage practically. From this it follows that in the transition layer between l_0 and d, the entire decrease of potential is caused by the positive space charge. With $l > d$ quasi-neutrality is restored, thus

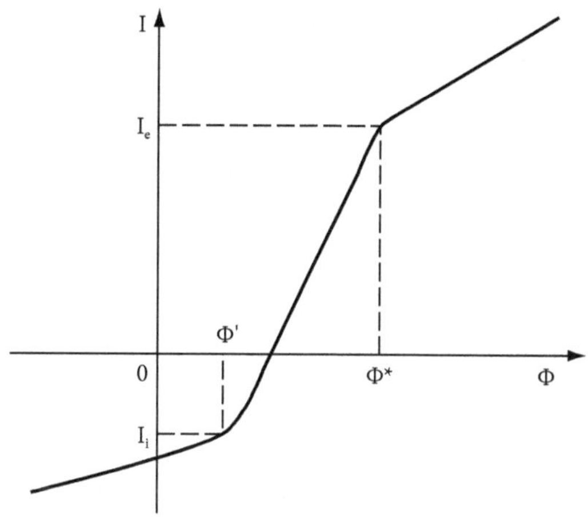

Fig. 4.17 Current/voltage characteristic of the electrode

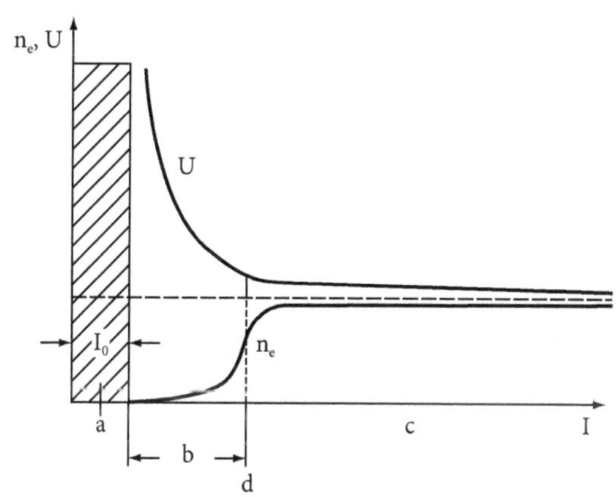

Fig. 4.18 Potential process and concentration distribution in the vicinity of the electrode

$n_i \sim n_e$, i.e., at point d we can use the saturation voltage $U_S \sim kT_e/e$.

The ions are accelerated under the influence of the same potential difference U_S and penetrate into the layer. If the condition $T_e \gg T_i$ is thus fulfilled, then the velocity of the ions that penetrate the layer is proportional to $(kT_e/m_e)^{0.5}$, and thus the ion saturation current is

$$i_i = en_i\sqrt{\frac{2kT_e}{m_i}}A. \quad (4.296)$$

If the electrode is evenly insulating – for example as a capacitor – then the total current must be zero. We then obtain the condition $|i_e| = |i_i|$. The electrode charges itself on the float voltage U_F. U_F results from Eq. (4.291) in combination with Eq. (4.296) as

$$U_F = \frac{kT_e}{e}\ln\sqrt{\frac{m_i}{4\pi m_e}}. \quad (4.297)$$

For argon we obtain a negative voltage at an electron temperature from $T_e = 5 \times 10^4$ K of 20 V.

If the high frequency voltage $u = U + \hat{u}\cos\omega t$ is applied on the electrode, then during the positive half-period at first a very large electron start up current will flow; in the negative half-period the ion saturation current flows. Mathematically, we obtain from Eq. (4.293) for $u < U$:

$$i_e = i_{eo}e^{\frac{U}{UT}}\left[l_o\left(\frac{\hat{u}}{U_T}\right) + 2\sum_{n=1}^{\infty} l_n\left(\frac{\hat{u}}{U_T}\right)\cos n\omega t\right], \quad (4.298)$$

where $U_T = kT_e/e$, and $l_n(x)$ represents the modified Bessel functions. A somewhat complicated expression results for $\hat{u} = U$. Since the ion saturation current is much smaller than the electron saturation current, the insulating electrode will charge itself on a DC voltage, which is higher than the peak voltage of the high frequency. From Eqs. (4.296) and (4.298), we have for the voltage U_I of the insulating electrode:

$$i_e = i_{eo}e^{\frac{U}{UT}}\left[l_o\left(\frac{\hat{u}}{U_T}\right) + 2\sum_{n=1}^{\infty} l_n\left(\frac{\hat{u}}{U_T}\right)\cos n\omega t\right]. \quad (4.299)$$

The electrode is, therefore, pre-charged so highly negative in the first periods after switching of the high frequency voltage by an excess of electrons that the total value of the current becomes zero.

The extent d of the dark range forming in front of the electrode can be calculated using the approach with the Boguslawsko–Langmuir law for infinitely expanded even electrodes.

The law is as follows:

$$j_i = \frac{4\sqrt{2}}{9}\frac{e_o}{d^2}\sqrt{\frac{e}{m_i}}U_1^{\frac{3}{2}}, \quad (4.300)$$

where

ε_0 is the absolute dielectric constant of the vacuum.

With the extent d of the dark range we can also determine the layer capacity $C = \varepsilon_0 A/d$. With the symmetrical structure of the two electrodes shape, the voltage distributes itself as U_1 on both dark layers.

Since the ion saturation current is constant everywhere, from Eq. (4.300) we obtain a relationship between voltage and layer thickness as

$$\left(\frac{U_1}{U_2}\right)^{\frac{3}{2}} = \left(\frac{d_2}{d_1}\right)^2, \quad (4.301)$$

and

$$\frac{C_2}{C_1} = \frac{\hat{u}_1}{\hat{u}_2} = \frac{A_2 d_1}{A_1 d_2} \quad (4.302)$$

is valid at the same time. With $\hat{u}_1 = U_1$ and $\hat{u}_2 = U_2$ we obtain the following allocation of voltage:

$$\frac{U_1}{U_2} = \left(\frac{A_2}{A_1}\right)^4. \quad (4.303)$$

References

1. Cap F (1970) Einführung in die Plasmaphysik I. Pergamon Press, Oxford
2. Linhart JG (1960) Plasma Physics. North Holland Publishing Co., Amsterdam
3. Chandrasekhar S (1960) Plasma Physics. University of Chicago Press
4. Maxwell JC (1954) A Treatise on Electricity and Magnetism (891). Dover Publications, Inc., New York
5. Mongomerry D, Tidman D (1964) Plasma Kinetic Theory. McGraw Hill, New York
6. Kunkel M (1966) Plasma Physics in Theory and Application. McGraw Hill
7. Dupree A (1961) Phys. Fluids 4, 696, 6, 1714 (1963)
8. Sears FW (1950) An Introduction to Thermodynamics, The Kintetic Theory of Gases and Statistical Mechanics. Addision-Weseley Press, Inc., Reading, Mass.
9. *CODATA Recommended Values.* National Institute of Standards and Technology, abgerufen am 20. Juni 2011. Wert für die Rydberg-Konstante
10. Cobine JD (1958) Gaseous Conductors. Dover Publications, Inc., New York
11. Phelps AV, Molnar JP (1963) Phys Rev 89:1202
12. Chuan RL (1957) A Note on the Adition of Heat to a Gas through Electrical Discharge, USCEC Rept. 68–202, January, 1957
13. Loeb LB (1956) The recombination of ions Handbuch der Physik, vol. XXI. Springer-Verlag, Berlin
14. Massey HSW, Burhop EHS (1952) Elektronic and ionic impact phenomena. Oxford University Press, Fair Lawn
15. Saha MN (1920) Ionization in the solar chromosphere. Phil Mag 40(238):472
16. Bond JW Jr (1954) The nature of shock front in argon. Los Alamos: Sci. Lab. Rept. LA-1693, Juli 1, 1954
17. Moore CE (1958) Atomic energy levels. Nat. Bureau of Standards Circ. 467 Vls. I,II and III. Washington: 1949, 1952 und 1958
18. Laporte O (1958) High temperature shock waves. In: Third AGARD Colloquium on Combustion and Propulsion. Pergamon Press, New York, p 499
19. Ohlsen HH (1961) Partition function cut-off and lowering of the ionization potential in an argon-plasma. Phys Rev 124(4):1703
20. Paschen F (1889) Ann.Phys. (4 Folge) 50, p 901
21. Penning FM (1957) Electical Discharges in Gases. The Macmillian Company, New York
22. Schulz P (1974) Elektronische Vorgänge in Gasen und Festkörpern. Verlag O. Braun, Karlsruhe
23. Davies WD, Vanderslice TA (1993) PhysRev 131:219
24. Westwood WD, Wilcox PC (1972) J Appl Phys 42:4099
25. Little PF, von Engel A (1954) The hollow cathode effect and the theory of glow discharges. Proc Roy Soc A 224:209
26. Musha T (1962) Cathode sputtering in hollow cathode discharges. J Phys Soc Japan 17:1440
27. Penache C (2002) Study of high-pressure glow discharges generated by micro-structured electrode (MS) arrays, Dissertation J.W. Goethe Universität Frankfurt Main
28. Mildner M, Korzec D, Engemann J (1999) 13.56 MHz hollow cathode jet matrix plasma source for large area surface coating. Surface and Coating Techn 112:366
29. Delcroix JL, Trinidade AR (1974) Advanc In Electronics and Electron Phys 35:87
30. Brown SC (1959) Basic data of plasma physica. John Wiley, New York
31. Wolf B (1995) Handbook of Ion Sources. CRC Press, Boca Raton
32. Sager O (1969) Frequency analysis of the probe current in an RF discharge. Proc IEEE 57:227

Gaseous Phase and Surface Processes

5

H. Frey

5.1 Elementary Gaseous Phase and Surface Reactions [1]

Consider the following reactions:

$$\begin{aligned} &\text{a)}\quad 2\,\mathrm{U} + 3\,\mathrm{V} \rightarrow \mathrm{W} + 2\,\mathrm{Z} \\ &\text{b)}\quad \mathrm{U} + \mathrm{V} \rightarrow \mathrm{W} + \mathrm{Z} \\ &\text{c)}\quad \mathrm{U} \rightarrow \mathrm{W} + \mathrm{Z@},\,, \end{aligned} \tag{5.1}$$

where U, V, W and Z are molecules. The reactions are denominated elementarily, if they occur in one step. This means all molecules in a), b) must collide at the same time, and/or in c) the decomposition "in a step" must take place. Reaction a) is in all probability not an elementary reaction, since five molecules must collide at the same time. For the reactions b) [2] an elementary reaction is conceivable:

$$\mathrm{O}^+ + \mathrm{O}_2 \rightarrow \mathrm{O} + \mathrm{O}_2^+\,;$$

likewise for reaction c):

$$\mathrm{A}^* \rightarrow \mathrm{A} + h\nu\,.$$

On the other hand, it is well known that the following reaction is not elementary:

$$\mathrm{Cl}_2 + \mathrm{H}_2 \rightarrow 2\,\mathrm{HCl}\,.$$

This reaction corresponds to b). From this it follows that on the basis of the stoichiometric equations a)–c) it cannot decided whether a reaction is elementary or not. However, chemical reactions can be divided into different basic steps:

1. Unimolecular reactions:

$$\mathrm{U} \rightarrow \text{final products.}$$

2. Bimolecular reactions:

$$\mathrm{U} + \mathrm{V} \rightarrow \text{final products.}$$

At high pressure triple molecular reactions can also occur:

$$\mathrm{U} + \mathrm{V} + \mathrm{W} \rightarrow \text{final products.}$$

In low pressure plasmas the reactions are usually complex. By means of the so-called stoichiometric coefficients the reaction rates can be defined. The reaction rates for reactants (U, V in Eq. (5.1)) are negative, for products positive:

$$R = \frac{1}{\alpha_j}\frac{\mathrm{d}n_j}{\mathrm{d}t}\,, \tag{5.2}$$

here

n_j is volume density of molecules at the jth component.

If we regard the reaction Eq. (5.1)a), then we obtain:

$$R = -\frac{1}{2}\frac{\mathrm{d}n_\mathrm{U}}{\mathrm{d}t} = -\frac{1}{3}\frac{\mathrm{d}n_\mathrm{V}}{\mathrm{d}t} = \frac{\mathrm{d}n_\mathrm{W}}{\mathrm{d}t} = \frac{1}{2}\frac{\mathrm{d}n_\mathrm{Z}}{\mathrm{d}t}\,.$$

In the case of surface reactions the particle volume densities n_j [cm^{-3}] must be replaced by surface densities n'_j [cm^{-2}]. Generally, R is a complicated function of the particle number

© Springer-Verlag Berlin Heidelberg 2015

H. Frey, H. R. Khan, *Handbook of Thin-Film Technology*, DOI 10.1007/978-3-642-05430-3_5

density of all reactants. In the case of elementary reactions, however, we distinguish the following simple forms:

$$\mathrm{Z} \rightarrow \text{products}$$

$$R = -\frac{\mathrm{d}n_\mathrm{Z}}{\mathrm{d}t} = K_1 n_\mathrm{Z}\,. \tag{5.3}$$

The quantity K_1 [s^{-1}] is called a rate constant of first order,

$$\mathrm{U} + \mathrm{U} \rightarrow \text{products}$$

$$R = -\frac{1}{2}\frac{\mathrm{d}n_\mathrm{U}}{\mathrm{d}t} = K_2 n_\mathrm{U}^2\,. \tag{5.4}$$

The parameter K_2 [$\mathrm{m}^3\,\mathrm{s}^{-1}$] is called a rate constant of second order. A reaction of second order can also be formed by two different reactants:

$$\mathrm{U} + \mathrm{V} \rightarrow \text{products}$$

$$R = -\frac{\mathrm{d}n_\mathrm{U}}{\mathrm{d}t} = -\frac{\mathrm{d}n_\mathrm{V}}{\mathrm{d}t} = K_2 n_\mathrm{U} n_\mathrm{V}\,. \tag{5.5}$$

For reactions of the third order we obtain:

$$\mathrm{U} + \mathrm{V} + \mathrm{W} \rightarrow \text{products}$$

$$R = -\frac{\mathrm{d}n_\mathrm{U}}{\mathrm{d}t} = -\frac{\mathrm{d}n_\mathrm{V}}{\mathrm{d}t} = -\frac{\mathrm{d}n_\mathrm{W}}{\mathrm{d}t} K_3 n_\mathrm{U} n_\mathrm{V} n_\mathrm{W}\,, \tag{5.6}$$

with K_3 in [$\mathrm{m}^6\,\mathrm{s}^{-1}$]. The rate constants are functions of temperature are but independent of the densities of the reactants involved.

5.1.1 Equilibrium Constant

We observe the following elementary reactions:

$$\mathrm{U} + \mathrm{V} \underset{K_{-2}}{\overset{K_2}{\rightleftarrows}} \mathrm{W} + \mathrm{Z}$$

with the forward rate $n_\mathrm{U} n_\mathrm{V}\, K_2$ and the backward rate $n_\mathrm{W} n_\mathrm{Z}\, K_{-2}$. In thermal equilibrium the rates are alike:

$$K_2 \bar{n}_\mathrm{U} \bar{n}_\mathrm{V} = K_{-2} \bar{n}_\mathrm{W} \bar{n}_\mathrm{Z}\,. \tag{5.7}$$

By division we obtain:

$$\frac{\bar{n}_\mathrm{W} \bar{n}_\mathrm{Z}}{\bar{n}_\mathrm{U} \bar{n}_\mathrm{V}} = \frac{K_2(T)}{K_{-2}(T)} = \bar{K}(T)\,. \tag{5.8}$$

Although Eq. (5.8) was derived from thermal equilibrium, this relation is also valid in a real system that is not in thermal equilibrium. The only condition is that the velocity of the collecting particles behaves as a Maxwell distribution with the temperature T. The reason is that the rate constants (also called rate coefficients) are independent of the particle densities. They depend on the activation cross-section and the collision dynamics:

$$K_\mathrm{UV}(T) + \langle \sigma_\mathrm{UV} v_\mathrm{R} \rangle = \int_0^\infty f_\mathrm{m} v_\mathrm{R} \sigma_\mathrm{UV} 4\pi v_\mathrm{R}^2 \mathrm{d}v_\mathrm{R}\,. \tag{5.9}$$

If the equilibrium constant is a rate coefficient (rate constant), then we can calculate the other constant from it. Using the example of collision excitation:

$$n_0 n_\mathrm{e} X(T_\mathrm{e}) = n^* n_\mathrm{e} Y(T_\mathrm{e})\,, \tag{5.10}$$

with the rate coefficient Y for excitation collision. *If* n_0 and n^* are in thermal equilibrium, then we obtain:

$$\frac{n_0}{n^*} = \frac{g_0}{g^*} \exp\left(-\frac{\Delta E}{kT_\mathrm{e}}\right)\,. \tag{5.11}$$

Thereby Y becomes X,

$$Y = X\frac{n_0}{n^*} = X\frac{g_0}{g^*} \exp\left[-\frac{\Delta E}{kT_\mathrm{e}}\right]\,. \tag{5.12}$$

5.2 Gaseous Phase Kinetics

Most processes are stationary [4], i.e. the entire process time, e.g. for etching trenches, is very much longer than reaction time constants or transportation time constants (Fig. 5.1) for the relevant gaseous phase species. A typical process

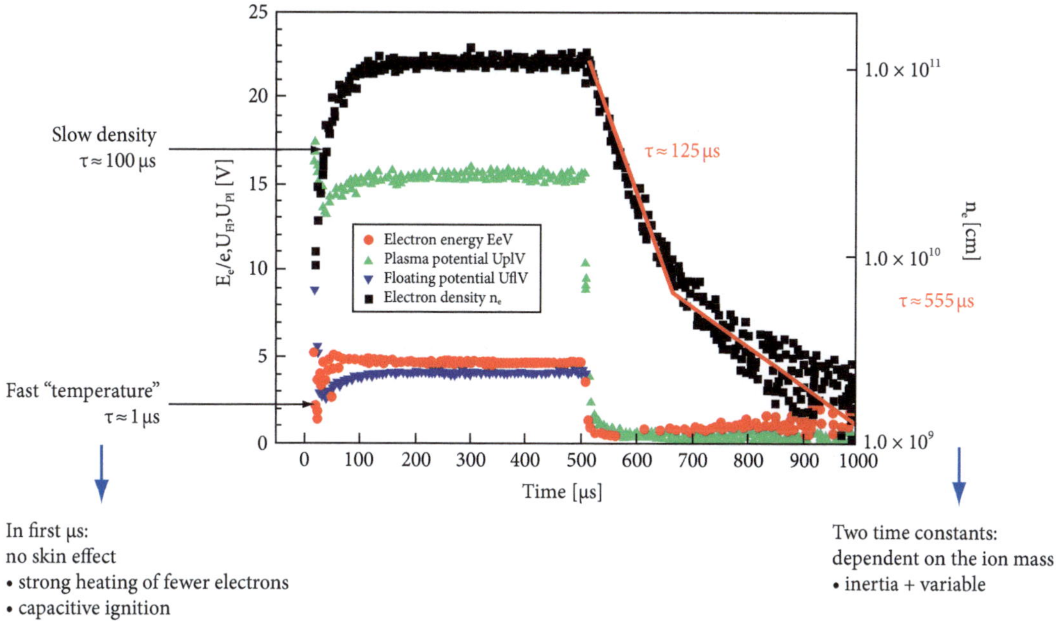

Fig. 5.1 Time constants in the pulsed inductively coupled RF-plasma

uses a constant gas flow in the reactor chamber (consisting of different gases), with constant power and constant pressure in the recipient. All gaseous species are evacuated from the surface.

For these reasons the particle densities of the gaseous species are constant, i.e. independent of the time. However, these densities cannot be determined based on the assumption of thermodynamic equilibrium, since this is not valid. However, if the rate coefficients are well known for the individual reactions, the appropriate rate equations can be calculated, which are often associated with one another.

5.2.1 Consecutive Reactions of First Order

We regard two consecutive time-varying reactions:

$$\mathrm{U} \xrightarrow{K_\mathrm{U}} \mathrm{V} \xrightarrow{K_\mathrm{V}} \mathrm{W} \tag{5.13}$$

without sources and sinks, i.e. the species U is not delivered from the outside and the species W does not disappear outwards.

The rate equations for this succession of reactions are:

$$\frac{\mathrm{d}n_\mathrm{U}}{\mathrm{d}t} = -K_\mathrm{U} n_\mathrm{U} \tag{5.14}$$

$$\frac{\mathrm{d}n_\mathrm{V}}{\mathrm{d}t} = K_\mathrm{U} n_\mathrm{U} - K_\mathrm{V} n_\mathrm{V} \tag{5.15}$$

$$\frac{\mathrm{d}n_\mathrm{W}}{\mathrm{d}t} = K_\mathrm{V} n_\mathrm{V}\,. \tag{5.16}$$

If we now begin now with an initial value $n_\mathrm{U} = n_\mathrm{U0}$, where $n_\mathrm{V} = 0$ for $t = 0$, then Eq. (5.14) can be integrated:

$$n_\mathrm{U} = n_\mathrm{U0} \exp(-K_\mathrm{U} t)\,. \tag{5.17}$$

If we apply (5.17) in (5.15), we obtain by integration:

$$n_\mathrm{V} = n_\mathrm{U0} \frac{K_\mathrm{U}}{K_\mathrm{V} - K_\mathrm{U}} [\exp(-K_\mathrm{U} t) - \exp(-K_\mathrm{V} t)]\,. \tag{5.18}$$

This procedure can be repeated again, in order to also calculate n_w [set (5.18) in (5.16)] and the following integration. Since this, however, concerns a final system,

$$n_\mathrm{U}(t) + n_\mathrm{V}(t) + n_\mathrm{W}(t) = n_\mathrm{U0} = \text{const.} \tag{5.19}$$

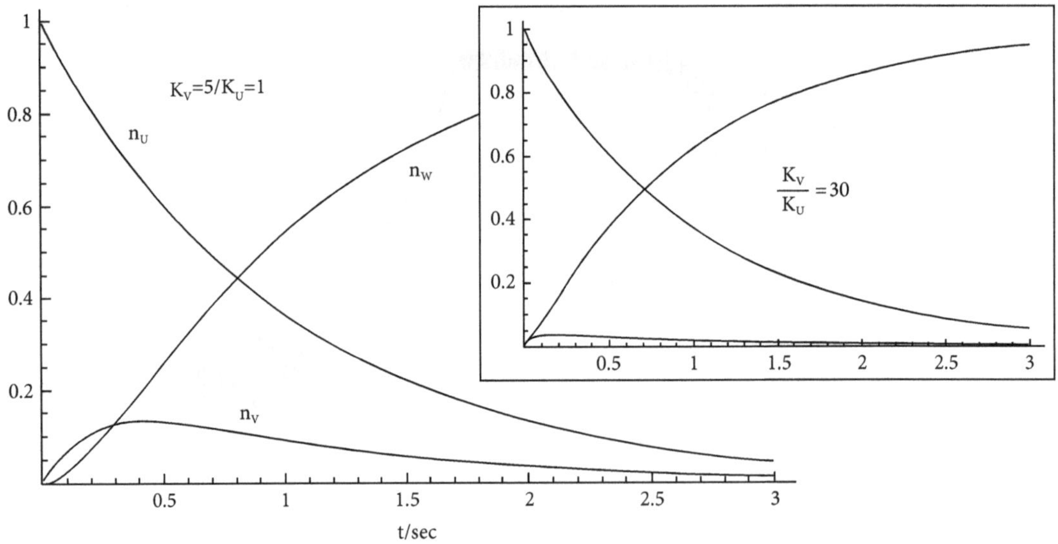

Fig. 5.2 Time-dependent reaction kinetics $K_U = 1$ and $K_V = 5$ of particle densities

is valid. After n_U and n_V are admitted, we obtain:

$$n_W(t) = n_{U0} - n_U - n_V\,. \tag{5.20}$$

From this we finally obtain:

$$n_W = n_{U0}\left[1 + \frac{1}{K_U - K_V}(K_V \exp(-K_U t) - K_U \exp(-K_V t))\right]. \tag{5.21}$$

If we regard two extreme cases:
case I $K_U \ll K_V$
case II $K_V \ll K_U$,
then the following are valid.

For case I:

$$n_W = n_{U0}(1 - \exp(-K_U t)) \tag{5.22}$$

and for case II:

$$n_W = n_{U0}[1 - \exp(-K_V t)]\,. \tag{5.23}$$

In both cases I and II the formation of the products is defined by the small rate constant. Since the reciprocal value of the rate constants in Eqs. (5.22) and (5.23) represent the time constants for the exponential time response, we obtain the formation of n_W by the slowest time constant limits, the so-called ***rate-limiting reaction step***.

In case I ($K_V \gg K_U$), the species V with a slow rate from the species U is produced. Subsequently, V is converted very fast into species W. For this reason, n_V disintegrates (after a short transition period) with the rate K_U, whereby $n_V \ll n_U$. Thus:

$$\frac{dn_V}{dt} \approx K_U n_V \ll K_U n_U \tag{5.24}$$

is valid. Therefore, n_V in Eq. (5.15) is set to zero.

This way we receive the following approximated solution for Eqs. (5.14)–(5.16):

$$n_U = n_{U0} \exp(-K_U t) \tag{5.25}$$

$$n_V \approx \frac{K_U}{K_V} n_U = \frac{K_U}{K_V} n_{U0} \exp(-K_U t) \tag{5.26}$$

$$n_W \approx n_{U0}[1 - \exp(-K_U t)]\,. \tag{5.27}$$

Here, species V is the reactive intermediate stage. Whereas

$$\frac{dn_V}{dt} \approx 0$$

is the stationary solution for the reactive intermediate stage V.

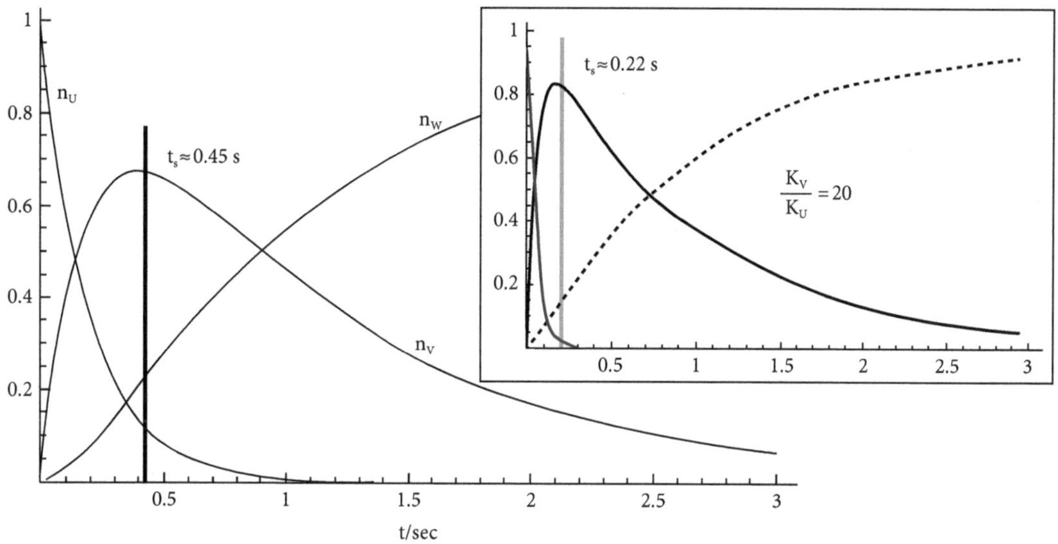

Fig. 5.3 Time-dependent reaction kinetics of the particle densities for $K_U = 5$ and $K_V = 1$

In case II ($K_U \gg K_V$) species V is produced faster by U than W is produced by V. Thereby two reactions are not developed from first order with the following solutions. For $t < t_s$ (separate ion time), we have:

$$\begin{aligned} n_U &= n_{U0} \exp[-K_U t] \\ n_V &\approx n_{U0} (1 - \exp[-K_U t]) \\ n_W &\approx 0 \end{aligned} \tag{5.28}$$

and for $t > t_s$, in contrast to this,

$$\begin{aligned} n_U &\approx 0 \\ n_V &\approx n_{U0} \exp[-K_V t] \\ n_W &\approx n_{U0} (1 - \exp[-K_V t]) . \end{aligned} \tag{5.29}$$

For the separate time, we have:

$$t_s = \frac{1}{\sqrt{K_U K_V}} , \tag{5.30}$$

which separate the fast (Eq. (5.28)) and the slow (Eq. (5.29)) time constants. Now with the consecutive reaction Eq. (5.13) during a stationary plasma process, we need a generation term G [$\mathrm{m}^{-3}\,\mathrm{s}^{-1}$] for the species U (for example, fresh gas feed), as well as a loss term for the species W (for example, by evacuating), which can be written as a reaction of first order.

Thus for Eqs. (5.14) to (5.16) we obtain:

$$\begin{aligned} \frac{dn_U}{dt} &= G - K_U n_U = 0 \\ \frac{dn_V}{dt} &= K_U n_U - K_V n_V = 0 \\ \frac{dn_W}{dt} &= K_V n_V - K_W n_W = 0 . \end{aligned} \tag{5.31}$$

If we solve the set of Eqs. (5.31), we obtain:

$$n_U = \frac{G}{K_U} \qquad n_V = \frac{G}{K_V} \qquad n_W = \frac{G}{K_W} . \tag{5.32}$$

5.2.2 Reactions Moving in Opposite Directions

Now we consider (5.13) with the respective reactions moving in opposite directions:

$$\mathrm{U} \underset{K_U}{\overset{K_U}{\leftrightarrows}} \mathrm{V} \underset{K_V}{\overset{K_V}{\leftrightarrows}} , \mathrm{W} \tag{5.33}$$

where $n_U = n_{U0}$ is set, without considering further sources or sinks. Then the following rate equations result:

$$\frac{dn_U}{dt} = -K_U n_U - K_{-U} n_V = 0 \tag{5.34}$$

$$\frac{dn_V}{dt} = K_U n_U - K_{-U} n_V - K_V n_V + K_{-V} n_W$$
$$= 0 \tag{5.35}$$
$$\frac{dn_W}{dt} = K_V n_V - K_{-V} n_W = 0\,. \tag{5.36}$$

With the associated solutions

$$n_V = \frac{K_U}{K_{-U}} n_{U0} = \bar{n}_V$$
$$n_W = \frac{K_V}{K_{-V}} n_V = \bar{n}_W\,, \tag{5.37}$$

where the solutions (Eqs. (5.37)) are valid for thermal equilibrium. However, if for the reactions (Eq. (5.33)) we consider a source term (G) and a loss term $(-K_W n_W)$ of first order, then we obtain

$$\frac{n_V}{n_U} = \frac{K_U}{K_{-U} + \frac{K_V K_W}{K_{-V}+K_W}}$$
$$\frac{n_W}{n_V} = \frac{K_V}{K_{-V} + K_W}\,, \tag{5.38}$$

where

$$n_W = \frac{G}{K_W}\,. \tag{5.39}$$

If we now compare the results of Eq. (5.38) with those of Eq. (5.37), then we see that the conditions of the particle densities of Eq. (5.38) lie below the equilibrium conditions of Eq. (5.37), as soon as sources and sinks are present. This is valid for almost all low pressure plasma processes that are not in thermal equilibrium.

We consider the following reaction:

$$U + V \rightarrow UV \tag{5.40}$$

if U and V are two atoms; this reaction cannot take place, since energy and momentum conservation would be invalid. However, there exist numerous so-called ***attachment processes***, such as, for example,

$$SF_6 + e^- \rightarrow SF_6^-\,.$$

In order to understand Eq. (5.40) an intermediate step must be considered:

$$U + V \xrightarrow{K_2} UV^*\,. \tag{5.41}$$

This process leads to an excited, but instable molecule. If the excitation energy is not transferred (very fast) in a following reaction step to another particle, the molecule dissociates immediately:

$$UV^* \xrightarrow{K_{-1}} U + V\,. \tag{5.42}$$

A possible mechanism for immediate loss of energy is radiation emission, which then leads to the following complex reaction:

$$U + V \underset{K_{-1}}{\overset{K_2}{\rightleftarrows}} UV^* \xrightarrow{K_1} UV + h\nu\,. \tag{5.43}$$

The stationary rate equations for Eq. (5.43) are:

$$\frac{dn_U}{dt} = \frac{dn_V}{dt} = -K_2 n_U n_V + K_{-1} n_{UV^*} + G$$
$$= 0$$
$$\frac{dn_{UV^*}}{dt} = K_2 n_U n_V - K_{-1} n_{UV^*} - K_1 n_{UV^*}$$
$$= 0$$
$$\frac{dn_{UV}}{dt} = K_1 n_{UV^*} - K_{1W} n_{UV} = 0\,, \tag{5.44}$$

where a source term G and a loss term $K_{1W} n_{UV}$ were used (in order to obtain a stationary system). Since the reaction Eq. (5.43) corresponds to Eq. (5.33), if K_{-V} is set to zero, the solution from Eq. (5.38) above can be used:

$$n_{UV^*} = \frac{K_2}{K_{-1} + K_1} n_U n_V\,. \tag{5.45}$$

The associated rate R *of the production* of n_{UV} is thus:

$$R = K_1 n_{UV^*} = \frac{K_1 K_2}{K_{-1} + K_1} n_U n_V\,. \tag{5.46}$$

Considering the reaction (Eq. (5.40): U + V → UV) as an elementary reaction (which it is not), we can write the associated rate constant as:

$$K_2' = \frac{K_1 K_2}{K_{-1} + K_1}\,. \tag{5.47}$$

For the estimation of this rate constant we use a typical dissociation time constant for an instable molecule UV* to $10^{-13} \leq UV^* \leq 10^{-12}$ s.

Thus we have $10^{12} \leq K^* \leq 10^{13}\,\mathrm{s}^{-1}$. Typical transition probabilities for spontaneous emission (A-values, Einstein coefficients) are of the order of magnitude $10^8 \leq K_1 \leq 10^9\,\mathrm{s}^{-1}$.

If we apply this in (5.47), we obtain:

$$K_2' \approx 10^{-5} - 10^{-3} K_2\,. \tag{5.48}$$

5.2.3 Three-Body Accretion Collisions

We consider the following type of collision, at which the species U and V react with one another and need a third constant collision partner M, in order to obtain the impulse and conservation of energy:

$$\mathrm{U} + \mathrm{V} + \mathrm{M} \rightarrow \mathrm{UV} + \mathrm{M}\,. \tag{5.49}$$

M can be any heavy particle (atom, molecule) in the system. From this equation (Eq. (5.48)) we recognize that the rate constant for two-body accretion collisions with radiation emission is small. For this reason, this process is unimportant in low pressure plasmas.

Since three-particle collisions are very rare in low pressure plasmas, the probability of a complex (i.e. not elementary) reaction is large:

$$\mathrm{U} + \mathrm{V} \underset{K_{-1}}{\overset{K_2}{\rightleftarrows}} \mathrm{UV}^*\,. \tag{5.50}$$

The particles V and U form a molecule, which is in an excited state. This then collides with the particle M, which is not changed:

$$\mathrm{UV}^* + \mathrm{M} \xrightarrow{K_{2\mathrm{M}}} \mathrm{UV} + \mathrm{M}\,. \tag{5.51}$$

In addition, the rate equations with sources and sinks are:

$$\frac{\mathrm{d}n_\mathrm{U}}{\mathrm{d}t} = -K_2 n_\mathrm{U} n_\mathrm{V} + K_{-1} n_{\mathrm{UV}^*} + G = 0\,,$$

as well as

$$\frac{\mathrm{d}n_{\mathrm{UV}^*}}{\mathrm{d}t} = K_2 n_\mathrm{U} n_\mathrm{V} - K_{-1} n_{\mathrm{UV}^*} - K_{2\mathrm{M}} n_{\mathrm{UV}^*} n_\mathrm{M} = 0$$

$$\frac{\mathrm{d}n_\mathrm{UV}}{\mathrm{d}t} = K_{2\mathrm{M}} n_{\mathrm{UV}^*} n_\mathrm{M} - K_{1\mathrm{W}} n_\mathrm{UV}\,. \tag{5.52}$$

In addition, the solutions are (with Eqs. (5.32) and/or (5.39)):

$$G = K_{1\mathrm{W}} n_\mathrm{UV} = K_{2\mathrm{M}} n_{\mathrm{UV}^*} n_\mathrm{M} \tag{5.53}$$

and

$$n_{\mathrm{UV}^*} = \frac{K_2}{K_{-1} + K_{2\mathrm{M}} n_\mathrm{M}} n_\mathrm{U} n_\mathrm{V}\,.$$

Thus with Eq. (5.53) we obtain the reaction rate of the production n_UV:

$$R = K_{2\mathrm{M}} n_{\mathrm{UV}^*} n_\mathrm{M} = \frac{K_2 K_{2\mathrm{M}} n_\mathrm{M}}{K_{-1} + K_{2\mathrm{M}} n_\mathrm{M}} n_\mathrm{U} n_\mathrm{V}\,. \tag{5.54}$$

From Eq. (5.54) we can see that the dependence on the collision partner M is complicated.

For low pressure ($K_{2\mathrm{M}} n_\mathrm{M} \ll K_{-1}$) we obtain the approximation:

$$R \approx \frac{K_2 K_{2\mathrm{M}}}{K_{-1}} n_\mathrm{U} n_\mathrm{V} n_\mathrm{M} \tag{5.55a}$$

and for high pressure ($K_{2\mathrm{M}} n_\mathrm{M} \gg K_{-1}$):

$$R \approx K_2 n_\mathrm{U} n_\mathrm{V}\,. \tag{5.55b}$$

For this reason, the reaction equation (Eq. (5.51)) looks like an elementary reaction with low pressure:

$$\mathrm{U} + \mathrm{V} + \mathrm{M} \xrightarrow{K_3'} \mathrm{UV} + \mathrm{M}\,, \tag{5.56}$$

with the rate constant

$$K_3' = \frac{K_2 K_{2\mathrm{M}}}{K_{-1}}\,. \tag{5.57}$$

If we consider Eqs. (5.50) and (5.51) as an elementary reaction:

$$\mathrm{U} + \mathrm{V} \xrightarrow{K_2'} \mathrm{UV}\,, \tag{5.58}$$

then in analogy to Eq. (5.47) and with $K_1 = n_\mathrm{M} K_{2\mathrm{M}}$ we obtain for the rate constant:

$$K_2' = \frac{K_2 K_{2\mathrm{M}} n_{2\mathrm{M}}}{K_{-1} + K_{2\mathrm{M}} n_{2\mathrm{M}}}\,. \tag{5.59}$$

Under the assumption of low pressure ($K_{-1} \gg K_{2M}n_M$) we obtain:

$$K_2' \approx K_2 \frac{K_{2M}n_M}{K_{-1}} . \quad (5.60)$$

With rate constants of $\approx 10^{-11} \le K_{2M} \le 10^{-10}\,\mathrm{cm^3\,s^{-1}}$ and $\approx 10^{12} \le K_{-1} \le 10^{13}\,\mathrm{s^{-1}}$ we obtain:

$$K_2' \approx K_2(10^{-24}\text{–}10^{-22})n_M . \quad (5.61)$$

If we set $p = 0.1\,\mathrm{mbar}$, $T = 300\,\mathrm{K}$, then we obtain:

$$n_M = 2.42 \times 10^{15}\,\mathrm{cm^{-3}}$$

and thus:

$$K_2' \approx 2.42 \times (10^9\text{–}10^7)K_2 . \quad (5.62)$$

For this reason, three-body accretion is weak, including neutral particle (M) in low pressure discharges. In high pressure discharges, however, with nM growth by more than four orders of magnitude, such reactions can become important, like, e.g. ozone production with pressure of more than 1 bar:

$$O_2 + O_2 + O \rightarrow O_3 + O_2 .$$

In this case, an oxygen molecule serves as the collision partner M.

5.2.4 Three-Body Positive–Negative Ion Recombination

Reaction between positive and negative charged ions of the species:

$$U^+ + V^- + M \rightarrow UV + M$$

can achieve large rate constants. It can be shown that these can play a role with pressures larger than 1 mbar. We are interested only in processes where the pressure range is clearly smaller than 1 mbar. A special case for this reaction class is the ***3-K electron ion recombination***:

$$e + U^+ + e \rightarrow U + e .$$

This reaction is not relevant for $T_e > 1\,\mathrm{eV}$ and $n_e < 10^{13}\,\mathrm{cm^{-3}}$.

5.3 Surface Processes

Etching of silicon (Si) treated with CF_4 is based on the fact that in the gaseous phase plasma F atoms are produced, which are transported to the surface (wafer). There they fluoridate the surface (i.e. they are involved with the solid body chemical reactions), as follows:

$$\mathrm{F\,(gas)} + \mathrm{Si} : \mathrm{F}_x\,\mathrm{(surface)} \Leftrightarrow \mathrm{SiF}_{x+1}\,\mathrm{(s)}$$
$$x = 0, 1, 2, 3 .$$

Essentially two species from the fluoridated surface diffuse back into the gaseous phase (because they are volatile) and are then evacuated as so-called corroding products [5]:

$$\mathrm{Si} : \mathrm{F}_x\,\mathrm{(s)} \Leftrightarrow \mathrm{SiF}_x\,\mathrm{(g)} \quad x = 2, 4 .$$

Further adsorption and desorption reactions of the type

$$\mathrm{F(g)} + \mathrm{Surface} \Leftrightarrow \mathrm{F} : \mathrm{Surface}$$

play an important role, whereby they affect the concentrations of the species in the gaseous phase. The discharge is affected by surface processes such as, for example, the neutralization of ions or the emission by secondary electrons.

In these plasmas the sets of equations of the gaseous phase and surface reactions are coupled. In the case of low pressure plasmas, this coupling is strong. For this reason, the most important surface reactions are treated in the following.

5.3.1 Neutralization of Positive Ions and Emission of Secondary Electrons

The exothermic neutralization reaction:

$$e + U^+ \rightarrow U$$

is forbidden in the gaseous phase, since momentum and conservation of energy are not ensured. However, at the surface (reactor wall, substrate, etc.) this reaction is very fast:

$$e + U^+ + \mathrm{Surface} \rightarrow U + S .$$

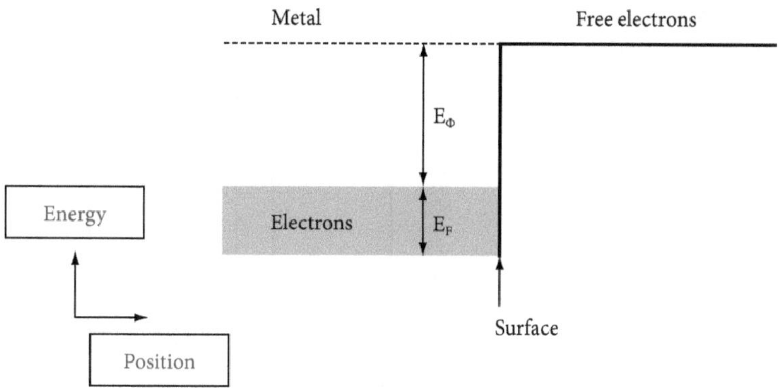

Fig. 5.4 Energy of the charge carriers as a function of position in the metal

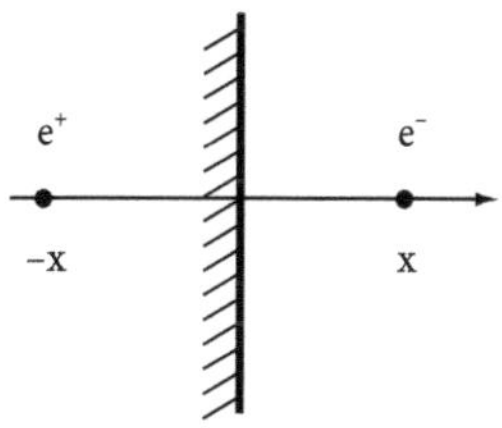

Fig. 5.5 Method of mirror images

"S" here and subsequently means "surface". For typical ion energies of 10–1000 eV all positive ions that strike the surface are neutralized immediately. In order for this to be understood fully, the inclusion of the electrons in the solid body will be discussed briefly.

In Fig. 5.4 the energy of the charge carriers is shown as function of the place in the metal. The bound electrons in the metal occupy energy levels lying very closely together in the so-called ***conduction band***. This is occupied up to an energy maximum, the so-called ***Fermi level***. The Fermi energy designates the energy difference E_F of the lower edge of the conduction band up to the Fermi level. This again lies around E_Φ below the potential energy $\Phi = 0$ for free electrons, where E_Φ is called the electron affinity is called. Thus the electrons are enclosed in the solid body by a potential barrier on a level with E_Φ.

E_Φ can be determined by a simple classical consideration of metals. We assume that eE_Φ is the work that can be performed, in order for an electron, which has a distance $x = a_{\mathrm{eff}}$ from a surface in the infinite, to transport (for $x < a_{\mathrm{eff}}$ *the* electron is still in the solid body). Here a_{eff} is the size of the atomic radius for any atom. Figure 5.5 shows that the force affecting the electron can be described by the method of mirror charges.

With the Coulomb law we obtain:

$$F_x = -\frac{e^2}{4\pi\varepsilon_0(2x)^2}\,. \tag{5.63}$$

The work performed by motion of the electron from a_{eff} to ∞ becomes:

$$eE_\Phi = -\int_{a_{\mathrm{eff}}}^{\infty} F_x \mathrm{d}x \tag{5.64}$$

and with it

$$E_\Phi = \frac{e}{16\pi\varepsilon_0 a_{\mathrm{eff}}}\,. \tag{5.65}$$

With $a_{\mathrm{eff}} = a_0$ equal the Bohr radius (0.53 Å = 0.053 nm) we obtain $E_\Phi \approx 6.8$ V, where the electron affinity for most materials is 4–6 V (it is lower for alkali and alkaline earth metals).

The atomic radius a_{eff} scales on average with:

$$a_{\mathrm{eff}} \approx a_0\sqrt{\frac{13.61\mathrm{V}}{E_\mathrm{i}}} \propto \frac{1}{\sqrt{E_\mathrm{i}}}\,. \tag{5.66}$$

Thus, it is valid for E_Φ:

$$E_\Phi \propto \sqrt{E_\mathrm{i}}\,, \tag{5.67}$$

where E_i is the ionization potential of the respective atom. In Fig. 5.6 the potential development

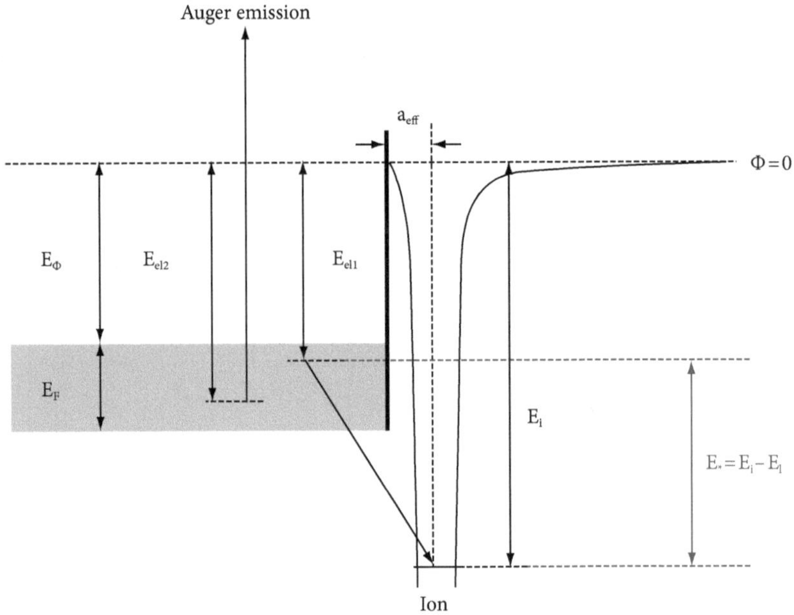

Fig. 5.6 Potential profile of an ion at a distance a_{eff} in front of a solid surface

of a positive ion is shown in the distance a_{eff} in front of a solid surface. This ion generates a deep potential well, which is separated from the surface by a narrow potential barrier for the electrons. The width of the barrier is approximately close to a_{eff}.

For neutralization two possibilities exit, whereby the first consists of the fact that the electron occupies an excited state:

$$e + \mathrm{U}^{+} + \mathrm{S} \rightarrow \mathrm{U}^{*} + \mathrm{S}\,, \tag{5.68}$$

with the excitation energy

$$eE^{*} \approx e(E_{\mathrm{i}} - E_{\mathrm{el}}) = h\nu\,. \tag{5.69}$$

If the excited state is not metastable, this becomes decomposed by radiation emission in the initial state (or a metastable state). This way, the neutralization of positive ions at a solid surface can produce metastable atoms or also recombination radiation. The second possibility for neutralization is based on the fact that the (first) electron drops back to the ground state of the atom and that a second electron from the conduction band of the solid body absorbs the energy, thereby becoming free. This is called ***Auger neutralization***. This is *non-radiating* and needs two electrons.

The (first) electron, which falls to the ground state, loses the energy,

$$\Delta eE = e(E_{\mathrm{i}} - E_{\mathrm{el1}})\,, \tag{5.70}$$

which the second electron in the solid body wins. If $\Delta eE < eE_{\mathrm{el2}}$, then this remains captured in the solid body. However, if $\Delta eE > eE_{\mathrm{el2}}$, then the electron from the solid body is released. One calls this procedure ***Auger emission*** or also ***secondary electron emission*** [6]. One frequently finds this in plasma technology, especially, e.g. if noble gas ions hit the surface of the recipient. Looking at Fig. 5.6, we can see that the condition for removing the second electron from the surface is most easily fulfilled, if both electrons are due to the upper edge of the conduction band, i.e.

$$E_{\mathrm{el1}} = E_{\mathrm{el2}} = E_{\Phi}\,. \tag{5.71}$$

The condition for emission then reads:

$$\Delta E = E_{\mathrm{i}} - E_{\Phi} \geq E_{\Phi} \quad \text{or} \quad E_{\mathrm{i}} \geq 2E_{\Phi}\,. \tag{5.72}$$

The maximally possible kinetic energy of the emitted electron is thus:

$$eE_{\max} = e(E_{\mathrm{i}} - 2E_{\Phi})\,. \tag{5.73}$$

Table 5.1 Sputtering yield as a function of the work of emission, the ion bombardment and the nature of the ions

Solid state body	Work of emission (eV)	Ion	Energy (eV)	γ_{se}
Si (100)	4.9	He^+	100	0.168
		Ar^+	10	0.024
			100	0.027
Ni (111)	4.5	He^+	100	0.170
		Ar^+	10	0.034
			100	0.036
Mo	4.3	He^+	100	0.274
		Ar^+	100	0.115
		N_2^+	100	0.032
W	4.54	He^+	100	0.263
		Ar^+	10	0.096
			100	0.095
		H_2^+	100	0.029
		N_2^+	100	0.025
		O_2^+	100	0.015

If both electrons of the lower end of the conduction band come, then

$$E_{el1} = E_{el2} = E_\Phi + E_F \tag{5.74}$$

is valid, and with it

$$\Delta E = E_i - (E_\Phi + E_F) \geq E_\Phi + E_F . \tag{5.75}$$

This way, one receives the minimum kinetic energy of the emitted electron:

$$eE_{min} \geq e(E_i - 2E_\Phi - 2E_F) \tag{5.76}$$

and/or zero, if (5.75) becomes negative. Thus it is clarified that (Eq. (5.76)) that secondary emission preferentially arises with noble gases because for these E_i is large and E_Φ is small for alkali and alkaline-earth solids.

There the time needed by an electron for tunnelling through the surface barrier in Fig. 5.6 is short in comparison to the time needed for ion collision with the surface; this is almost independent of the secondary emission process of the ion energy (i.e. before the ion can transfer the energy to the surface, the secondary electron has already been emitted). Therefore the secondary emission depends only on the nature of the gas (E_i) and on the wall material (E_Φ). Although the neutralization of ions and the emission of secondary electrons was only described for metals (see above), a comparable process takes place also on the surface of semiconductors and insulators.

The emission of secondary electrons is described with the coefficient γ_{se}; this indicates the number of secondary electrons emitted per incident ion. A frequently used empirical formula is:

$$\gamma_{se} \approx 0.016 \times (E_i - 2E_\Phi) \tag{5.77}$$

under the condition that $E_i > 2E_\Phi$.

In particular, metastable neutral particles (e.g. Ar^* or He^*) produce secondary electrons very efficiently under the condition $E^* > E_\Phi$. Additionally, secondary electrons can also be produced by ion energies $eE_i > 1$ keV. Such high energies only appear, if at all, in capacitive coupled machines with a large area ratio. Equation (5.77) only gives a rough estimation of the secondary emission coefficient. This actually also depends on the morphology, impurities and surface allocations. Some measured values E_Φ and γ_{se} for ions hitting a clean surface are given in Table 5.1.

Independently of neutralization and secondary emission, heavy particles (ions, neutral ones) have a similar behaviour if they hit on surfaces.

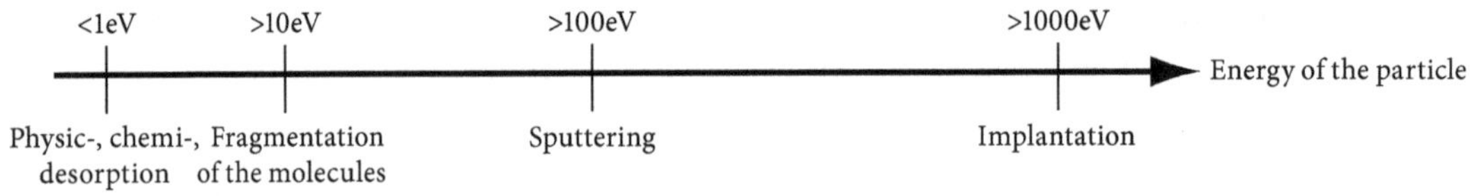

Fig. 5.7 Physical processes, depending on the energy of the bombarding ions

With low (= thermal) energies ($< 1\,\text{eV}$) physisorption, chemisorption and desorption can occur. With higher energies (some 10 eV) molecules are divided into fragments (= fragmentation). If the energies are within the range of some 100 eV, solid atoms are sputtered (= sputtering) at the surface and with energies of thousands of eV the implantation becomes important [7] (see Fig. 5.7).

5.3.2 Adsorption and Desorption

Both adsorption and desorption (removal of adsorbent particles) influence plasma-technical processes critically, since multiplying one or both is step limiting for surface processes. The adsorption is a reaction with the surface, which leads to a binding:

$$\mathrm{U} + \mathrm{S} \rightarrow \mathrm{U} : \mathrm{S}\,. \tag{5.78}$$

Desorption is the reaction in the opposite direction.

Adsorption is based on attraction between the arriving molecule (ion) and the surface. One differentiates two different types of adsorption: physisorption and chemisorption. Physisorption is based on van der Waals reciprocal effect between the molecule and the surface.

If we consider a Bohr atom in the environment of a surface, then this can be described in accordance with Fig. 5.8 as an oscillating dipole. The component (time-dependent) perpendicular to the surface of the dipole moment reads:

$$P_{\mathrm{d}x} \approx e a_0 \cos \omega_{\mathrm{at}} t\,, \tag{5.79}$$

where e is the elementary electric charge, a_0 the distance and ω_{at} the atomic angular frequency.

The force between two dipoles can be found with the method of mirror charges, where for the dipole with $+x$ a dipole moment $+p_{\mathrm{d}x}(t)$ is accepted and a rectified dipole $-x$ with the dipole moment $+p_{\mathrm{d}x}(t)$ exists at the same time t. Therefore, both dipoles attract each other (near field approximation) with the force:

$$F_x = -\frac{6\left\langle p_{\mathrm{d}x}^2(t)\right\rangle}{4\pi\varepsilon_0(2x)^4} \approx -\frac{3a_0^2e^2}{4\pi\varepsilon(2x)^4}\,. \tag{5.80}$$

The interaction potential after van der Waals also results from the force:

$$\begin{aligned}\left|\vec{F}\right| = q\left|\vec{E}\right| &= -e\frac{\mathrm{d}V}{\mathrm{d}x} \Rightarrow V(x)\\ &= -\frac{a_0^2 e}{64\pi\varepsilon_0 x^3}\,.\end{aligned} \tag{5.81}$$

However, if the distance between atom and surface becomes very small ($0.1 \le d \le 0.3\,\text{nm}$), a repulsive (Coulomb) force forms between the electron cloud of the atom and the solid electrons.

If the transition between attraction and repulsion is with approximately 0.1–0.3 nm, then we can measure the potential depth with (5.81):

$$\begin{aligned}V &\approx \frac{(0.53\times10^{-10})^2\times1.6\times10^{-19}}{64\pi\times8.85\times10^{-12}(1\times10^{-10})^3}\\ &\approx 0.26\,\text{V}\,.\end{aligned}$$

For a distance of 0.3 nm the potential amounts to about 0.01 V. With this fact it becomes clear that the energy necessary to remove a physisorbed particle from a surface within the range of 0.01–0.26 eV is appropriate. These energies are clearly smaller than typical ion energies at the surface. Since 0.01 eV corresponds to about 116.0 K (1 eV = 11,604.5 K; ambient temperature: 293 K), this is the sufficient ambient temperature in order for physisorbed particle to be replaced. Energies of 0.01–0.25 eV ensue in enthalpies $|\Delta H| \approx 1\text{–}25\,\text{kJ/mol}$.

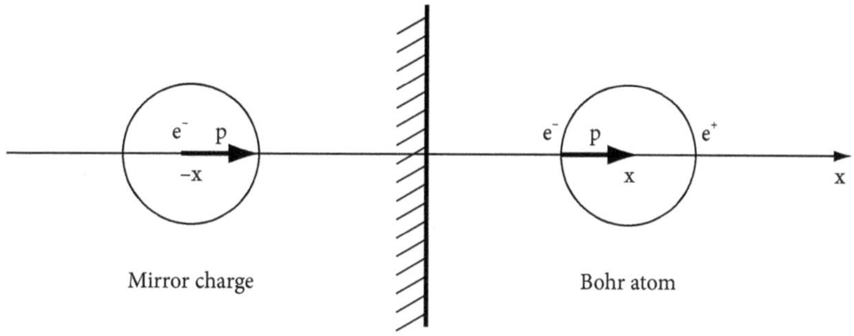

Fig. 5.8 Atom model of Bohr near a surface (oscillating dipole)

Physisorption reaction is exothermic, i.e. it becomes free with the physisorption of a mole particle species (6.022×10^{23} particles) at an energy of 1–25 kJ/mol. Contrary to physisorption, chemisorption is based on a chemical bonding between the respective particle and the surface [8].

Also the chemisorption reaction is exothermic, however, the enthalpy values are clearly larger, i.e. $|\Delta H| \approx$ 40–400 kJ/mol. This corresponds to a potential depth of $E_{\text{chem}} \approx$ 0.4–4 eV, whereby the minimum of the potential well with approximately 0.1–0.15 nm is located in front of the surface. If a molecule consisting of two atoms is bound over a double bond, then chemisorption can take place, as a bond arm breaks open and binds with the surface:

$$\text{U} = \text{V} + \text{S} \rightarrow \text{UV} : \text{S}\,. \tag{5.82}$$

Simply bound molecules are often dissociated, if they chemisorb at the surface:

$$\text{UV} + \text{S} \rightarrow \text{U} : \text{S} + \text{V} : \text{S}\,. \tag{5.83}$$

This process is called dissociative chemisorption and needs two free surface bond arms (dangling bonds).

Both nature of the adsorption exist often at the same time. If we consider the molecule U–V (simply bound), then first the physisorption potential well removed from the surface (smaller depth), in order to reach afterwards the chemisorption potential well which was more near to the surface (larger depth), Fig. 5.9.

The potential develops with a positive upward gradient called ***attraction, negative repulsion***. If the value of the potential is negative,

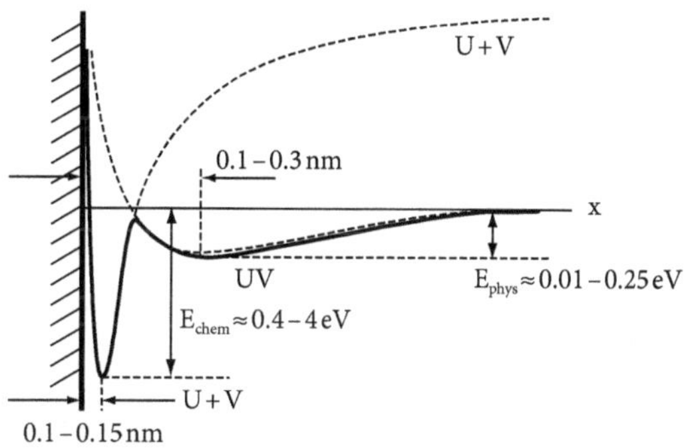

Fig. 5.9 Energy curve for physisorption and chemisorption

the molecule binding the molecule (atom) at the surface takes on a positive value, the molecule is free. With dissociative chemisorption the molecule is UV and/or is always bound to the two atoms U, V, i.e. the total developing potential (thick line, Fig. 5.9) is always negative.

In the case of the Fig. 5.10 there results a barrier of the height of E_{ads} in the developing potential, which prevents the molecule firstly from chemisorption. An incident molecule UV is thus physisorbed, in order to later change with not too high a barrier into the dissociative chemisorbed condition. In both cases (Figs. 5.8 and 5.10), the potential processes of UV and U + V have an intersection, i.e. the original molecule UV is chemisorbed (Fig. 5.9), in order to arrive at an energetically more favourable state. This is also valid, Fig. 5.10, if the barrier E_{ads} is overcome (e.g. by energy entry from the outside). These two cases correspond in principle to Eq. (5.83). The reaction in Eq. (5.82) is shown in Fig. 5.11.

Particles on a surface impact can only be adsorbed, if they lose energy with the collision. The normal component of the loss of energy must be

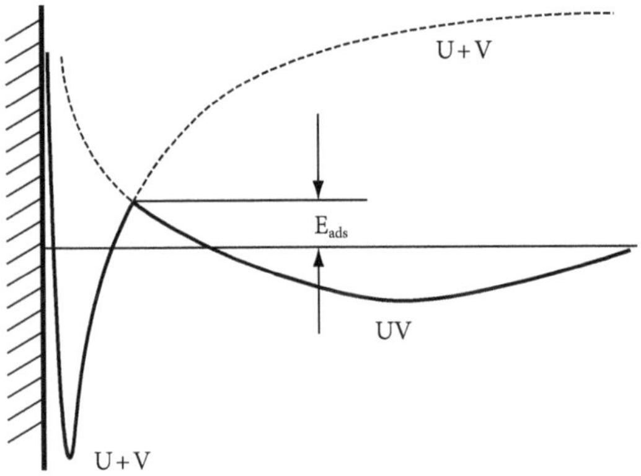

Fig. 5.10 Potential curve for the chemisorption

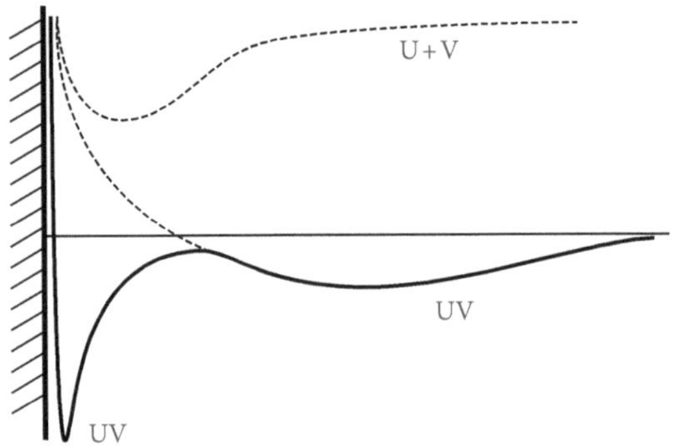

Fig. 5.11 Response curve for the chemisorption

suitable, in order to hold the particle in the adsorption pot. That is, the energy must be larger than any potential barrier and may not to be too large, so that other processes (than adsorption) do not take place. If particle species U is in the plasma, then this has a flow density at the surface j_U, for which

$$j_U = \frac{1}{4} n_U \bar{v}_U \,, \tag{5.84}$$

with the mean velocity v_U and the particle density n_U (Eq, (5.84)), is valid. This concerns a molecular chemisorption (developing UV). We do not consider the energetically more unfavourable state of dissociated chemisorption (developing U + V) here.

For the flow density (often only called flux) of the particles that are adsorbed, we have:

$$j_{ads} = s j_U = \frac{1}{4} s \bar{v}_U n_U \,, \tag{5.85}$$

with the so-called ***sticking coefficients***. These are in the general case a function of the surface coverage θ, as well as the gas and surface temperatures. The surface coverage θ marks the portion of the surface covered with adsorbate. That is, $\theta = 1$ means that the entire surface is covered and $\theta = 0$ that the surface is free from adsorbate.

A generally used approximation (in accordance with Langmuir kinetics) of the sticking coefficient for non-dissociative adsorption is:

$$s(\Theta, T) = s_0(T)(1 - \Theta) \,, \tag{5.86}$$

where s_0 is the sticking coefficient with $\theta = 0$ and $1 - \theta$ describes the portion of the surface that is not covered with adsorbate.

Generally it is valid that chemisorption stops if at a surface all active (i.e. free) surface bonds are occupied; in the case of the deposition process is means that new dangling bonds are free etched. This corresponds to approximately a monolayer of the coating. The subsequent adsorption then takes place only on the basis of the physisorption, which exhibits substantially weaker bonds (see above). In this way many mono-situations can have physisorption, i.e. the continuous adsorption and/or condensation of adsorbates can take place. For non-active surfaces (e.g. reactor walls, flanges, windows, ...) it is usually valid that physisorption and desorption are in equilibrium, i.e. the particle net flux to these surfaces is zero.

The temperature variation of s_0 (sticking with degree of coverage zero) depends on whether the chemisorption possesses an energy barrier (Fig. 5.12) or not (Fig. 5.13).

If no barrier is present (Fig. 5.9), then $s_0 \approx 1$ is obtained for low temperatures. Against it s_0 decreases if T increases, since the parts of particles and/or molecules that lose a suitable energy quantity for chemisorption decrease. However, if a barrier on the level of E_{ads} is present (Fig. 5.10), then the sticking coefficient is small; s_0 has an ***Arrhenius form***:

$$s_0 = s_{00}(T) \exp(-E_{ads}/kT) \,, \tag{5.87}$$

where s_{00} decreases with rising T, as this applies to the case without barrier. Equation (5.87)

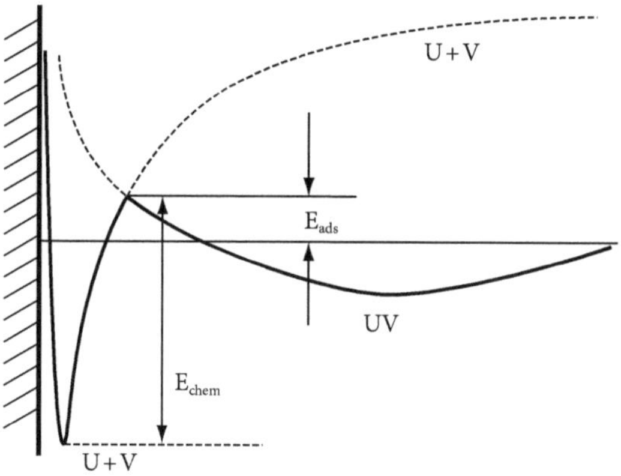

Fig. 5.12 Sticking coefficient > 0

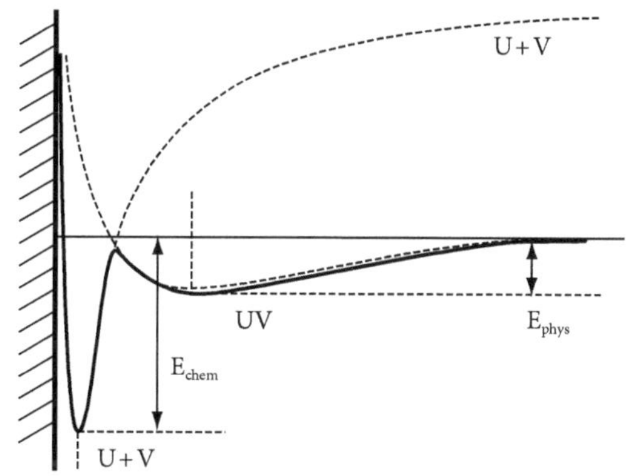

Fig. 5.13 Sticking coefficient = 0

shows that with increasing T the exp-term goes towards 1 and with decreasing T towards zero.

The measured sticking coefficients vary at 300 K (= 26 meV) by $10^{-6} - 1$ and depend strongly on crystal orientation and surface roughness, whereby s_0 likewise increases with increasing roughness. With many surfaces the active bonding places, which cause the sticking coefficient, at surface defects are stages, free spaces, dislocations, etc. Chemically reactive gases, in particular radicals, do not exhibit sticking within the range of $0.1 \leq s_0 \leq 1$ at metals such as Fe, Ni, ... Often, sticking coefficients for other surfaces are smaller. For example, $s_0 \approx 1$ for H on Si, but $s_0 \approx 0.01$ for H_2 on Si, or $s_0 \approx 10^{-4}$–10^{-3} for CO_2 on Si.

Desorption is the process moving in directions opposite to adsorption:

$$\mathrm{U} : \mathrm{S} \rightarrow \mathrm{U} + \mathrm{S} . \tag{5.88}$$

Both processes balance themselves in thermal equilibrium. One can show that the desorption rate (first-order) also possesses an Arrhenius form:

$$K_{\mathrm{des}} = K_0 \exp(-eE_{\mathrm{des}}/kT) , \tag{5.89}$$

where E_{des} designates the depth of the barrier (E_{chem} or E_{phys}).

A theoretical estimation gives:

$$K_0 \approx 6 \times 10^{12} \bar{s}(T) \frac{\bar{g}_{\mathrm{esc}}}{\bar{g}_{\mathrm{ads}}} \quad \left[\mathrm{s}^{-1}\right] , \tag{5.90}$$

with the statistical weights for "escaping" and "-caught" molecules.

In accordance with Eq. (5.87) the thermal sticking coefficient also has an Arrhenius form. In the case of physisorption it is $10^{14} \leq K_0 \leq 10^{16}\,\mathrm{s}^{-1}$ and for chemisorption $10^{13} \leq K_0 \leq 10^{15}\,\mathrm{s}^{-1}$.

5.3.3 Fragmentation

Impacting ionized or neutral molecules with sufficient high energy on the surface can break into atoms, which are adsorbed or reflected at the surface. The threshold energy of the fragmentation is in the range of the binding energy of the molecules. With collision energies, which correspond four to five times to the binding energy, more than half of the molecules are fragmented. There molecular bonds lie between 1 and 10 eV (Table 5.2) and ion bombardment energies on surfaces are often clearly higher than the associated binding energy (in particular on the smaller electrode of capacitive driven HF-discharges), very often molecular ions fragment as soon as they impact a surface.

The same is valid also for neutral particles, which result from charge reversal collisions close to the surface, where their energy likewise can exceed the typical binding energy (Table 5.2).

In Fig. 5.14 the associated wavelengths of the light emitted in the plasma, which correspond to the respective binding energies, are given. From this it follows that the wavelengths associated to

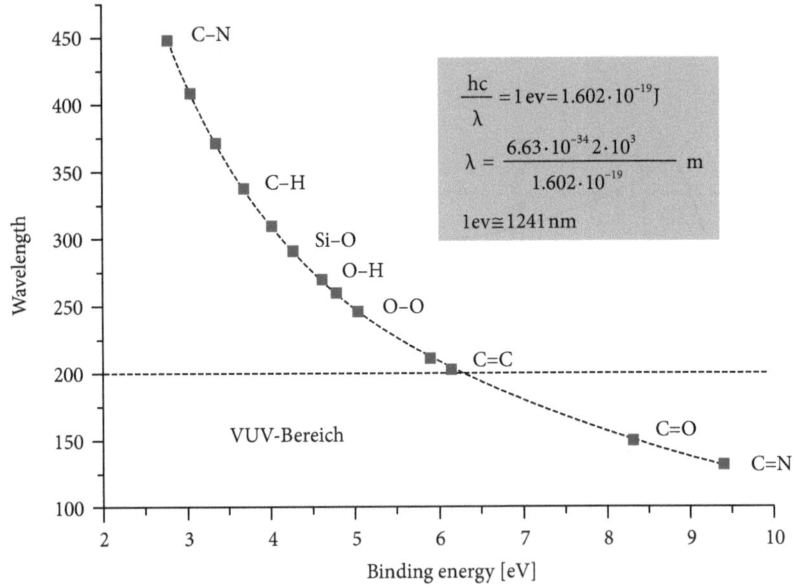

Fig. 5.14 Wavelength generated by reloading collisions in the plasma emitted photons

Table 5.2 Binding energies close to the wall by reloading collisions

Binding	Energy [eV]
C=N	9.41
C=O	8.31
C=C	6.14
H–F	5.90
O–O	5.14
C–F	5.04
O–H	4.78
Si–O	4.61
H–Cl	4.46
C–H	4.27
N–H	4.02
C–C	3.69
C–Cl	3.36
C–O	3.34
Si–O	3.05
C–N	2.78

the two largest binding energies lie in the vacuum UV (VUV, $\lambda \leq 200\,\text{nm}$). Since technical plasmas wavelengths can radiate within the range under 200 nm (as a function of the used gas mixture), it is clear that also the light emission of the plasma in the UV and VUV spectral regions can produce free surface bonds.

References

1. Schlichting H (1990) Methoden und Mechanismen der thermischen Desorption: Adsorptions-, Desorptions-Kinetik, Epitaxie und Ordnung von Edelgasschichten auf Ru(001). Im Fachbereich Physik der Technischen Universität München eingereichte Dissertation. München
2. Christmann K (1991) Introduction to Surface Physical Chemistry. Springer, New York
3. Demtröder W (2000) Atome, Moleküle und Festkörper Experimentalphysik, vol. 3. Springer, Berlin, Heidelberg, New York, Barcelona, Hongkong, London, Mailand, Paris, Singapur, Tokio
4. Henzler M (1994) Oberflächenphysik des Festkörpers. Teubner, Stuttgart
5. Bean FE (1978) Anisotropic etching of silicon, IEEE Trans. Electron.Devices ED-25, No. 10, pp. 1185–1193, Okt. 1978
6. McCracken RJ (1975) The behaviour of surfaces under ion bombardement. RepProgPhys 38(2):241–327
7. Behrisch R, Eckstein W (eds) (2007) Sputtering by Particle Bombardment: Experiments and Computer Calculations from Threshold to MeV Energies. Springer, Berlin
8. Duke CB, Park RL (1972) Surface structure – An emerging spectroscopy. Phys Today 25(8):23–28

6 Cathode Sputtering

H. Frey

6.1 Introduction

Cathode sputtering [1, 2] is more frequently used in thin film technology than evaporation. The reasons for this are the following:

- High melting material or dielectrics with a high frequency generator/source can be sputtered from a solid target.
- Sputtering is a ballistic process in which the target remains relatively cold. The composition of the released particle flux corresponds to the stoichiometry of the sputtering target.
- Sputtering with a reactive gas or a mixture of gases generates films of chemical compounds with a defined stoichiometry from elementary target material.
- The sputtering process can be used to deposit large areas with very high lateral homogeneity.
- By ion bombardment, the substrates can be cleaned before coating, and the properties of the sputtered film can be influenced, e.g. adhesion, strength, structure, etc.

The sputtering process is represented in Figs. 6.1–6.3.

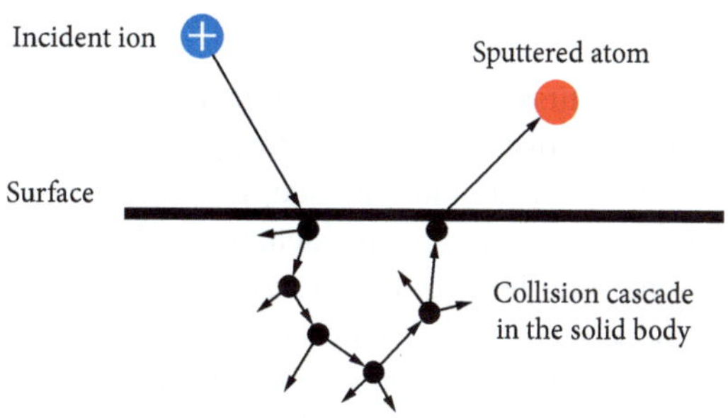

Fig. 6.1 The sputtering principle

6.2 Sputtering

At energies of approximately 30 eV and more, particles knock out atoms from a solid surface [3]. Generally, particles with sufficient energy are ions. The number of atoms N per unit area per second leaving the target is given by:

$$N = J^+/ge\,Y(E_i, A, B)\,, \qquad (6.1)$$

where J^+ is the current density of the bombarding ions, g is the number of electronic charges per ion, and Y is the sputter yield in atoms per incident ion, which is a function of the ion energy E_i, the ion species A, and the target material B.

The sputtering principle is shown in Fig. 6.1.

Within this collision cascade, by the energy transfer of the bombarding particle, secondary particles of the energy E_{sec} in a solid angle $d\Omega_{sec}$ are created (see Fig. 6.2).

Thus the number of atoms impacted in the solid body (recoil), whose energy is $E_{sec} \geq E_0$, can be described by

$$N(E, E_0) \approx \Gamma \frac{E}{E_0}\,, \qquad (6.2)$$

where Γ is a function of the interaction potential between the incident ion and the target atom

© Springer-Verlag Berlin Heidelberg 2015
H. Frey, H. R. Khan, *Handbook of Thin-Film Technology*, DOI 10.1007/978-3-642-05430-3_6

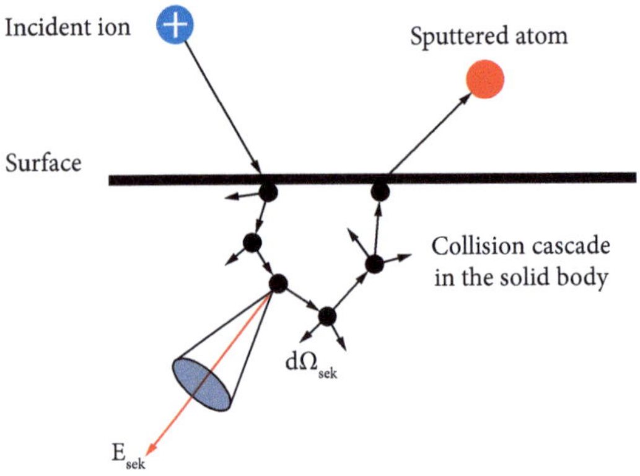

Fig. 6.2 Energy transfer

(for the first collision), in general, between all further collisions between projectile and target (ion/atom) within the cascade, E is the energy of the 1st collision of the incident ion with the atom in the solid body or the energy of the impacting solid atom (projectile) for all further collisions.

Equation (6.2) is only valid if the energy E of the incident recoils multiplied by the kinematic factor (i.e. the maximum portion of energy transfer possible) is larger than E_0,

$$\gamma_{\mathrm{kin}} E \gg E_0 \quad \text{with} \quad \gamma_{\mathrm{kin}} = \frac{4 m_1 m_2}{(m_1 + m_2)^2} . \tag{6.3}$$

The number of collisions with energies between E_0 and $E_0 + \mathrm{d}E_0$ is then:

$$\frac{\mathrm{d}N(E, E_0)}{dE_0} \approx \Gamma \frac{E}{E_0^2} = F(E, E_0) . \tag{6.4}$$

The number of particles that are impacted at the surface in a time $\mathrm{d}t_0$, resulting from the flux density of the projectiles (equal to the flux density of the incident ions for the first collision) multiplied by the number of recoils N and by time is described in Eq. (6.5)

$$G(E, E_0)\,\mathrm{d}E_0 = \Phi \times N(E, E_0)\,\mathrm{d}t_0 . \tag{6.5}$$

Thus $G(E, E_0)$ is the number of impacted particles with energies greater than E_0, since each of these particles passes through the interval between E_0 and $E_0 + \mathrm{d}E_0$ by energy release in the solid body.

Formally, we replace $\mathrm{d}t_0$ by:

$$\mathrm{d}t_0 = \frac{\mathrm{d}E_0}{|\frac{\mathrm{d}E_0}{\mathrm{d}t}|_{t=t_0}} = \frac{\mathrm{d}E_0}{\frac{\mathrm{d}x}{\mathrm{d}t}|\frac{\mathrm{d}_0}{\mathrm{d}x}|_{t=t_0}} = \frac{\mathrm{d}E_0}{v_0|\frac{\mathrm{d}E_0}{\mathrm{d}x}|} , \tag{6.6}$$

where $\mathrm{d}t_0$ is the (mean) time, in which a recoil cools down from $E_0 + \mathrm{d}E_0$ to E_0 and v_0 is the velocity related to E_0.

Thus, from Eq. (6.6) with Eq. (6.4) and Eq. (6.5), we obtain:

$$G(E, E_0)\,\mathrm{d}E_0 = \Phi\Gamma \frac{E}{E_0} \times \frac{\mathrm{d}E_0}{v_0|\frac{\mathrm{d}E_0}{\mathrm{d}x}|} . \tag{6.7}$$

Now calculating the flux density J at particles that are impacted toward to the substrate surface, this is given in accordance with Fig. 6.3 by projection of the individual velocity vectors from the solid angle $\mathrm{d}\Omega_0$ on the surface-normal by

$$\begin{aligned} J(E, E_0)\mathrm{d}E_0\,\mathrm{d}\Omega_0 &= G(E, E_0)\mathrm{d}E_0\,\mathrm{d}\Omega_0 v_0 \cos\Theta_0 \\ &= \Phi\Gamma \frac{E}{E_0} \frac{\mathrm{d}E_0}{v_0|\frac{\mathrm{d}E_0}{\mathrm{d}x}|} v_0 \cos\Theta_0\,\mathrm{d}\Omega_0 . \end{aligned} \tag{6.8}$$

An emission of a particle from the surface, resulting from the last impact, is contrary to a barrier from a perpendicular velocity component that the level E_{SB} (surface binding) opposes. The energy component thereby becomes

$$m \frac{(v_0 \cos\Theta_0)^2}{2} = E_0 \cos^2\Theta_0 .$$

For the energy in the perpendicular and parallel directions to the surface we obtain:

$$\text{vertical} \qquad E_0 \cos^2\Theta_0 - E_{\mathrm{SB}} = E_1 \cos^2\Theta_1 \tag{6.9}$$

$$\text{parallel} \qquad E_0 \sin^2\Theta_0 = E_1 \sin^2\Theta_1 . \tag{6.10}$$

According to the (energy) flux conservation at the surface, we obtain:

$$E_0 \cos\Theta_0 \mathrm{d}\Omega_0 = E_1 \cos\Theta_1 \mathrm{d}\Omega_1 , \tag{6.11}$$

transformed to

$$\cos\Theta_0 \mathrm{d}\Omega_0 = \frac{E_1}{E_0} \cos\Theta_1 \mathrm{d}\Omega_1 .$$

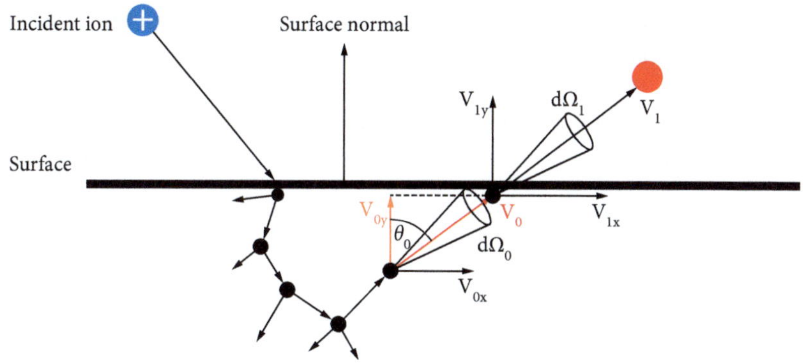

Fig. 6.3 Contribution to the flux j in the direction of the surface

Using Eq. (6.8), we obtain the flux yield as a relationship of the sputtered current density J to the irradiated flux density θ

$$Y = \frac{J}{\Phi} = \Gamma \frac{E}{E_0} \frac{\mathrm{d}E_1}{|\frac{\mathrm{d}E_0}{\mathrm{d}x}|} \frac{E_1}{E_0} \cos \Theta_1 \mathrm{d}\Omega_1 \,. \quad (6.12)$$

Adding the left-hand and right-hand sides of the Eqs. (6.9) and (6.10), we obtain:

$$E_0 - E_{\mathrm{SB}} = E_1 \Rightarrow E_0 = E_1 + E_{\mathrm{SB}}\,,$$

which applied to (6.12) gives

$$\frac{\mathrm{d}Y}{\mathrm{d}E_1 \mathrm{d}\Omega_1} = \Gamma \frac{E}{|\frac{\mathrm{d}E_0}{\mathrm{d}x}|} \frac{E_1}{(E_1 + E_{\mathrm{SB}})^2} \cos \Theta_1 \,, \quad (6.13)$$

where $|\mathrm{d}E_0/\mathrm{d}x|$ instead of $E_0 = E_1 + E_{\mathrm{SB}}$ is determined.

Equation (6.13) indicates the number of sputtered particles scattered in the solid angle $\mathrm{d}\Omega_1$ with energy E_1 in relation to the impacted particle flux θ. This is the so-called ***Thompson distribution***. The ***Thompson distribution*** has a maximum for plane surfaces:

$$E_1 = E_{1\,\max} = \frac{E_{\mathrm{SB}}}{2(l - m)}\,, \quad (6.14)$$

with $m \approx 0.2$–0.3 where $E_1 \approx 1\,\mathrm{keV}$ or $m = 0$ for $E_1 = E_{\mathrm{SB}}$. Furthermore, Eq. (6.13) shows that the maximum power in the spectrum of sputtered atoms depends only on the target, not on a specific ion or its energy. Thus, this precondition is the fundamental approach. For the determination of the sputter yield as a function of angle and energy, a Monte Carlo code [5, 6] should be used (TRIM code [3]).

The sputter yield corresponds approximately with:

$$Y = \gamma_{\mathrm{sput}} \propto \frac{1}{E_{\mathrm{SB}}} \frac{m_{\mathrm{ion}}}{m_{\mathrm{SB}} + m_{\mathrm{target}}}\,, \quad (6.15)$$

where E_{SB} in this equation can be equalized to the evaporation enthalpy.

As an approximation the ***Bohdansky equation*** is used, were the sputter yield is a function of the ion energy E_{ion} of the ions impacting the surface at the incident angle α ($\alpha = 0°$: perpendicular incidence), and we can write

$$Y(E_{\mathrm{ion}}, \alpha = 0°) = Q s_{\mathrm{a}}(\varepsilon) \left(1 - \left(\frac{E_{\mathrm{th}}}{E_{\mathrm{ion}}}\right)^{2/3}\right) \left(1 - \frac{E_{\mathrm{th}}}{E_{\mathrm{ion}}}\right), \quad (6.16)$$

with the effective cross-section (for stopping):

$$s_{\mathrm{n}}(\varepsilon) = \frac{3.441\sqrt{\varepsilon}\ln(\varepsilon + 2.718)}{1 + 6.355\sqrt{\varepsilon} + \varepsilon(6.882\sqrt{\varepsilon} - 1.708)}\,, \quad (6.17)$$

and the empirical parameter Q, attained from experimental sputtering data, then for the standardized energy ε, we obtain

$$\varepsilon = \frac{E_{\mathrm{ion}}}{E_{\mathrm{TF}}}\,, \quad (6.18)$$

with E_{TF},

$$E_{\mathrm{TF}} = \frac{Z_1 Z_2 e^2}{a_{\mathrm{L}}} \frac{m_1 + m_2}{m_2}$$

where E_{ion} is the energy of the incident ion, $Z_{1,2}$ the nuclear charge value of the projectile and the target atom, and the mass of both, $m_{1,2}$. The quantity a_L in (6.18) designates an effective Bohr radius with:

$$a_L = \left(\frac{9\pi^2}{128}\right) a_{bohr} \left(Z_1^{2/3} + Z_2^{2/3}\right)^{-0.5} = \frac{0.4685}{\sqrt{Z_1^{2/3} + Z_2^{2/3}}} \ [\text{Å}] , \quad (6.19)$$

where a_{bohr} is the Bohr radius of the hydrogen atom. The threshold energy E_{th} is differentiated according to the mass ratio of the projectile and the target atom.

For the projectile mass $m_1 < 0.2m_2$,

$$E_{th} = \frac{E_{SB}}{\gamma_{kin}(l - \gamma_{kin})} \quad (6.20)$$

is valid, with the „kinematic factor" and the surface bounding energy in accordance with γ_{kin} (6.3) and E_{SB} refers to (6.19a).

$$\gamma_{kin} = \frac{4m_1m_2}{(m_1 + m_2)^2} . \quad (6.19a)$$

For larger projectile masses $m_1 > 0.2m_2$

$$E_{th} = 8E_{SB}\left(\frac{m_1}{m_2}\right)^{0.4} \quad (6.21)$$

is valid. If the sticking coefficients of the different components are similar, then the deposited film and the target have the same compound. Under the assumption, that sputter atoms impact the substrate, the physical sputter deposition rate is then:

$$R_{deposition} = \frac{\gamma_{sput}\Gamma_{ion}}{n_{film}} \frac{A_{target}}{A_{substrat}} \quad [\text{cm/s}] , \quad (6.22)$$

where Γ_{ion} is the ion flux in $[1/\text{cm}^2\,\text{s}]$, γ_{sput} the sputter yield density of the deposited film n_f in $[1/\text{cm}^3]$, the target, and A_t and A_f the substrate surface.

By using, for example, 1 kV argon ions with $\gamma_{sput} = 0.83$ Al (Fig. 6.7), we obtain a film density of $n_f = 5 \times 10^{22}\,\text{cm}^{-3}$ and an ion current density of $1\,\text{mA/cm}^2$ (i.e. $\Gamma_{ion} = 6.25 \times 10^{15}\,\text{cm}^{-2}\,\text{s}^{-1}$), with $A_t/A_f = 1$:

$$R_{dep} = \frac{0.83 \times 6.25 \times 10^{15}}{5 \times 10^{22}} \approx 100 \frac{\text{nm}}{\text{min}}$$

if the process gas pressure is 10^{-3} mbar (typical), then the argon concentration at a gas temperature of 300 K is:

$$\begin{aligned} p = nkT \Rightarrow n &= \frac{p}{kT} \\ &= \frac{0.1}{1.38 \times 10^{-23} \times 300} \frac{1}{\text{m}^3} \\ &= 2.42 \times 10^{19} \frac{1}{\text{m}^3} \\ &= 2.42 \times 10^{13} \frac{1}{\text{cm}^3} . \end{aligned}$$

With the gas-kinetic cross-section for Al $\approx 5 \times 10^{-15}\,\text{cm}^2$ in argon, we obtain for the mean free path of the aluminum atoms:

$$\begin{aligned} \lambda = \frac{1}{n\sigma} &\approx \frac{1}{2.42 \times 10^{13} \times 5 \times 10^{-15}} \frac{\text{cm}^3}{\text{cm}^2} \\ &\approx 8.3\,\text{cm} . \end{aligned}$$

Therefore, for this example it is necessary that the distance from the target and substrate does not become larger than approximately 8 cm. The energy distribution of the sputtered atoms is proportional to:

$$f(E) \propto \frac{E_t E}{(E_t + E)^3} , \quad (6.23)$$

which can be seen in Fig. 6.4a,b. E_t is the surface bounding energy of the target material. In Fig. 6.4a, $E_t = 1$ eV, and in Fig. 6.4b $E_t = 4$ eV. Most sputtered atoms have energies with $0.5E_t$, whereby typical E_t-*values* lies between 1–4 eV.

If the atoms impact the substrate at those energies, this leads to mixing and diffusion of the substrate material with the incident atoms. This results in an improvement of the bonding and a better adhesion of the film on the substrate.

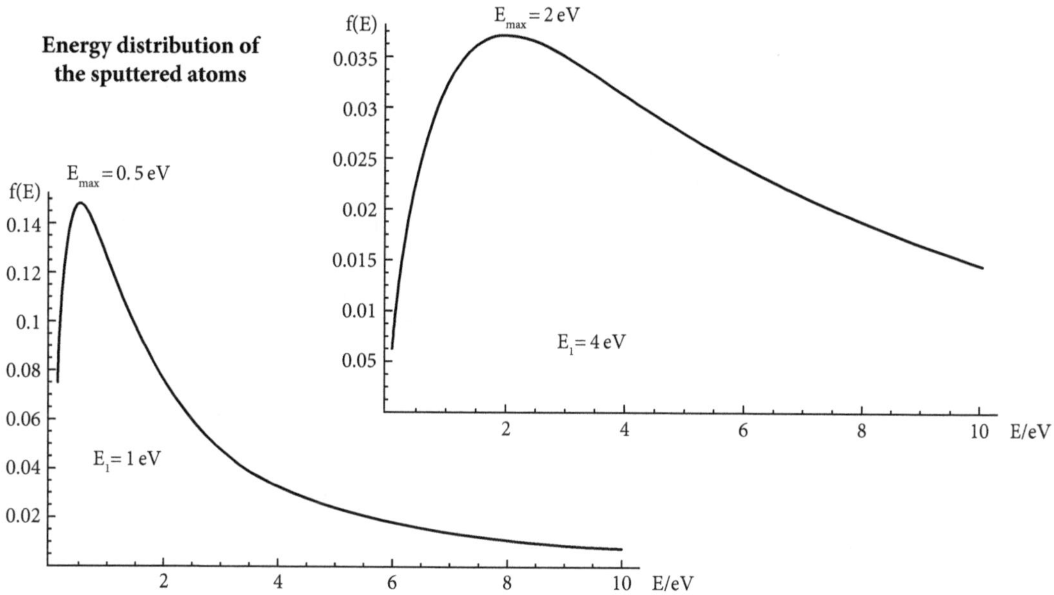

Fig. 6.4 Energy distribution of the sputtered atoms (Ar+ bombarding particles)

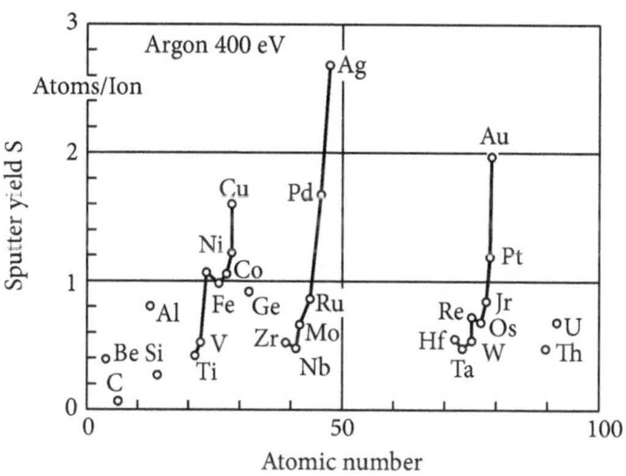

Fig. 6.5 Plot of the atomic number of the target materials against sputter yield for perpendicular bombardment by argon ions at 400 eV [7]

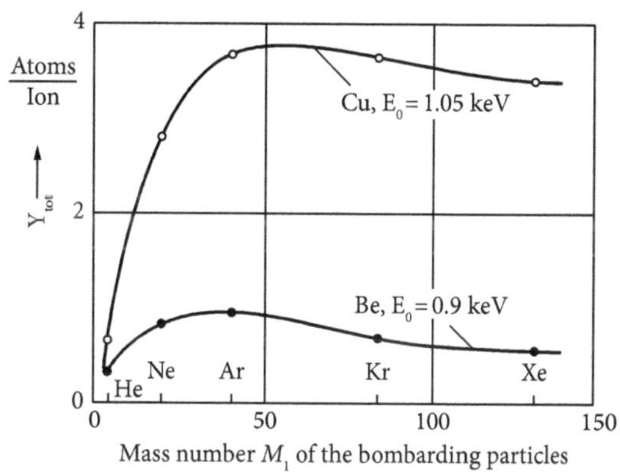

Fig. 6.6 Total sputter yield Y_{tot} of polycrystalline Be and Cu against the perpendicular bombardment of ions from noble gases at approximately 1 keV

Figure 6.5 shows Y_{tot} over the atomic number Z_2 of the target elements [7]. The dependence M_1 of the bombardment particles is essentially determined by the energy, which will transfer during the deceleration of the projectile at the first collisions with the target atoms.

Figure 6.6 shows the total sputter yields Y_{tot} for target materials Be and Cu as a light and moderately heavy element for the perpendicular bombardment by noble gases ions [8].

Figure 6.7 shows the sputter yield curve of different elementary targets by ion bombardment as a function of the incident angle θ to the normal surface [9].

More detailed analysis of the dependence of the bombardment angle of the yield Y_{tot} shows that up to a certain energy value, the cascade theory with the momentum redistribution of the bombarded particles does not exact explain the results.

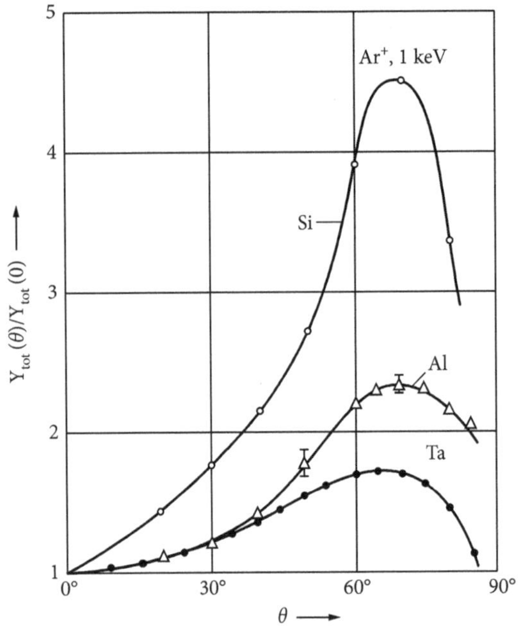

Fig. 6.7 Dependency of standardized sputter yield Y_{tot} as function of angle θ at Ar-ion bombardment with 1 keV for different polycrystalline targets

For an estimation of the optimal bombardment angle θ_{opt} for the maximum yield $Y_{tot}(\theta)$, we obtain

$$Y_{tot}(\Theta) = Y_0 + \Delta Y(\Theta) . \quad (6.24)$$

The increase in $\Delta\gamma(\theta)$ in the range $\theta < \theta_{opt}$ is nearly

$$\Delta Y(\Theta) = c\sigma_1 \frac{\gamma E_o}{d^2 U_o} \Theta^2 / \Theta^2_{opt} , \quad (6.25)$$

where σ_1 is the distance from the center of the incoming projectile with energy E_0 to the target atom and d the target atom radius. At 1 keV, σ_1 for Ar-ions is about 2-2-5 Å and only varies a little with the target materials. The constant c is 2.5×10^{-2} according to the measured values. The optimal bombardment angle for maximum $Y_{tot}(\theta)$ is approximately in range of 60° to 70°, depending on the surface roughness and is little influenced by the bombardment conditions. The influence of the bombardment angle θ can be calculated with Eq. (6.25). Figure 6.8 shows this for several ion/target combinations.

If the sputtered current contains different types of particle X, then the partial sputter yield Y_X is a part of the total sputter yield. We then obtain:

$$Y_{tot} = \Sigma Y_X . \quad (6.26)$$

During the sputtering of an elementary target X applies to the released atoms and during the molecules developed during the actual sputtering process. With investigations of sputtered metals, for example, by bombardment with a perpendicular 1 keV Ar ion beam, Me_2 dimer or Me_3 trimer arise. The partial yields are in the order of magnitude of $Y_{Me_2} = 10^{-1}$–10^{-2} and $Y_{Me_3} = 10^{-3}$–10^{-4} [10].

For alloy targets at the sputter equilibrium we have

$$Y_X^{(Leg)} = c_X Y_{tot} , \quad (6.27)$$

where c_x is the atomic volume concentration of the alloy's component X.

During the sputtering of assemblies such as oxides, nitrides, etc., Eq. (6.27) is no longer valid. Figure 6.9 [11] shows appropriate examples of sputtered oxides.

6.2.1 Angle Distribution of Sputtered Particles

A perpendicular incident ion beam focused at a specific point on the target surface leads to a cosine distribution of the (ejected) atoms or molecules. The cosine distribution correlates with the isotropic sputtering cascade theory. At higher bombardment energies the particle flux increases into the solid angle range around the surface-normal. The associated angle distribution can then be approximated by the relationship $\cos^n \varphi$ with $n > 1$. Thereby φ is the (polar-)angle measured against the surface plane.

For amorphous or texture (topography/topology)-free polycrystalline targets the angle distribution shows a rotationally symmetrical distribution based on the surface plane. At bombardment energy below 1 keV, the angle distribution of the sputtered particles changes to the range of $\cos^n \varphi$ with $n < 1$. With decreasing bombardment energy, the number of the particle collisions leading

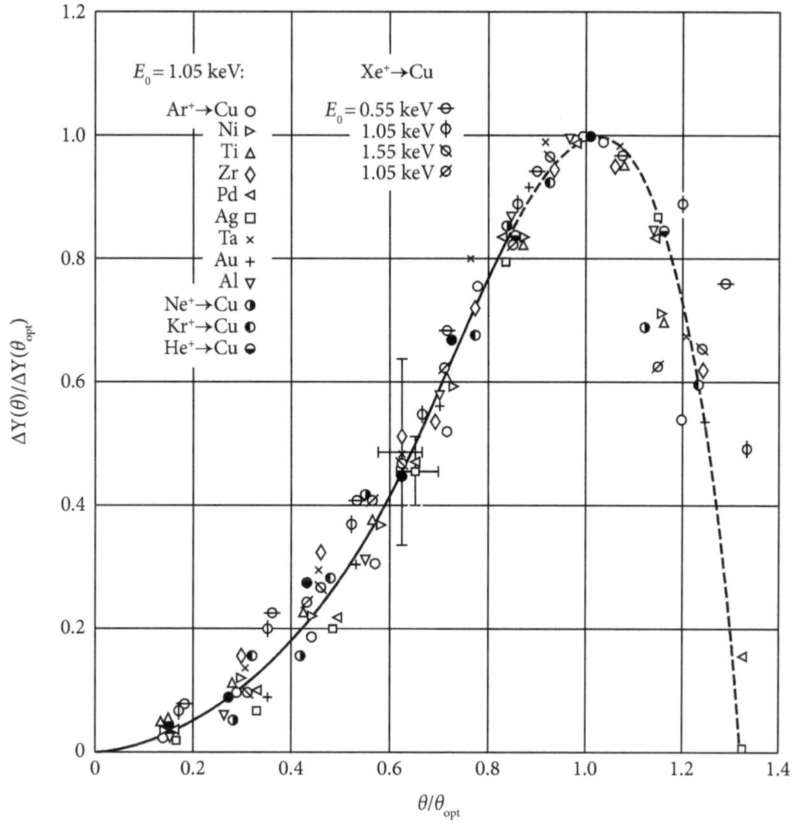

Fig. 6.8 Measured values against the bombardment angle [9]

to the emission of a surface particle always becomes smaller.

Thus there is an increase in the probability of a particle emission in the lateral direction, particularly with target materials that have a high surface energy barrier. The angle distribution of the sputtered particles assumes a torus-like form with a clear depression in the environment of the surface plane at bombardment energies around 100 eV, Fig. 6.10.

Figure 6.11 shows a comparison between measured angle distribution with angular ion bombardment and simulation values [13].

6.2.2 Sputtering Effects with Single-Crystal Solids

The regular structure of the atoms in the bulk material and at the surface of crystalline sputter targets affects the interaction between the bombardment particles and the atoms of the solid lattice and also the properties of the atomic impact cascades which lead to the emission of surface particles. If a single-crystal target is bombarded with a particle beam of sufficiently small divergence parallel to the so-called channel direction (Fig. 6.12), and/or a low indicated lattice direction, then a part of these particles penetrates into the channel between the tightly packed atomic rows, and the transfer of the bombardment energy takes place at a larger lattice depth. Since the energy transfer within the cascade process is smaller at the uppermost atomic layer, the sputter yield Y_{tot} clearly decreases with the bombardment along the channel direction.

The channeling property of the crystalline substrate depends on the angle of the entrance Φ of the individual ions and on the ion and substrate characteristics. If the entrance angle Φ is large, then the amplitude of oscillation will be large, and the ion will not be channeled (trajectory a in Fig. 6.12). There will be a maximum permissible amplitude at which the ion will re-

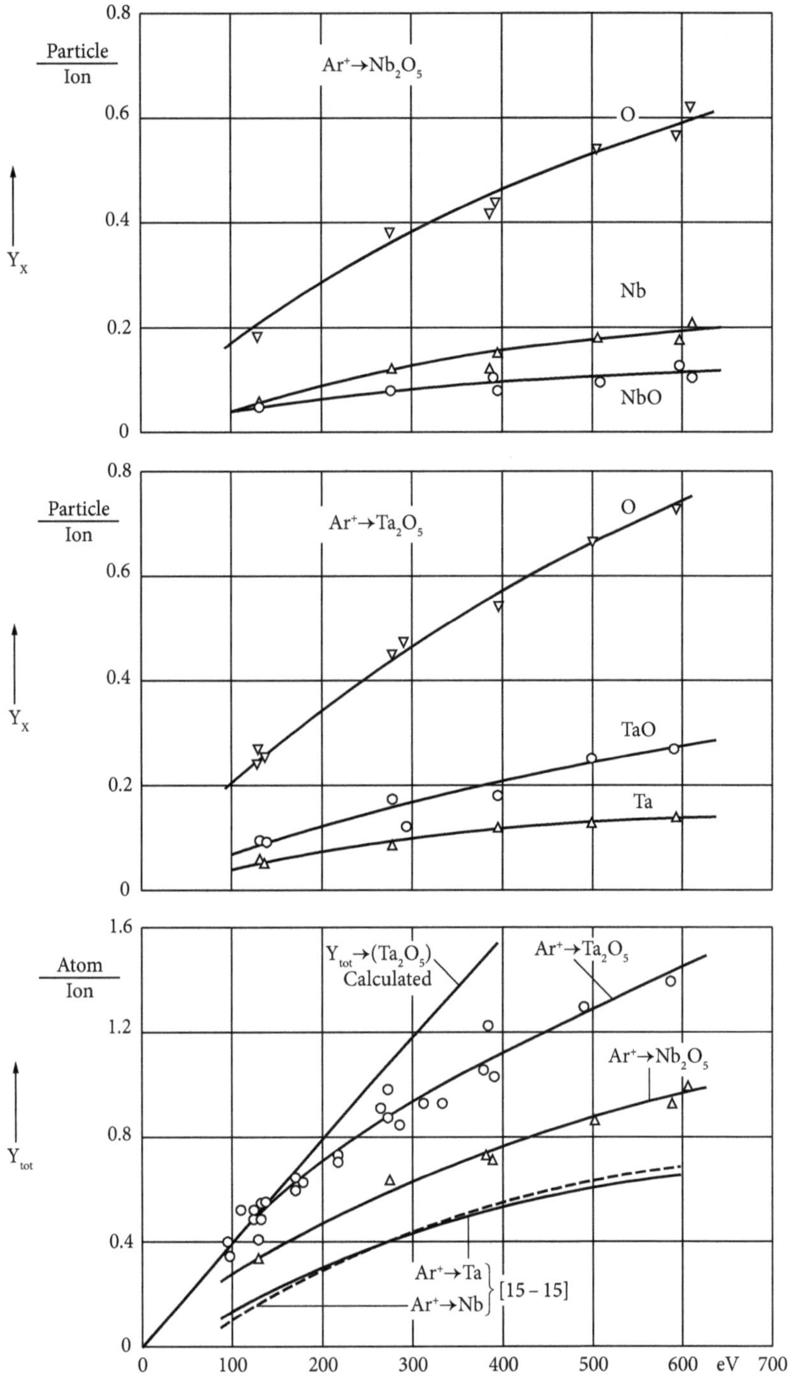

Fig. 6.9 Total and partial sputter yields for the perpendicular bombardment of Ta_2O_3 and Nb_2O_3 by Ar-ions with 600 eV

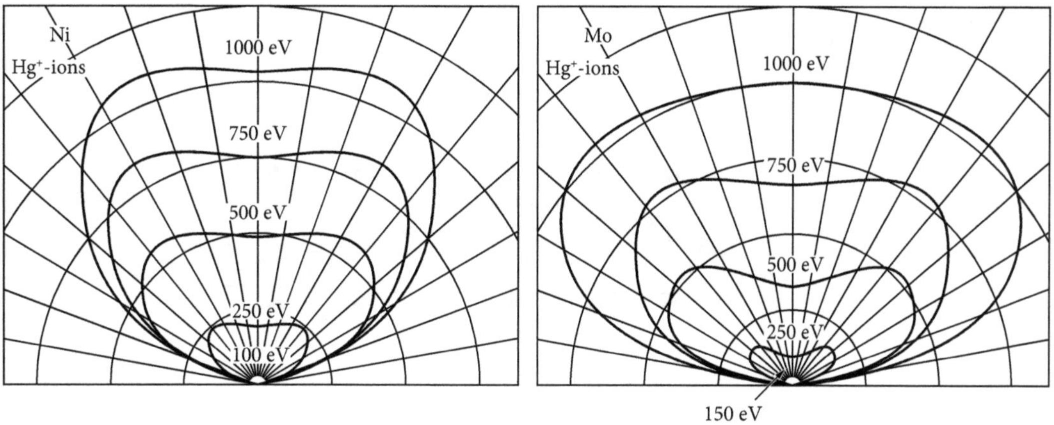

Fig. 6.10 Angle distribution of the sputtered particles with perpendicular bombardment of polycrystalline Ni and Mo with Hg ions of different energy [12]

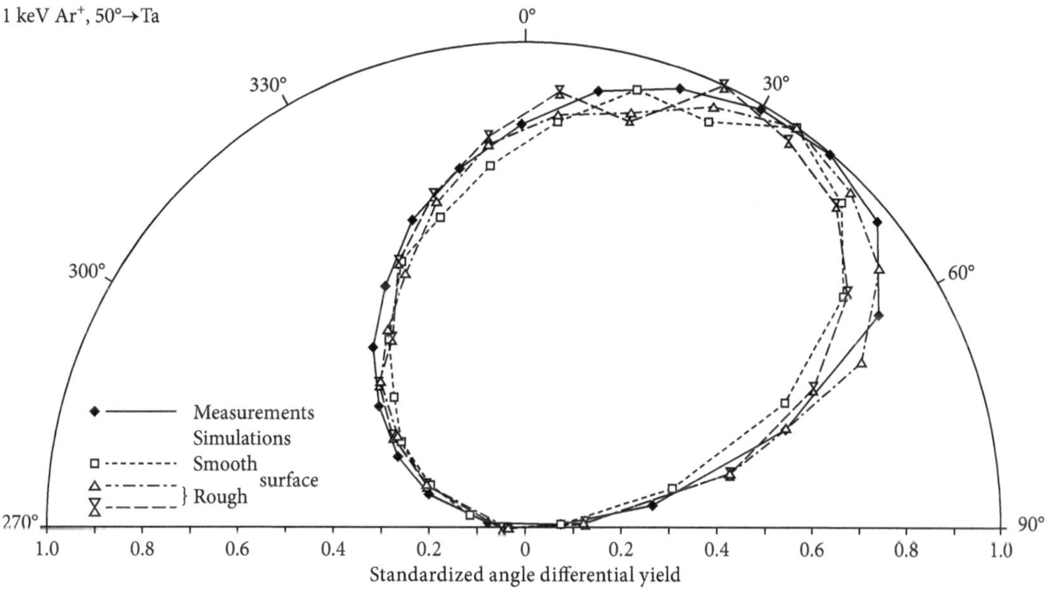

Fig. 6.11 Angle distribution for measured and simulated values for the sputtering of polycrystalline Ta bombardment with Ar-ions of 1 keV less than 50° against the target plane. Different surface roughness are simulated

main channeled (trajectory b in Fig. 6.12); Φ_c is the critical angle that produces this maximum amplitude trajectory.

This critical angle for channeling can be approximated [14] by

$$\Phi_c = (2Z_1Z_2e^2/4\pi\varepsilon_0 Ed)^{0.5}. \quad (6.28)$$

It is clear from this equation that as the ion energy E is increased, Φ_c decreases, and has more difficulty remaining in a stable trajectory. On the other hand, for closely packed atomic rows d decreases and channeling is easier.

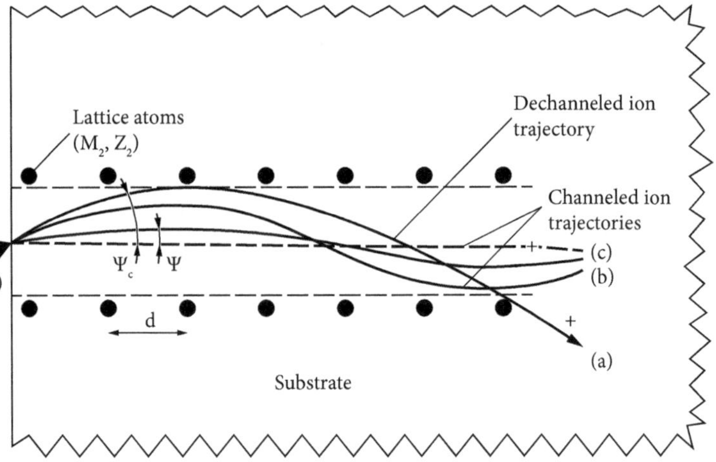

Fig. 6.12 Trajectories of channeled particles in a crystallographic direction

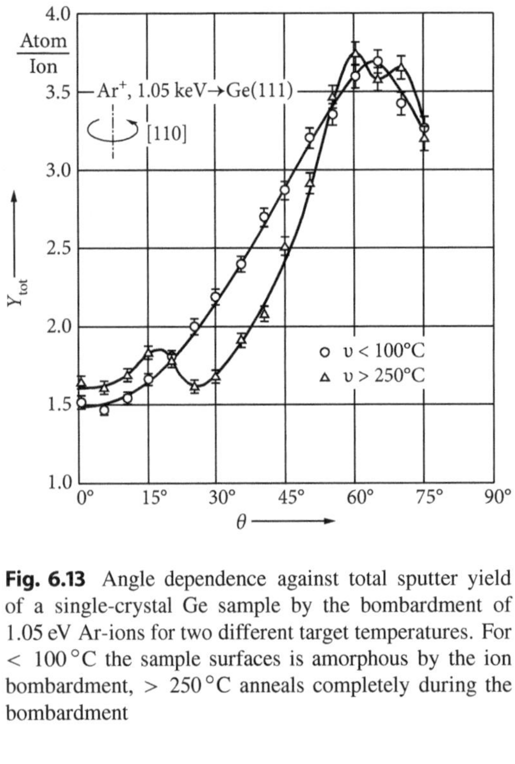

Fig. 6.13 Angle dependence against total sputter yield of a single-crystal Ge sample by the bombardment of 1.05 eV Ar-ions for two different target temperatures. For < 100 °C the sample surfaces is amorphous by the ion bombardment, > 250 °C anneals completely during the bombardment

An appropriate example is shown in Fig. 6.13. The figure describes the bombardment of a Ge surface with 111 orientation with 1.05 eV Ar ions [15]. At a temperature above 250 °C the atomic structure at the surface of the semiconductor crystal remains uniform, since with thermal annealing he bombardment-induced defect generation prevails at the uppermost atomic layers.

A 110 orientated substrate has a crystal structure with the closest atomic packing and thus the highest transparency of a single-crystal lattice. By ion bombardment, a large penetration depth is evident due to the channeling effect. The channeling effect leads to a reduction of the sputter yield up to an incident beam angle of 35°. A further increase in the incidence ion angle with respect to the surface plane as shown in Fig. 6.13 at the angle of 70° – as for $\theta = 0°$ – leads to a further (111) orientation.

Since the associated lattice channels are far less than in the 110 orientation, the decreasing of the sputter yield is less pronounced. Figure 6.13 shows the curve of $Y_{tot}(\theta)$ for the Ge-sample measured at temperatures below 100 °C. The uniform curve of the bombardment angle dependence is to be attributed to the fact that semiconductors, Ge or Si, become amorphous under bombardment below an annealing temperature. The annealing temperatures for Ge and Si are approximately 250 °C and 500 °C.

Semiconductors show a characteristic temperature dependence on the sputter yield; an increase of approximately 10 % Y_{tot} is observed when the annealing temperature is exceeded. A further effect, apart from lattice channeling, is the so-called '*focusing collision*' sequences in tightly packed lattices [16].

In a certain energy range there is a focusing of the transported particle momentum by atomic collisions along a tightly packed atomic

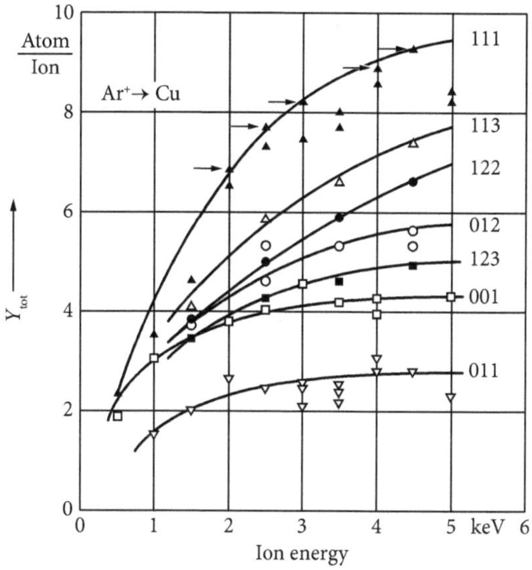

Fig. 6.14 The sputter yield against the perpendicular bombardment energy of Ar-ions for differently orientated single-crystal Cu. *Arrows* on the 111 orientated curve show the measured values for the electropolished surface

chain. Surface atoms at the end of such atomic chains are preferentially emitted along the appropriate low indicated lattice directions. Therefore, a strong anisotropy effect of the sputtered particles along the angle distribution is possible, whereby the always existing contribution from statistic sputtering cascades are overlaid with further maxima along the tightly packed lattice directions (so-called Wehner spots) [17]. Similarly, the focusing of the surface atoms after emission from the surface can be affected; they may recombine with the bombardment ion and other emitted atoms [18]. Figure 6.14 shows the sputter yield with perpendicular bombardment with Ar-ions for differently orientated single crystals Cu [19]. The characteristics of the sputter yield of the Cu polycrystalline targets are indicated.

6.3 Coating with Sputtering

The major advantage of sputtering is its generality. In principle, any vacuum-compatible material can be coated by this method [20, 21]. Both DC (direct current) and RF (radio frequency) gas discharges in a variety of configurations have been used. Power is required to generate the plasma from which the useful ions are to be extracted and accelerated to the target. In the case where the target material is insulating, means must be found to neutralize the surface charge built up on it by the impinging ions. The substrate is bombarded by inert gas molecules, electrons, photons, and negative ions from the plasma, as well as the desired atoms from the target. Means to avoid deleterious effects from these extraneous sources must be found. Ion bombardment and discharge pressure lead to internal film stress. Examples of sputtered films and their applications are given in Table 6.1.

In principle, a transition from a high to a lower level of internal stress is possible by increasing the argon pressure. This is attributed to a decrease in the ion energies (during the increase in the pressure).

6.4 DC Diode Sputtering

Figure 6.15 shows the cathodic sputtering method with DC-diode assembly. In the chamber there are two parallel-plate electrodes, one composed of the target material and the other of the substrate that is to be coated. In the case of DC discharge, by applying a potential of several kilovolts (kV) between planar electrodes in an inert gas atmosphere at pressures between 10^{-1} and 10^{-3} mbar, a glowing plasma is formed, which is separated from each electrode by a dark space. The plasma is essentially at a constant positive potential with respect to the cathode.

Electrons are drawn out of the boundary to the plasma close to the anode and accelerated across the anode dark space. The electrons then strike the anode, where the energy they carry is dissipated. Ions are drawn out of the other boundary of the plasma, accelerated across the cathode dark space (***Crooke's zone***), and strike the target cathode, causing sputtering. The plasma is replenished by electron–ion pairs formed by the collision of neutral molecules with the secondary electrons generated at the cathode surface, accelerated across the cathode dark space, and injected

Table 6.1 Application of sputtered layers

Application		Material
Electronic:	Electrodes/interconnects	Au, Al, Cu, Cr, Ti, Pt, Mo, W, Al/Si, ...
	Resistances	Cr, Ta, Re, TaN, NiCr, $SnO_2, \dots$
	Dielectric materials	HfO_2, PbO, AlN, Si_3N_4, Al_2O_3, $SiO_2, \dots$
	Insulators	Si_3N_4, Al_2O_3, SiO_2, TiO_2, $Ta_2O_5, \dots$
	Superconductors	Nb, NbN, La-Sr-Cu-O, Y-Ba-Cu-O, ...
	Semiconductors	Ge, Si, Se, Te, SiC, ZnO, ZnSe, CdSe, ...
	Passivation	Si_3N_4, SiO, SiO_2
Optic:	Deposition	SiO_2, TiO_2, SnO_2, In_2O_3
Other:	Hardening	Cr, TiN, SiC, WC
	Decoration	Al, Zn, Cd, Cr, Ti, Ta, W, TiN, TiC, SiC, Ag, Au ...

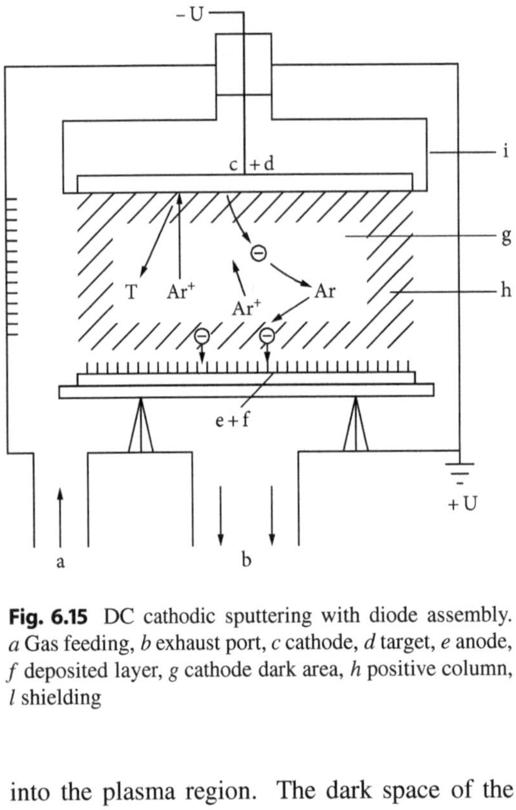

Fig. 6.15 DC cathodic sputtering with diode assembly. *a* Gas feeding, *b* exhaust port, *c* cathode, *d* target, *e* anode, *f* deposited layer, *g* cathode dark area, *h* positive column, *l* shielding

into the plasma region. The dark space of the cathode has a thickness about the same as that of an electron–ion collision. The dark space of the anode has a thickness determined by space-charge-limited flow of electrons from the plasma to the anode (see Sect. 4.6).

This mode of DC sputter operation works only for electric conducting materials. Insulators cannot be sputtered with the DC-mode; the positive charges on the cathode surface cannot discharge from the plasma ions. The consequence would be damage to the cathode potential, and then no more plasma ions would be accelerated toward the target, resulting in the automatic termination of the sputter process. In order to prevent this, the back part of the cathode or the electrical feed line must be shielded against sputtering.

Between cathode and substrate there is always a shutter installed for pre-sputtering in order to prevent a film that is contaminated on the substrate.

For the impurity degree of C_j, by undesirable adsorption films on the substrate surface, the following relationship is valid:

$$C_j = \frac{\alpha_j N_j}{\alpha_j N_j + S}, \quad (6.29)$$

where

α_j is adsorption coefficient of the j_{th} residual gas component,

N_j the number of impacting residual gas molecules for each cm^2/s, and

S the number of sputtered particles.

Contrary to active residual gas atoms, an integration of inactive sputtering gas atoms in crystalline metallic films is marginal, i.e. with the cathodic sputtering the chamber should be evacuated into the high vacuum range and in addition high purity sputtering gas ($\sim$ 99.999 %) should be used.

The integration of argon or krypton on interstitial places is not possible because of the small interatomic distances. For cubic-face-centered (FCC, Face Cubic Center) metals like gold, the free radius of an octahedron gap is $0.414 \times R_{\text{Au}}$,

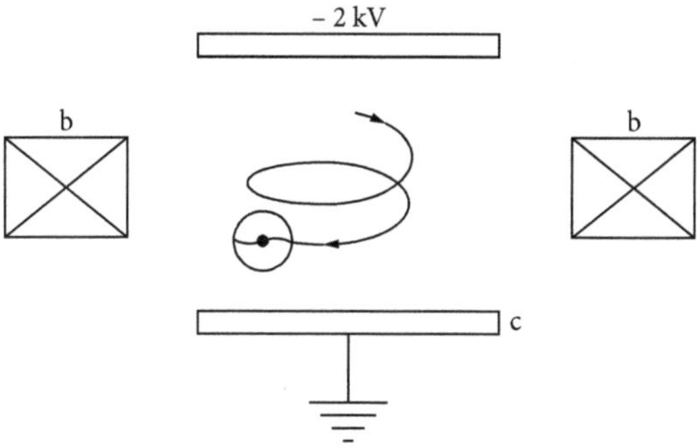

Fig. 6.16 Schematic structure of a DC sputtering system with incorporation of a magnetic field. *a* Cathode, *b* magnetic coil, *c* anode

and $0.732 \times R_{Nb}$ for cubically body-centered (BCC, Body Cubic Center) metals.

For Nb ($R = 0.147\,\text{nm}$), the calculated free radius of a possible foreign atom on a cubic space-centered surface is:

$$R = 0.108\,\text{nm}\,.$$

This only permits the entrapment of the small noble gas atoms ($r_{He} = 0.093\,\text{nm}$, $r_{Ne} = 0.112\,\text{nm}$) on interstitial sites. The entrapment of the larger Ar- or Kr-atoms would lead to a lattice destruction/deformation. The entrapment of sputter gas atoms in a deposited film is marginal, amounting to a maximum of 10^{-2} at. % in Ti-films.

Amorphous metallic films can be generated by quench condensation [22]. Metals such as copper condense at a temperature of 4 K, obtaining a nanocrystalline structure (crystallite size ~ 3 nm). However amorphous conditions can be obtained by alloying with other metals. From [23] it is possible to obtain amorphous alloys, if:

- The difference of the atomic diameters ≥ 15 %.
- The concentration ranges of binary systems extend to approx. 20 to 80 %.

The quantity of the sputter gas entrapped in the film can be reduced by reducing the gas pressure or in some cases by heating the substrate [23]. In order to achieve high sputtering rates with low pressure, the ionization degree of the sputtering gas must be increased [24]. This is possible by:

- Increasing the probability of ionization of the existing atoms.
- Increasing the number of ionizing electrons.
- Using an ion beam source.

By using a magnetic field $\vec{B}$ (5×10^{-3} to 10^{-2} T), we can increase the degree of ionization in the direction of the electron motion. The electrons then move helically around the electrical stream lines of the field (see Fig. 6.16).

The current density j on the target surface can be increased to $5 \times 10^{-3}\,\text{A cm}^{-2}$ with a sputter gas pressure of 10^{-3} mbar. Coating rates of 500 nm/min are then reachable.

6.5 Triode Sputtering

With the triode configuration (Fig. 6.17) the ionization level of the plasma is increased by electrons emitted from a hot cathode and accelerated in an electrical field. The electrical field is formed by an auxiliary anode.

The cathode is generally a tungsten filament heated to 2500 °C or a hollow cathode plasma source. The anode is a disk or tube with approximately +100 V against bias to earth. In addition

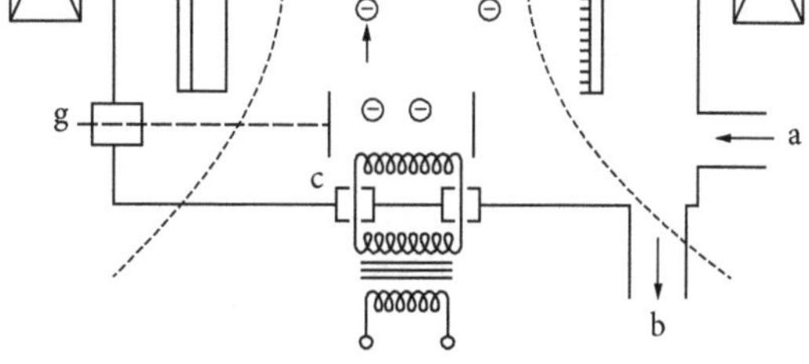

Fig. 6.17 Schematic structure of a triode arrangement. *a* Gas feeding, *b* vacuum port, *c* glow cathode, *d* target, *e* anode, *f* deposited layer, *g* auxiliary anode, *h* magnetic coil

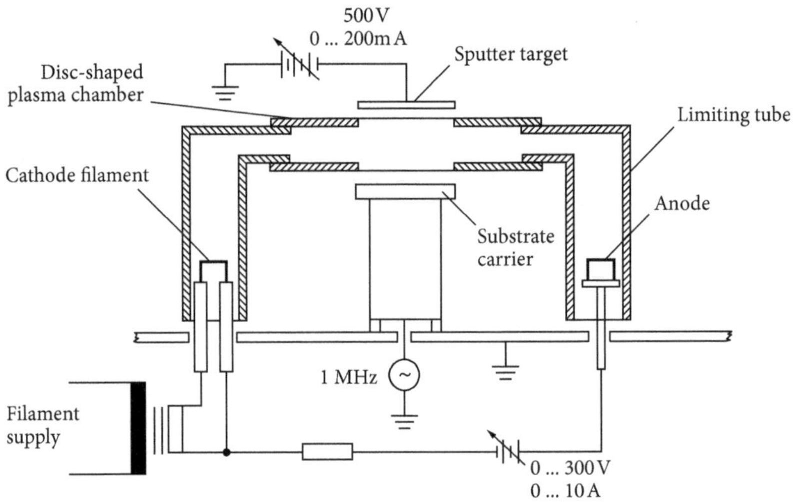

Fig. 6.18 Layout of a sputter system according to the triode principle with RF-excitation

to the electrical field, a magnetic field is usually overlaid. The arrangement of the heating filament-anode-magnet forms a plasma generation unit that is independent of the substrate and target. The potential of the target does not influence the maintenance of the discharge, which means that smaller discharge voltages, 500–000 V, are needed. The substrate can also be arbitrarily arranged and configured. Figure 6.18 shows the concept of triode sputter equipment with plasma excitation in combination with an electron cyclotron wave resonance system [25].

6.6 RF (Radio Frequency) Sputtering

Direct current discharges have difficulties with initiation of the discharge and with the sputtering of insulating materials. These disadvantages are eliminated by using high frequency arc discharges (1–30 MHz). If one of the electrodes is capacitively coupled to an RF generator, it develops a negative DC bias with respect to the other electrode. This is because electrons are much more mobile than ions, which hardly respond to the applied RF field at all, When the RF discharge is initiated, the capacitively coupled electrode becomes negatively charged as electrons are deposited on it. However, this charge is only partially neutralized by positive ions arriving during the negative half-cycle [26]. Since no charge can be transferred through the capacitor, the electrode surface retrains a negative bias such that only a few electrons arrive at the surface during the next positive half-cycle, and the net current averaged over the next full cycle is zero. The RF voltage only serves for the generation of the plasma (Fig. 6.19).

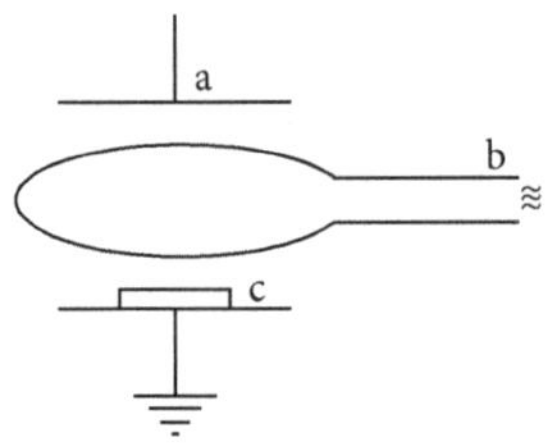

Fig. 6.19 Schematic of a high frequency sputtering system. *a* Cathode, *b* high frequency coil, *c* substrate

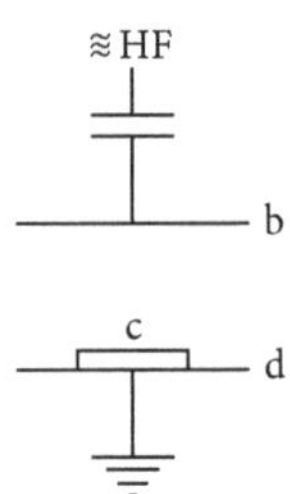

Fig. 6.20 Schematic structure of a high frequency sputtering arrangement in a diode circuit

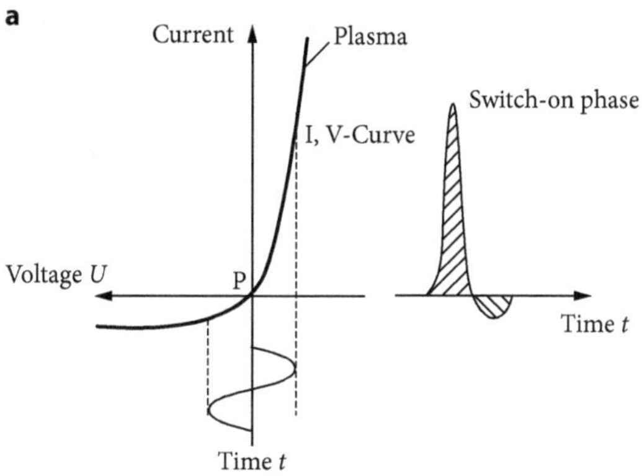

Fig. 6.21 Structure of negative DC voltage at the cathode in high frequency sputter equipment by different velocities of electrons and ions. **a** Start up state with electron excess, **b** shift of the operating point by structure of negative DC voltage at the cathode, so that the total value of the current disappears

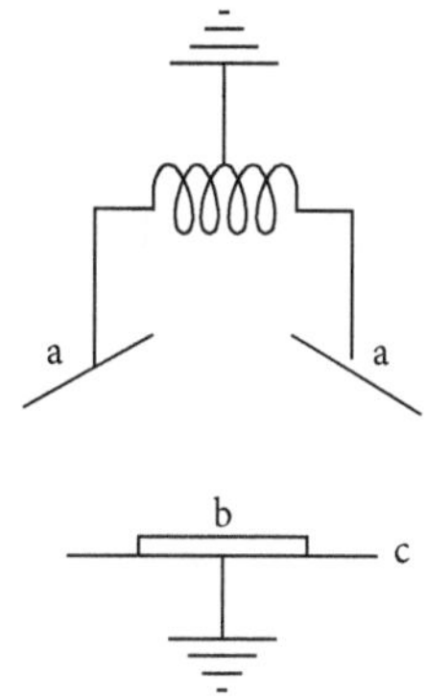

Fig. 6.22 Schematic of a high frequency sputter system with a double cathode. *a* Cathode, *b* substrate, *c* anode

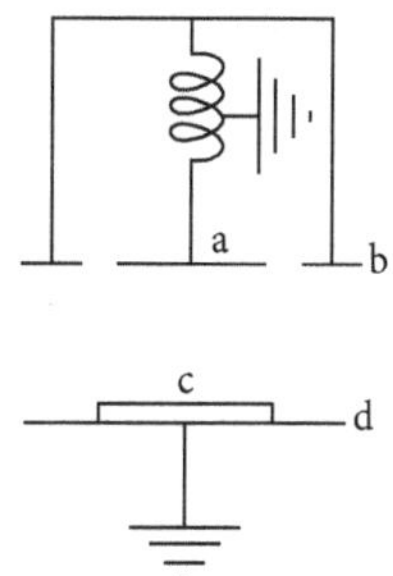

Fig. 6.23 Schematic of a high frequency sputter system with an axial symmetrical double cathode. One electrode is a circular plate, the other a ring format. *a* Internal cathode, *b* ring cathode, *c* substrate, *d* anode

Ions diffusing from the plasma behave as if only the negative DC bias voltage existed and sputter the negatively charge electrode (Figs. 6.20 and 6.21). The conducting anode, which is directly connected to the power supply, receives pulses of energetic electrons accelerated essentially to the full amplitude of the RF field.

In the case of gas-discharge sputtering, where the substrate is an electrode of the system, the energy dissipated in the sample by electrons from the plasma can be substantial. A small amount of heating can be advantageous, since the coating is continuously annealed, which results in a stress-free film. However, excess heating can cause distortion of the film material, and eventually evaporation of the thin material. Heating of the deposited film from electron bombardment limits the rate of the film growth that can be utilized in sputtering, even before the target heating effect. The so-called '***double cathode***' arrangement (Figs. 6.22, 6.23), has the additional benefit of increasing the efficiency of generated ions.

In this system the RF frequency voltage is applied to two cathodes in the recipient, so that the ions per half-wave fly to the appropriate cathode and thus both electrodes are sputtered. RF sputtering permits the deposition of film from arbitrary material, conducting or non-conducting. The system has two, three, or possibly four RF targets. Figure 6.24 shows the attainable coating rates. The diagram shows that the deposition rate is directly proportional to the power applied to the target.

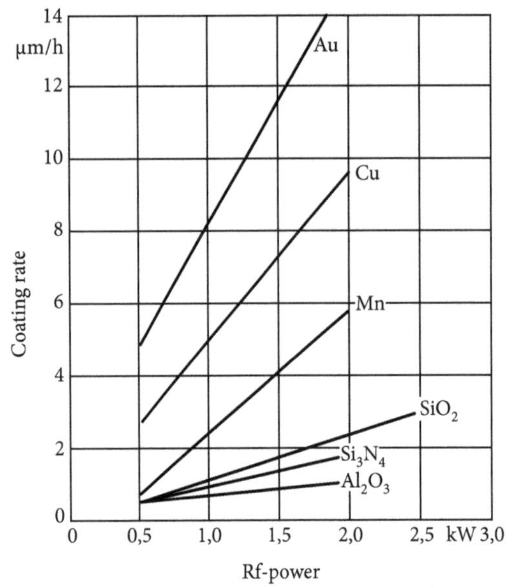

Fig. 6.24 Coating rate as a function of the RF power, target surface area $100\,cm^2$, target/substrate separation 25 mm

With a current density on the target surface of $3\,mA/cm^2$ and a distance of 25 mm between target and substrate, a coating rate of $6\,\mu m/h$ is possible. With a power density of around $1\,W/cm^2$ and a distance of 25 mm between substrate and target, the substrate temperature can reach up to 400 °C.

High frequency sputtering is usually operated with a frequency of 13.56 MHz. Also systems with 1 MHz, 2 MHz, 27.12 MHz, 40.68 MHz and with even higher frequencies are operated.

Industrial RF generators contain a quartz-stabilized oscillator and a power setting enabled by coupled transistor amplifiers up to approximately 5 kW. For high power operation, vacuum tubes (triodes) are used. The RF outputs of such generators are designed with a reflection-free coaxial conducting cable with a characteristic impedance (Z) of 50 Ω, Fig. 6.25.

In the generators output cables are reflectometers, which supply signals for the delivered forward power P_V and the reflected power P_r.

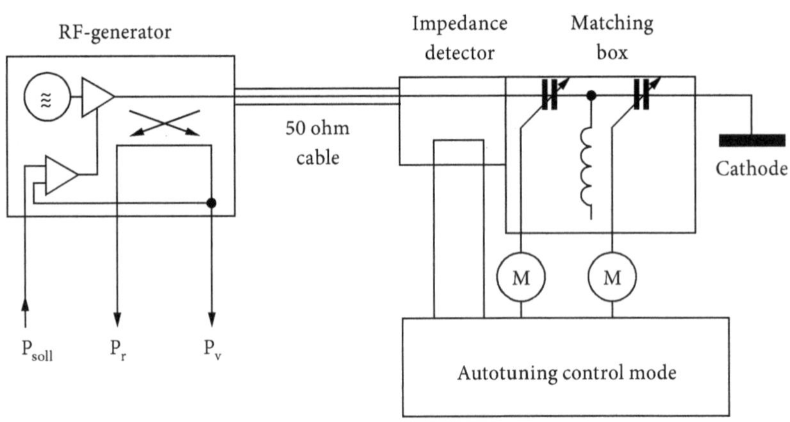

Fig. 6.25 RF frequency supply for a sputtering cathode

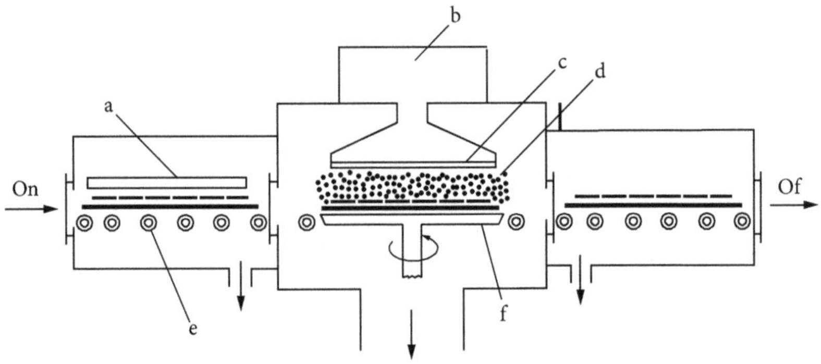

Fig. 6.26 Three-chamber sputtering system with RF-diode assembly. *a* Glow, heat, *b* match element, *c* target, *d* plasma, *e* transport device, *f* substrate plate

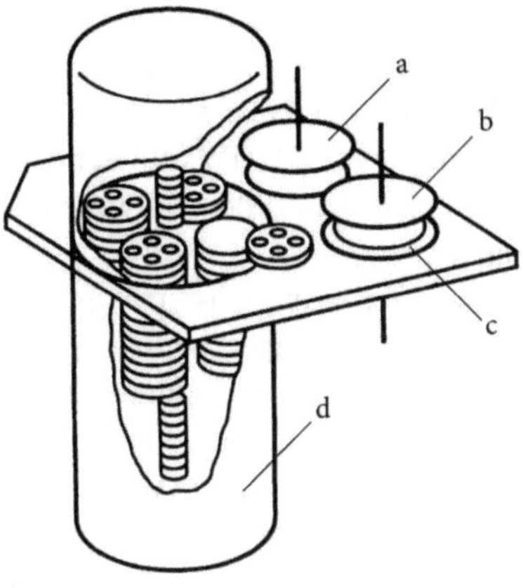

Fig. 6.27 RF diode sputtering system with substrate magazine. *a* Target, *b* heater, *c* corroding electrode, *d* substrate magazine

The reflected power shows the deviation of the load impedance from the nominal impedance. At good adjustment $P_r/P_V \leq 2\,\%$, the released power is given through the forward power P_V and is regulated over these. Accuracies of approximately $\pm 5\,\%$ are reached.

Figure 6.26 shows the schema of a multi-chamber system. In order to degrease the process chamber pumping time and reduce the time required for the pre-treatment of the substrate, a pre-chamber (load lock) is often used for pre-pumping. The substrates arrive over a load lock into the main vacuum chamber. Afterwards the substrates come into the ventilation chamber. The principle of a completely automated sputter system is shown in Fig. 6.27. The system possesses a substrate magazine containing a large number of substrates, which are introduced successively into the sputter chamber. In a preliminary stage, the substrates are heat-treated and ion-etched. Figure 6.28 shows an RF sputter system for SiO_2 deposition. The target disk consists of quartz tube coils.

Fig. 6.28 Sketch of a sputtering system with bilateral deposition. *a* Transmitter connection, *b* pit, *c* coupling capacity, *d* inductance, *e* current feed, *f* flange, *g* adjustment capacity, *h* substrate abrasion, *i* screen, *j* substrate, *k* substrate electrode, *l* target, *m* magnet shoe, *n* pole shoe

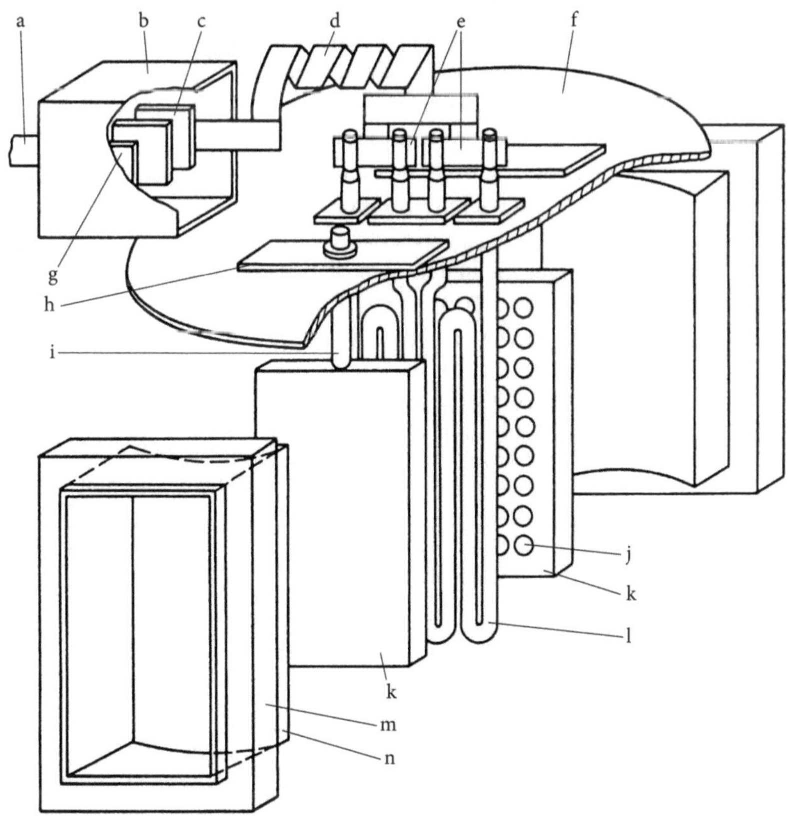

6.7 Bias Sputtering

If the substrate is isolated from earth and a negative voltage is applied to the substrate, then the deposited film is constantly bombarded with ions from the plasma, Fig. 6.29. The deposition film is continually cleaned by adsorbent gas particles during the process.

Bias sputtering can also be combined with an AC source (alternating current) sputtering [27].

6.8 Reactive Sputtering

If an active gas such as oxygen, water, or hydrogen sulfide is intentionally added to the inert working gas, it can react with the target material either at the target surface or in the gas phase, resulting in the deposition of a new compound [28]. By varying the pressure of the active gas, control can be gained on the stoichiometry of the film.

The products of the dissociation react with the target chemically in addition to the bombardment of the high energy ions. The film deposited at the substrate is a combination of the materials from the sputtered target and the reactive gas components. Typical applications are deposited films, whose components possess different vapor pressures and hence different sticking coefficients at the substrate. For example, to generate a SiO_2 film with a SiO_2 target, Si increases in the deposited SiO_x film (i.e. $x < 2$). With the addition of oxygen (O_2) to the inert working gas (Ar), the correct stoichiometry of $Si : O = 1 : 2$ is obtained. Likewise, a pure Si target can be used to deposit SiO_2 films only with the addition of

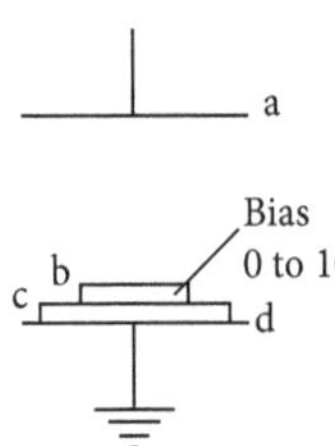

Fig. 6.29 Schematic of a bias sputtering system. *a* Cathode, *b* substrate, *c* insulator, *d* anode

oxygen gas to the recipient. Silicon dioxide films have many applications, e.g. in microelectronics as insulator films and also as a mask material. In food technology, PET (Polyethylenterephthalat) bottles are coated in order to improve the diffusion barrier.

Another application is the generation of high temperature superconducting films like YBaCuO, using a YBaCuO target and sputtering with an O_2 additive. The admixture of the following additives (often to the working gas Ar) by dissociation creates the appropriate reactive gas components:

- O_2 and H_2O $\Rightarrow$ O-atoms
- N_2 and NH_3 $\Rightarrow$ N-atoms
- CH_4 and C_2H_2 $\Rightarrow$ C-atoms
- SiH_4 $\Rightarrow$ Si-atoms.

In principle, ceramics could also be used as targets but they are sensitive to high power fluxes (they break with inhomogeneous heat gradients). Therefore, metal targets are usually used in combination with a reactive gas quantity.

In principle there are two operation modes to deposit a multi-component film with a metallic target ("Compound"):

1. ***Low ion flux and large gas flow*** $\Rightarrow$ ***target with compound film covered*** $\Rightarrow$ ***covered modus***.
2. ***Large ion flux and low gas flow*** $\Rightarrow$ ***target remains metallic*** $\Rightarrow$ ***metallic modus***.

In the metallic mode larger deposition rates are obtained than in the covered mode. If we keep the ion flux constant and vary the gas flow, we receive a transition of one to another mode. This transition shows a hysteresis (Fig. 6.30), which starts in the metallic mode and increases the gas flow, then the plasma "jumps" with a certain gas flow $\Gamma_{m\to b}$ into the covered mode, whereby the deposition rate clearly becomes smaller. If we now again reduce the gas flow, the transition to the metallic mode takes place at a smaller gas flow $\Gamma_{b\to m}$:

$$\Gamma_{m\to b} > \Gamma_{b\to m} .$$

In stationary condition the deposition rate of compound material at the target is equal to the erosion rate by sputtering of the compound mate-

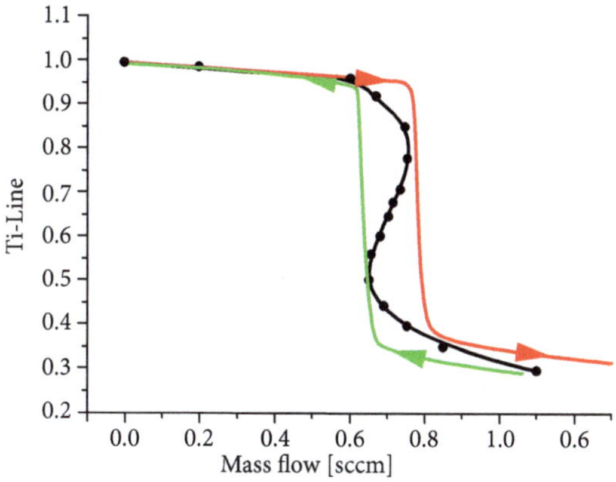

Fig. 6.30 A Ti-line compensated hysteresis with the reactive sputtering deposition of TiN layers (measured values)

rials from the target:

$$n'_{\mathrm{T}}\frac{\mathrm{d}\Theta_{\mathrm{T}}}{\mathrm{d}t} \stackrel{!}{=} 0 = i\Gamma_{\mathrm{rea}}s_{\mathrm{rea}}(1-\Theta_{\mathrm{T}}) - \Gamma_{\mathrm{ion}}\gamma_{\mathrm{C}}\Theta_{\mathrm{T}}\,. \tag{6.30}$$

For the description a simple model is used, with following denotations:

- A_{T} and A_{S} are the surfaces of the target and the substrate.
- Θ_{T} and Θ_{S} are the surface degrees of coverage from target and substrate with compound material.

γ_{M} and γ_{C} are the sputtering rates (target) of the metal and the compound materials at the target.
Γ_{ion} is the ion flux.
Γ_{rea} is the flow of the reactive gas molecule.
s_{rea} is the sticking coefficient of these gas molecules on the metallic target.

Here i is the number of radicals created in the plasma by dissociation (e.g. O-radical from O_2: $i = 2$, C-radical from CH_4: $i = 4$). The first term describes the flow of the reactive gas molecules (Γ_{rea}), which are dissociated in the plasma (i) adhering to the target naturally only on a still free surface $(1-\Theta_{\mathrm{T}})$.

The second term in Eq. (6.30) designates the sputtering of the compound film through bombarding ions (Γ_{ion}) with the yield γ_{C}, where this can naturally only take place with an already covered surface of the target (Θ_{T}).

We now consider the coverage of the substrate Θ_{S} in the stationary condition; those reactive gas molecules striking the metallic part of the substrate $(1-\Theta_{\mathrm{S}})$ are increased. Further this is increased by the flux of compound material, which is sputtered from the target and strikes on the free, i.e., metallic, surfaces of the substrate $(1-\Theta_{\mathrm{S}})$.

The coverage of the substrate is decreased by part of Θ_{S} of the flux of sputtered metal atoms $Y_{\mathrm{M}}\Gamma_{\mathrm{ion}}$, which come from the target $(1-\Theta_{\mathrm{T}})$, which strikes on the substrate surface covered with compound material (Θ_{S}). If we now still consider the relationship from target to substrate surface ($A_{\mathrm{T}}/A_{\mathrm{S}}$), then we obtain:

$$\begin{aligned} n'_{\mathrm{S}}\frac{\mathrm{d}\Theta_{\mathrm{S}}}{\mathrm{d}t} &\stackrel{!}{=} 0 \\ &= i\Gamma_{\mathrm{rea}}s_{\mathrm{r}}(1-\Theta_{\mathrm{S}}) + \Gamma_{\mathrm{ion}}\gamma_{\mathrm{C}}\Theta_{\mathrm{T}}\frac{A_{\mathrm{T}}}{A_{\mathrm{S}}}(1-\Theta_{\mathrm{S}}) \\ &\quad - \Gamma_{\mathrm{ion}}\gamma_{\mathrm{M}}(1-\Theta_{\mathrm{T}})\frac{A_{\mathrm{T}}}{A_{\mathrm{S}}}\Theta_{\mathrm{S}}\,. \end{aligned} \tag{6.31}$$

If we describe the total number of reactive gas molecules that are used per second in order to generate a compound film on the substrate, then we obtain:

$$\frac{\mathrm{d}N_{\mathrm{rea}}}{\mathrm{d}t} = \Gamma_{\mathrm{rea}}s_{\mathrm{r}}[(1-\Theta_{\mathrm{T}})A_{\mathrm{T}} + (1-\Theta_{\mathrm{S}})A_{\mathrm{S}}]\,, \tag{6.32}$$

since the reactive gas flux goes equally to the substrate and to the target. The flux by which the target is sputtered, that is that the target flows off after the sputtering of the metallic part and the compound fraction, is:

$$\Gamma_{\mathrm{sput}} = \Gamma_{\mathrm{ion}}[\gamma_{\mathrm{M}}(1-\Theta_{\mathrm{T}}) + \gamma_{\mathrm{C}}\Theta_{\mathrm{T}}]\,. \tag{6.33}$$

Now Eq. (6.30) can be resolved after covering of the surface of the target:

$$\Theta_{\mathrm{T}} = \frac{i\Gamma_{\mathrm{rea}}s_{\mathrm{r}}}{\Gamma_{\mathrm{ion}}\gamma_{\mathrm{C}} + i\Gamma_{\mathrm{rea}}s_{\mathrm{r}}}\,. \tag{6.34}$$

Likewise (6.31) can be resolved after the covering of the surface of the substrate, and θ_{T} inserted after (6.34):

$$\Theta_S = \frac{1 + \frac{A_T}{A_S}\frac{\Gamma_i \gamma_C}{\Gamma_i \gamma_C + i\Gamma_r S_r}}{1 + \Gamma_i \frac{A_T}{A_S}\left[\gamma_C \frac{1}{\Gamma_i \gamma_C + i\Gamma_r S_r} + \gamma_M \left(\frac{1}{i\Gamma_r s_r} - \frac{1}{\Gamma_i \gamma_C + i\Gamma_r S_r}\right)\right]} . \quad (6.35)$$

If we insert Eqs. (6.34) and (6.35) into (6.32) and in (6.33), we have the hysteresis problem that was addressed before.

With reactive sputtering many interfaces can be produced. Thus, a multiplicity of hard material combinations are created on the basis of carbides, nitrides and oxides and/or their sputtered mixtures of metals. They are frequently abrasive wear-preventing and anticorrosive [25, 29, 30].

6.9 Magnetron Discharges

The high flexibility of sputtering techniques is also based on various magnetically confined discharge configurations. By means of the $E \times B$, drift plasmas of high density within a closely limited range can be concentrated [31, 32].

6.9.1 Structure

A typical magnetron configuration is shown in Fig. 6.31. It consists of electrodes that are connected with a DC voltage V_{DC} (above the cathode, under the anode). Above the cathode a system of permanent magnets is arranged so that the cylinder-symmetrical chamber makes a circular penetration of the magnetic field possible. The

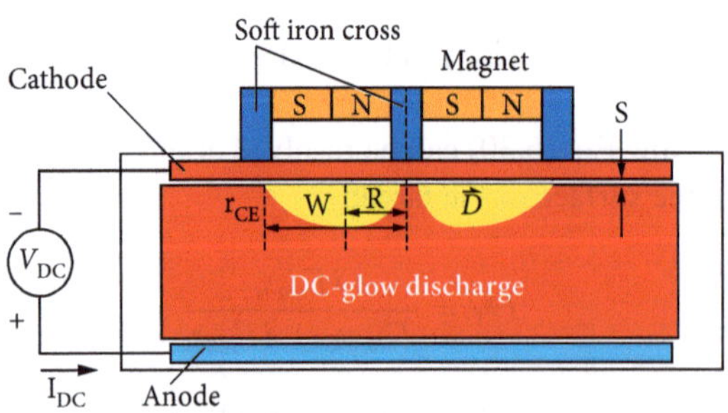

Fig. 6.31 Planar magnetron. Electrons originating at the cathode surface are confined by the fields of permanent magnets placed behind the cathode

center radius of the ring is R, the width of the ring w and the height r_{ce}. If we apply a voltage V_{DC} and ignite a glow discharge, we obtain a high density, brightly shining plasma within the range of the magnetic field. Outside of the circular field the plasma density is lower.

In the environment of the cathode a skin layer forms with the thickness $s \approx 1$ mm.

6.9.2 Function of the Magnetron

Positive inert gas ions attain kinetic energy in the cathodic skin layer and bombard the (cold) cathode, which is configured as a target. On bombardment the target material sputters and diffuses into the discharge. The mean free paths of these sputtered particles are in the range of the distance between anode and cathode. If a substrate is on the anode, this is coated with the target material. Additionally, the ion bombardment of the cathode causes the emission of secondary electrons from the target. These are repelled by the cathode and penetrate into the bulk plasma, where they are held by the magnetic field lines. The electrons collide with the available neutral particles and ionize these.

A typical magnetron discharge parameter is as follows:

$B_0 \approx 200\,\text{G} = 0.02\,\text{T}$ (magnetic flow density with the radius R).

$p \approx 0.3$–0.7×10^{-2} mbar .

$j_{ion} \approx 20\,\text{mA/cm}^2$ (over the torus averaged ion current density).

$V_{DC} \approx 800\,\text{V}$.

The loss of energy of the electrons pro electron ion pair is E_C.

Electrons in the discharge lose energy through:

- Ionization
- Excitation
- Elastic collisions (elastic scattering).

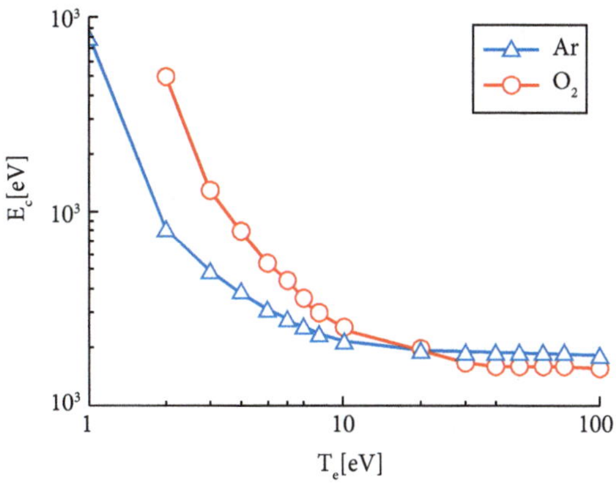

Fig. 6.32 Necessary electron energy E_C per electron–ion pair T_e [eV] (loss of energy of the electrons per electron–ion pair)

At high electron temperatures the loss of energy E_C generates an electron–ion pair that corresponds, for instance, to the double ionization energy E_{ion}, that is $E_C \approx 2E_{ion}\,f.\ T_e \geq 100\,\text{eV}$.

For electron temperatures below E_{ion}, E_C increases steeply, since the excitation increases strongly in relation to the ionization.

In argon ($E_{ion} = 15.8\,\text{eV}$) with $T_e \approx 4\,\text{eV}$ a total energy, for instance, $E_C \approx 40\,\text{eV}$ is needed (Fig. 6.32), in order to produce an electron–ion pair.

Molecule gases need additional energy (CO_2 in Fig. 6.31), in order to excite the vibratory and rotatory oscillations of the molecules. Further energy is used by dissociation processes. In these cases, E_C at the same temperature is approximately two to ten times higher than for noble gases.

6.9.2.1 Estimation of the Magnetron Voltage V_{DC}

When nearly the total voltage V_{DC} in the cathodic skin zone drops, the ion energy is, for instance, V_{DC}. If an ion strikes the target with an energy of approximately 200–1000 eV, then in the case of argon ions striking an aluminum target, an electron is emitted per 10 ions that is $\gamma_{se} \approx 0.1$.

This secondary electron flies back to the plasma and becomes captured by the more or less horizontal streamlines of the field. The secondary electron produces pairs of N electrons–ions, whereby per pair the energy E_C is used:

$$N \approx \frac{eV_{DC}}{E_C}\,. \tag{6.36}$$

If the secondary electrons have energies between 200–1000 eV (not to be compared with temperatures!), then E_C is about 30 eV per electron–ion pair. Because of the nearly horizontal gradient of the streamlines of the field not all electrons become captured but some make a gyration in order to fly back again to the cathode, the effective secondary electron emission γ_{eff} is smaller than γ_{se}. Estimations give:

$$\gamma_{eff} \cong 0.5\gamma_{se} \cong 0.05\,. \tag{6.37}$$

The discharge is self-sustaining; if the secondary electron flying into the plasma has sufficient energy to create many ions, which again generate sufficient secondary electrons, then:

$$N\gamma_{eff} = 1\,. \tag{6.38}$$

Inserting Eqs. (6.36) and (6.37) into (6.38), we obtain:

$$\frac{1}{2}\gamma_{se}\frac{eV_{DC}}{E_C} = 1\,. \tag{6.39}$$

For the supply voltage of the magnetron we then obtain:

$$V_{DC} \approx \frac{2E_C}{e\gamma_{se}}\,. \tag{6.40}$$

For $E_C = 30\,\text{eV}$ and $\Gamma_{se} = 0.1$ we obtain $V_{DC} = 600\,\text{V}$. The secondary electrons emitted by the target gyrate around the magnetic streamlines of the field of B_0 with the gyration radius r_{CE}. In the horizontal area of the streamlines of the field, the vector of the electron velocity is perpendicular to the streamlines of the field. For the gyration radius we obtain:

$$v = \omega \times r \Rightarrow r_{CE} = \frac{v_\perp}{\omega_{CE}} = \frac{v_e}{\omega_{CE}}\,. \tag{6.41}$$

The electrons receive their kinetic energy by acceleration in the skin layer, which means:

$$eV_{DC} = \frac{1}{2}m_e v_e^2 \Rightarrow v_e = \sqrt{\frac{2eV_{DC}}{m_e}}\,. \tag{6.42}$$

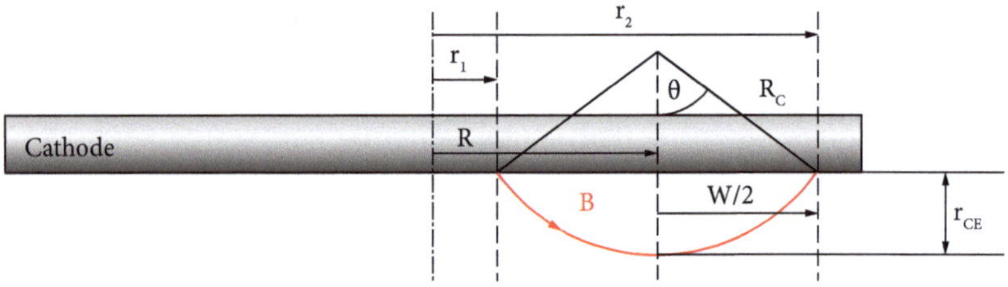

Fig. 6.33 Schematic illustration of the magnetron principle at the cathode surface

To determine the angular frequency, we used the equilibrium of forces between the centrifugal force and the Lorentz force:

$$qv \times \vec{B} = m\frac{v^2}{r} \Rightarrow qvB \equiv e\omega rB = m\omega^2 r$$
$$\Rightarrow \omega = \frac{eB}{m}. \tag{6.43}$$

Inserting Eqs. (6.42) and (6.43) into (6.41), we obtain:

$$r_{CE} = \frac{1}{B_0}\sqrt{\frac{2m_e V_{DC}}{e}}, \tag{6.44}$$

for $B_0 = 200\,\mathrm{G}$ and $V_{DC} = 600\,\mathrm{V}$ $r_{CE} \approx 5\,\mathrm{mm}$. Under the same conditions for argon ions we obtain gyration radius of $r_{Ar} \approx 1.3\,\mathrm{m}$. For an approximation of the ring width w, we must take 6.33 into account.

Figure 6.33 shows a part of the cathode from Fig. 6.31. The mean part of the plasma ring under the cathode corresponds, for instance, to the gyration radius r_{CE} of the electrons around the horizontal streamlines of the magnet field (more in the center, less at the edge).

For further calculations, it is accepted that the thickness of the skin layer is much smaller than the gyration radius, i.e. $s \ll r_{CE}$. If secondary electrons are emitted, these keep up with the streamlines of the magnet field and travel along them ($v \neq 0$). Thus, they arrive at the edge of the skin layer with r_1 and r_2, where they are reflected back due to the large barrier potential. Thus, they move between the two radii. If we accept the radius of curvature of the B-lx ines of flux to be R_C, and the height of the ring as r_{CE} and the width as $w \approx r_2 - r_1$, then we obtain:

$$\sin\theta = \frac{w/2}{R_C} \Rightarrow R_C \sin\theta = w/2. \tag{6.45}$$

Furthermore:

$$r_{CE} + R_C\cos\theta = R_C \Rightarrow R_C\cos\theta = R_C - r_{CE}. \tag{6.46}$$

After squaring Eqs. (6.45) and (6.46) and adding the two equations together, we obtain:

$$R_C^2(\sin^2\theta + \cos^2\theta) = (R_C - r_{CE})^2 + \frac{w^2}{4}. \tag{6.47}$$

With $\sin^2\theta + \cos^2\theta = 1$ we obtain for w:

$$w^2 = 4\left(2R_C r_{CE} - r_{CE}^2\right). \tag{6.48}$$

Then inserting into Eq. (6.47) the value already calculated for $r_{CE} = 0.5\,\mathrm{cm}$ and a typical value of $R_C = 4\,\mathrm{cm}$, the ring width is:

$$w \approx 3.8\,\mathrm{cm}.$$

The pressure for magnetron discharges is very small, thus there are no collisions in the outer zone. At the edge, the layer thickness, s, is small and there is a relatively high voltage (V_{DC}) drop. Therefore, the ion current density of the ring at the cathode can be calculated with "the $U^{3/2}$" from the ***Child-Langmuir law***:

$$\bar{j}_i = \frac{4}{9}\varepsilon_0\sqrt{\frac{2e}{m_{ion}}}\frac{V_{DC}^{3/2}}{s^2}, \tag{6.49}$$

with the dielectric constant ε_0 and the ion mass m_{ion}. At the same time, the ion current density corresponds to the entire discharge current divided by the ring surface:

$$\bar{j}_i = \frac{I_{DC}}{\pi(r_2^2 - r_1^2)} = \frac{I_{DC}}{\pi(r_2 - r_1)(r_2 + r_1)}. \tag{6.50}$$

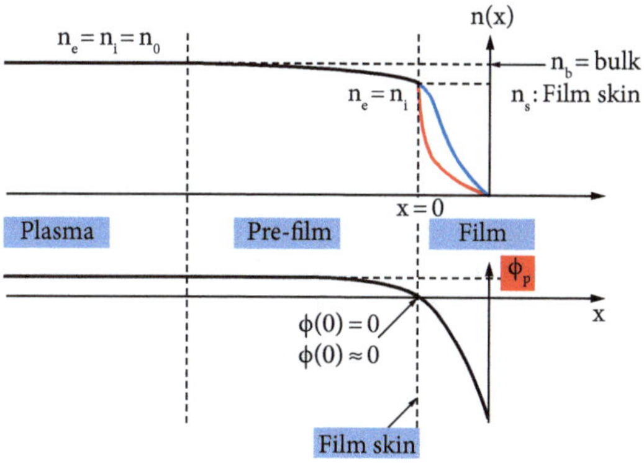

Fig. 6.34 Course of the plasma density on the magnetron surface

For $r_2 - r_1 = w$ and for $r_2 + r_1 \approx R + w/2$ we obtain:

$$\bar{j}_i \frac{I_{DC}}{\pi(Rw + w^2/2)} . \qquad (6.51)$$

Using the calculated value of $R = 4\,\text{cm}$ and $w = 3.8\,\text{cm}$ and a typical discharge voltage of $I_{DC} = 5A$, we obtain:

$$\bar{j}_i = \frac{5\,\text{A}}{70.4\,\text{cm}^2} \approx 71 \frac{\text{mA}}{\text{cm}^2} .$$

Applying this value in Eq. (6.48), we obtain for the thickness of the edge layer:

$$s \approx 0.6\,\text{mm} .$$

The above assumption that $s \ll r_{CE}$ ($= 5\,\text{mm}$) is justified.

6.9.3 Estimation of Plasma Density

To determine the plasma density in the magnetron, we consider the plasma skin layer [33].

The plasma is divided into bulk, pre-layer and layer edge. The plasma density at "the interface", between bulk and pre-layer is designated by bulk density n_b. The density at the layer edge, i.e. "the interface" between pre-layer and layer, is designated by n_s. The potential with n_b is F_p, which with n_s is set to zero, i.e. $F_s = 0$. The ions are accelerated in the pre-film with the ***Bohm*** velocity v_B, i.e.

$$v_B = \sqrt{\frac{kT_e}{m_{ion}}} \qquad (6.52)$$

is the velocity of the ions at the film edge (Fig. 6.34). When the ions move through the pre-film potential, the ion acquires the energy:

$$\frac{1}{2} m_{ion} v_B^2 = e\Phi_p . \qquad (6.53)$$

With Eq. (6.43), we obtain:

$$kT_e = 2e\Phi_p . \qquad (6.54)$$

In order to determine the plasma density of the edge layer n_s, the Boltzmann equation in regard to the bulk density: n_b is required:

$$n_s = n_b \exp(-e\Phi_p/kT_e) . \qquad (6.55)$$

So extrapolating from (6.54), we obtain:

$$n_{i,s} = n_s n_b \exp(0.5) \approx 0.61 n_b . \qquad (6.56)$$

Now the ion current density can be expressed:

$$\bar{j}_i = e n_{i,s} v_s = 0.61 \times e n_b v_B , \qquad (6.57)$$

and/or with (weak) T_e dependence the ion density can be determined:

$$n_i = \frac{\bar{j}_i}{0.61e} \sqrt{\frac{m_i}{kT_e}} . \qquad (6.58)$$

For $T_e \approx 3\,\text{eV}$ and a calculated current density of $j_i \approx 70 \times 10^{-3}\,\text{A/cm}^2 = 700\,\text{A/m}^2$, we obtain for the ion density:

$$n_i = n_e = \frac{700}{0.61 \times 1.6 \times 10^{-19}} \times \sqrt{\frac{40 \times 1.67 \times 10^{-27}}{1.38 \times 10^{-23} \times 3 \times 11604}} = 2.67 \times 10^{18} \frac{1}{\text{m}^3} = 2.67 \times 10^{12} \frac{1}{\text{cm}^3} .$$

6.9.3.1 Determination of the Sputter Rate

The calculation of the sputter rate now only requires the sputter yield, which is given in Fig. 6.5 with $\gamma_{\text{sput}} = 0.83$.

With the atomic density of aluminum in a solid body $n_{\text{Al}} \approx 6 \times 10^{22}\,\text{cm}^{-3}$, we obtain:

$$R_{\text{sput}} = \gamma_{\text{sput}} \frac{\bar{j}_{\text{i}}}{e} \frac{1}{n_{\text{Al}}} = 0.83 \frac{70 \times 10^{-3}}{1.6 \times 10^{-19}} \frac{1}{6 \times 10^{22}}$$
$$= 6 \times 10^{-6} \quad [\text{cm/s}]\,.$$

A sputter erosion of approx. 0.5 cm results in a 24 h period. The discharge carves a channel into the target in the form of a ring and the diameter is in accordance with the distribution of the ion current density over the ring width w.

An expensive target of highly pure material, which has a bad utilization represents a substantial cost factor. With mobile magnets or with new high frequency concepts this disadvantage can be reduced.

6.9.4 Discharge Power

The electrical supply power into the discharge is:

$$P_{\text{mag}} = I_{\text{DC}} U_{\text{DC}} = 5 \times 600\,\text{W} = 3\,\text{kW}\,.$$

Since the power is essentially consumed by the cathode, the cathode must be cooled.

6.9.5 Types

One differentiates between cylindrical and planar magnetrons. An example is the coaxial magnetron configuration shown in Fig. 6.35. It consists of electrodes made in the form of concentric cylinders, the target cathode being the innermost electrode and the substrates placed on the inside surface of the outer cylinder. The magnetic field is axial. Electrons from the cathode surfaced are trapped by the crossed electric and magnetic fields in the cathode dark space and can escape only by losing energy in ionizing collisions. The loss of energy results in the electrons moving in a tighter spiral path as they drift towards the anode. As this process continues, the energy carried by the electrons becomes smaller and the spiral path becomes tighter. This process results in a greater efficiency of ionization and a substantial lowering of the power dissipated by the electrons at the anode. The configuration can be used for DC or RF discharges.

The plasma concentration results from the E×B drift. The magnetic pressure $P_{\text{mag}} = B^2/2\mu_0$ increases the concentration of the plasma in the environment of the cathode surface. Planar magnetrons have greater importance in thin film technology than cylindrical magnetrons [35, 36]. The planar magnetron, Fig. 6.36, is a simple extension of the parallel plate discharge in which magnets (permanent or electromagnets) are placed behind the cathode. The magnetic field lines enter and leave normal to the cathode surface. The transverse component of the magnetic field in front of the target is arranged to be in the range 200–500 G. Electrons are trapped in the crossed electronic and magnetic fields, both enhancing the efficiency of the discharge and reducing the energy dissipated by electron arriving at the anode.

The target is bonded on a water-cooled disk in order to dissipate the heat caused from the

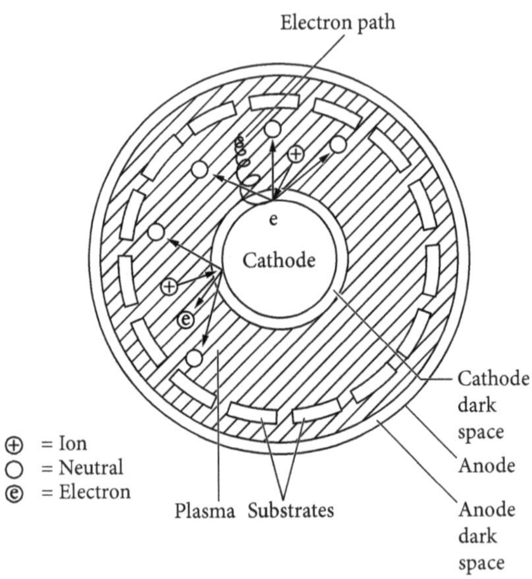

Fig. 6.35 Coaxial magnetron [34]

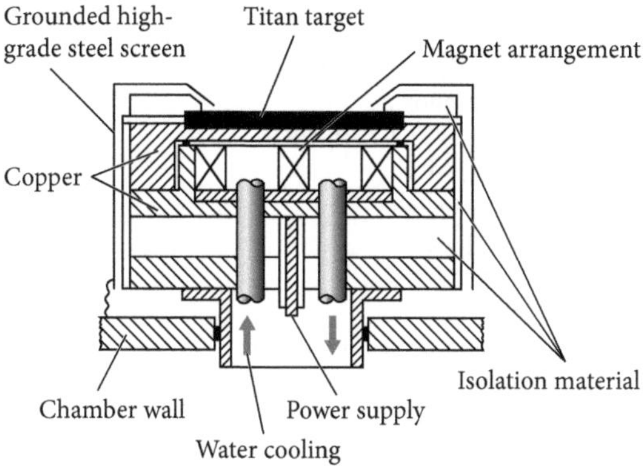

Fig. 6.36 Intersection through a planar magnetron

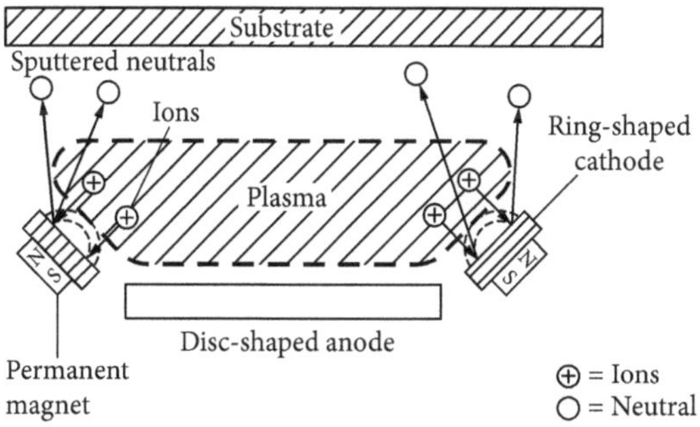

Fig. 6.37 Circular magnetron. Electrons originating at the cathode are confined by fields from permanent magnets placed behind the cathode. They lose energy in spiral paths in the plasma and are collected by the disk anode. The substrate, not being an electrode of the system, collects only sputtered neutrals

ion bombardment [37, 38]. Circular magnetrons (sputter guns or "S" guns) in which the cathode is in the form of a ring that surrounds a planar disk-shaped anode are also popular (Fig. 6.37).

Magnets behind the cathode enhance the discharge and the electrons are collected by the anode. A large fraction of the sputtered material from the target cathode is ejected in a forward direction and deposited on a substrate that no longer needs to be an electrode of the system. This way, very high power densities (50 W/cm^2) in the erosion zone of the target can be dissipated. The main disadvantage of planar and circular magnetrons is the fact that the deposition rate is not uniform across the region where the substrates are placed. This, as in evaporation, is overcome by an appropriate mechanical motion of the substrate table so that each substrate is exposed to the same average flux.

6.9.6 Cathodes

Table 6.2 shows some examples of sputter technology application, including the film functions and the materials used, and in Table 6.3 the respective requirements of the arranged targets are shown [37]. Targets are manufactured in vacuum by induction or electron-beam melting with subsequent casting into the desired form. Targets of high melting materials or from high melting compounds that are technically not producible can be manufactured metallurgically by pressing of powders followed by sintering or hot-pressing. Rectangular targets are used especially for manufacturing films on large surfaces, such as paper, plastic foils, windowpanes or other large-scale substrates. For DC voltage sputtering magnetrons, cathodes up to 350 cm in length are used for the production of thin films. Figure 6.38 shows some forms of cathodes.

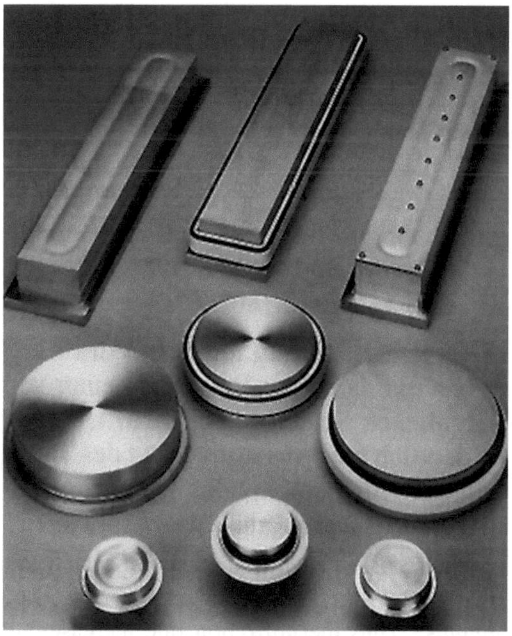

Fig. 6.38 Different target forms before and after use

Table 6.2 Examples of application of cathodic sputtering

Sector	Thin film function	Material
Microelectronic	Circuit path and contacts Buffer layers	Al, Al-alloys Ti, W, WTi, Cr silicide ($TaSi_2$, $MoSi_2$) Noble metals
Data storage	Magnetic storage layer for rigid disks Optic and magneto-optic memory	Co-alloys (CoNiCr) Cr, C FeCo-rare-earth-alloys Al, Si
Deposition of glass	Thermal insulation layer Coloration of glass Heated windows	Sn, InSn, ITO (In_2O_3/SnO_2) Ag, NiCr
Display technology	Electrodes for liquid crystal readouts	ITO (In_2O_3/SnO_2)
Hard material layers	Wear reducing layers for tools	TiN, TiAIN, HfN
Decorative coatings	Wear resistant decorative coatings (gold imitation, anticorrosive)	TiN, Au-alloys

Table 6.3 Requirements at targets

Application	Requirements/problems
Microelectronic (Al, Cr, WTi, Silizide)	Purity (> 5 N) Avoidance of alkali impurities Avoidance of alpha emitters (U, Th) Avoidance of precipitations uniform structure low O_2 content
Magnetic layers (Co-Basis, Ni-Basis)	Low magnetic permeability Avoidance of precipitations Purity Structure
Coating of glass (Sn, InSn, ITO)	Homogeneity (up 3 m) Avoidance of particle formation Cost
Hard materials layers Corrosion protection layers	Avoidance of particle formation Cost

The largest erosion ditch at the magnetron target surface is formed in the range where the electrical field and the magnetic field are perpendicular to each other. To increase the utilization rate, different possibilities are available. The method most frequently applied with round cathodes moves the magnets in form of Lissajous figures. Thus the utilization rate increases, the erosion of the target is more evenly distributed and hence there is an improved homogeneity of the film thickness.

Under continuous production conditions high target utilization rates with constant sputtering conditions are necessary. This is only possible by stabilization of the discharge parameters, with respect to the erosion profile, Fig. 6.39.

Figure 6.40 shows the dependence of the magnetic field force for different target thicknesses and the associated reduction in the discharge voltage for two different intensities of current. The operating point of the discharge is not only regulated by the power of the current supply. A reproducible coating process is only possible with stabilization of the magnetic field force at the target surface [38].

A constant magnetic field force can be attained by adjusting the height of the magnetic system according to the target erosion, Fig. 6.41 [38].

The lifting system also has the advantage that there is around a 5 to 10 % increase of the target utilization. The target utilization can be also improved substantially with an interpol target, Fig. 6.42 [39]. The magnetic field lines deviate parallel to the target surface; target erosion is more homogenous. So that the electrons do not escape out of the plasma, the pole pieces have a floating potential. Interpol targets are available in a round or rectangular form. A round cathode

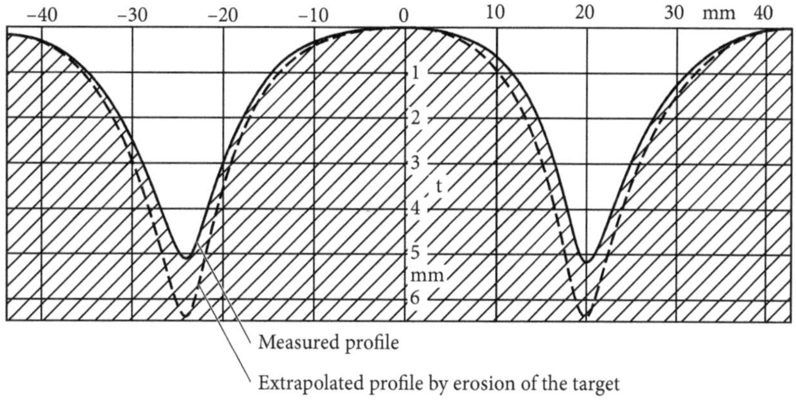

Fig. 6.39 Sectional cross-section through an eroded target during a typical utilization of 25–30 %

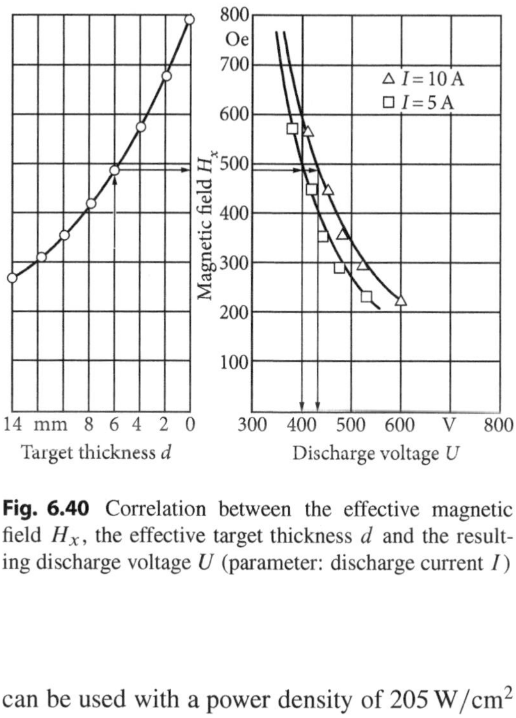

Fig. 6.40 Correlation between the effective magnetic field H_x, the effective target thickness d and the resulting discharge voltage U (parameter: discharge current I)

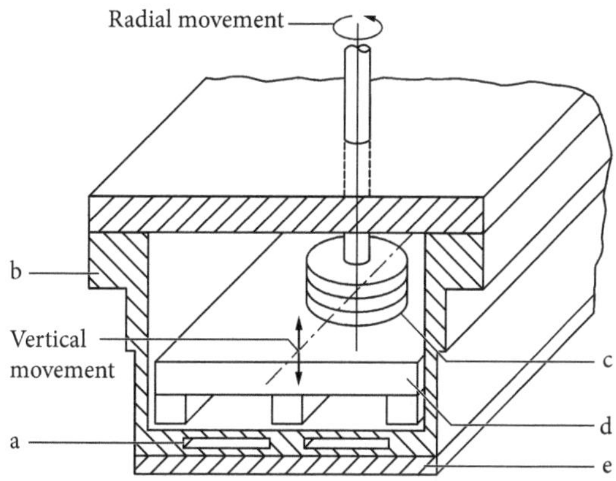

Fig. 6.41 Representation of the lifting mechanism in the sectional view. *a* Cooling, *b* retaining plate, *c* raising system, *d* magnetic cross-head, *e* target

can be used with a power density of 205 W/cm^2 in continuous operation. At a substrate distance of 50 mm, it is possible to obtain a coating rate of 201 nm/s. The density of the sputtered particles is so high that a sufficient number of the particles are ionized, and hence the sputter process can be maintained without noble gas discharge. With copper, the density of the ions is already high enough so that with a discharge power of 80 W/cm^2 no argon is necessary [39].

Reactive sputtering with magnetrons in the direct current modus has the disadvantage that arcing is a frequent occurrence. The reason is shown in Fig. 6.42.

Within the internal ranges of the erosion ditch, the ion current density becomes so high that no oxide or other insulator material is generated. On the sides of the discharge ditch a strong drop-off of the ion current density occurs. This finally results in fast creation of oxides and nitride films. These films lead to arc discharges. With the high discharge current at the points of breakthrough material evaporates, and subsequently particles strike the substrate, which leads to irregularities in the film. A reduction of the discharge arcs can be reached with an improvement of the interpol magnetron, Fig. 6.43.

A shutter between target and substrate and the gas inlet in the vicinity of the substrate cause a reduction of reactive gas transport to the target and also a reduction of the condensation rate. An AC

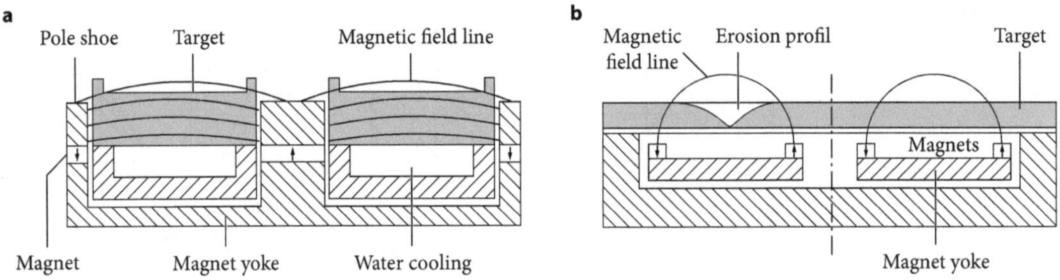

Fig. 6.42 **a** Schema of an interpol magnetron, **b** Erosion profile of an interpol magnetron coating rate: 60 nm/s. Utilization of target: 60 %

magnetic field that is attached laterally in front of the poles of the magnet has two advantages:

- The magnetic field extends deeper into the discharge area and in closer vicinity to the gas inlet and to the substrate.
- With directly opposing magnetic fields, a broadening effect on the sputter ditch of the target occurs.

Only a small zone remains, in the millimeter range, a poorly conducting layer develops. Therefore, the number of arcing discharges is substantially reduced at the direct current magnetron with the result that the insulation surface range is significantly larger [39]. Rotating magnetrons have good target utilization and are stable in operation; three different types of systems exist [40].

Figure 6.44 presents the system were the target rotates around a fixed magnetic system.

Extensive cooling of the target is possible. This type is effective for sputtering of low melting point materials. With the use of target rotation, nearly the entire target surface in the active zone is exposed to an intensive ion bombardment. A stable and long-term mode of operation in a reactive gas atmosphere is attainable. A reverse type of fixed target and rotating magnets shows similar attributes, Fig. 6.45.

This system is suitable for sputtering of alternating materials of different types. A further system (Fig. 6.46, [40]), which utilizes a rotating magnetic field, can be a cylindrical target sputtered all over, without long-term operation leading to instabilities.

The Twin MAG sputter system offers high condensation rates, large target utilization rate and arc-free long-term operation, Fig. 6.47. Two identical, standard magnetrons lying closely side

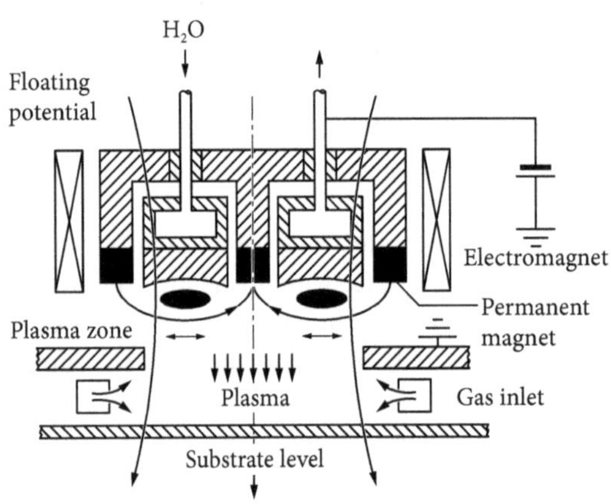

Fig. 6.43 Schema of an interpol magnetron with an additionally overlaid electromagnetic AC field

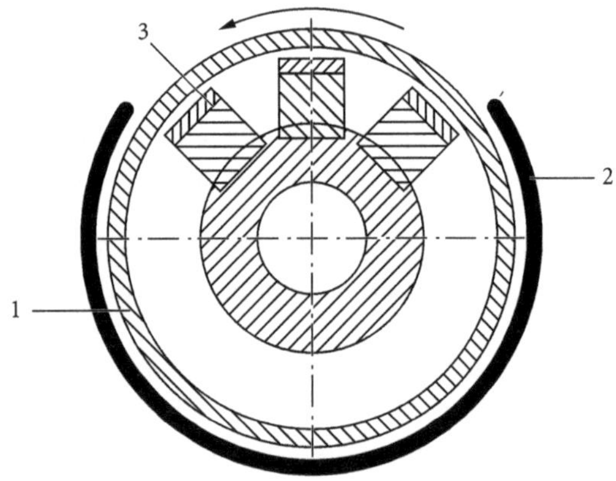

Fig. 6.44 Pattern of a cylindrical magnetron with a rotary target. *1* Target, *2* dark space shields, *3* fixed permanent magnets

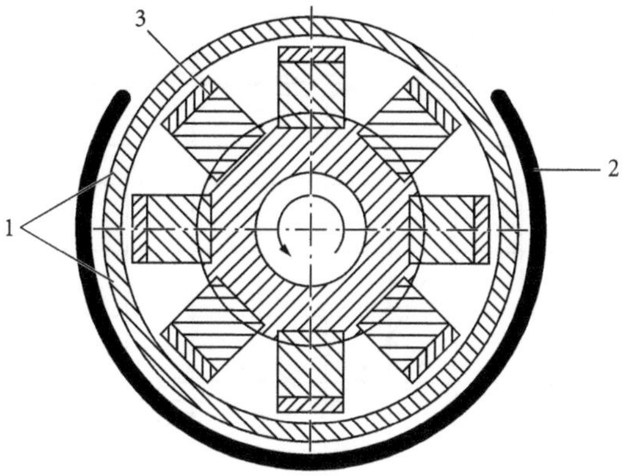

Fig. 6.45 Schema of a cylindrical magnetron with rotary magnets. *1* Target materials, *2* dark space shields, *3* rotary permanent magnets

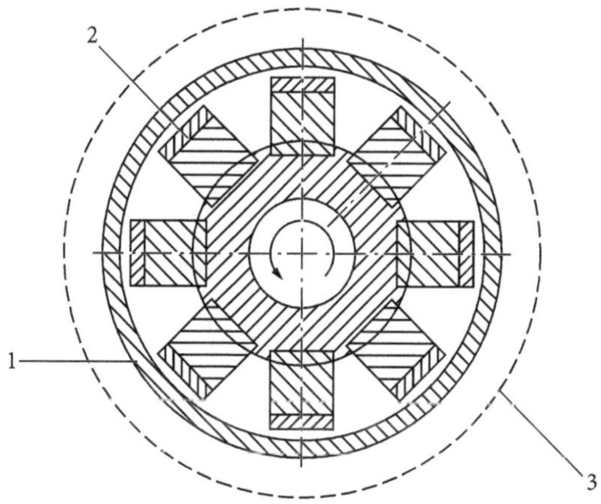

Fig. 6.46 Schema of a cylindrical magnetron with sputtering of the entire lateral surface. *1* Target, *2* rotary permanent magnets, *3* anode

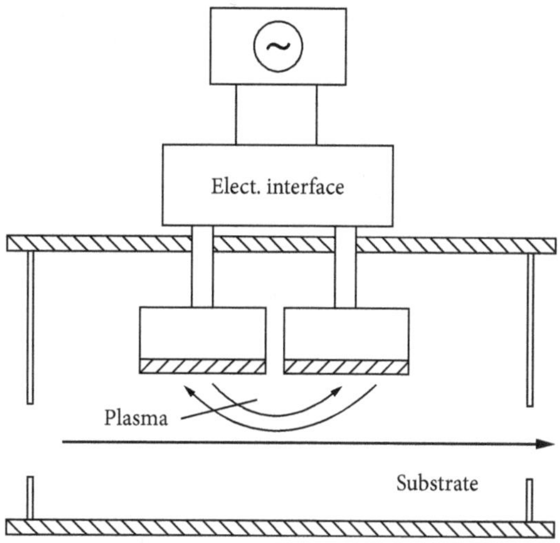

Fig. 6.47 Schema of a twin-MAG sputtering system

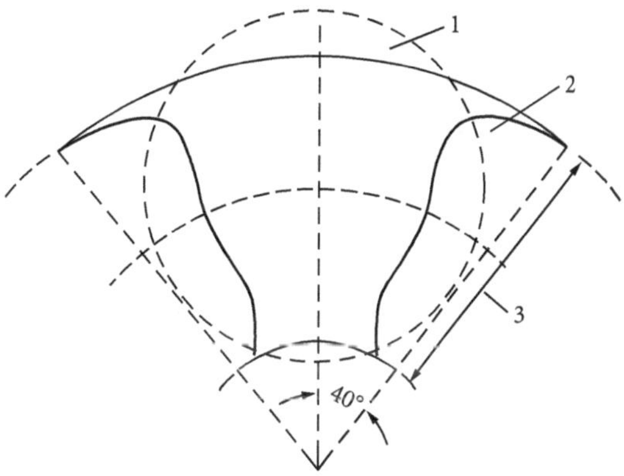

Fig. 6.48 Distribution shutter for a magnetron cathode of 200 mm in diameter. *1* Cathode, *2* distribution shutter, *3* circular ring provable with substrates

by side form a unit. On both cathodes an AC voltage is applied with a frequency of some kHz. At one-half periods, one magnetron serves as the cathode, the other as the the following half-wave the polarity is inverted. The consequence is that any generation of an insulating layer on the target edge will be uncharged in the second half-period as it switches anode polarity. Therefore, no high voltage develops, which could lead to a breakthrough.

6.9.7 Shutters

At start-up or after a long operation time, the recipient must ventilate or the working gas must be changed. Thereby, impurities or out-gassing products from the target must be eliminated. In large manufacturing facilities, a pre-sputtering conditioning process with no test substrate or test substrates is used, or in the case of smaller systems, a shuttering screen is used. Substrate cleaning by particle bombardment or structuring of target layers by etching must likewise involve a shutter to protect the target surface against impurities. When substrate coating with a round cathode is used, a shutter for homogeneous distribution of the film thickness is necessary. Figure 6.48 shows an example of a distribution shutter.

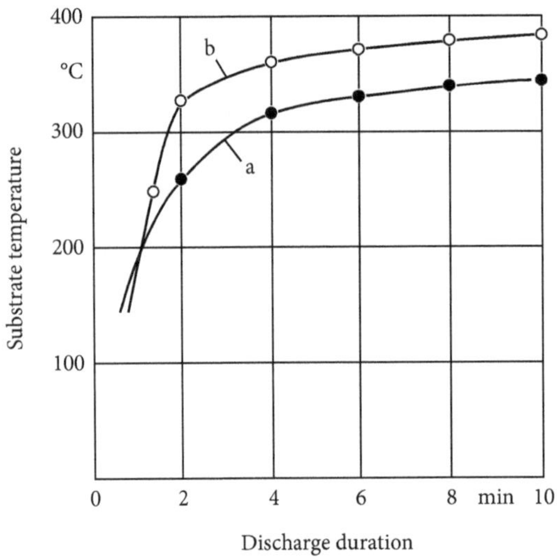

Fig. 6.49 Substrate heating up with a DC voltage discharge; a = $0.74\,W/cm^2$, b = $1\,W/cm^2$

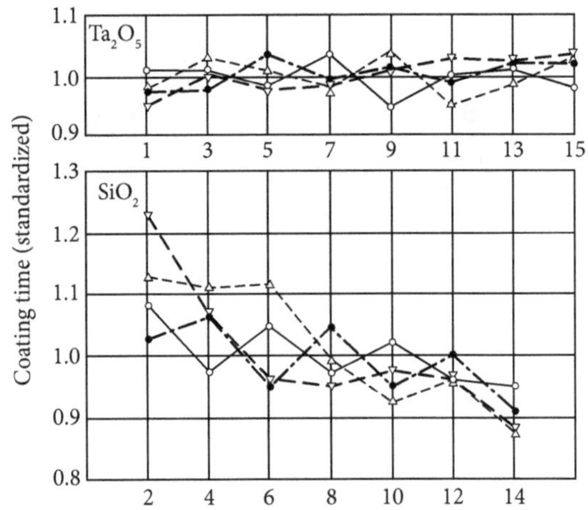

Fig. 6.50 Standardized representation of the coating duration of the individual $\lambda/4$-layer at the production of four equal edge filters

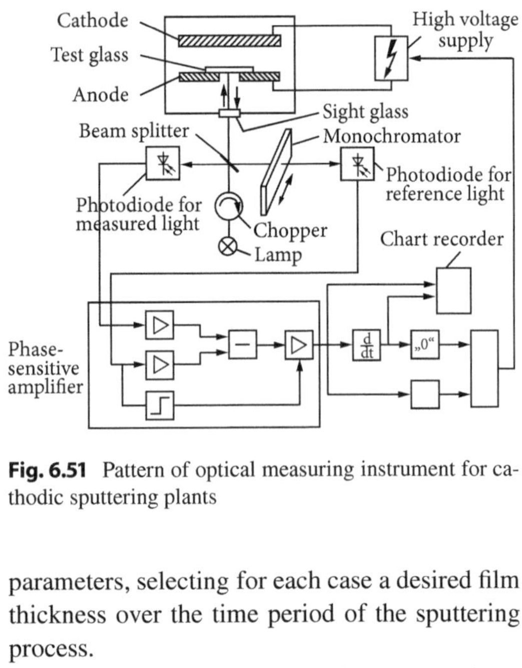

Fig. 6.51 Pattern of optical measuring instrument for cathodic sputtering plants

6.9.8 Infrared Heating

For homogeneous film deposition, not only constant discharge parameters should be taken into consideration but a constant substrate temperature is also necessary. The coating process usually begins with a lower substrate temperature and subsequently increases due to particle bombardment. Until temperature equilibrium is reached, a difference in film morphology and, therefore, in film properties will be occur. Figure 6.49 illustrates this, whereby all three thermal factors, heat conductivity and heat transition coefficients, influence the curve progression. Infrared heating before starting the coating process improves the homogeneous structure of the film.

6.9.9 Rate and Film Thickness Measuring Instruments

Whereas with vacuum evaporation the evaporation rate increases exponentially with the source temperature, the coating rate during cathodic sputtering and the condensation rate are far less critically dependent on the discharge parameters.

One can usually tune in rate or film thickness measuring instruments generated with discharge parameters, selecting for each case a desired film thickness over the time period of the sputtering process.

For determination of the coating rate as a function of the process parameters the use of a quartz crystal oscillator is useful. Without interruption of the vacuum (in situ) all relevant parameters can be determined. There are exceptions when exact control of the film thickness and optically effective films with high precision demands are necessary during the coating procedure. Sputtering of a SiO_2-target is very sensitive to the residual gas. According to Fig. 6.50 [41], the

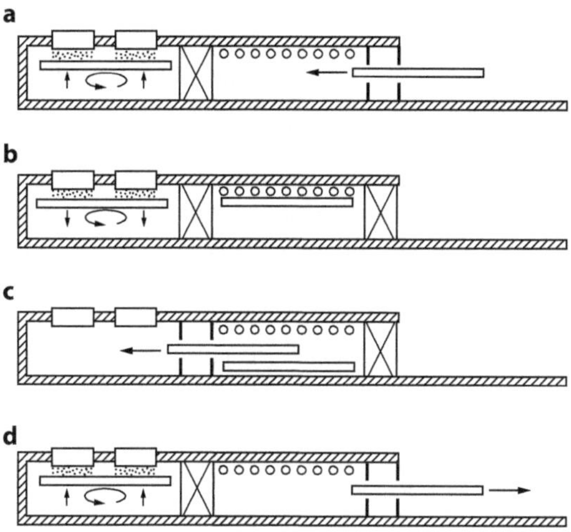

Fig. 6.52 Process steps in a cathodic sputtering plant with air-lock; **a** Introducing the substrates to be coated in the load lock, **b** Pretreatment of the substrates, **c** Introducing the substrates into the coating chamber, **d** Discharge of the substrates from the load lock, the substrates coating

deposition duration necessary for generating the first SiO_2 film is substantially longer than for the following films where the inside of the equipment is better conditioned.

Figure 6.51 [42, 43] shows the schema of an optical measuring instrument that uses the minimum/maximum method to obtain an exact signal for a given film thickness. The condensation rate for depositing a SiO_2 film is so constant that it is not necessary to use a PID (proportional-integral-derivative) controller. The optical instrumentation serves as a thickness monitor only.

6.9.10 Two-Chamber Equipment

The film properties are only reproducible if the residual gas does not have a noticeable influence during the deposition. This requires a high effective pump speed and sufficient recipient conditioning. Moisture, recipient temperature and cleanliness after each ventilation or conditioning procedures will result in different partial pressures of the residual gases. With a load lock arrangement in which the coating chamber remains under constant vacuum, these disadvantages do not arise. The simplest form of load lock arrangement is shown in Fig. 6.52

The chamber and the load lock are separated by a valve. During the recovery of the coated substrate and the evacuation of the load lock this valve remains closed. The residual gas conditions in the coating chamber can be kept constant, which saves long conditioning times. These simple load lock installations do not offer optimal use of the substrate transport and the pump down time. With a double transport system this disadvantage is eliminated. Two substrate carriers can be treated at the same time, since the load lock chamber possesses two transportation levels. While the processes are executed in the coating

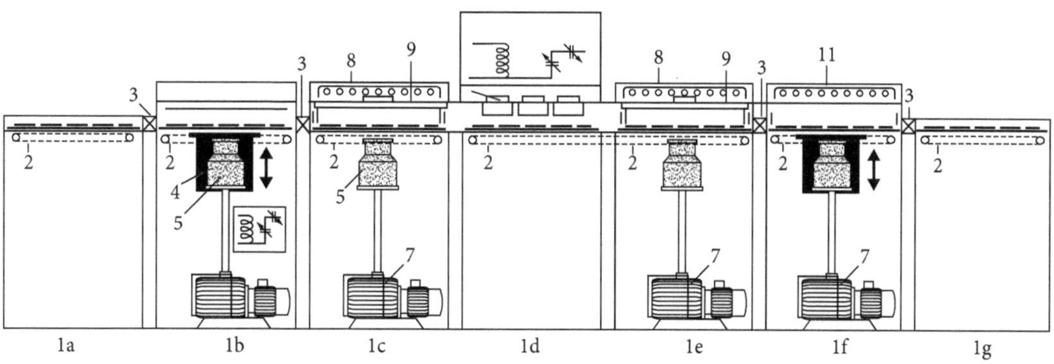

Fig. 6.53 Schema of a sputtering plant with seven modules. *1a* Loading module, *1b* lock-in module, *1c* process module, *1d* process module, *1e* process module, *1f* lock-out module, *1g* reloading module, *2* transport system, *3* valve, *4* lifting device for HF-etching or for cooling, *5* turbo-molecular pump or cryopump, *6* electrical matching elements, *7* pre-vacuum pump, *8* substrate heating, *9* DC or RF cathode, *10* DC or RF magnetrons, *11* cooling station

chamber, the second substrate carrier is loaded into the load lock (Fig. 6.53*a*) and then evacuation and pre-treatment steps are possible (heating, etching) (Fig. 6.53*b*). After coating, there is an exchange of processed and non-processed substrates. Subsequently, the coated substrate is removed and the sputtering process is only interrupted briefly.

6.9.11 Multi-Chamber Equipments

Multi-chamber systems are arranged in a modular construction, which offers large production capacity and flexibility, in addition to optimum reproducibility for production of different film systems. All pre and post-treatments, including different sputtering options with any number of integrated cathode arrangements, are possible. Figure 6.53 shows a seven-module design.

References

1. Westwood WD (2003) Sputter Deposition AVS Education Committee Book Series, vol. 2
2. Ziegler JF, Biersack JP, Littmark U (1984) The Stopping and Range of Ions in Solids Stopping and Ranges of Ions in Matter, vol. 1. Pergamon Press, New York
3. Ortner K, Birkholz M and Jung T (2003) "Neue Entwicklungen beim Hohlkatoden-Gasflusssputtern". *Vac. Praxis* (in German) **15**: 236–239 M. Birkholz and T. Jung (2003). *Vac. Praxis* (in German) **15**: 236–239
4. Sigmund P (1987) Mechanisms and theory of physical sputtering by particle impact. Nucl Instr Meth Phys Res B 27:1
5. Binder K (1992) The Monte Carlo Method in Condensed Matter Physics. Springer, Berlin [u. a.], ISBN 3-540-54369-4; 2. Auflage, 1995, ISBN 3-540-60174-0
6. Binder K, Landau DP (2000) *A Guide to Monte Carlo Simulations in Statistical Physics*. Cambridge University Press, Cambridge [u.a.], ISBN 0-521-65366-5; 3. Auflage, 2009, ISBN 978-0-521-76848-1
7. Laegreid N, Wehner GK (1961) Sputtering yields of metals for Ar^+ and Ne^+ ions with energies from 50 to 600 eV. J Appl Phys 32, pp 365/369
8. Oechsner H (1974) Ergebnisse zur Festkörperzerstäubung durch Ionenbeschuss Metalloberfläche. Ange. Elektrochemie 28 Jahr., pp 449/455
9. Yamamura Y, Mössner C, Oechsner H (1987) The bombarding-angle dependence of sputtering yields under various surface conditions. Rad Eff 103, pp 25/43
10. Oechsner H (1990) SNMS-investigations on the formation of sputter-generated molecules by atomic combination. Int J Mass Spectr Ion Proc 103, pp 32/43
11. Oechsner H, Schoof H, Stumpe E (1978) Sputtering of Ta_2O_3 by Ar^+-ions at energies below 1 keV. Surface 76, pp 343/354
12. Wehner GK, Rosenberg D (1960) Angular distribution of sputtered material. J Appl Phys 31, pp 177/179
13. Yamamura Y, Mössner C, Oechsner H (1984) Angular distribution of sputtered atoms from ionbombarded surfaces. Rad Eff 165, pp 31/41
14. Linhard J (1965) Influence of cristal lattice on motion of energetic charged particles. K Dan Vidensk Selzk Mat Fys Medd 34(14):64
15. Gesang WR (1979) Struktureffekte bei der Zerstäubung von Germanium und kubisch flächenzentrierten Metallen unter Beschuss mit Ar^+-ionen von 1 keV. Diss. TU Clausthal
16. Silsbee RH (1957) Focusing in collision problems in solids. J Appl Phys 28:1246–1250
17. Andersen GS, Wehner GK (1960) Atom ejection patterns in single crystal sputtering. J Appl Phys 31:2305–2313
18. Lehmann G, Siegmund P (1966) On the mechanism of sputtering. Phys Stat Sol 16, pp 507/512
19. Southern AL, Willis WR, Robinson MT (1963) Sputtering experiments with 1- to 5- keV Ar^+ions. J Appl Phys 34, pp 153/163
20. Glocker DA, Ismat Shah S (eds) (2002) *Handbook of Thin Film Process Technology* (2 vol. set). Bristol, U.K.: Institute of Physics Pub
21. Mattox DM (1998) Handbook of Physical Vapor Deposition (PVD) Processing: Film Formation, Adhesion, Surface Preparation and Contamination Control. Noyes Publications, Westwood, N.J.
22. Kishinevsky LM (1973) Estimation of electron potential emission yield dependence on metal and ion parameters. Rad Eff 19, pp 23/27
23. Cuomo JJ, Gambino RJ (1977) Incorporation of rare gas in sputtered amorphous metal films. J Vac Sci Techn 14, Nr. 1, p 152
24. Nowick AS, Mader SR (1965) A hard-sphere model to simulate alloy thin films. IBM journal 9(5/6):358
25. Bunshah, Roitan F (eds) (1994) Handbook of Deposition Technologies for Films and Coatings: Science, Technology and Applications, 2nd edn. Materials science and process technology series. Noyes Publications, Park Ridge, N.J.

26. Oechsner H (1974) Electron cyclotron wave resonances and power adsorption effects in electrode less low pressure HF plasma with a superimposed static magnetic field. Plasma Phys 16, pp 835/844

27. Fredrichs R (1962) J Appl Phys 33:1898

28. Depla D, Mahieu S (2008) Reactive Sputter Deposition. Springer, Berlin

29. Nakamura K Magawa K, Topuruda K, Komiya S (1977) Application of war resistant thick films formed by physical vapor deposition processes. Thin Solid Films 40, pp 155/167

30. Haefer RA (1987) Oberflächen- und Dünnschicht-Technologie (Teil 1). Springer-Verlag, Berlin, Heidelberg, New York

31. Vossen JC, Kern W (eds) (1978) Thin Film Processes. Academic Press, New York, pp 76–170

32. Butler HS, Kino GS (1963) Plasma sheath formation by radiofrequency fields. Phys Fluids 6(9):1346

33. Window B, N (1986) Charged particle fluxes from planar magnetron sputtering sources. J Vac Sci Technol A4, pp 196/202

34. Window B, W (1986) Unbalanced dc magnetron as sources of high ion fluxes. J Vac Sci Technol A4, pp 453/456

35. Münz ED (1991) The unbalanced magnetron: Current status of development. Surf Coat Technol 48, pp 81/94

36. Stjerna B, Granquvist GG (1990) Optical and electrical properties of SnO thin films made by reactive rf. magnetron sputtering. Thin Solid Film 193/194(1990):704–711

37. Peukert B, Scjmitty SU, Weigert (1991) Sputtertargets Metall, Forschung und Entwicklung. Degussa, Frankfurt/M, pp 255–272

38. Kukla R, Krug T, Ludwig R, Wilmers K (1990) A highest rate self-sputtering magnetron source. Vacuum 41:1968–1970

39. Krug TG, Beisswenger S, Kukla R (1991) High Rate Reactive Sputtering with a New Planar Magnetron. 34 th. Annual SVC Conference Philadelphia, March

40. Wright M, Beardow KT (1986) Design advances and application of the rotable cyclindrical magnetron. JVacSciTechnol A 4(3):388–392

41. Kienel G (1981) Optical Layers Produced By Sputtering. Thin Solid Films 77, pp 213/224

42. Kienel G, Stengel W (1978) Herstellen optischer Schichten durch Kathodenzerstäubung. Vak Techn 27, pp 204/211

43. McBridge, Michael W (1990) Optical Coaters. Breaking out of the box. Lasers & Optronics 3, :34–38

Plasma Treatment Methods

7

H. Frey

7.1 Plasma Polymerization

7.1.1 Introduction

Plasma polymerization is a contactless coating method with monomers of polymers in a plasma discharge [1, 2]. The monomers polymerize in the plasma and are deposited on the substrate.

Thin polymer coatings have other properties than polymer foils, which are manufactured by conventional methods, e.g. by evaporation of solvents from the polymer solution or by lamination from the polymer melt. A condition for this process is the presence of chain-forming atoms such as carbon, silicon or sulphur in the process gas. Since the monomer molecules are fractured in the plasma to reactive particles, the chemical structure of the outlet gas in the polymer coating remains at most partial, which entails cross-linking and unordered structuring (Fig. 7.1).

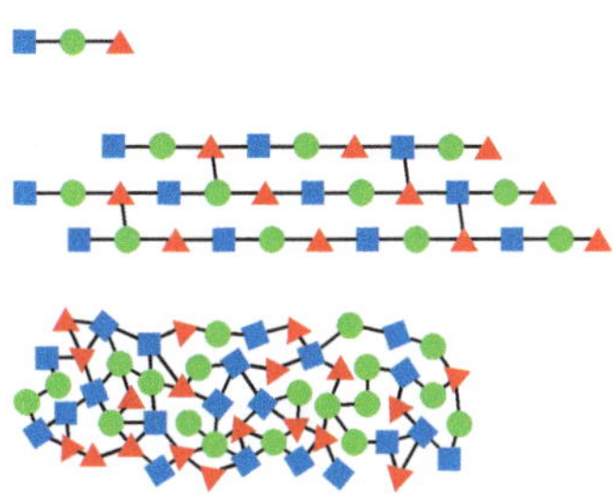

Fig. 7.1 Structure between chemically polymerized and plasma polymerized films

Structural preservation and cross-linking degrees can be controlled with process parameters such as pressure, process flow and electrical power, so that one can also develop so-called gradient films, which, e.g. exhibit a cross-linking degree that increases over the thickness [3–7].

The coating rate by plasma polymerization is between 5 and 500 nm/min. The coating rate as well as the physical and chemical properties of the film can be adjusted within wide ranges by variation of the process parameters such as pressure, flow, plasma power and frequency of the applied voltage. With the method of plasma polymerization, membranes are manufactured for seawater desalination, containment for medicines depot, antireflection coatings of optical components, protection of capacitors and semiconductor components against moisture.

With textiles both the breathing activity of polyester fabric and dirt deflecting films can be improved.

Properties of plasma polymerization are:

- There are no problems with drying process or solvent.
- The process is fast and adjustable.
- Only small heating up of the substrate is required.
- The film composition changes continuously (during the coating).
- There are insoluble, nearly pinhole-free layers with good adhesion.
- Batch or in-line processing can be used.

© Springer-Verlag Berlin Heidelberg 2015

H. Frey, H. R. Khan, *Handbook of Thin-Film Technology*, DOI 10.1007/978-3-642-05430-3_7

7.1.2 Physics and Chemistry of Plasma Polymerization

Monomer gas is excited in the range of a glow discharge at a plasma pressure between 0.1 and 10 mbar. Energy sources are the ions in the range of the cathode fall and the electrons. These can reach energies around 10 eV, which is enough in order to break chemical bonds and start up reactions, which would otherwise only run off at substantially higher temperatures. The cold glow discharges the plasma supply – similarly to the irradiation by electrons, gamma quanta or UV radiation –, at ambient temperature high energies in a very effective form for chemical reactions.

The gas discharge in monomer vapours can excite both by direct current (DC) and by alternating current (AC) [8]. At direct current discharge one of the two electrodes is heated. The deposited film can thereby be destroyed. The chemical composition of the films is, therefore, not homogeneous. With AC current these difficulties are reduced with rising frequency.

A plasma-polymerized film develops because the high energy ions and the electrons existing in the plasma take up energy in the electrical field. Thus excited molecules, ions and radicals [9] develop. At inelastic impacts the particle energy is to deliver to the monomer gas and/or to feed gas atoms. The actual polymerization process is only occurred on the substrate surface [10, 11]. If molecules impact the substrate surface to be coated, then the molecules are partly adsorbed.

If the adsorption energy is higher than the condensation energy and the saturation vapour pressure of the monomer is very much larger than the process pressure, then in the equilibrium a monomolecular coverage of the coating surfaces develops. Impacting monomer ions introduce the polymerization process; thus the diffusion equilibrium is disturbed, and new molecules can add themselves.

Time and area related number N of the impacting molecules is:

$$N = \frac{1}{6} n \overline{v}, \tag{7.1}$$

whereby

$\overline{v}$ the middle thermal velocity and n the monomer gas density, and

N for cyclohexane by an assumed temperature of 300 K is $4.35 \times 10^{20}\,\mathrm{cm^{-2}\,s^{-1}}$.

A current density of $2\,\mathrm{mA/cm^2}$, as corresponds, for instance, to the plasma polymerization conditions, is identical to a flux of 12.4×10^{15} elementary charges for each square centimetre and second.

From approximately 10^5 molecules that strike the substrate surface, there is only one ion, so that the thermal diffusion process on the substrate surface take place practically undistributed.

Schematically the polymerization process in the adsorption film of the solid surface takes place about as follows:

$$R\cdot_n + M \overset{K_1}{\rightarrow} R\cdot_{n+1}, \tag{7.2}$$

$$R\cdot_m + R\cdot_n \overset{K_2}{\rightarrow} P_{n+m}, \tag{7.3}$$

$$R\cdot_n \overset{K_3}{\rightarrow} R\cdot_n \quad \text{(adsorbed)}, \tag{7.4}$$

where K_i are the reaction rates, R free radicals, M an adsorbent monomer gas molecule, and P the polymerized final product.

The different product quantity P depends on the binding energy in the molecular structure of the monomer used. It is needed for:

Opening of a C=C bond nearly 2.6 eV,
disconnecting of a CC bond (aliphatic) nearly 3.7 eV,
disconnecting of a C–H bond nearly 4.3 eV,
disconnecting of a C–F bond nearly 5.4 eV.

As an example, some reactions with the monomer ethylene are considered:

$$\begin{array}{ll} & C_2H_4^* + e^- \\ C_2H_4 + e^- \rightarrow & C_2H_4^+ + 2e^- \\ & C_2H_3\cdot + 2H\cdot + e \\ C_2H_3 \cdot C_2H & C_4H_4\cdot \rightarrow -(C_2H_4)_{\overline{n}}\,. \end{array}$$

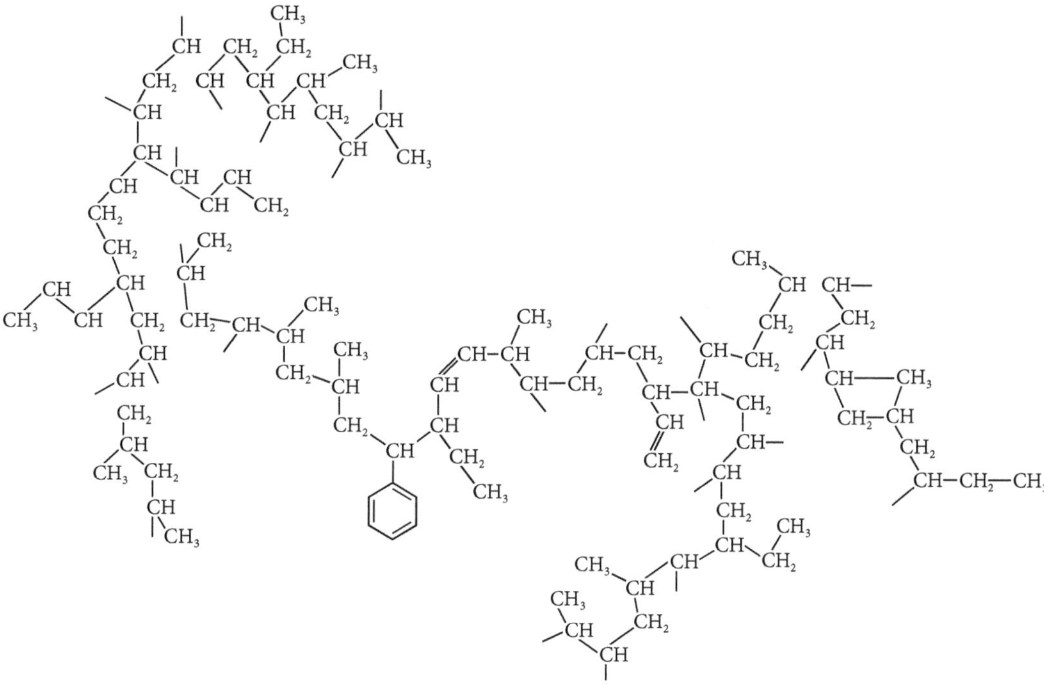

Fig. 7.2 Structural formula of a plasma-polymerized ethylene film

The structural formula of the final product is shown in Fig. 7.2.

The polymerization procedure occurs in steps. We can differentiate them roughly as follows:

- Excitation in the gaseous phase: forms highly reactive molecules with up to seven carbon atoms.
- Adsorption: excited molecules like monomer molecules adsorb on the solid surface.
- Polymerization on the surface.

Figure 7.3 shows the principle of plasma polymerization equipment with capacitive coupling of electric power. The recipient with the substrate to be coated, must be arranged in the inside in such a way that the monomer gas flows laminarly over the substrate surface. By variation of the process parameters such as pressure, monomer flow, plasma power and frequency of the voltage, the physical and chemical properties of the films can be changed [12]. As monomers for the plasma polymerization nearly all organic gases and liquids come into consideration whose

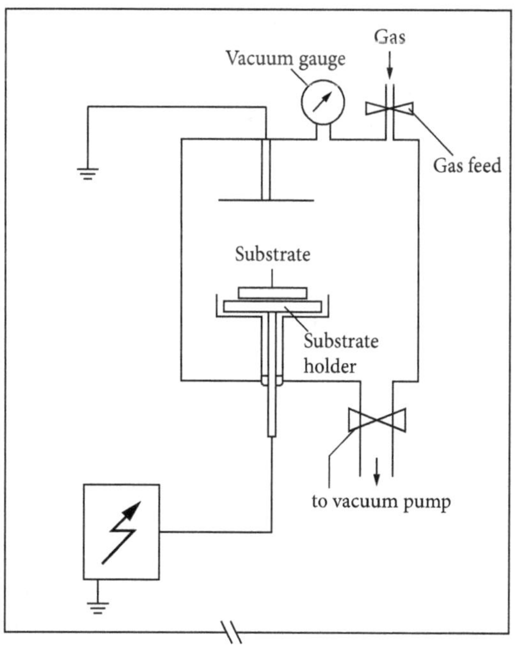

Fig. 7.3 Schematic structure of equipment for plasma polymerization

Table 7.1 Monomers

Styrol	
Acrylonitrile	
Ferrocene	
Acrylic acid	
Vinyl ferrocene	
Ethylene	Hexamethyldisiloxane
Propylene	Triethylsilane
Acetylene	Tetraethylsilane
Butadiene	Pyridine
Vinyl acetate	
Propylene oxide	
Methyl ethylene	
Cyclohexane	
Benzene	
1,3,5 trichlorobenzene	
Toluol	
Aniline	
Divinylbenzene	
Xylene	
Tri fluorine ethylene	
Tetra fluorine ethylene	
Hexafluoropropane	
Hexafluoropropylene	

Table 7.2 Polymerization effectiveness of monomers

Monomer	Effectiveness [g/kWh]	mol [kWh]
Vinyl ferrocene	300	1.4
1,3,5 trichlorobenene	190	1.05
Chlorobenzene	75	0.67
Styrol	69	0.66
Ferrocene	67	0.36
Picoline	65	0.70
Naphtaline	62	0.48
Pentamethylbenzene	61	0.44
Nitrotoluene	55	0.40
Acrylonitrile	55	1.04
Toluol	38	0.41
Aniline	38	0.41
Tetra fluorine ethylene	12	0.12
Ethylene	11	0.39
Acetylene	9	0.35
Thiophene	13.5	0.16
1,2,4-trichlorobenzene	5.5	0.03
p-xylene	45	0.42

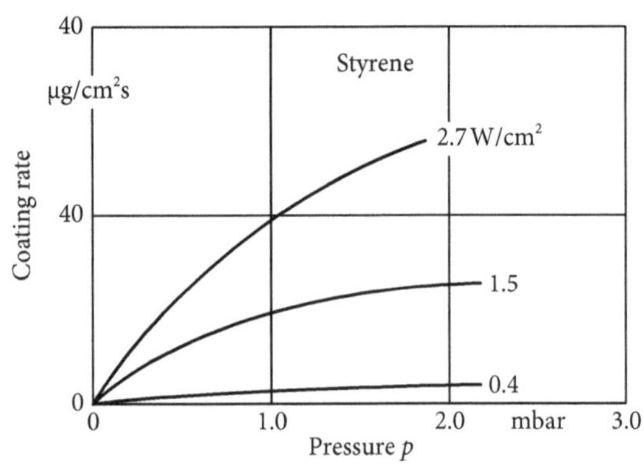

Fig. 7.4 Growth rate as a function of pressure and power

vapour pressure is at an ambient temperature higher than 1 bar. Relatively fast growing films are developed with styrene, acrylonitrile, ferrocene and acrylates; with siloxane, silazane, ethylene, propylene, benzene, toluene and xylene, however, they polymerize more slowly [13]. Table 7.1 shows some common monomers.

Table 7.2 shows the polymerization effectiveness of monomers created in a parallel reactor.

The coating quality and the growth rate are determined primarily by the selected monomer and/or the monomer mixture. Processing parameters such as pressure, flow rate, electrical power and frequency, and the position of the coating materials are less important.

By adding inert gases to the monomers, in many case one can increase the velocity of the polymerization. For the development of a plasma-polymerized film several monomers must be selected and then the optimal processing parameters must be determined by experimental tests.

After taking the polymerized coating out of the coating chamber, the film surface reacts with oxygen and hydrogen in air. This leads to a further decrease of the solubility and an increase of the specific weight. This reaction shows that radicals are still present in polymerized films. In air the radicals are converted into peroxide radicals, which can further be reacted to hydro peroxides or diperoxides. These molecules lead to subsequent secondary reactions, which result in rigid and hard films. Typical influence curves of the

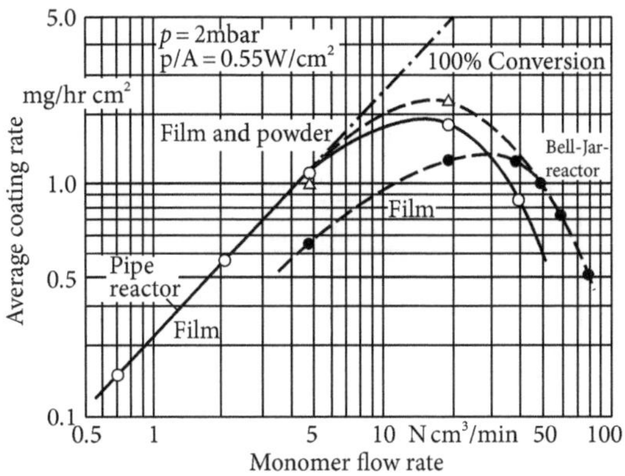

Fig. 7.5 Growth rate as function of monomer gas flow for two different reactor types (pressure 2.5 mbar, power 0.55 W/cm^2)

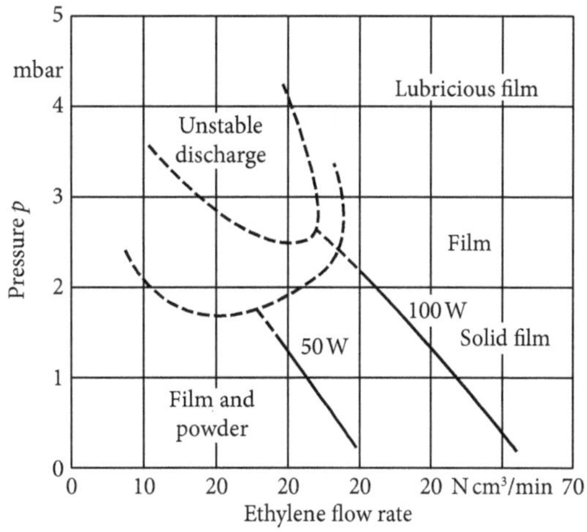

Fig. 7.7 Characteristic curves of plasma polymerization of ethylene

parameters pressure, power, flow and frequency are shown in Figs. 7.4 and 7.5.

The growth rate rises with increasing pressure and plasma power, the film quality is reduced up to an optimum. This is only valid for the flow (Fig. 7.5).

Figure 7.6 shows the influence of the frequency. However, we should not only see the growth rate but we should also consider the film quality and the process engineering aspects of the different frequency ranges. The parameter flow rate, pressure and electrical power feed into the plasma are plotted in characteristic curves in Fig. 7.7.

7.2 Plasma Diffusion Treatment

The industrial use of the plasma diffusion treatment methods began in the sixties. Today nitrogen, carbon, boron and metallic alloying elements can be diffused with diverse procedures into the edge zones of workpieces. A classification of plasma diffusion treatment in relation to other plasma procedures is shown in Fig. 4.1.

7.2.1 Description of the Method

The principle of the plasma diffusion method is shown in Fig. 7.8.

The workpieces that will be treated are isolated by being hung up at a charge rack, which locks in a recipient. The recipient is evacuated up to a pressure of 1 to 5 mbar. Between the charge rack and the recipient wall an electrical DCvoltage between 300 to 600 V is applied, so that the process gas is ionized in the recipient. A glow discharge is develops. The charge rack potential

Fig. 7.6 Growth rate of ethylene as a function of the discharge frequency

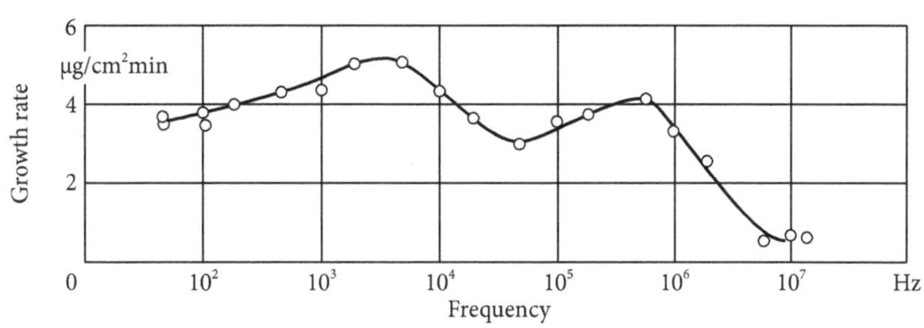

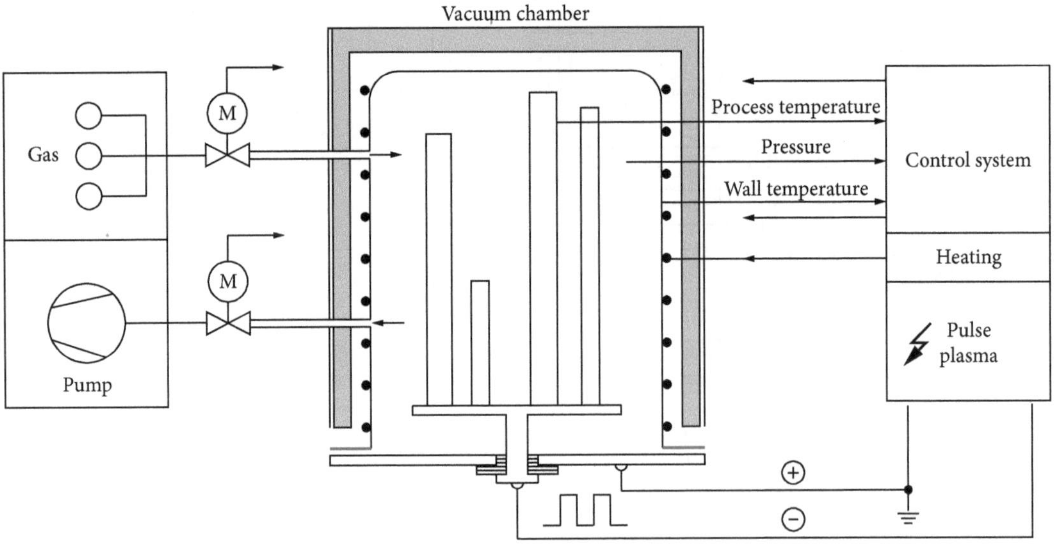

Fig. 7.8 Schema of equipment for plasma diffusion treatment with a cold wall recipient

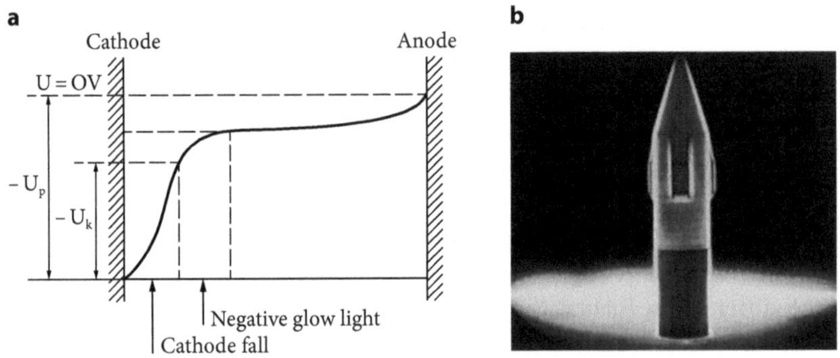

Fig. 7.9 **a** Potential process between anode (recipient), **b** glow discharge on a workpiece and cathode (workpiece), U_p process voltage, U_k: cathode fall

is negative. Positive ions are accelerated to the charge rack. In the range of the cathode fall of the glow discharge the work piece contours will be post-formed very well (Fig. 7.9).

The positive gas ions take up kinetic energy when they fly through the cathode fall and impact the surface of the workpieces. The energy that is transferred thereby leads to heating of the workpieces. Depending on the process gas mixture, the chemically reactive ions and excited atoms execute a chemical interconnection with the material of the workpieces. As a function of the workpiece surface temperature, particles diffuse into the substrate material. During the treatment process the process gas pressure must be held constant in the recipient. For nitrating or nitrocarburizing nitrogen and hydrogen gas mixtures, ammonia and/or nitrogenous gases with carbon additives such as methane are used with the material of the workpieces [14–16]. In the range of the outermost edging zone of the workpiece it comes to formation of a diffusion zone or to enrichment with nitrogen and/or carbon.

7.2.2 Fundamental Processes

The fundamental interaction processes between material surface and plasma are described in Fig. 7.10 .

Thereby the following processes occur:

- Heating of the workpiece
- Dislodge of atoms from the surface, whereby impurities of the surface are removed
- Penetration of gas ions into the workpiece surface

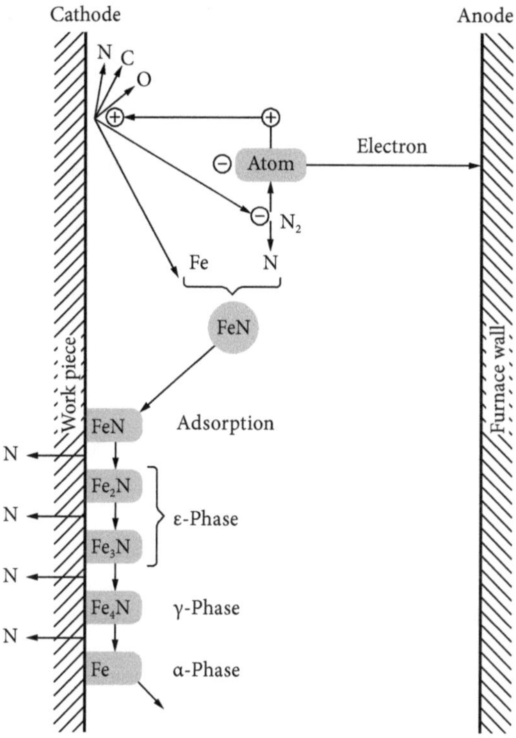

Fig. 7.10 Interaction processes between workpiece surface and plasma

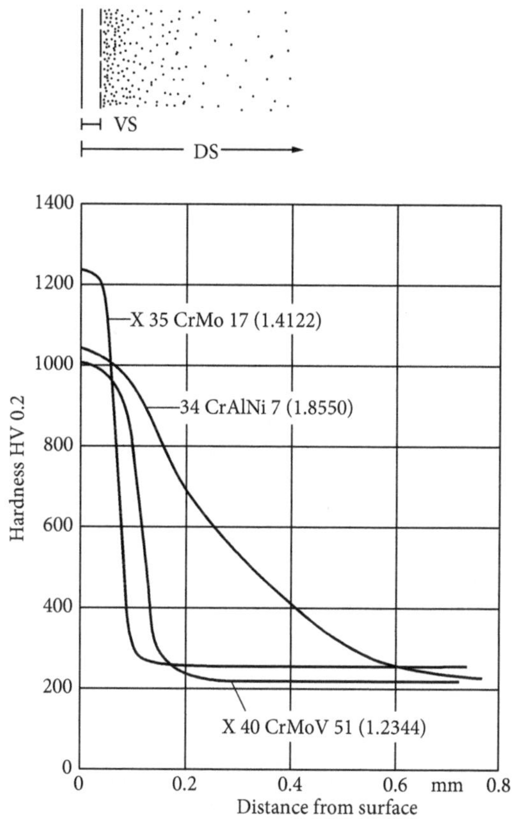

Fig. 7.11 Plasma of edge zones of nitrated workpieces from steel with the hardness–depth profile

- Release of electrons necessary for the further existence of the independent gas discharge.

7.2.3 Correlations Between Different Plasma Parameters

Aspects of plasma diffusion treatment are as follows:

- The ranges of voltage and current density for the perpetuation of an abnormal glow discharge are small. They depend on:
  - stability of the plasma
  - development of sparking.
- The current density in the plasma on the workpiece surface is large, which means:
  - Intense heating of the workpieces, in particular with a large surface/volume relationship
  - Intense sputtering of the workpiece surface by ion bombardment.
- The expansion of the glowing corona around the workpieces is low, which means:
  - Good post-forming by complex workpiece geometry
  - Good penetration in holes and screw threads of workpieces.

At the start of the process one begins with low temperature at low pressure. If the workpieces are heated only by the glow discharge, the heating-up time is long. To increase the temperature of the work pieces the gas pressure in the recipient is increased. The transition of the elements in the process gas to the material inside the workpiece is not yet completely resolved. Some models explain some phenomena, however contradictions due to experimental results remain. The ***model of Kölbel*** [17] is commonly used.

Figure 7.11 shows the curves of the hardness–depth profile of different samples from alloyed steels [18–20].

Table 7.3 Comparison of different plasma-supported diffusion procedures for the treatment of steel

Plasma diffusion method	Applicator gas	Treatment temperature	Time of treatment	Separation layer		Depth of diffusion
				Sort	Thickness	
		(°C)	(h)		(μm)	(μm)
Nitrogen hardening	NH_3, N_2	300–600	0.2–100	Fe_4N, $Fe_{2-3}N$	0–10	100–800
Nitro carbonization	NH_3, N_2, $+ CH_4$, CO_2	500–600	0.5–30	Fe_4N	2–20	100–800
Carbonization	CH_4, C_3H_8	800–1000	0.5–3	–	–	500–1500
Carbo nitriding	CH_4, C_3H_8, $+ N_2$	800–1000	0.5–3	–	–	500–1500
Borating	BCl_3, BF_3, B_2H_6	800–1000	1–10	FeB, Fe_2B	20–200	100–1000
Alloying	Al-, Cr-, Mo-, Ti-, W-Steam	900–1050	3–6	–	–	10–300

Table 7.4 Relation of different plasma-supported diffusion procedures for treatment of nonferrous metals

Plasma diffusion method	Applicator gas	Treatment temperature	Time of treatment	Separation layer		Depth of diffusion
				Sort	Thickness	
		(°C)	(h)		(μm)	(μm)
Nitrogen hardening of titanium	N_2	800–1000	1–20	TiN	1–5	20–80
Nitro carbonization of titanium	N_2, CH_4	800–1000	1–20	TiC-N	1–5	20–80
Carbonization of titanium	CH_4, C_3H_8	800–1000	1–20	TiC, a–C	1–5	20–80
Nitrogen hardening of aluminium	NH_3, N_2	450–500	1–10	AIN	1–5	1–5

7.2.4 Plasma Nitration of Titanium

In order to apply wear-resistant titanium nitride films on titanium surfaces, the titanium workpieces can be directly nitrated in the plasma [21–24]. Because of the large affinity of oxygen, the oxygen content in the recipient must be as small as possible during the treatment.

- The temperatures during the treatment must be between 800 and 1100 °C.
- The incorporated film thickness reaches values between 2 and 5 μm.
- The diffusion zone is between 20 and 80 μm.
- The hardness of the surfaces is appropriate between 1000 and 2000 HV.

7.2.5 Plasma Nitration of Aluminium

Aluminium alloys are usually made wear-resistant by anodic oxidation at the surface. This method can be substituted by surface nitration. Because of there always exist rugged oxides on aluminium surfaces, nitration without removal of the oxides from the substrates is not possible. A method to do this is plasma etching. The aluminium surface will not be damaged thereby. Aluminium is nitrated at 450 and 500, °C. The hardness of the nitrated layer amounts to about 1000 HV. The thicknesses of the films reach values between 4 and 5 μm.

7.3 Plasma Spraying

At plasma spraying a powdered coating material is transferred under use of a plasma flame from the solid into the liquid state. Afterwards it is generally deposited onto a nonheating substrate surface [25]. The principle of plasma spraying is shown in Fig. 7.12.

The energy source is an electric arc, which is ignited by means of a high DC-voltage spark or a high frequency spark. The arc burns between intensively water-cooled electrodes, a finger-shaped tungsten cathode and a copper anode are designed as a nozzle. The plasma gas is led into this nozzle, which consists of argon or a mixture of argon, helium, nitrogen and hydrogen. The electrons emitted from the cathode dissociate and ionize the gas atoms and gas molecules through collisions.

The arc base point rotates due to electromechanical effects and so a too high local heating of cathode and anode is avoided. The arc is constricted by thermal effects (thermal pinch) and by its own magnetic field (magnetic pinch) in the cross-section. At the nozzle opening, after leaving the arc, recombinations of ions to atoms and molecules in the plasma take place. By the released recombination energy, temperatures of up to 20,000 K arise in the core of the plasma. The energy content and the temperature of the plasma depend on the gas or gas mixture used. The enthalpy of the monatomic gases argon and helium is substantially lower than that of the diatomic gases hydrogen and nitrogen. On the other hand, the temperature is substantially higher in the plasma than with hydrogen and nitrogen [26].

With the feed gas stream, powder can be injected into the high energy plasma beam. After injection the powder particles are transferred into the liquid state and accelerated. The axial velocity component of the powder particles depends on the flow rate of the plasma gas and the volumetric expansion of the gas by the transition to the plasma. At a high particle velocity the flight phase of the powder particle to the impact on the substrate surface is only few milliseconds, on this

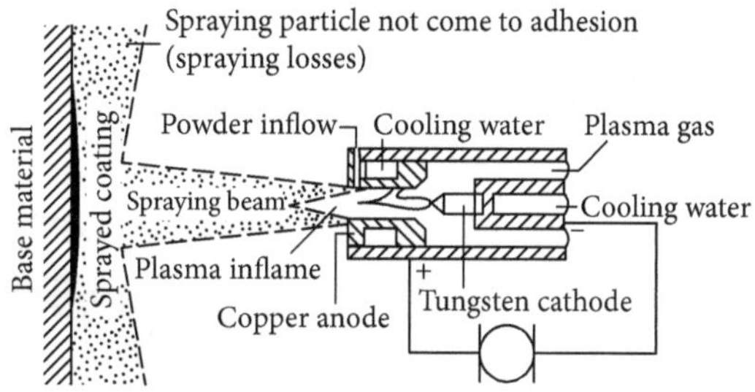

Fig. 7.12 Principle of plasma spraying

the particles solidify and form gradually a thick, compact bonding coating, Figs. 7.13 and 7.14.

Due to the construction space used for the plasma burner, coatings are possible only on freely accessible surfaces as well as on rotation symmetrical components. An internal coating of pipes with atmospheric plasma starting from an inside diameter of 30 mm is possible, whereby the coating depth is limited due to the nozzle construction. A coating of undercuts is, in principle, not possible.

An advantage of plasma spraying is given also by the simultaneous processing of more than only a powdered material to a film. The different powder materials are supplied to the plasma over separate nozzles [26]. The composition of the powder substance, and thus the properties of the film, can be changed by controlling the process. In order that the plasma-sprayed films adhere to the workpiece surface, the surface must, for example, be irradiated with corundum [27].

Plasma spraying is a chaotic process. The melted powder particles anchor themselves onto the surface, and with increasing film thickness into one another, and with atmospheric plasma spraying cavities arise. Therefore the treated surface has many lattice defects such as dislocations, microcavities or lattice deformations. The melted powder particles penetrate because of their high kinetic energy and the capillary effect in unevenness and undercuts. There they solidify very fast and it comes to mechanical anchoring. During the coating of metallic materials with metallic bulk material it also comes to metallurgical interactions between spraying particles and bulk material. At very high powder particle velocities

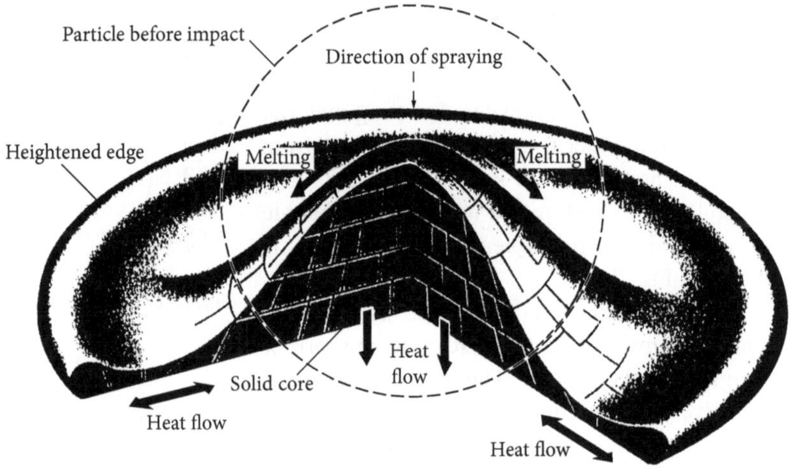

Fig. 7.13 Microscopically small droplet melted coating material

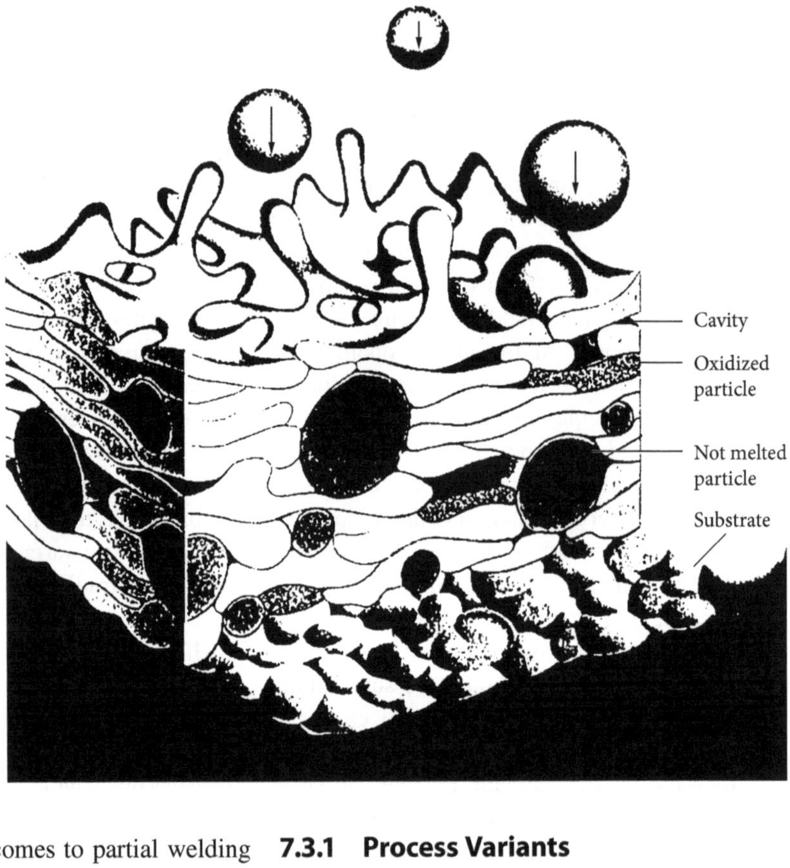

Fig. 7.14 Development of a plasma-sprayed layer

in a sprayed coating it comes to partial welding of the spraying particles among themselves. In addition, in the transient range between film and bulk material it can also come to microwelding. The effects of interaction can be seen in Fig. 7.14.

7.3.1 Process Variants

The number of variables influencing plasma spraying is very large. Their interactions are illustrated in Fig. 7.15.

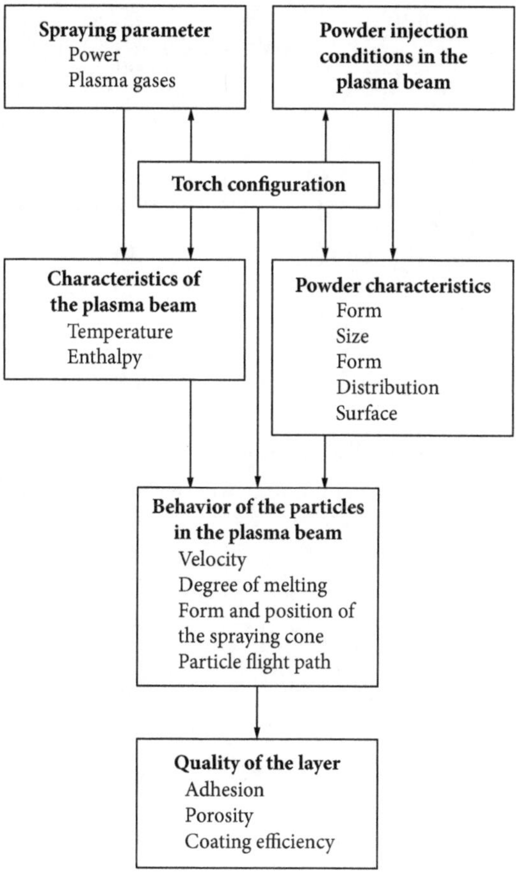

Fig. 7.15 Interaction effects of the most important plasma spray parameters

7.3.1.1 Processes in the Atmosphere

We differentiate between two processes: ***atmospheric plasma spraying*** and ***high speed plasma spraying***.

7.3.1.1.1 Atmospheric Plasma Spraying (APS)

The economically most important variant of plasma spraying is atmospheric plasma spraying [28]. The spraying particles can thereby react with the surrounding atmosphere, the materials of which are limited to such materials that have no or only a small disposition to oxidize. The oxide fraction that develops during the coating process is also introduced into the produced sprayed coating. The spraying distance from the nozzle to the workpiece amounts to a function of the material and the coating parameters between 100 and 150 mm. Typical materials for the APS procedure are ceramic(s), metals and also alloys.

The porosity of the films is between 1 and 5 percent. Atmospheric plasma spraying equipment is installed in chambers to protect the environment against noise and radiation, and have efficient exhaust and filtering units for the removal of spray dust. Frequently, a robot takes over the nozzle guidance in order to be able to coat complex geometry.

7.3.1.1.2 High Speed Plasma Spraying (HPPS)

High speed plasma spraying (HPPS) is an upgrade of previous conventional atmospheric plasma spraying. In particular, with the processing of high melting point ceramic materials, the kinetic energy of the particles represents a determining factor in the case of collision with the substrate, and the film quality regarding adhesion and porosity.

7.3.1.2 Processes in Controlled Atmosphere

7.3.1.2.1 Inert Gas Plasma Spraying (IPS)

Inert gas plasma spraying (IPS) takes place in chambers. With this version the plasma beam and the powder jet are projected by a surrounding protective gas that shields against the atmosphere, i.e. inert gas plasma shielding gas (SPS).

With inert gas plasma spraying in closed chambers, the process chamber must first be evacuated to a residual gas pressure of $< 10^{-2}$ mbar, so that ventilation can take place after the spraying process with an inert gas (argon). With small oxygen partial pressure, which is determined by the quality of the gases and the evacuation process, oxidation of easily reactive materials is prevented. During spraying one can also work with high cooling quantities of inert gas, so that the substrate temperature can be kept just as low as with the APS process. This of particular importance withe the processing of hard and brittle materials with small thermal expansion coefficients. When using conventional substrate materials (ferritic and austenitic steel) a workpiece temperature that is too high can lead to the sprayed plasma coating splitting off due to different coefficients of expansion.

The process of inert gas jacket plasma spraying or shrouded plasma spraying (SPS) does not take place in a closed chamber. In principle, it is an ***APS*** process, by which the plasma beam is protected by an inert gas flow, which the plasma coaxially envelopes, and so an interaction of the spraying particles and the plasma with the surrounding atmosphere is prevented to a large extent. The additional gas flow is not ionized and cools the substrate.

7.3.1.3 Vacuum Plasma Spraying (VPS)

With vacuum plasma spraying (***VPS***) the coating process takes place in a vacuum chamber. After evacuating the chamber (pressure $< 10^{-2}$ mbar) the plasma spraying process in inert gas takes place at pressures between 50 and 400 mbar. In order to keep the pressure in the chamber constant, the inflow plasma gases must be exhausted. In the inert atmosphere the liquid spraying particles are protected. An advantage over the ***APS*** procedure is the possibility of using an electric arc for substrate surface cleaning [29].

Because of the larger gas expansion in the vacuum, the plasma beam is more stable because of the weakened interaction effect with the surrounding atmosphere. The use of a Laval nozzle amplifies this effect [30]. With the smaller number of gas molecules in the low pressure atmosphere the plasma beam becomes longer and the gas velocity higher than under atmospheric pressure. A turbulence of the plasma beam arises only with a larger distance of the nozzle. The hot zone of the plasma beam thereby becomes longer [31]. The larger expansion of the plasma beam is connected with a decrease of the power density.

With vacuum plasma spraying also the basic material is heated to higher temperatures, since the lack of cooling and the smaller number of surrounding molecules means that less warmth can be dissipated. In addition, this has the advantage that at the lower cooling speed thermal stresses between layer and substrate material remain small. The higher temperatures also increase the diffusion between layer and basic material, with the result of improved film adhesion. The spraying distance with the ***VPS*** process lies between 250 and 350 mm. Films with porosities of less than 1 % are typical.

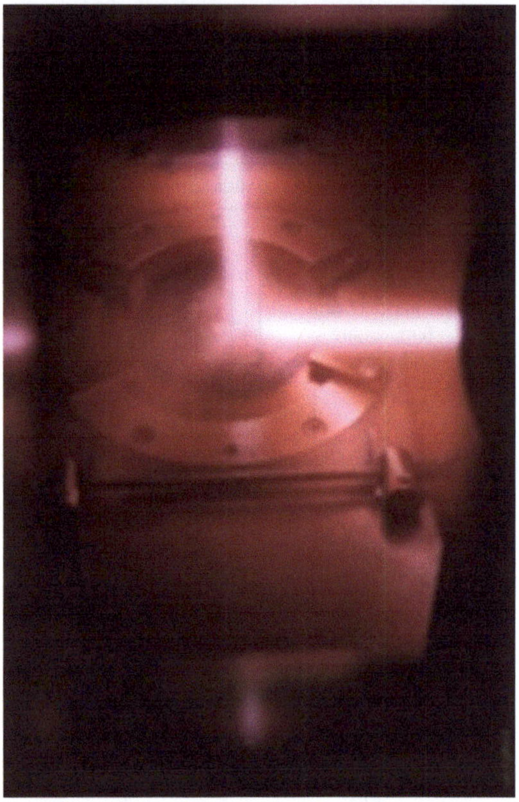

Fig. 7.16 View into the vacuum chamber of two-beam RF-plasma spraying equipment during the manufacture of $YBa_2Cu_3O_{7-x}$-layers

7.3.1.4 Electrodeless Plasma Spraying

Electrodeless plasma spraying was developed for the production of very pure films. Figure 7.16 shows the view inside two-beam equipment for the deposition of high temperature superconductor films of $YBa_2Cu_3O_7$ [32]. Argon was used as the working gas. The pressure during the coating amounted to 4 mbar, the RF-frequency was 2 MHz and the power was between 600 W and 1.2 kW. The powder stream was about 1 g/min and the gas flow 2 l/min. An advantage over the RF plasma spraying process is that the power coupling can be adjusted exactly along the discharge tube. The film prosperities can, therefore, be adjusted very exactly.

References

1. Hartmann R (1990) Beschichtung durch Plasmapolymerisation. TECHNIKA 19:151–154
2. Yasuda H (1981) Glow discharge polymerization. In: Journal of Polymer Science: Macromolecular Reviews 16:199–293
3. Linder EG, Davies AP (1931) J Phys Chem 35:3649
4. Gregor LV (1965) In: Thun R, Haas G (eds) Physics of Thin Films. New York, Academic Press 1065, Bd. 3 pp 113
5. Holland L (1956) Vacuum Deposition of Thin Films. Chapman and Hull, London
6. Shen M (ed) (1976) Plasma Chemistry of Polymers. Marcel Deker
7. Shen M, Bell AT (1979) Plasma Polymerization ACS Symposium Series, vol. 108.
8. Hollahan JR, Bell T (eds) (1974) Techniques and Applications of Plasma Chemistry. Wiley, New York
9. Poll HV, Tiller HJ, Wickleder KM (1978) Polymere dünne Schichten. In: Hamann G (ed) Organische Festkörper und anorganische dünne Schichten. Akademische Verlagsgesellschaft, Berlin, pp 399–490
10. Poll HV (1970) Zur Angew Phy 29:260
11. Kieser J (1989) Plasmapolymerisation als industriell einsetzbare Technik. In: VDI-Technologiezentrum (ed) Plasmagestütztes Verfahren der Oberflächentechnik. VDI Verl. p 106/131
12. Grünwald H (1992) Polymerisation zur Lösung von Adhäsionsproblemen. In: plasmatec '92. Universität Duisburg
13. Topics in Current Chemistry. Vol. 89 und 90, Springer-Verl. 1980
14. Flis J, Mankowski J, Rolinski E (1989) Corrosion behaviour of stainless steel after plasma and ammoniac nitriding. Surface Engineering 5(2):151
15. Briglia T, Terwagne G, Bodart F, Quaeyhaegens C, D'Haen J, Stals LM (1996) Study of nitrogen-implanted stainless steels by CEMS and TEM. Surf Coat Technol 80:105–108
16. Berg M, Budtz-JÆrgensen CV, Reitz H, Schweitz KO, Chevallier J, KringhÆj P, BÆttiger J (2000) On plasma nitriding of steels. Surf Coat Technol 124:25–31
17. Kölbel J (1965) Die Nitridschichtbildung bei der Glimmentladung Forschungsbericht d. Landes Nordrhein-Westfalen, vol. 1555. Westdeutscher Verlag, Opladen
18. Dearnley PA, Namvar A, Hibberd GGA, Bell T (1989) Some observations on plasma nitriding austenitic stainless steel Proc. 1st Int. Conference on plasma surface engineering, Garmisch-Partenkirchen. DGM Informationsgesellschaft, Oberursel, p 219
19. Camps E, Muhl S, Romero S, Garcia JL (1998) Microwave plasma nitride austenitic AISI-304 stainless steel. Surf Coat Technol 106:121–128
20. Flis J, Mankowski J, Rolinski E (1989) Corrosion behaviour of stainless steel after plasma and ammoniac nitriding. Surface Engineering 5(2):151
21. Rie KT, Eisenberg S (1987) Mikrostruktur von Titan und Titanlegierungen. Härtereitechn Mitt 42(6):344–349
22. Rolinski E, Karpinski T (1987) Wear Restistance and Corrosion Behaviour of Plasma nitride Titanium Alloys. Proc. Of the Intern Seminar on Plasma Heat Treatment. PYC Edition, Paris, Senlis/France, pp 255–263
23. Souchard JP, Jacquot P, Coll B, Buvrun M (1990) Ionic Nitriding and Ionic Carburizing of pure Titanium and its Alloys. Proc of the 2nd Inter. Conf on Ion Nitriding and Ion Carburizing, Cincinnati, Ohio/USA, 1989. ASM
24. Eisenberg S (1988) Plasmadiffusionbehandlung von Titan und Titanlegierungen. VDI-Verlag
25. Suryanarayanan R (1993) Plasma Spraying: Theory and Applications. World Scientific Pub Co Inc., p 211
26. Papyrin A, Kosarev V, Klinkov S, Alkhimov A, Fomin V (2007) Cold Spray Technology. Elsevier, Oxford
27. Lugscheider ER, Agethen R, Knepper M (1989) Löten von Keramik-Metall-Verbindungen durch Aufbringen von gradierten Zwischenschichten. Verbindungstechnik in der Elektrotechnik (2):82–85
28. Pawlowski L (1995) The Science and Technology of Thermal Spray Coatings. Wiley, New York
29. Steffens HD, Brandl W, Dvorak M, Wewel M, Rucinski D (1991) Vacuum Plasma and Vacuum Arc Sprayed Reactive Materials in Comparison. Proc. 2nd Plasma Technik Symposium, Luzern CH, pp 263–276
30. Jäger D (1989) Spritztechnische Verarbeitung von Refraktärmaterialien und Refraktärhartstoffen für Korrosions- und Verschleißschutzanwendungen. Ed.: Lugscheider E, Techn-Wissensch. Berichte Nr. 22.19.1.89 1989, RWTH Aachen
31. Heimann RB (1996) Plasma spray coating principles and applications. VCH, Weinheim, New York, Basel
32. Jäger H, Schulze K, Frey H (1990) Preparation of $YBa_2Cu_3O_{7-x}$ Films by Low-Pressure RF-Plasma Deposition. J. of. The Less-Common Metals, 164 u. 165, pp 486–492

8 Particle Beam Sources

H. Frey

8.1 Introduction

In this section we consider the fate of energetic ions (1–100 keV) incident on a solid surface. The 10 ways in which ions can interact with a surface are illustrated in Fig. 8.1. An incoming ion can be backscattered by an atom or group of atoms in the bombarded sample (1). The backscattering process generally results in a deflection of the ion's incident path to a new trajectory after the encounter and an exchange of energy between the ion and the target atom. The energy exchange can be either elastic or inelastic, depending on the constituent particles and the energy of the ions. The momentum of an ion can be sufficient to dislodge a surface atom from a weakly bound position on the sample lattice and cause its relocation on the surface in a more strongly bound position (2).

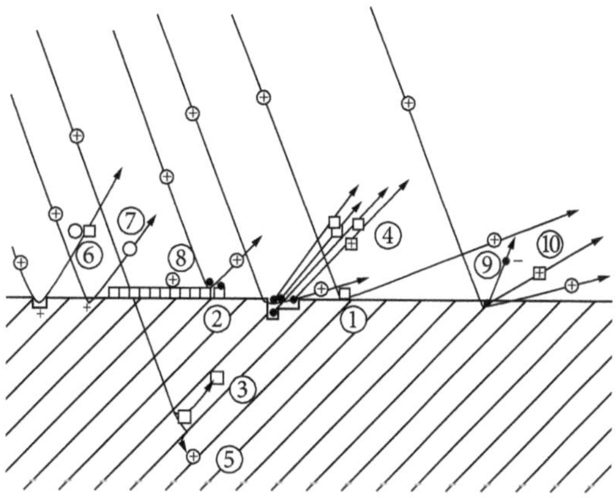

Fig. 8.1 Ion–solid interactions

This process is called atomic dislocation. Ions with greater energies can cause internal dislocations in the bulk of the sample (3). Physical sputtering (4) results when ions strike the sample surface and transfer enough momentum to entirely free one or more atom. Ions can penetrate into the lattice and become trapped as their energy is expended (ion implantation) (5). Chemical reaction of the ions with the ions with the surface atoms can occur, and, as a result, new compounds can be formed on the sample surface, or the outermost layer of atoms may leave as a gas (chemical sputtering) (6). The bombarding positive ion can gain an electron from the surface by Auger neutralization and be reflected as a neutral atom (7). Ions can become bound to the surfaces or the sample as adsorbed ions (8). Secondary electron emission (9) occurs under suitable conditions of ion bombardment of metal surfaces. Secondary ion emission (10) results when surface atoms are excited to ionized states and emitted from the sample [1].

Ions are a tool for the modification of films near the surface. This is a very flexible tool due to the various process parameters of ion radiation. These parameters can usually be adjusted independently and reproducibly, i.e. the tool of the ion beam is very well controllable. Because

© Springer-Verlag Berlin Heidelberg 2015

H. Frey, H. R. Khan, *Handbook of Thin-Film Technology*, DOI 10.1007/978-3-642-05430-3_8

of the variety of possible mechanisms of the ion radiation the effects are not predictable in most cases.

8.2 Basic Processes of Ion–Solid Interaction

8.2.1 Loss of Energy

To analyze the energy loss between two major energy loss processes it is convenient to differentiate between electronic collisions and nuclear collisions.

The first process is the interaction of the fast ion with the lattice electrons, resulting in excitation and ionization. Since the density of the electrons in the target is high and the collisions are so numerous, the process, similarly to the electron energy loss, can be regarded as continuous.

The nuclear loss results from collisions between the screened nuclear charges of the ion and the target atoms. The frequency of these collisions is lower, so they may be described by elastic two-body collisions. At high energies they are accurately described by Rutherford scattering [2], at medium energies by screened Coulomb scattering, while at low energies the interaction becomes even more complex.

In addition, there is a contribution to the energy loss resulting from the charge exchange between the moving ion and target atom [3]. This energy loss reaches a maximum when the relative velocity is comparable to the Bohr electronic velocity ($2 \times 10^6\,\mathrm{m\,s^{-1}}$). The total energy loss $-\mathrm{d}E/\mathrm{d}x$ can thus be regarded as the sum of three components, nuclear, electronic, and charge exchange,

$$\mathrm{d}E/\mathrm{d}x = (\mathrm{d}E/\mathrm{d}x)_\mathrm{n} + (\mathrm{d}E/\mathrm{d}x)_\mathrm{e} + (\mathrm{d}E/\mathrm{d}x)_\mathrm{ch} . \tag{8.1}$$

At low ion energies nuclear stopping dominates and is responsible for most of the angular dispersion of an ion beam. At high energies electronic collisions are more important. A useful rule of thumb is that the energy $<$ lost to the lattice by nuclear collisions becomes predominant at energies less than A keV, where A is the atomic weight of the incident ion. At intermediate energies the charge-exchange contribution may rise to roughly 10 % of the total. The variation of the energy loss with ion energy is shown in Fig. 8.2.

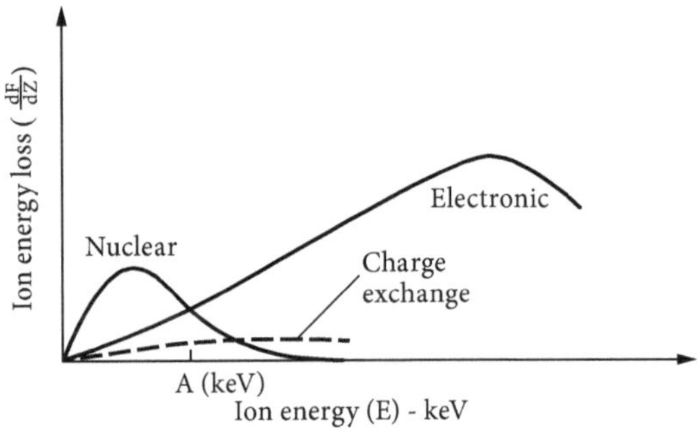

Fig. 8.2 Variation of ion energy loss with ion energy

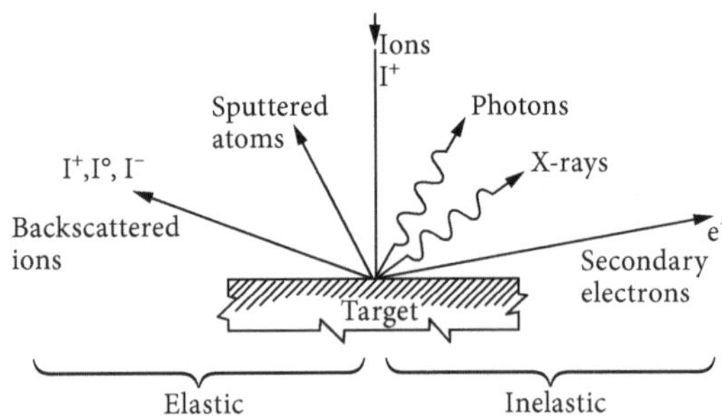

Fig. 8.3 Schematic representation of ion–target interactions

Inelastic interactions with the target electrons lead to secondary electron emission, emission of characteristic X-rays, and optical photon emission; elastic interactions lead to the displacement of lattice atoms, the formation of defects, and surface sputtering, as is illustrated schematically in Fig. 8.3.

The energy spectrum of ions [4] with initial energy E_0 scattered from a solid target is represented schematically in Fig. 8.4. There is a broad low-energy hump (10–30 eV) corresponding to the emitted neutral atoms (sputtered atoms) and a high-energy hump distributed about the incident ion energy E_0 (elastically scattered ions).

If we assume that independently nuclear and electronic interaction occurred, the entire stopping cross-section consists of an electronic part

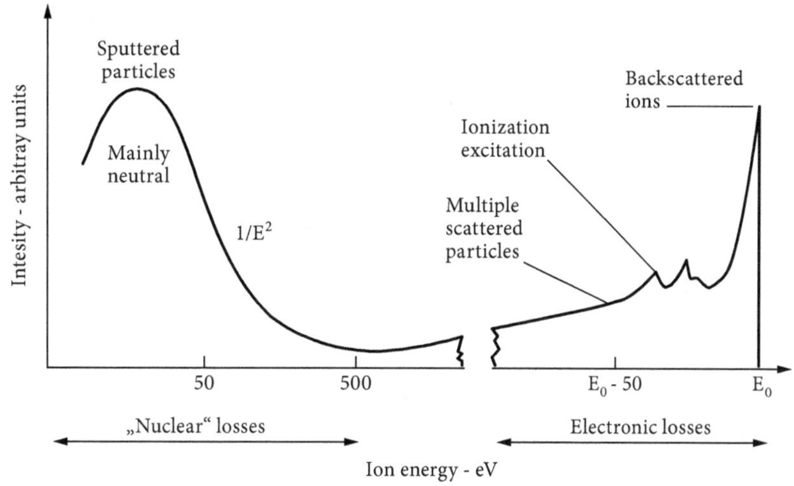

Fig. 8.4 Energy spectrum of ions scattered from a solid target

and a nuclear part:

$$S = S_e + S_n \,. \tag{8.2}$$

Both parts can be calculated from the differential effective cross-section of the single scattering process, which depends on the energy transfer Γ_e:

$$S = \int T_e \, d\sigma \,. \tag{8.3}$$

For nuclear interaction the energy transfer is:

$$T_n = \frac{4\, m_1 m_2}{(m_1 + m_2)^2} E \, \sin^2\left(\frac{\theta_s}{2}\right) , \tag{8.4}$$

whereby θ_s is the deflection angle (in the gravity center system), $m_{1,2}$ the masses of the nucleus (the indices 1 and 2 designate the projectile and the collision partner). The nuclear collision cross-section can be calculated from the laws of classical mechanics for interatomic interaction potential. An overview is given in [5]. Such potentials can be written in the form of a shielded Coulomb potential

$$V = \frac{Z_1 Z_2 e^2}{4\pi\varepsilon_0 r} \phi\left(\frac{r}{a}\right) , \tag{8.5}$$

with the nuclear charge numbers of $Z_{1,2}$, the internuclear distance r, the shielding function ϕ, and the shielding length α. For analytic calculations it is meaningful to use universal approximations.

The shielding length after ***Lindhard*** et al. [6] for the ***Thomas Fermi*** potential $\phi = \phi_{TF}$ can be calculated in accordance with

$$a_{TF} = \frac{0.8853\, a_0}{\left(Z_1^{\frac{2}{3}} + Z_2^{\frac{2}{3}}\right)^{\frac{1}{2}}} , \tag{8.6}$$

whereby a_0 is the first ***Bohr radius*** $a_0 = 5.59 \times 10^{-9}$ cm.

With new systematic calculations for many ion–atom interactions [6] a universal shielding function can be written as an exponential series:

$$\begin{aligned} \phi_U(y) &= 0.182 e^{-3.2y} + 0.51 e^{-0.942y} \\ &\quad + 0.28 e^{-0.403y} + 0.0282 e^{-0.202y} , \end{aligned} \tag{8.7}$$

with the universal shielding length

$$a_U = \frac{0.8853\, a_0}{Z_1^{0.23} + Z_2^{0.23}} . \tag{8.8}$$

For the universal calculation of the stopping data we use dimensionless variables for the energy [7]:

$$\varepsilon = \frac{4\pi\varepsilon_0 a\, m_2}{Z_1 Z_2 e^2 (m_1 + m_2)} E \tag{8.9}$$

and the distance

$$\varrho = \pi a^2 n \frac{4 m_1 m_2}{(m_1 + m_2)^2} x \,. \tag{8.10}$$

The reduced nuclear stopping power for a ***Thomas Fermi*** potential is in good approximation given by [8]

$$-\left(\frac{d\varepsilon}{d\varrho}\right)_n = \frac{3.44\sqrt{\varepsilon}\log(\varepsilon + 2.718)}{1 + 6.35\sqrt{\varepsilon} + \varepsilon(6.882\sqrt{\varepsilon} - 1.708)} . \quad (8.11)$$

The electronic deceleration must be treated quantum-mechanically for high projectile velocities. In the ***Thomas Fermi*** model

$$\overline{v} = v_0 Z_2^{\frac{2}{3}} , \quad (8.12)$$

whereby $v_0 = c/137$ is the velocity of the electrons at the first Bohr' orbit (velocity of light).

For the ***Bethe*** formula we obtain [9]:

$$S_e^> = \frac{2\pi Z_1^2 Z_2 e^4}{(4\pi\varepsilon_0)^2} \frac{m_1}{m_e} \frac{1}{E} \log\left(4\frac{m_e}{m_1}\frac{E}{I}\right) . \quad (8.13)$$

In Eq. (8.13) e is the elementary charge, m_e the mass of the electron, and I the average ionization potential.

For the other limiting case of small velocities, ***Lindhard and Scharff*** [10] found by means of the classical scattering theory of electronic stopping power, with the projectile velocity $v = \sqrt{2E/m_1}$ proportional to the electronic stopping power,

$$S_e^< = \frac{2e^2 a_0}{\varepsilon_0} \frac{Z_1^{7/6} Z_2}{(Z_1^{2/3} + Z_2^{2/3})^{3/2}} \frac{v}{v_0} . \quad (8.14)$$

The intermediate range of average velocities is only partly understood and thus describable. Therefore from an interpolation formula we use

$$\frac{1}{S_e} = \frac{1}{S_e^<} + \frac{1}{S_e^>} , \quad (8.15)$$

with $S_e^< \sim \sqrt{E}$ and $S_e^> \sim E^{-1}\log E$, as given in Eqs. (8.13) and (8.14).

Semi-empirical fit functions describe the electronic stopping over large energy ranges and for many ions and substances [7–10].

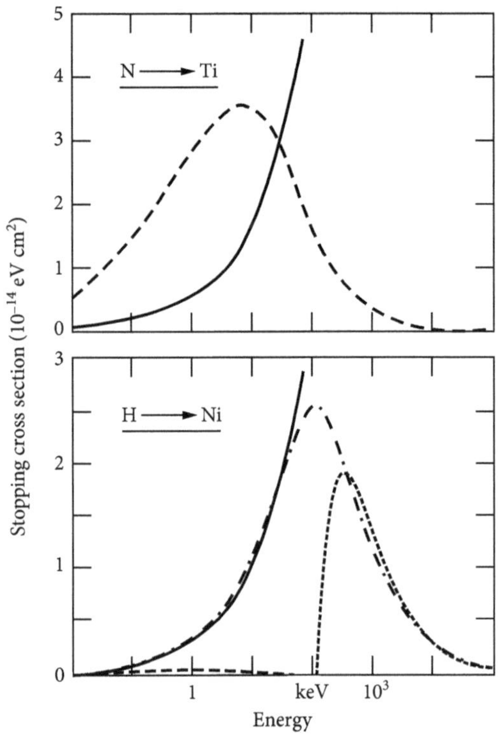

Fig. 8.5 Contributions to the stopping cross-section for two different projectile target combinations. Low-energy electronic stopping (Eq. (8.14), *continuous line*), high-energy electronic stopping (Eq. (8.13), *broken line*), nuclear stopping (Eq. (8.11), *dashed line*), electronic stopping after Anderson and Ziegler (*dash-broken line*)

Figure 8.5 shows the theoretical forecasts for two ion material combinations. For hydrogen in nickel the semi-empirical adaption after ***Andersen and Ziegler*** [11] is shown in Fig. 8.5.

For the lightest ions the nuclear part of the stopping is negligible; it dominates for moderately heavy ions up to energies of approximately 10 keV. The ion stopping was only considered simple in monatomic species. In multi-component materials it can be accepted approximately that ion stopping is not influenced by chemical effects.

For a substance of two elements A_yB_z ***Bragg's rule*** of summation of the stopping cross-section is valid:

$$S_{A_yB_z} = yS_A + zS_B . \quad (8.16)$$

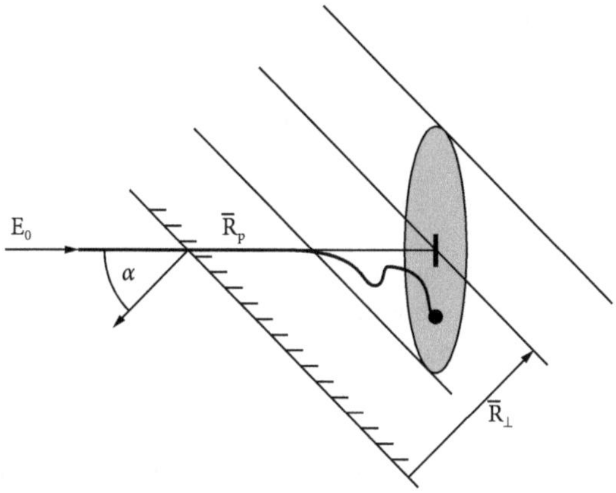

Fig. 8.6 Schematic representation of the ion trajectory and the average ranges

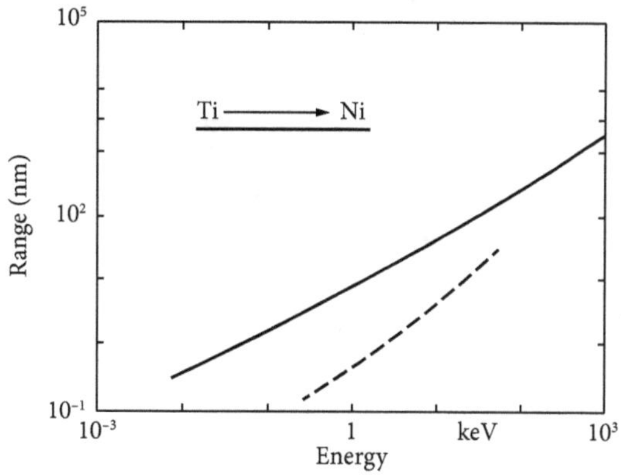

Fig. 8.7 Total distance of titanium ions in nickel (*solid line*) and projected distance after Eq. (8.19)

From the literature there are well-known materials that have intense covalent bounds; those from ***Bragg's rule*** deviate by up to 40 % [10–16]. Examples are oxides, carbides, and amorphous hydrocarbons films.

8.2.2 Range Distances

If the total stopping cross-section S is well-known, the average total path of an ion with the injection energy E_0 can be calculated in accordance with

$$\overline{R}_t = \frac{1}{n} \int_0^{E_0} \frac{\mathrm{d}E}{S(E)} . \qquad (8.17)$$

Because of the statistic nature of multiple scattering a probability can be defined for each individual ion. The ions are stopped within a discus-shaped spatial distribution, then their average projected distance is defined after Fig. 8.6.

For heavy ions in light substances multiple scattering is small. In raw approximation and in the energy range, in which electronic stopping can be neglected, the ratio from total distance to projection range in case of $m_1 \geq m_2$ [17] is valid:

$$\frac{\overline{R}_t}{\overline{R}_p} \approx 1 + \frac{m_2}{3m_1} . \qquad (8.18)$$

For light ions ***Schiøtt*** [18], assumed that electronic stopping dominates $\lambda < 1$,

$$\frac{\overline{R}_t}{\overline{R}_p} \approx \frac{1}{\lambda(1-2\lambda)} ; \quad \lambda = \frac{m_1}{m_2} \frac{S_e(E_0)}{S_n(E_0)} . \qquad (8.19)$$

For many ions within the beam spot an implantation zone results around the average implantation depth $\overline{R}_\perp$; for those in the case of an inclined ion incidence the average projected range in accordance with (Fig. 8.3) is

$$\overline{R}_\perp = \overline{R}_p \cos(\alpha) . \qquad (8.20)$$

Relationships become more complicated when for light ions in heavy substances, particularly with small energy or even injection, a large part of the ions are out-scattered (reflected) from the surface. Figure 8.7 shows the total and projected distances (after correction Eq. (3.19)) for titanium in nickel.

The treatment for small energies is particularly problematic. Distances in about 1 Å are physically absurd. The principle of the average distance becomes irrelevant by strong multiple scattering in connection with ion reflection.

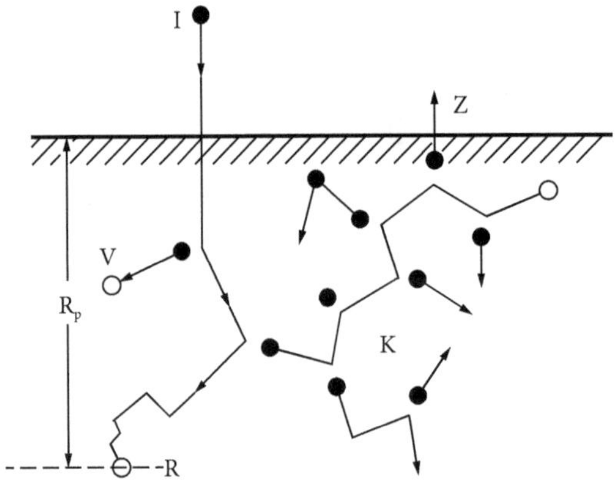

Fig. 8.8 Schematic representation of the effects of nuclear collisions by an incident ion that stopped at the place R and its projected distance R_p. In the collision cascade K dislocations (V) and sputtering incidence (Z) can take place

8.2.3 Collision Cascades and Radiation Damage

The effect of an incident ion on the atoms of the irradiated material is illustrated schematically in Fig. 8.8.

Along the path the ion will transfer the energy Γ_n by elastic collisions after Eq. (8.4). The binding energy of an atom in its lattice is some eV; therefore the atom strikes out with very soft collisions at its place. Then it can be regarded as a fast particle. If its initial energy Γ_n and distance is sufficient low, it will return through elastic interaction with the lattice to its place. In order to produce a stable defect, a ***Frenkel*** pair from an interstitial atom and a lattice vacancy, and threshold energies from $\sim$ 10 to 50 eV are necessary. With crystalline materials this threshold energy depends on the direction of the impacted atom.

With high energy transfer a primarily adjoining atom can produce a collision cascade and more stable ***Frenkel pairs*** by further collisions. Using the approximation (Fig. 8.8) the power density is relative low, so that fast atoms essentially only collide with resting atoms.

Within this range of the so-called linear cascade the number of Frenkel defects per primary collision results when using the modified ***Kinchin-Pease*** formula [22]

$$
v_F(T_n) = \begin{cases} 0 & \text{for } T_n < E_d \\ 1 & \text{for } E_d < T_n;\ T_n^* < 2.5\,E_d \\ \dfrac{0.8}{2E_d} T_n^* & \text{for } T_n^* > 2.5 E_d\,. \end{cases} \tag{8.21}
$$

T_n^* is that part of the primary energy transfer, which is not dissipated by the electronic interaction effect.

$T_n^* \sim 0.9 T_n$ is a good approximation for light to moderately heavy ions [23]. No defects are created, if the energy is so small that at a central collision the minimum dislocation energy is exceeded. This is about half as large as the effective dislocation energy in raw approximation. Thus we obtain the threshold energy for the generation of defects:

$$
E_0^s \approx \frac{E_d}{2\gamma}\,, \tag{8.22}
$$

whereby γ is the energy transfer function (see Eq. (8.4))

$$
\gamma = \frac{4 m_1 m_2}{(m_1 + m_2)^2}\,. \tag{8.23}
$$

At very high densities for the deposited energy the model of the linear cascade is no longer valid (typically for heavy ions with energies around 100 keV). Ranges develop that contain only fast atoms, which collide among themselves and occur during a short time in a thermalized ensemble.

In a simple model of this heat pulse (thermal spike) [24–26] we assume a straight ion trajectory, along which the energy is deposited in a line shape. From the theory of heat conduction, for the temporal temperature gradient in the distance r from the ion trajectory we have:

$$
T(r,t) = \frac{n\,S_n}{4\pi\lambda t} \exp\left(-\frac{\varrho c r^2}{4\lambda t}\right)\,, \tag{8.24}
$$

whereby λ designate the heat conductivity; ρ the density, and c the specific warmth of the target material. The number of dislocations per ion is accordingly

$$
\frac{\mathrm{d}N_d}{\mathrm{d}x} = \frac{k\,n^2 S_n^2 v_0}{24\pi E_d^2 \lambda}\,, \tag{8.25}
$$

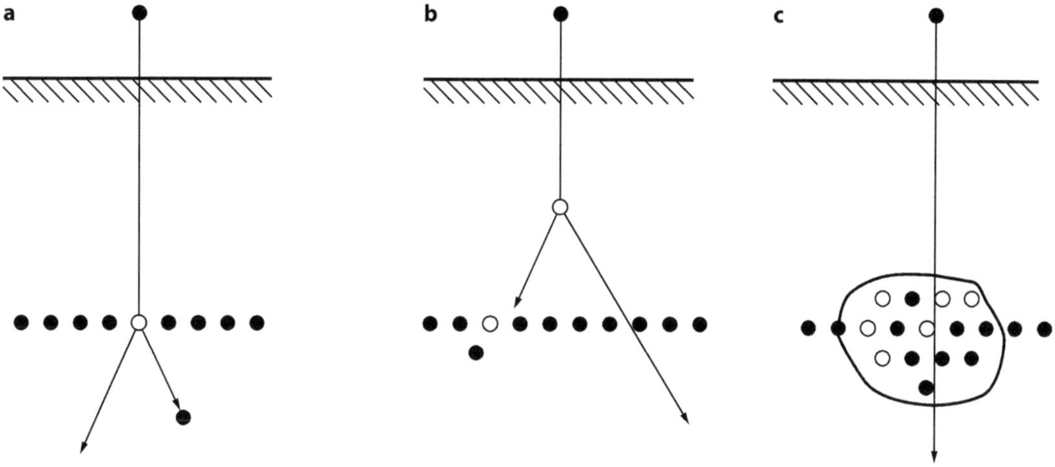

Fig. 8.9 Schematic representation of the different elementary processes of atomic mixing by ion bombardment. **a** Collisions with marker atoms, **b** collisions with matrix atoms, **c** cascade mixing

whereby $v_0 \sim 10^{13}$/s is a typical frequency of the lattice vibrations.

In metals the maximum concentration of Frenkel pairs lies at some 10^{-3}. Further Frenkel pairs are stable only at very low temperatures if both the interstitial atoms and the lattice vacancy are immovable. In insulators the electronic interaction can also contribute substantially to the creation of radiation damage [27, 28].

8.2.4 Ion Mixing

The ion mixing is based on the spatial dislocation of atoms within a collision cascade. Several processes contribute to mixing by nuclear collisions (Fig. 8.9).

Sigmund and Grass-Marti [29] found the following for cascade broadening:

$$\Omega^2 = \frac{2\gamma}{\pi^2}\alpha\left(\frac{m_2}{m_1}\right) S_n(\overline{E})\frac{(R(E_e))^2}{E_e}\Phi\,, \quad (8.26)$$

where $\overline{E}$ is the average energy of the ions at the place of the marker atoms, E_e the minimum energy of an atom for the dislocation, and $R(E_e)$ the assigned average distance. E_e is generally significantly lower than the dislocation energy E_d for the creation of a ***Frenkel*** pair, if the bulk material is already strongly damaged with high bombardment fluencies. Ion mixing can be affected by further temperature-dependent processes. Such processes are chemical in nature in connection with the thermodynamics of the material mixture [30], or correlated with the mobility of defects [31].

8.2.5 Irradiation Amplified Diffusion

By a solid body we distinguish, in principle, two mechanisms of mass transport. Foreign atoms, which are located at interstitial locations, can migrate at sufficiently high temperature between these locations. The diffusion of foreign atoms on lattice sites and thus the self-diffusion of the matrix atoms require the presence of lattice vacancy, which by their migration exchange places with the atoms. For both processes a diffusion coefficient can be indicated as a temperature-independent factor D_0 and an activation energy U_m

$$D = D_0 e^{-\frac{U_m}{kT}}\,. \quad (8.27)$$

In the simple model for stochastic migration we can write:

$$D_0 = \frac{v_0 d^2}{6}\,, \quad (8.28)$$

whereby v_0 indicates the oscillating frequency in the lattice ($\sim 10^{13}$/s) and d the hopping width of a single atom.

The activation energy for interstitial lattice diffusion generally has values in the order of 1 eV. For the lattice vacancy mechanism the activation energy is the sum of lattice vacancy energy of formation and lattice vacancy migration energy $U_m = \breve{U}_f + \breve{U}_m$, which typically have some eV. The lattice vacancy mechanism begins in non-irradiated material only at very high temperatures, if sufficient lattice vacancies are produced in the thermal way.

In a general definition we are speaking of ***radiation-increased diffusion*** if the internal material transport in relation to thermal transport is increased by irradiation. Ion beam mixing can, therefore, also be regarded as a radiation-increased diffusion. Further, in particular for interstitial atoms, thermal diffusion induction of a thermal pulse of an ion can be increased temporarily, which leads to an intensified diffusion. The lattice vacancy mechanism thereby plays a substantial role; its activation energy is then only given by the migration energy of the lattice vacancy (about 1 eV).

In the case of irradiation, the balance equations of the defects ($c_{i,v}$ and $D_{i,v}$ atomic concentrations and diffusion coefficients of the interstitial atom (i) and the lattice vacancy (v)) are:

$$\frac{\partial c_i}{\partial t} = D_i \frac{\partial^2 c_i}{\partial x^2} - 4\pi\, D_i\, a_{iv} n\, c_i\, c_v + F_d(x) \quad (8.29)$$

and

$$\frac{\partial c_v}{\partial t} = D_v \frac{\partial^2 c_v}{\partial x^2} - 4\pi\, D_i\, a_{iv} n\, c_i\, c_v + F_d(x)\,. \quad (8.30)$$

$a_{iv} \approx n^{-1/3}$ is the critical distance for recombination of interstitial atoms with lattice vacancy and F_d the local production rate of Frenkel pairs.

The boundary conditions are: $c_i(x = 0) = c_v(x = 0) = 0 = c_i(x \to \infty) = c_v(x \to \infty)$ at the surface and deep in the bulk material; neglected are internal sinks such as dislocations or vacancy lattice complexes.

The migration of interstitial atoms at sufficiently low temperatures occurs substantially faster than with those of the lattice vacancy. Except within fewer atomic layers in the proximity of the surface the recombination with lattice vacancies is very probable due to their high concentration. Interstitial atoms in the bulk material have a higher probability of survival. Thus both defect types contribute to the intensive, long ranges of diffusion. To date there is neither a model nor a calculation method to describe the procedures completely.

8.2.6 Surface Texture and Growth Mode

The morphology of an increasing film is substantially determined by surface processes. Surface defects and irregular packing growing into the film can be neutralized partly by ion bombardment.

The particle energies E are indicated in units of the ***Lennard Jones*** potential depth ε and have values between 0.1 eV up to some eV. Thermal oscillations were not considered.

A defect-free film (***Frank van der Merwe***-growth) morphology with layer on layer growth depends on the energetic bombardment with ions. Without bombardment, however, a morphology with many cavities (***Volmer-Weber***-growth) results from rather an island-like growth (Fig. 8.10). This effect of the ion bombardment becomes explicit, if more realistic conditions are used. Figure 8.11 shows the results of simulations for ion beam deposition of silver layers at different energies [33].

The increased tendency to ***Frank-van-der-Merwe*** growth under ion bombardment is shown by experimental results [34]. It is based on the fact that the ion beam, on the one hand, increases the nucleus density and, on the other hand, dissolves again by lateral dislocation islands. This is represented schematically in Fig. 8.12. Both partial mechanisms require the creation of surface defects.

After this conception if a defective-free morphology is to be possible, for example at epitaxial growth, the generation of surface defects should be maximized, while volume defects should be minimized at the same time. Quantitatively this view was examined in various observations of the

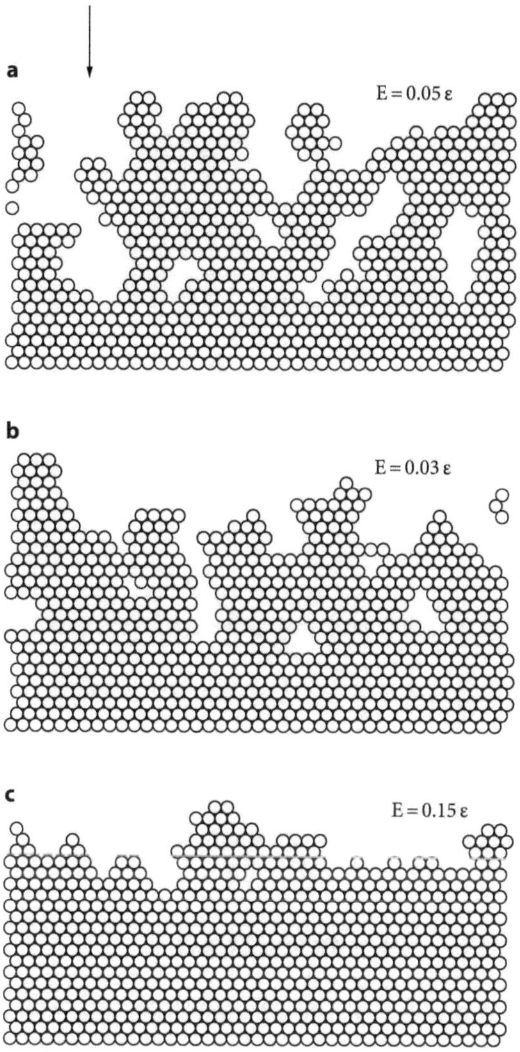

Fig. 8.10 Two-dimensional molecular-dynamic simulation of film growth by atomic deposition with the help of a ***Lennard Jones*** potential [32]

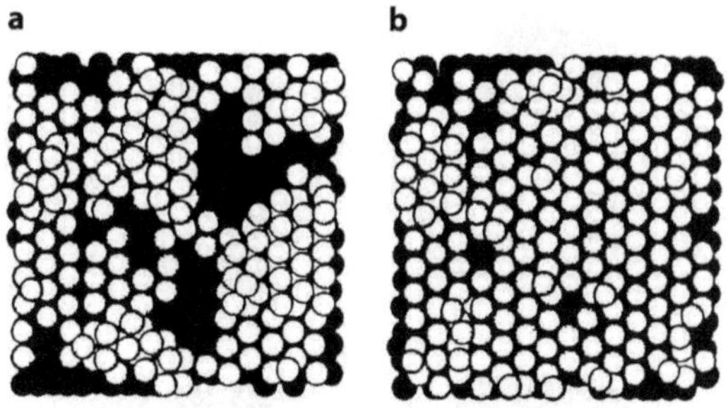

Fig. 8.11 Molecular-dynamic simulation of the growth of silver layers at ambient temperature with an EAM potential for incidence energies of 0.1 eV (**a**) and 10 eV (**b**)

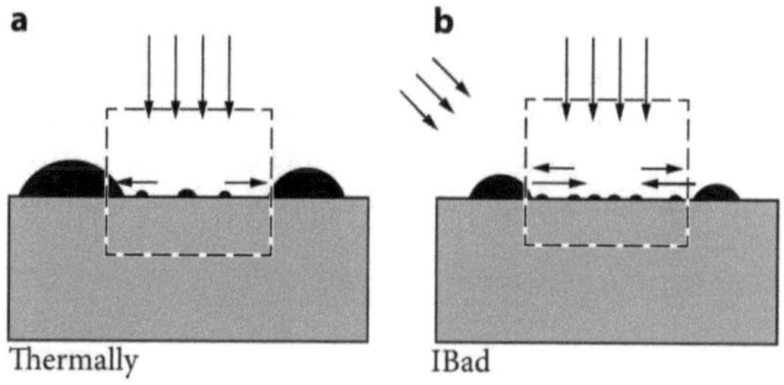

Fig. 8.12 Schematic representation for the suppression of island growth by ion bombardment

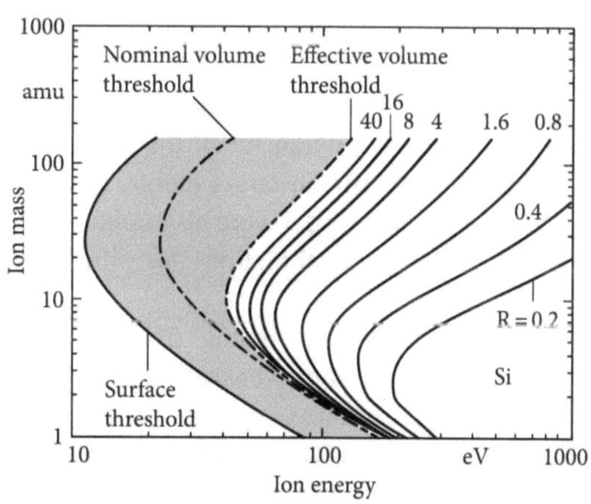

Fig. 8.13 Isocontour lines of the relationship R of surface defects to volume defects as function of ion energy and ion masses for the bombardment of silicon. Further volume thresholds and surface thresholds are drawn

ion bombardment by silicon and germanium with different ions [35].

Quantitatively this view was examined in various observations of ion bombardment by silicon and germanium with different ions [35]. Figure 8.13 shows isocontour lines for constant relationship from surface dislocations to volume dislocations for different ion energies and ion masses.

This relationship becomes infinitely large during the approach to the effective volume dislocation threshold energy. The threshold energy after (8.22), calculated for surface dislocation energy was accepted to be half as large as the volume dislocation energy.

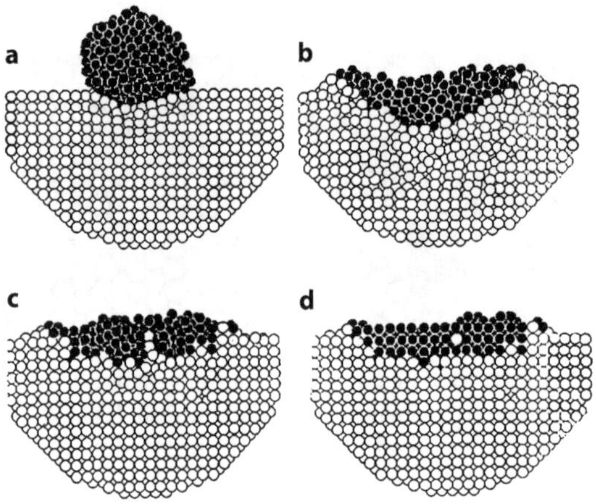

Fig. 8.14 Molecular-dynamic simulation of the integration of an energetic molybdenum cluster into a molybdenum layer. The energy of 5 keV distributes itself on 1024 atoms of the cluster. The time differences between the snapshots of simulation amount to **b** 2.25, **c** 9, and **d** 17.25 ps [37]

With ion energies under the effective volume dislocation threshold only surface dislocations are created, until finally, with very low energies, no more dislocations are possible. The form of the isocontour lines and limiting curves result from the energy transfer of the incident ion (Eq. 8.4) on the silicon target atoms. The general trends of this model were also confirmed by molecular-dynamic simulations for silicon layers [35].

New aspects for modeling of film growth bring in the deposition by energetic ionized clusters [36]. With clusters of the order of magnitude of 1000 atoms, with high particle energies of several keV, incidence energies between 1–10 eV per atom can also be obtained (Fig. 8.14).

The cluster first penetrates with very high transient temperatures and pressures into the surface and thereby spreads out laterally. A film of approximately 10 atoms thick on the substrate is initially disturbed. It is hoped to get a film with a good epitaxial property, a sufficient mixture with the basis, and only low defects in the film and in the substrate. This simulation qualitatively agrees with experimental analyses.

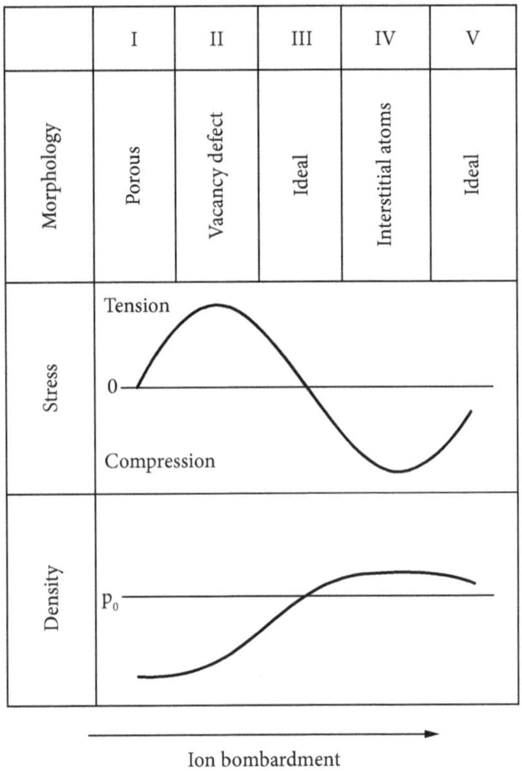

Fig. 8.15 Schematic representation of the relationship between film morphology, density, and internal stress as a function of ion bombardment

8.2.7 Density and Stress

With the model of ***Thornton***, Fig. 6.15, it becomes explicit that ion bombardment affects the morphology and the density of the film. Figure 8.15 shows a qualitative pattern for low substrate temperatures. Without ion bombardment (regime I) a vapor-deposited or a sputtered film becomes porous. Their density is relatively low: the porous structure can only take absorb stresses.

With increasing ion bombardment (increasing ion energy or a higher ion/neutral relationship) the cavities are filled, so that there remain especially small lattice vacancy clusters or a single lattice vacancy: an internal tensile stress develops. With increased ion bombardment, atoms that are displaced from the surface in the volume fill the lattice vacancy (regime III), so that with determined conditions the ideal density of the unstressed material results. Accordingly, highly strengthened bombardment preferentially leads to interstitial atoms (or lower accumulations of interstitial atoms) and thus to increased density and internal compressive stress.

At relative low energies, up to a few 10 eV, within the ranges I–IV there are especially single collision processes. It is only with higher energies in regime V that it comes to extended cascades, within those by dislocation or recoordination of interstitial atoms the compression strength is reduced and the density decreased. The dependence of tension with the use of ion bombardment is semi-empirically described in [38].

The *range III* with the structure of compressive strengths during the film deposition is described by sputtering [38]. The relative volume variation, which is connected over ***Hooke' law*** with the tension, is assumed proportionally to the dislocation rate. The latter is calculated similarly as with ***Sigmund sputtering theory***.

For the tension in the film we obtain

$$S = C\, j\, E^{1/2} \frac{M}{1-v}, \tag{8.31}$$

whereby
C indicates a constant,
J the ion flux,
E the ion energy,
M the modulus of elasticity, and
v the Poisson number.

The transition from regime IV to regime V is characterized by the fact that interstitial atoms are deposited by direct implantation or recoil implantation, but compete by dislocation or structural change in expanded collision cascades to be again removed. A simple analytic model of the supplantation can be determined from the flux balance. For an ion beam sputtering process it is assumed that a fraction P (probability of penetration) of the striking ions penetrates into the surface and forms interstitial atoms, while the remainder on the surface is deposited and the density of the undisturbed material assumed.

For the relationship of interstitial atoms (atomic concentration c_i) and lattice atoms (c_0) for an incident ion flux j it is then valid:

$$\frac{c_i}{c_0} = \frac{j\,P - j\,Y_c c_i}{j(1-P) + j\,Y_c c_i}, \tag{8.32}$$

with the conversion rate of yield Y_c. From this it follows that:

$$c_i = 1 - c_0 = \frac{P}{1+Y_c}. \tag{8.33}$$

If we continue to assume the fact that interstitial atoms do not cause a distortion then for the resulting number density n we obtain:

$$\frac{1}{n} = \frac{c_0}{n_0}, \tag{8.34}$$

whereby n_0 is the density of the defect-free material. Thus it follows that:

$$\frac{n}{n_0} = \frac{1+Y_e}{1+Y_e-P}. \tag{8.35}$$

For approaches for the identification of Y_e we use the model of the thermal pulse in the form of a point source.

For amorphous carbon we obtain a conversion rate of ***yield***

$$Y_e \approx 0.016 \left(\frac{E_0}{E_c}\right). \tag{8.36}$$

The results are shown in combination with experimental results in Fig. 8.16.

8.2.8 Phases in Material Films

We obtain hard material films with good mechanical properties if we prefer deposition processes that create hard material phases. Tetragonal phases are excellent according to cubic phases (diamond-like carbon, c-BN) and hexagonal phases (graphite, h-BN). The hexagonal phases are energetically more favorable, so that the deposition requires non-equilibrium processes, for example, by ion bombardment at low temperatures. In all these cases it was found experimentally that the part of the tetragonal phase from the

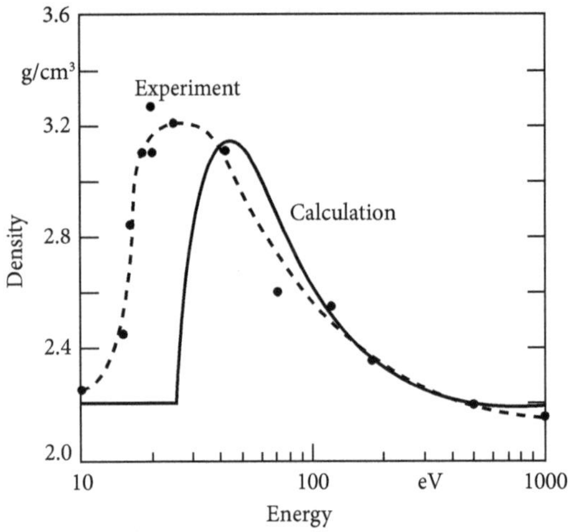

Fig. 8.16 Density of an amorphous carbon film as a function of ion energy from experiments and model calculations [39]

energy of the ions depends on striking the sample during deposition; we find one for a certain material and for a certain deposition procedure only in each case an optimal ion energy.

8.3 Particle Beam Sources

8.3.1 Ion Beam Generation

In this section, we are concerned with the creation and extraction of ions [40]. We will first discuss the different ways in which ions can be created and then the properties of ion sources used in microelectronic applications. This will be followed by a discussion of the space-charge effects that play an important role in the ion-beam formation used for ion milling, ion implantation, and ion microscopy.

In general, each application of ion beams requires a somewhat different source and set of beam parameters. The source performance parameters of greatest importance are the following:

- The ion species produced
- The ion current produced by the source
- The brightness of the source
- The energy spread.

An ion beam consists of a large number of charged atoms or molecules, which have approximately the same velocity, energy, and direction. Their trajectory can be manipulated due to their charge by electrical and magnetic fields. Ion beam technology needs high vacuum or ultrahigh vacuum, since otherwise with the scattering with neutral atoms and molecules, the ion movement with respect to energy and direction changes or causes a complete deterioration of the beam.

Characteristic sizes of an ion beam are the state of charge and energy of the ions as well as the ion current density and the lateral distribution [41].

To illustrate the different characteristics of the ion sources, three classes of ion sources are discussed below: electron-impact, field-ion, and liquid-metal field-ion sources.

8.3.1.1 Electron-Impact Ionization Sources

Electron-impact ionization is the most used technique for generating ions for implantation and ion processing. In this ionization process the energy transferred to a molecule from an energetic electron exceeds the ionization energy E_i, where E_i/e is the ionization potential for that molecule (see Sect. 4.5.4). Electrons for ionization can be created by thermionic or cold cathode emission or can result from the discharge itself. These electrons are accelerated by the use of DC or RF fields and confined by the use of magnetic fields.

Energetic electrons can lose energy to gas particles through elastic collisions, the dissociation of gas particles, and the excitation of gas particles. The momentum transfer to the gas particles is usually negligible because of the large mass imbalance; thus the increase in kinetic energy of the gas particles is negligible. When an incident (primary) electron has energy in excess of the ionization energy, this excess energy can appear as the kinetic energy of a scattered incident electron, as the kinetic energy of an ejected (secondary) electron, as the multiple ionization of a gas particle, the excitation of a gas particle, or any combination of these effects.

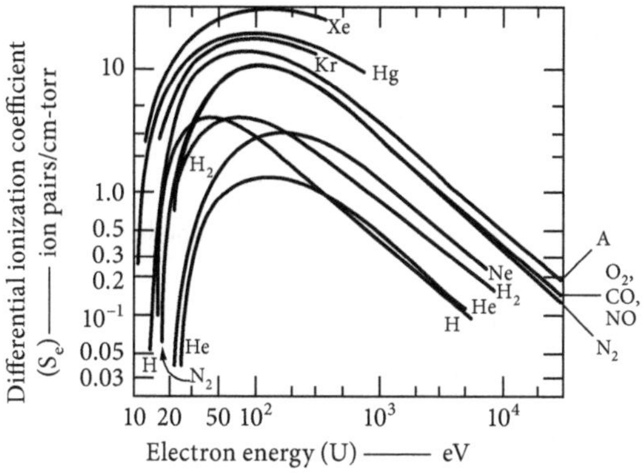

Fig. 8.17 Differential ionization coefficients S_e as a function of electron energy E

The number of ionizing collisions suffered by an electron in passing through a gas per unit path length per unit pressure is called the differential ionization coefficient S_e; the value of this coefficient depends on the electron energy E_e and the gas species. The differential electron-impact ionization coefficients for several gas species are shown in Fig. 8.17. It can be seen that electrons with energies much less than or much greater than the atomic and molecular ionization energies do not ionize these particles very effectively; electron energies a few times the ionization energies are optimum for electron-impact ionization.

The number of ions per square centimeter per second $\acute{n}_i$ produced by an electron current density j (A/cm^2) passing a distance l (cm) through a gas of pressure p (mbar) is given by

$$\acute{n}_i = (j - zplS_e)/e\,. \qquad (8.37)$$

As an energetic electron loses energy from ionizing and scattering collisions, its ionizing effectiveness changes. When its energy decreases below the lowest ionization energy of any atom in the volume, it no longer causes ionization and is called an ultimate electron. Energetic electrons can cause ionization efficiently only if they are maintained in the ionizing volume for a time long enough for them to lose their energy by ionizing collisions. A confining force resulting, for example, from a magnetic field or an oscillating RF field is often employed to increase the path length and electron lifetimes. These concepts are illustrated in Fig. 8.18a–c.

In the hot and cold cathode ion sources the anode is maintained at a positive potential with respect to the two end plates, so that electrons emitted by the cathode are injected through the plasma sheath in front of the cathode into a region where the potential varies with both the radial and axial distances.

The application of an approximately axial magnetic field prevents the electrons from moving directly to the outer cylinder. In this crossed electric and magnetic field configuration, the electrons spiral around the magnetic field lines and oscillate longitudinally. As a result of ionizing collisions between the electrons and the atoms of the source gas, the plasma will be supplied continuously with ions and electrons. The primary electrons lose energy and diffuse across the magnetic field lines to the outer cylinder, where they are collected, causing an electric current to flow from the anode to the cathode. The plasma essentially fills the discharge chamber. In principle, this discharge mechanism can be used to ionize any gas (Sect. 4.6).

In the RF ion source the ever-present electrons formed in the discharge tube fill the chamber and form a plasma sheath along the chamber wall and the extractor electrode. The wall potential is given by [42]

$$U_w = kT_e/2e \, \ln\{T_e m_i / T_i m_e\} \qquad (8.38)$$

for an insulator surface, where T_e and T_i are the electron and ion temperatures, and m_e and m_i are the electron and ion masses, respectively [43]. This wall potential causes the plasma boundary S_1 to move away from the wall surface S_2, as shown in Fig. 8.19.

The measure of plasma withdrawal is given by λ_D, where λ_D is the Debye shielding length (see Sect. 4.3). If an aperture larger than 2 λ_D is made in the chamber wall, the plasma will balloon out through this opening, as shown in Fig. 8.19a. In the opening of the extractor electrode the value of the negative potential applied to this electrode determines the shape of the plasma

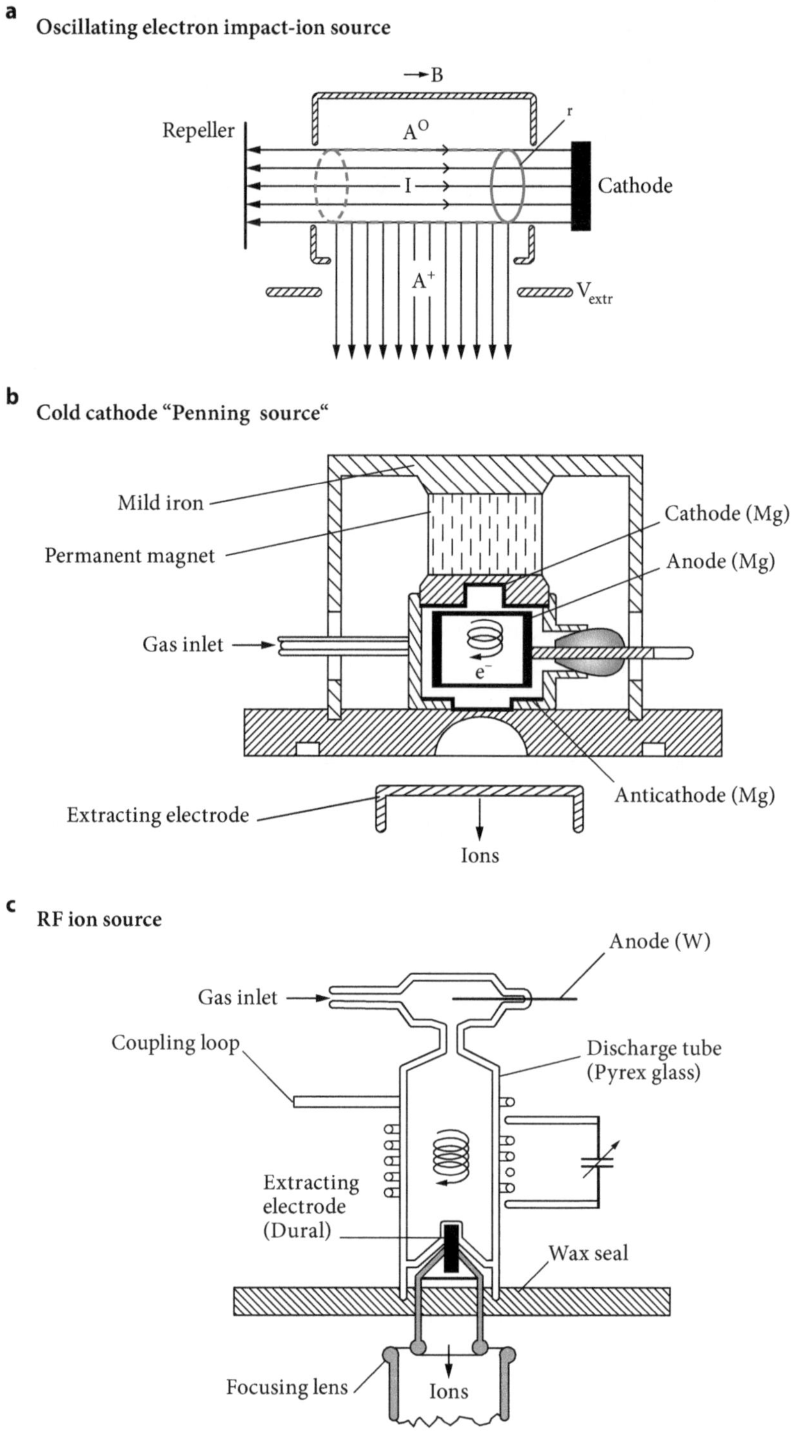

Fig. 8.18 Electron-impact ionization sources. **a** Oscillating electron impact-ion source, **b** cold cathode "Penning source", **c** RF ion source

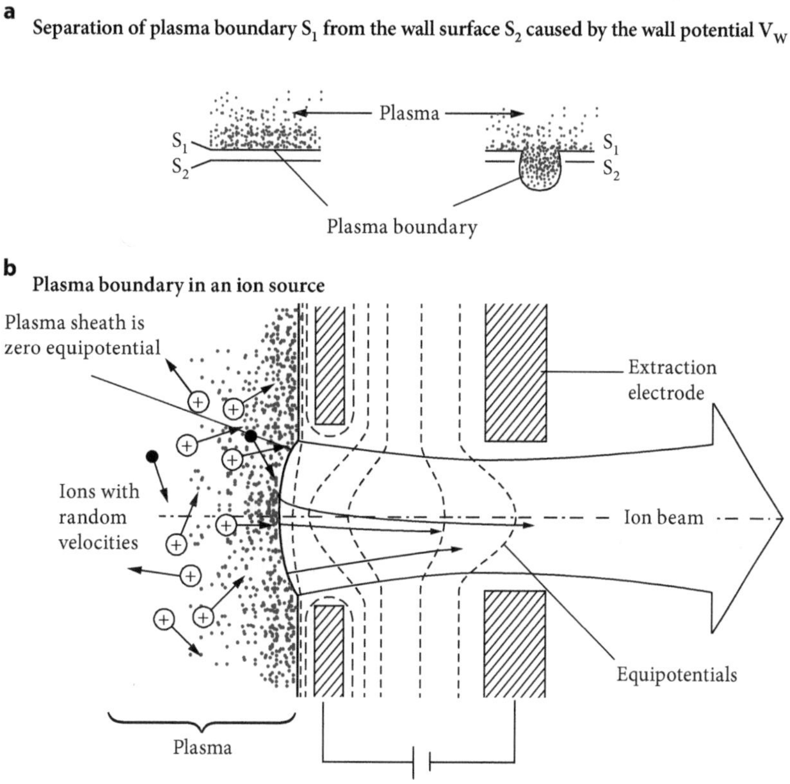

Fig. 8.19 Beam extraction from plasma sources. **a** Separation of plasma boundary S_1 from the wall surface S_2 caused by the wall potential V_w, **b** plasma boundary in an ion source

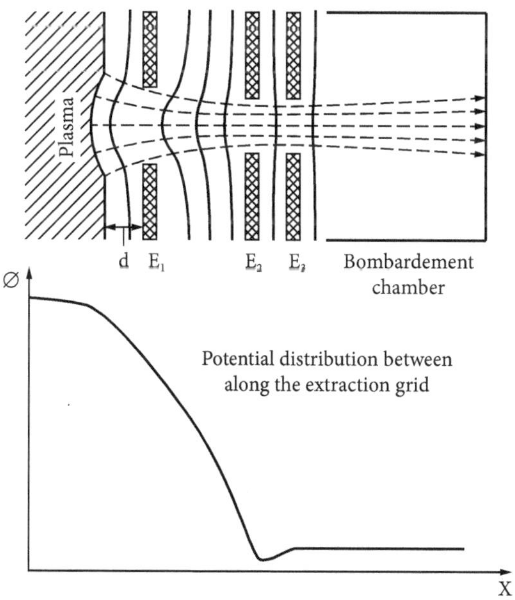

Fig. 8.20 Schema of an ion extraction optics with three electrodes and the potential distribution ($\phi(x)$) between plasma and bombardment area. Potential distribution between along the extraction grid

surface (Fig. 8.19b) because the plasma boundary must be an equipotential surface and the normal component of the electric field must be zero. This means that if the negative potential is increased, the plasma surface is forced back so that a curved surface will be the ion emitter forming a convergent beam of ions.

The shape and position of the ion-emitter surface of the plasma are thus dependent on the value of the negative potential applied to the extracting electrode. This means that the ion-optical system composed of the extracting electrode and the plasma surface can have different geometries determined by the value of the negative potential applied to the extracting electrode. This imposes limitations on the extracting voltage so that the beam intensity cannot be varied beyond a given value by varying the extracting voltage.

Under normal operating conditions the shape of the plasma boundary adjusts itself so that the ion extraction rate is equal to the supply rate from the plasma. In this case, the extracted ion current

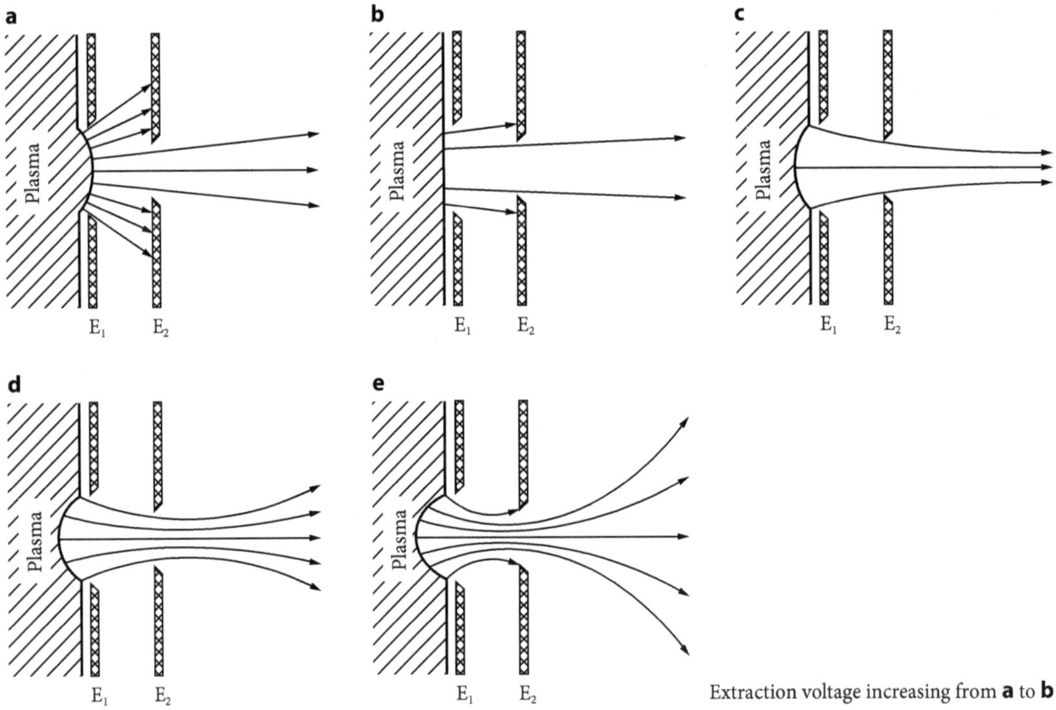

Fig. 8.21 Influence of the extraction voltage on the form of the plasma edge layer and the ion beam. Case **c** corresponds to the optimal operating point.

density is given by the Child–Langmuir equation:

$$j_i = 4\varepsilon_0/9(2e/m_i)^{0.5}(U_{ext})^{1.5}/d^2\,, \quad (8.39)$$

where U_{ext} is the extractor potential, d is the extractor electrode–plasma boundary distance, and m_i is the ion mass.

Often a third electrode is used, which is again weakly positive on E_2 weakly positive (Fig. 8.20). Thus it can be prevented that electrons, which were caught by the space charge of the ion beam, flow back into the plasma. By a suitable choice of the voltage applied at E_3, the electron flux from the beam volume in the plasma and thus the focusing of the ion beam can be controlled. The difference of potential between plasma or electrode E_1 and the extraction electrode E_2 determine the form of the ion-emitting plasma boundary surface (Fig. 8.21).

At too low extraction voltages a convex boundary surface is formed. The predominant part of the extracted ion beam in this case strikes the extraction electrode E_2 and sputters it. The extracted ion current density is smaller than its maximally possible value. In the case of an increase of voltage the beam is ever more focused (Fig. 8.21b), until finally its diameter at the level of E_2 accepts the opening diameter of this electrode (Fig. 8.21c) and only very few ions strike E_2.

The transition from divergent to convergent beam distribution is accompanied by an appropriate change in the form of the ion-emitting plasma boundary surface. With a further increase of the extraction voltage the plasma boundary surface is ever more concave and the ion-emitting surface (Fig. 8.21d,e) increases. This process is accompanied with an increase of the ion current extracted from the plasma. With further deformation of the plasma boundary a stronger focusing of the ion beam arises, whereby its waistline is shifted closer to the extraction system, until finally it is between the electrodes (e). Under current increase larger repulsive space charge forces

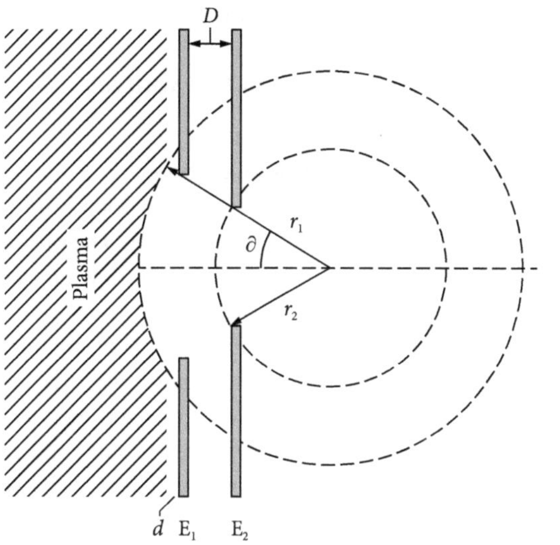

Fig. 8.22 Sectional view by an extraction lens. The sizes indicated determine the perveance of the arrangement

in the beam waistline arise, which lead to an intensified expansion of the beam in the bombardment area. Thus, the current density at the substrate is, however, again reduced. If the focus lies between the electrodes, then the expansion of the beam can already be so large in the level of E_1 that a part again strikes E_2 and with the now high ion energies leads to a substantial erosion of the electrode material.

The maximal ion current extractable by an ion optics is limited by its own space charge. Generally, for the space charge-limited total current by an extraction system it is valid that

$$I = P \times U^{3/2}\,, \tag{8.40}$$

where P is the perveance of the ion-optical extraction system and U the ion-accelerating potential difference between the starting point of the ions and the exit electrode.

If we regard a concentric extraction system with round grid openings of suitable size as a spherical sector (Fig. 8.22) with the half opening angle δ, then we obtain for the perveance:

$$P = \frac{8\sqrt{2}}{9} \times \pi\varepsilon_0\sqrt{\frac{e_0}{M}} \times \frac{1-\cos\delta}{F(r_1/r_2)}\,, \tag{8.41}$$

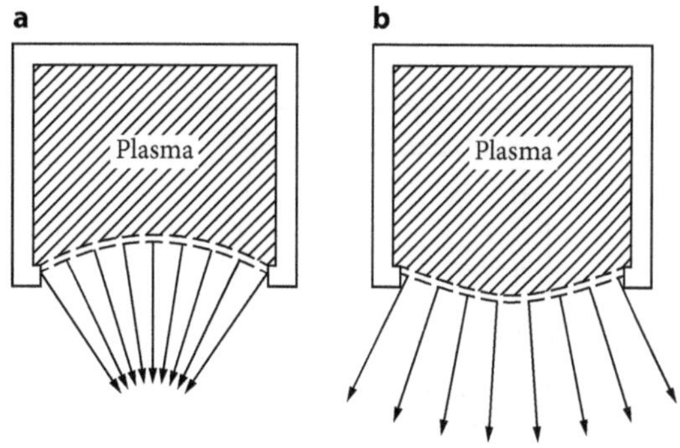

Fig. 8.23 Ion extraction with curved grid structures in the form of a spherical recess. **a** Focusing high current density. **b** Defocusing low current density

where $F(r_1/r_2)$ is a function of the two sphere radii r_1 and r_2 (as those shown in Fig. 8.22 in tabulated form) [44].

For the case represented in Fig. 8.22 for the potential difference U in Eq. (8.40), it is valid that

$$U = U_{P_1-E_1} + U_{E_1-E_2}\,. \tag{8.42}$$

The maximally extractable ion current is the product of the ion saturation current density j_{ion} from the plasma and the surface A of the respective ion-emitting plasma boundary surface. Because of the variation of A with the extraction voltage U the relationship r_1/r_2 and concomitantly the perveance changes with U. This means that the geometry of an extraction system must be adapted to the desired ion current density, ion energy, and beam divergence. Detailed theoretical views for the calculation of extraction optics can be found in [45].

For the generation of a large-scale ion beam a large number of grid openings must be used. Large area ion beam sources for thin film technology have the following parameters:

- Ion energy: 0.1–20 keV
- Ion current density: 0.5–20 mA/cm^2
- Energy spread: approximately 10–20 eV
- Beam homogeneity: approximately 5 %.

In order to keep changes of electrode geometry as small as possible by thermal deformation during the operation, extraction grids are made of graphite or of high-melting metals (Mo, Ta), which are covered with ceramic layers at the sur-

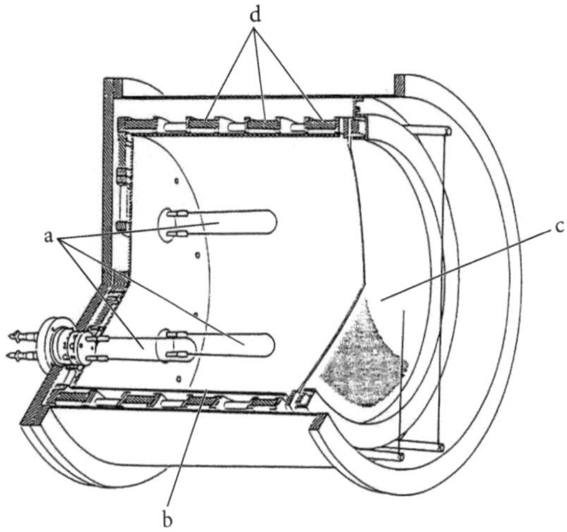

Fig. 8.24 Large area ion gun [48]. *a* cathodes, *b* anode, *c* two-electrode extraction lattice, *d* permanent magnets (multipole arrangement)

face. Frequently also extraction arrangements in curved form are used (Fig. 8.23).

Such concave or (convex) formed extraction systems can then also contribute to the focusing (or defocusing) of ion beams [46]. Electrons injected into the ion beam can likewise decrease beam divergence due to the space charge of the ion beams.

8.3.2 Large Area Sources

8.3.2.1 Direct Current Ion Sources

8.3.2.1.1 Ion Sources with Hot and Cold Cathodes

Figure 8.24 shows a large area ion source based on a direct current discharge [47, 48]. The plasma is maintained by means of electrons, which are emitted from glow cathodes.

Behind the anode wall permanent magnets are arranged in a so-called *multipole* arrangement. Under the influence of crossed electrical and magnetic fields in the wall environment there develops a circular electron flux with a drift velocity,

$$\vec{v} = \frac{\vec{E} \times \vec{B}}{B^2} . \tag{8.43}$$

This leads to an extension of the electron trajectories and the duration of the stay of the electrons. This increases the probability of the electron collision ionization of neutral gas atoms in the wall environment. The spatial homogeneity of the plasma density in the discharge chamber is thereby improved. The source can be operated with argon at a pressure between 10^{-3} and 10^{-4} mbar with discharge voltages < 50 V.

Low electron energies have the advantage that only a few percent of multiply ionized gas atoms

Fig. 8.25 Freeman ion source

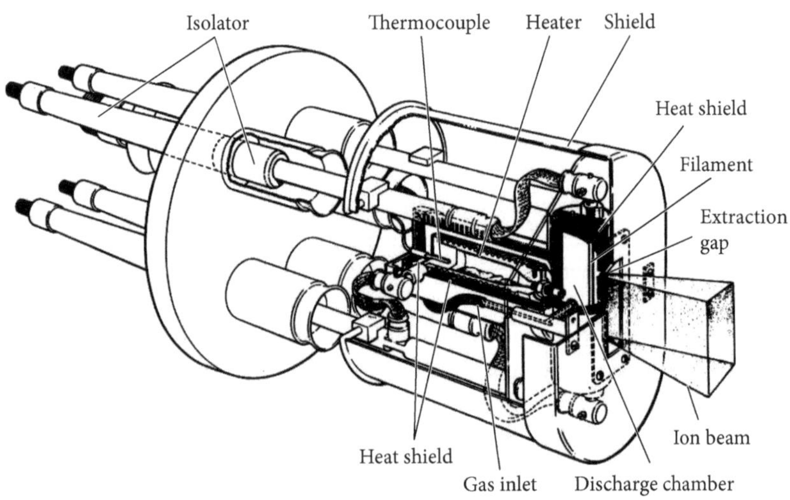

develop. Despite their simple design, sources with hot cathodes have disadvantages, which limits their application. Apart from the thermal evaporation of the cathode wire, operation with reactive gas leads to fast deterioration of the cathodes.

Freeman Source

Similarly to the source of Fig. 8.24, the ***Freeman source***, Fig. 8.25, is a low-voltage gas discharge with hot cathodes.

An outer axial magnetic field serves to increase the probability of ionization. The ions are extracted not at the end of the discharge chamber, but laterally by a slot (e.g. 40 mm × 2 mm). The accelerating electrode is generally in a large variable distance of 1 to 3 cm in front of the source. The ions can be extracted from the plasma with relatively large accelerating voltages of some 10 kV [49].

Freeman sources are particularly used in ion implantation equipment. Figure 8.26 shows a source used in high current implantation equipment. The source is suitable for the production of P^+ and As^+ ions. The attainable ion beam currents are, for instance, 10 mA. Figure 8.27 shows an evaporator unit for the Nova–Freeman ion source.

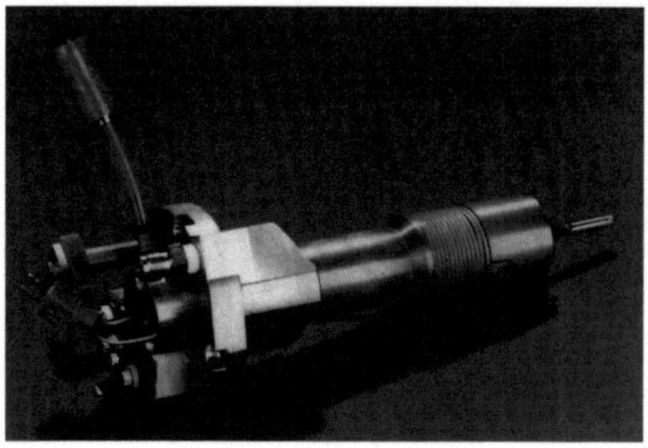

Fig. 8.27 Evaporator equipment for the Nova–Freeman-source (company photo Eaton)

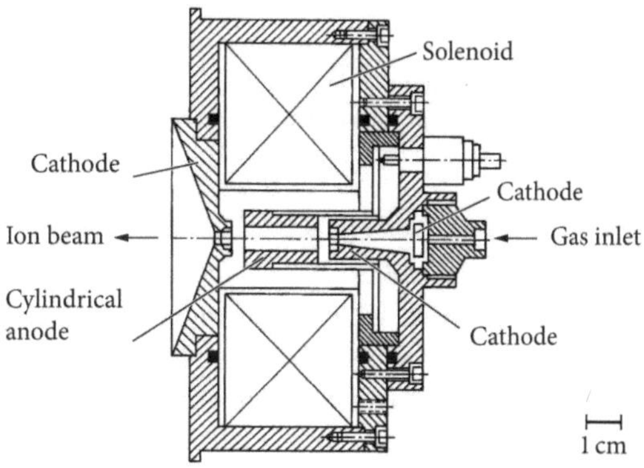

Fig. 8.28 Cross-section of a ***Bethge–Baumann*** source

Fig. 8.26 Nova–Freeman ion source

8.3.2.2 Electrodeless High Frequency Ion Sources

For operation with long maintenance intervals, electrodeless high frequency or microwave ion guns are better alternatives. With these sources also ions from reactive gases and metals can be generated.

The ***Bethge–Baumann*** source [50] works according to the Penning principle. Figure 8.28 shows a schema of this source.

8.3.2.2.1 Sources with Ring Discharges

The simplest design of electrodeless high frequency ion sources by inductive excitation is to use a high frequency ring discharge, Fig. 8.29.

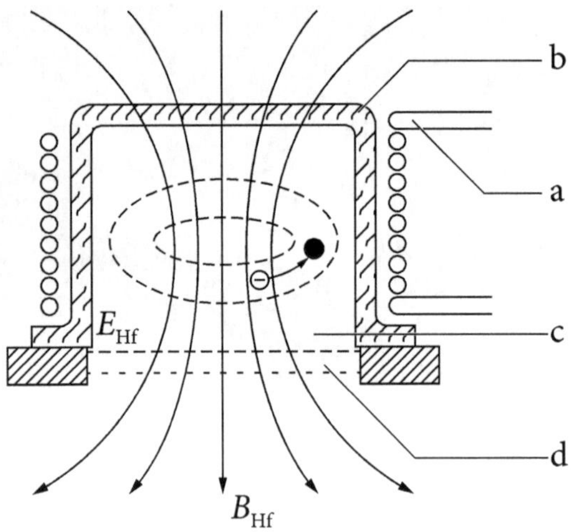

Fig. 8.29 Schema of a ring discharge ion source. *a* High frequency coil, *b* plasma chamber (dielectric), *c* discharge area, *d* two-electrode extraction grid

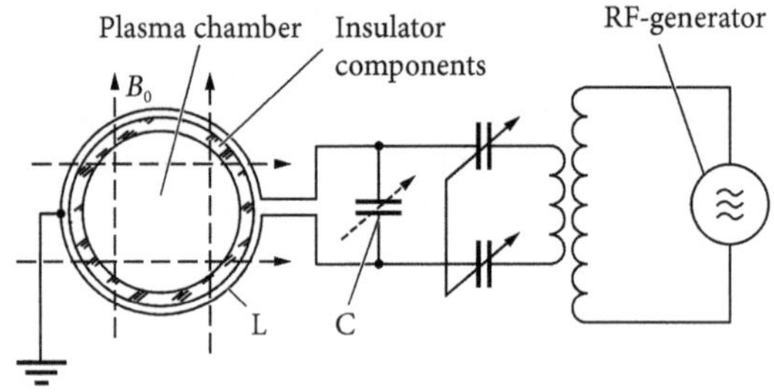

Fig. 8.30 Arrangement for excitation for plasma by means of electron cyclotron wave resonance

The plasma chamber consists of a cup-shaped dielectric material, which is surrounded by a high frequency solenoid. The high frequency axial magnetic field B_{HF} induces an electrical high frequency cyclic field E_{HF}. E_{HF} increases with r. The electrons in the environment of the wall are accelerated more intensively, they form a torus-shaped plasma density distribution (ring discharge). Without collisions with gas particles in the plasma chamber the electrons in all RF-ion sources have a phase difference of $\pi/2$. They move in the direction to the electrical field and from there in the temporal average, no energy is taken up. Only the phase difference is changed the electrons taken up from the electrical field sufficient energy, in order to maintain by collision ionization the plasma discharge.

Typically used frequencies for sources with ring discharge are some 100 kHz up to few MHz.

8.3.2.2.2 ECWR Ion Sources

This type of source uses the electron cyclotron resonant frequency for plasma generation [51]. The overlapping of a weak direct magnetic field B_0 perpendicularly to the direction of the high frequency B-magnetic field can develop electron cyclotron resonance and thus the ionization increases the degree of the inductively excited plasma. The schema of a ***EWCR*** (electron cyclotron resonant frequency) source is shown in Fig. 8.30.

The plasma chamber usually consists of a glass tube, which is limited by two flat walls. The cross-sectional shape of the plasma chamber is almost arbitrary and can be adapted to the operation purposes. Around the glass tube is put a one-turn high frequency coil equipped with a lengthwise slot, which covers the glass tube nearly completely. Located with an adjustable capacity the coil forms a tunable electrical circle, in which the high frequency is inductively coupled in.

To use the electron cyclotron resonance the plasma discharge perpendicular to the symmetry axis of the plasma chamber and/or to the direction of the high frequency magnetic field is superposed by a static homogeneous magnetic field, which is created, e.g., by a pair of ***Helmholtz*** coils. In the case of resonant excitation the ionization degree in ***ECWR*** plasmas can reach some percent. Thus plasma densities of 10^{10}–10^{11} ions and/or electrons per cm^3 result in the case of process gas pressures from 10^{-4} to 10^{-3} mbar. In the case of resonance, the impedance of the arrangement from the load circuit coil and plasma becomes real, and thus the coupling in of the injected high frequency power is maximal. Due to the high density and temperature of the electrons in ***ECWR*** plasmas, they can be used also after ionization of neutral particles.

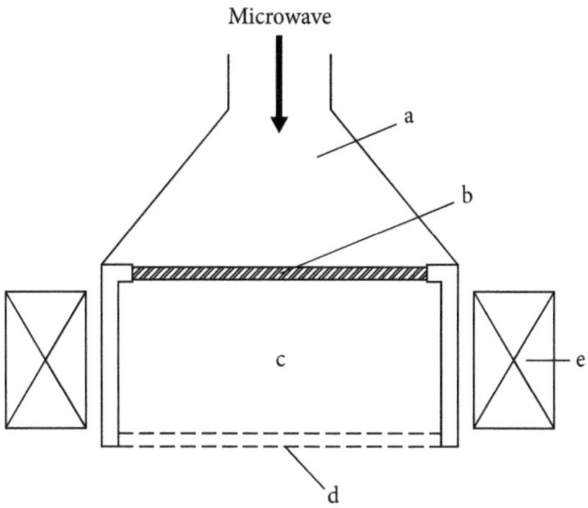

Fig. 8.31 ECR ion source. *a* Waveguide, *b* quartz window, *c* discharge area, *d* two-electrode extraction grid, *e* magnetic field coil

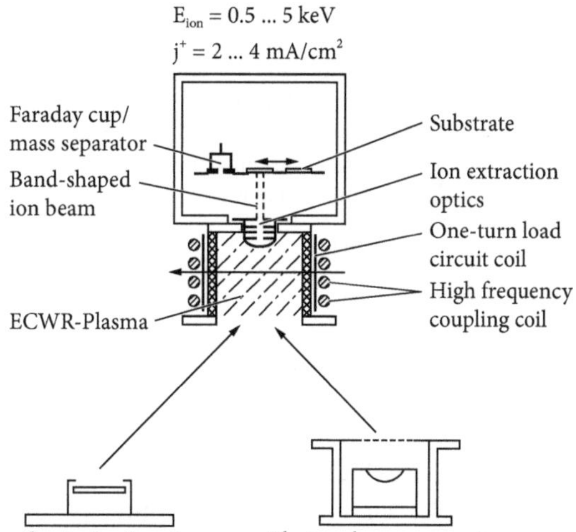

Fig. 8.32 Schematic structure of an ECWR ion gun for the generation of band-shaped ion beams of arbitrary composition. *a* Sputtering module, *b* evaporator unit

8.3.2.2.3 ECR Ion Sources

The electron cyclotron resonance (ECR) – ion sources [52] are usually operated with a microwave frequency of 2.45 GHz. The microwave is fed into the plasma chamber over a waveguide and a quartz window. The chamber form must be adapted to the plasma, which is generated by means of the microwave, in order to achieve an even propagation of the transversal electromagnetic waves in the discharge chamber. The plasma chamber is surrounded with one or more magnetic field coils to adjust the resonance condition ω (microwave) $= \omega_e$ (electrons). At a microwave frequency of 2.45 GHz and a magnetic field overlaid by 875 Gauss the electron cyclotron resonance is $\omega = 2\pi \times 2.45$ GHz. ***ECR*** sources supply pressures between 10^{-4} and 10^{-5} mbar and ion current densities up to some mA/cm^2. Figure 8.31 shows the schematic structure of an ***ECR*** source.

8.3.2.3 Metal Ion Beams from Plasma Ion Sources (Plasmatrons)

For the creation of ion beams with a metallic component and of pure metal ion beams, metal atoms are injected into the source plasma in two ways (Fig. 8.32), i.e.:

a) Sputtering of a target in the plasma chamber;
b) evaporating metal atoms into the plasma chamber.

The two modules for the injection of metal atoms into the source plasma can alternatively be flange-mounted to the bottom of the plasma chamber.

By sputtering of a target in the discharge chamber a negative DC voltage is applied on the target with several kV, whereby ions of the source plasma are accelerated to the target, and released atoms are ionized partially by electron collisions.

Since the sputtered and after-ionized metal atoms on an average possess a higher kinetic energy than the plasma ions, the relationship from metal to process gas ions in the ion beam can be varied purposefully by the appropriate deceleration potential with the ion extraction. This method always must use a process gas. If metal atoms are evaporated into the plasma chamber inside, then the realising probability is larger than with sputtered particles because of their smaller kinetic energy. With this method, the relationship of metal and process gas ions in the ion beam is adjustable over the metal atomic current evaporated into the plasma as well as the pressure of the process gas.

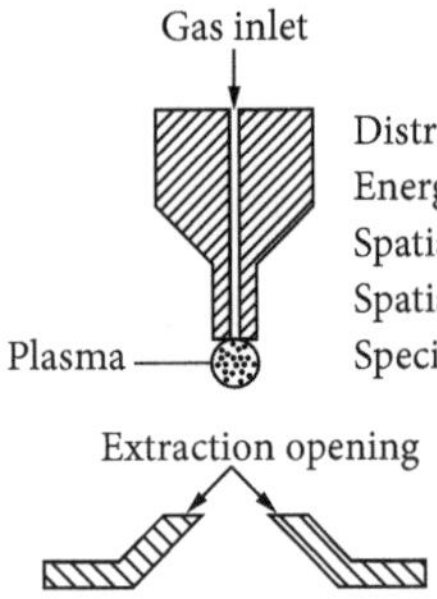

Distribution of intensity	10^3 Acm^{-2} sr^{-1}
Energy dispersion	5 eV
Spatial current density	3 mA/sr
Spatial current density	50 µm
Species of ions	H, He, Ar

Fig. 8.33 Principle of a fine-focusable plasma source

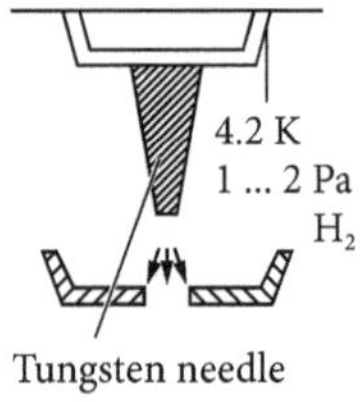

Distribution of intensity	10^5 Acm^{-2} sr^{-1}
Energy dispersion	1 eV
Spatial current density	10 ... 60 µA/sr
Beam diameter	< 10 Å
Species of ions	H_2^+

Fig. 8.34 Principle of a field emission source

8.3.2.4 Fine Focus Sources

In principle, an ion source for directly writing fine focus systems, with which structural widths under 0.5 µm are to be generated, must supply high current densities with little energy uncertainty out of as small an emission area as possible, i.e., only sources with a very high directional beam value can be used.

8.3.2.4.1 Plasma Fine Focus Sources

By plasma sources the emission area from the diameter of the extraction opening is determined. The extractable ion current is fixed by the ion density in the plasma. The size of the emission area can be lowered to a diameter of approximately 50 µm; the discharge then burns at the tip of a capillary (Fig. 8.33) [53].

The extractable current thereby drastically decreases so that the directional beam value remains limited to:

$$R \leq 10\,\mathrm{A\,cm^{-2}\,sr^{-1}}\,.$$

The spectral brightness R_S reaches only maximum values:

$$R_S = R/\Delta E \quad \text{of} \quad 10\,\mathrm{A\,cm^{-2}\,sr^{-1}\,eV^{-1}}\,.$$

Plasma sources from there are not suitable for fine focus ion beam systems of high current density.

8.3.2.4.2 Field-Ionization Sources

At this type of source (Fig. 8.34), in the environment of a cooled tungsten-needle a gas pressure of typically 10^{-2} mbar is adjusted.

At the tungsten-tip the gas is adsorbed. At an extraction opening with a tip radius $\leq$ 1 µm, which is put on negative potential, electrical field forces around 20 V/nm can be reached. In this electrical field the gas molecules and/or gas atoms are ionized. Effective source sizes of 0.5 nm are expected in the future [54].

Since ionization energies of atoms and molecules are usually higher than the work functions of solid surfaces, the field required to extract the electron is usually an order of magnitude greater than for field demission from solid surfaces. Typically, fields of about 10^8 V/cm or 1 V/Å are required for field ionization. Such fields can be simply generated around sharp points of a radius of about 1000 Å. Since ionization occurs only an angstrom or so from the surface, the energy with which the electron strikes the tip is always small. The ions, on the other hand, are accelerated radially from a region close to the point, which forms a tiny virtual source. At very high field forces the ion current i produced in a small volume element close to the tip is given by

$$i = e\acute{n}\,, \tag{8.44}$$

where $\acute{n}$ is the number of particles arriving in the volume element and e the elementary charge.

If A_0 is the area of the tip where the field is high, from kinetic theory we obtain:

$$\acute{n}_0 = A_0 p/(2\pi m k T)^{0.5}\,, \tag{8.45}$$

where p is the pressure, m the mass, k the Boltzmann constant, and T the absolute temperature.

However, because of the high field gradient around the point, molecules are drawn toward it from a relatively large distance, where the fields are sufficient to polarize the molecule. This gives

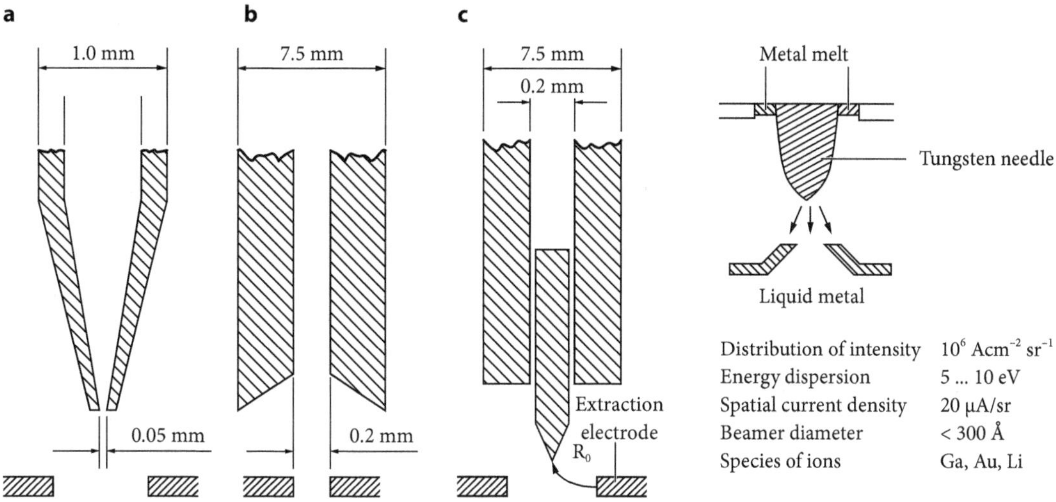

Fig. 8.35 Liquid field ionization sources; **a** nozzle configuration, **b** tube configuration, **c** ring configuration

rise to a much larger effective area of capture A given by

$$\frac{A}{A_0} = 1 - \frac{2}{3}\frac{V_0}{kT}, \quad (8.46)$$

where $-V_0 = \mu E_0 + 0.5\alpha E_0^2$ is the potential energy of the particles just outside the tip, of average permanent dipole moment μ in the field direction, and of polarizability α, and E_0 is the field at the tip. The effective area of capture can exceed that of the tip by a factor of 10–100, giving rise to correspondingly higher field-ionization currents. For ion-beam lithography and microscopy, where a high brightness is required, the use of field-ionization sources is very attractive. With simple optical systems a resolution of $< 1000\,\text{Å}$ can be achieved with monatomic gases (H, He, Ar, Xe).

A gas field emission source can generate a total current of 10^{-8}–10^{-7} A, with a virtual source size of 10 Å, and brightness:

$$\beta \sim 10^8\,\text{A/cm}^2\,\text{sr}\,.$$

The source extractor resolution for gas field emission source systems is limited by chromatic aberration due to the wide energy spectrum (4–15 V) of the extracted ions. The ultimate limit of the emitted density is set by the kinetic supply of ionizable gaseous particles to the high field ($> 2.2\,\text{V/Å}$) region of the conically shaped emitter. This limit can be removed if a liquid film reservoir of the ionizable material is formed on the high field region of the emitter.

8.3.2.4.3 Liquid Field Ionization Sources

A liquid metal ion source [55] typically consists of a fine tungsten capillary (0.02–0.002 cm in diameter) or of a capillary with a tungsten needle projecting through it and an extraction electrode. The operation of the source involving feeding liquid metal to the tip of the capillary tube and applying a voltage between the tip and the extractor electrode is shown in Fig. 8.35.

The interaction of the electrostatic and surface-tension forces causes the liquid-metal meniscus to form a sharply peaked cone of small radius (a ***Taylor cone***). The application of the critical Taylor voltage on a liquid-metal cone extracts ions or electrons, depending on the polarity. This critical cone-forming voltage is given by:

$$V_c = 1.43 \times 10^3 \gamma^{0.5} R_0^{0.5}\,, \quad (8.47)$$

where R_0 is the electrode spacing (cm) and γ is the surface tension (dyn/cm). To understand V_c, let us assume $R_0 \sim 0.1$ cm, and $\gamma = 700$ for liquid Ga; then we find that $V_c = 11$ kV.

It was shown by Taylor [56] that at the critical voltage, where the onset of cone formation occurs, the cone half-angle is 49.3°. As the cone forms the apex the radius decreases sufficiently to enable field emission. If a negative field is applied to the emitter, field electron emission occurs when the local apex field reaches typical field-emission values (0.1–0.5 V/Å), while field desorption as ionization occurs when the polarity is reversed and the field reaches values > 1 V/Å. At higher voltages considerable heating takes place, resulting in the vaporization of the liquid metal, followed by field ionization of the thermally evaporated atoms. Without a negative feedback mechanism the liquid cone radius r_a (ion source size) would decrease without obvious limit with increasing apex field. The apex field (E_a) is given by [57]:

$$E_a = 332(\gamma/r_a)^{0.5}\,\text{V/cm}\,. \tag{8.48}$$

The formation of Ga ions via field desorption requires a field of 1.6×10^8 V/cm. With this E_a and $\gamma = 700$ dyn/cm Eq. (8.48) gives a r_a value of 30 Å. The ion emission is extremely stable owing to space-charge effects in the emission process, even at low currents (< 1 μA). In the case of electron emission from liquid Ga cathode current pulses as high as 250 A with a 2 to 3 ns rise time have been observed at a repetition rate of 40–8000 pulses/s. The mechanism involves the formation of a field stabilized zone of the liquid cathode, which forms a sufficiently small apex radius (r_a) that a regenerative field electron current initiates an explosive emission process. Runaway is not observed for ion emission because there is always sufficient negative feedback present in all regimes of ion emission. This is provided by space charge and by the fact that a decrease in the apex field leads to an increase in the radius of the apex, hence a further decrease in the field and thus decreases in emission, which in turn reduces the space charge, thus leading to an increase in the field, and so on. Space charge is effective in ion emission but not in electron emission because the primary charge carriers in ion generation are massive and the countercharge carriers (which reduce charge) are light, whereas the opposite situation occurs in electron emission.

Typical onset currents in liquid-ion sources are 10^{-5} A, which is a factor of 10^3 times larger than the equivalent current of the mass flow or the supply of liquid in the capillary. Thus the conclusion is that, in contrast to the case of gas field-ionization sources, for liquid-ion sources the atom supply is adequate for the formation of a space-charge stabilized ion current. Probe diameters obtained from liquid-ion sources are limited by the chromatic aberration at the extractor lens. Spherical aberration is negligible in the region of half-angles between 1 and 6 mrad. The liquid metal source represents a major breakthrough in high-brightness ion-source technology. Its superior performance can be used to form focused ion beams with probe sizes and currents comparable to those used for scanning electron microscopy.

8.4 Application Areas of Ion Beams

Applications of ion beams, depending upon the kinetic energy of the ions (Fig. 8.36), are divided into:

⇒ ion beam deposition ***IBD***
⇒ ion beam etching ***IBE***
⇒ ion implantation ***II***.

In the range of low particle energies, at which the product of the striking rate and adhesion probability is larger than sputtering caused by the particle bombardment, it comes to the development of surface films consisting of beam

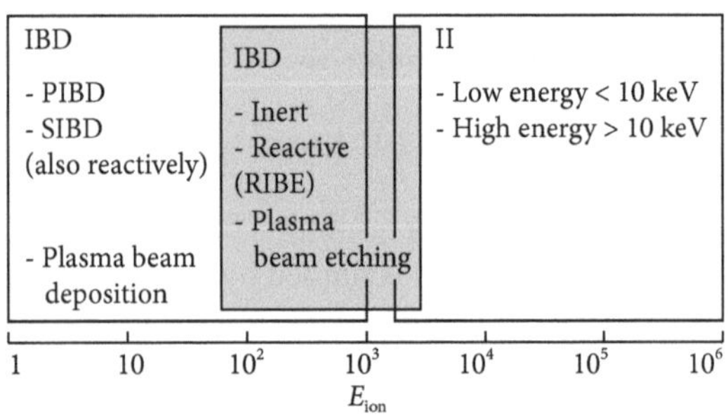

Fig. 8.36 Application areas of ion beam methods as an energy function of the impacting particles

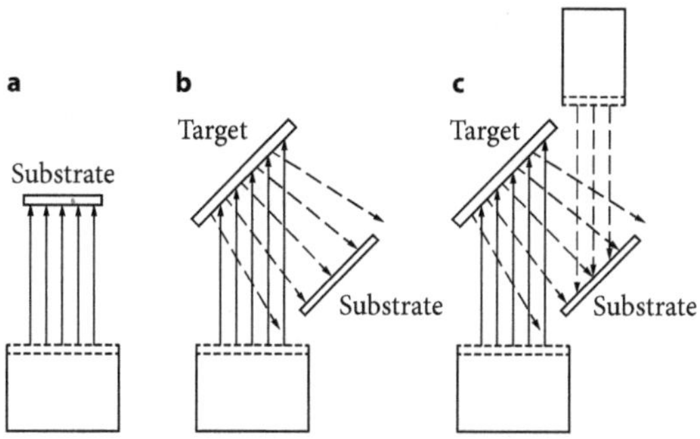

Fig. 8.37 Application areas of the different ion beam procedures. **a** Application of individual beams for – primary ion beam deposition ***PIBD***, – ion beam etching (***IBE***, ***RIBE***), – ion implantation (***II***), – plasma beam deposition, plasma beam etching. **b** Single beam procedure during secondary ion beam deposition (***SIBD***) – ion-(plasma-)beam sputtering (***IBS***). **c** Two-beam procedure during the secondary ion and/or plasma beam deposition (***SDIBD***)

particles. This method is called ***primary ion beam deposition*** (***PIBD***), see Fig. 8.37a.

An increase of the ion energy causes increased sputtering of the bombarded surface. Thus, films can be generated by condensation of sputtered target atoms on substrate surfaces by ion beam sputtering (***SIBD***), Fig. 8.37b.

PIBD and ***SIBD*** differ in substantial points. With ***PIBD*** predominantly ions with controlled energy reach the substrate. With ***SIBD*** predominantly neutral atoms with a middle energy of a few eV and a certain energy distribution, depending on the target material, reach the substrate.

Frequently with ***SIBD*** the substrate is, therefore, bombarded additionally by means of a second ion gun (Fig. 8.37c), with the goal to influence the creation of the film by means of additional energy and electrical charge.

Ion beam etching, with energies between 100 eV and 1 keV, is used in microstructure technology [58] or to dilute substrates for transmission electron microscopy (Fig. 8.37a). With this method of ***ion beam milling*** also surfaces can be smoothed or polished atomically, if one or more ion beams under flat angles are directed toward a rotary target. For ion energies $>$ 1 keV and perpendicular bombardment, we reach the region of the ion implantation extending into the MeV range. Here we must differentiate between low energy implantation ($<$ 10 keV) and high energy implantation.

With implantation of low energies only regions near the surface, up to some 10 nm, are modified by direct irradiation with simultaneous, relatively high sputtering yield. At higher energies the sputtering yield generally decreases again, and the ions penetrate, depending upon energy and material, up to several μm deeply into the surface (Fig. 8.37a).

8.4.1 Primary Ion Beam Deposition (PIBD)

For ***PIBD*** the film consists of particles extracted from the ion source. The kinetic energy of the particles results in a better adhesion tension of the developing film with the substrate surfaces. Compared with evaporation or sputtering methods where the striking particles have only small energies, ***PIBD*** leads to a density increase. The increased energy of ions in the beam that are bombarding the substrate surface also causes a changed nucleation behavior due to increased surface mobility and surface migration of the impacting particles.

Observations [59] of the ion beam deposition of thin In-films on Si_3N_4 showed that in comparison to the molecular beam epitaxy (***MBE***) an increased surface diffusion of In-atoms takes place at ambient temperature. During ***MBE*** many small indium islands develop by the nucleation of the substrate, and the increased particle energy ($<$ 300 eV) of the In-ions supports a faster growth of larger and homogenous islands with simultaneous circumvention of further secondary nucleation.

With the ***PIBD*** method possible adherence of low substrate temperatures at the deposition of the films is sometimes important, in order to avoid unwanted diffusion or segregation processes with doped substrates or with doping during the coating. By means of the ***PIBD*** method also completely new, thermodynamically not producible alloys with new material properties can be manufactured.

Since the substrate and film temperatures are low compared with sintering procedures, it does not come to the flocculation of individual components, which is typical with metal mixtures. Rather, with the ion beam processes the respective mixing proportions of the species involved become frozen.

Direct particle beam deposition by means of a plasma beam source is particularly suitable for the production of electrically high-isolating layers [60]. Thus it is possible to manufacture [61] films with diamond-like characteristics at low substrate temperatures ($< 50\,°C$) by plasma beam deposition.

A range, which is both of technical and of physical interest, is the deposition of thin stoichiometric layers of oxidic high temperature superconductors. ***H. Frey*** et al. [62] developed equipment with which by means of a high frequency plasma beam, purifying powders were injected into a rf-plasma beam, which can produce superconducting $YBa_2Cu_3O_{2-x}$ films.

8.4.2 Secondary Ion Beam Deposition (SIBD)

With the ***SIBD*** method, the influence of the ion energy and the position of the substrate relative to the sputtering target are important for the film characteristic obtained [63]. Depending upon energy and angle adjustment strong changes of the reflecting power and the internal tensions of the deposited films can be achieved.

8.4.3 Ion Beam Etching (IBE, Reactive RIBE)

In microelectronics and microsystem technology anisotropic etching techniques are needed, by which the etching process takes place depending on the direction [58, 64].

With ***ion beam etching*** it often comes to local charging effects. The ions can be recombined by electrons from an electron gun, so that only neutral particles impact the substrate surface. An alternative to ion beam etching is ***plasma beam etching***. The plasma beam sources supply a quasi-neutral plasma beam in the low energy range ($< 150\,eV$). If we use reactive gases for plasma generation, selective anisotropic structures can be manufactured [65].

8.5 Ion Implantation

8.5.1 Introduction to Implantation

The most successful and widespread surface modification technique in semiconductor technology is *ion implantation*. Most integrated circuits are fabricated using this process. Electrical dopants are introduced directly into a semiconductor surface film by bombarding it with energetic ions. Ion implantation allows excellent control over the number and distribution of atoms that can be injected, and it is undoubtedly this feature which has made the process an indispensable part of semiconductor technology [66, 67].

With ion implantation the dopant concentrations of thin films can be stopped more exactly than with diffusion processes. Furthermore, the concentration maximum is with diffusion at the surface and with the implantation inside the base material (Fig. 8.38).

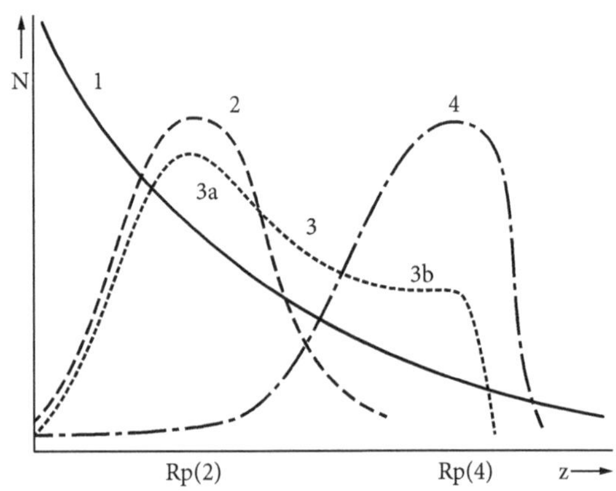

Fig. 8.38 Schematic dopant concentration distributions of N (profile) inside a solid body as a function of the doping depth z. *1* Generated by diffusion; *2* generated by low energy implantation; *3* generated by channeling; *4* generated by high energy ion implantation; R_p distance of the object surface the projected range

Since the implantation is an extreme nonequilibrium method, normal solubility is by far exceeded. One does not generate only conventional alloys in very thin films, but also new metastable compact alloys can be created (e.g. lead (Pb) in iron (Fe), gold (Au) in iron (Fe), or tantalum (Ta) in copper (Cu)). Metastable systems are converted at higher temperatures into finely dispersed precipitations, while amorphous systems are inclined to recrystallize.

Recapitulation: the advantages of the ion implantation are the following:

- Doping levels can be controlled precisely, since the incident ion beam can be measured accurately as an electrical current.
- Doping uniformity across a surface can be accurately controlled.
- The depth profile can be regulated by choice of the incident ion beam energy.
- It is a low-temperature process. This feature is important for some compound semiconductors that are unstable at high temperatures.
- Extreme purity of the dopant can be guaranteed by mass analysis of the ion beam.
- Particles enter the solid as a directed beam, and since there is little lateral spread of the beam, smaller and faster devices can be fabricated.
- Dopants can be introduced that are not soluble or diffusible in the base material.

8.5.1.1 Qualitative Features of Ion Implantation

In the process of ion implantation [68] atoms of the desired doping element are ionized and accelerated to high velocities and then made to enter a substrate lattice by virtue of their kinetic energy (or momentum). After the energetic ion comes to rest and equilibration has occurred the implanted atom may be in a position in which it serves to change the electronic, resistance and abrasion properties of the substrate lattice; that is, doping occurs. Lattice defects caused by the energy loss process of the ions may also change the properties of the substrate [69].

Some ions from a well-collimated beam that is directed toward a channel (certain crystallographic directions in which open spaces exist among the rows of atoms) are able to penetrate deeply into the crystal lattice before coming to rest in interstitial sites (Fig. 8.39). The directions that allow large amounts of this ion channeling are limited. When the crystal is examined from directions other than along a channel or plane, the atoms appear more randomly oriented, roughly as in a dense atomic gas. However, even when ions are injected along a nonchanneling direction, it is difficult to prevent channeling completely.

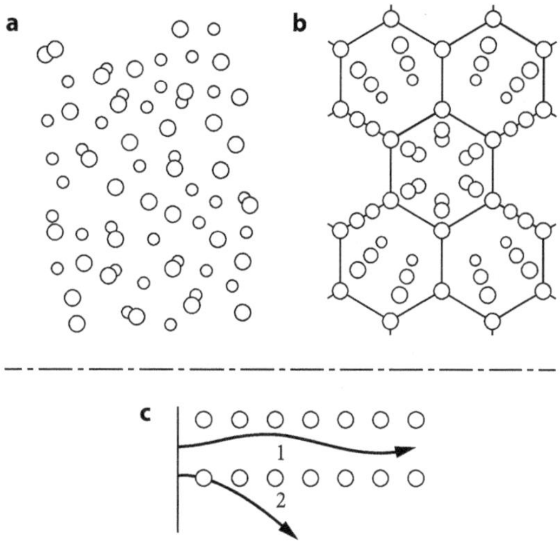

Fig. 8.39 Schematic representation of channeling. **a** Crystal in arbitrary direction; **b** crystal seen in channeling orientation; **c** side view: channeling

The penetration depth and final distribution of the ions in the crystal depend on the ion energy, the crystal and ion species, and the angular alignment of the ion beam with the crystal axis. The characteristics do not depend strongly on the crystal axis, nor the crystal temperature. One can distinguish two primary classes of ion–crystal interaction that yield quite different penetration depths and ion density distributions. Ions penetrate a target that is either amorphous or crystalline, but if the ion path is misaligned from any crystal axis, they will suffer collisions that reduce their inward motion in such a way that the resulting density profile and depth is roughly ***Gaussian***. If the ion dose is low and the trajectories are directed precisely in an open crystallographic direction, the ions can penetrate deeply into the crystal and will stop rather abruptly at the end of their range. Some fraction of the ions traveling along a channel leaves the channel prematurely. Channeled profiles cannot be maintained at high dose levels, since energetic ions striking the surface of a crystal will displace the crystal atoms, resulting in a near amorphous condition close to the surface. Although the ion dose required to create this amorphous layer varies somewhat with the ion and crystal species, it is usually about 10^{14} ion/cm^2 or more. Thus, with high dose levels, the implanted distribution will always be near Gaussian.

8.5.1.2 Range Theory

It is customary to assume that there are two major forms of energy loss for an ion entering a target (see Sect. 8.2). These are interactions of the ion with the electrons in the solid and collisions of the ion with the nuclei of the target. Therefore the total energy loss can be written as the sum

$$-\frac{\mathrm{d}E}{\mathrm{d}z} = N[S_\mathrm{n}(E) + S_\mathrm{e}(E)]\,, \tag{8.49}$$

where $-\mathrm{d}E/\mathrm{d}z$ is the average rate of energy loss with distance along R (eV/cm), E is the energy of the ion at point z along R (eV), $S_\mathrm{n}(E)$ is the nuclear stopping power (eV cm^2), $S_\mathrm{e}(E)$ is the electronic stopping power (eV cm^2), and N the target atom (density $\sim 5 \times 10^{22}/\mathrm{cm}^3$ for Si).

The total distance that the ion travels before coming to rest is called *its range* (R). The projection of this distance onto the direction of incidence is called *the projected range* (R_p). The distribution of the stopping points for many ions in space is *the range distribution.*

If $S_\mathrm{n}(E)$ and $S_\mathrm{e}(E)$ are known, then the *total range* can be calculated:

$$R = \int_0^R \mathrm{d}z = \frac{1}{N}\int_0^{E_0} \frac{\mathrm{d}E}{S_\mathrm{n}(E) + S_\mathrm{e}(E)}\,. \tag{8.50}$$

R is the average *total range* and one should expect a distribution of R in the direction of incidence (ΔR_p) around R_p, the *projected range*, and a transverse distribution (ΔR_T) around R_T, the *transverse range*. Generally, the shape of the distribution depends on the ratio between the ion mass M_1 and that of the substrate atoms, M_2. The relative width $\Delta R_\mathrm{p}/R_\mathrm{p}$ of the distribution depends on the ration of M_1 to M_2. (For light ions such as B, $\Delta R_\mathrm{p}/R_\mathrm{p}$ is large. For heavy ions such as As and Sb $\Delta R_\mathrm{p}/R_\mathrm{p}$ is small.)

8.5.1.3 Useful Approximations

The expression for the nuclear stopping power can be written in the following form:

$$S_\mathrm{n} - 2.8 \times 10^{-15}\frac{Z_1 Z_2}{Z^{1/3}}\frac{M_1}{M_1 + M_2}\,\mathrm{eV} \times \mathrm{cm}^2\,, \tag{8.51}$$

where

S_n is the nuclear stopping power (independent of E),

Z_1 is the ion atomic number,

M_1 is the atomic mass,

Z_2 is the substrate atomic number (14 for Si) and $Z^{1/3} = (Z_1^{2/3} + Z_2^{2/3})^{1/2}$.

The electronic stopping power can be approximated as

$$S_\mathrm{e}(E) = kE^{1/2}\,. \tag{8.52}$$

Here, E is the energy of the ion and k is a constant that depends on both the ion and the substrate and is given by

$$k = Z_1^{1/6}\frac{0.0793\, Z_1^{1/2} Z_2^{1/2}(M_1 + M_2)^{3/2}}{(Z_1^{2/3} + Z_2^{2/3})^{3/4} M_1^{3/2} M_2^{1/2}}\frac{C_\mathrm{R}}{C_\mathrm{E}^{1/2}}\,, \tag{8.53}$$

where

$$C_\mathrm{R} = \frac{4\pi a^2 M_1 M_2}{(M_1 + M_2)^2} \tag{8.54}$$

and

$$C_\mathrm{E} = \frac{4\pi\epsilon_0 a\, M_2}{Z_1 Z_2 q^2 (M_1 + M_2)}\,, \tag{8.55}$$

where ε_0 is the permittivity *of free space* $\varepsilon_0 = 8.85 \times 10^{-11}/\mathrm{F\,cm}^{-2}$, a is the Bohr radius $a = 0.529 \times 10^{-8}$ cm, q is the elementary charge $q = 1.602 \times 10^{-9}$ C for an amorphous Si substrate, and k is a constant independent of the type of ion.

Equation (8.53) reduces to:

$$k = 0.2 \times 10^{-15}\,\mathrm{eV}^{1/2} \times \mathrm{cm}^2\,.$$

Table 8.1 Projected range corrections R_p/R (Si substrate)

Ion	R_p/R Values				Rule of thumb value
	20 keV	40 keV	100 keV	500 keV	$(1 + M_2/3M_1)^{-1}$
Li	0.54	0.62	0.72	0.86	0.4
B	0.57	0.64	0.73	0.86	0.54
P	0.72	0.73	0.79	0.86	0.77
As	0.83	0.84	0.86	0.89	0.89
Sb	0.88	0.88	0.89	0.91	0.93

Note that S_n is independent of E (Eq. (8.49)) and S_e increases with E. At some critical energy E_c, the electronic and nuclear stopping powers are equal, that is, at $E = E_c$

$$S_n(E_c) = S_e(E_c)$$

and

$$\sqrt{E_c} = \frac{S_n^0}{k},$$

hence

$$\sqrt{E_c} = 14\frac{14Z_1}{(14^{2/3} + Z_1^{2/3})^{1/2}}\frac{M_1}{M_1 + 28}. \tag{8.56}$$

For amorphous Si the critical energy is given as:

$$E_c \cong 10\,\text{keV} \quad \text{for B} \quad (Z = 5, M = 10)$$
$$E_c \cong 200\,\text{keV} \quad \text{for P} \quad (Z = 15, M = 30)$$
$$E_c > 500\,\text{keV} \quad \text{for As and Sb}.$$

Thus B tends to be stopped by electronic interactions; P, As, and Sb tend to be stopped by nuclear collisions.

If $E \ll E_c$, then:

$$\frac{dE}{dz} = kS_n$$

and from Eq. (8.48)

$$R = (0.7\,\text{Å})\frac{Z^{1/3}}{Z_1Z_2}\frac{M_1 + M_2}{M_1}E_0. \tag{8.57}$$

Equation (8.55) is useful for heavy ions such as As, Sb, and sometimes P. If $E \gg E_c$, then:

$$\frac{dE}{dz} = NkE^{1/2}$$
$$R = 20\sqrt{E_0}\,\text{Å} \quad \text{for } E_0 \text{ in eV}. \tag{8.58}$$

Equation (8.56) is *useful for boron* if channeling does not take place. In Eqs. (8.57) and (8.58) R is the total range, not the projected range R_p. For the general case when the above limiting cases do not apply:

$$R = \frac{1}{N}\int_0^{E_0}\frac{dE}{S_n + kE^{1/2}}. \tag{8.59}$$

8.5.1.4 Projected Range

The previous equations were in terms of R, (the *total range*), but we really require R_p (the *projected range*). It can be shown that [68]

$$\frac{R}{R_p} = 1 + b\frac{M_2}{M_1}, \tag{8.60}$$

where $b \approx 1/3$ for nuclear stopping and $M_1 > M_2$, that is, for Sb and As. Although b is smaller for electronic stopping (B, P), $1/3$ is still correct for the first order. For example, As : M = 75, therefore $R/R_p \approx 1.12$. For other mass and energy combinations the projected range can be estimated using the R_p/R calculations in Table 8.1.

The dopant concentration distribution as a function of the distance in the target is shown in Fig. 8.34, where:

$$R_p = \frac{R}{1 + bM_2/M_1},$$

and the standard deviation:

$$\Delta R_p \cong \frac{2}{3}\frac{\sqrt{M_1M_2}}{M_1 + M_2}R_p$$
$$= \frac{\text{half-width at}(1/2)N_{max}}{(2\ln 2)^{1/2}}$$

$$N_{max} \cong \frac{N_D}{2.5\Delta R_p},$$

Table 8.2 Projected range and standard deviation in projected range for various ions in silicon in Å

Ion	(keV)	20	40	60	80	100	120	140	160	180	200
B	R_p	714	1413	2074	2695	3275	3802	4289	4745	5177	5588
	(ΔR_p)	276	443	562	653	726	793	855	910	959	1004
N	R_p	491	961	1414	1847	2260	2655	3034	3391	3728	4046
	(ΔR_p)	191	312	406	479	540	590	633	672	710	745
Al	R_p	289	564	849	1141	1438	1737	2036	2335	2633	2929
	(ΔR_p)	107	192	271	344	412	476	535	591	644	693
P	R_p	255	488	729	976	1228	1483	1740	1998	2256	2514
	(ΔR_p)	90	161	228	291	350	405	495	509	557	603
Ga	R_p	155	272	383	492	602	712	823	936	1049	1163
	(ΔR_p)	37	64	88	111	133	155	176	197	218	238
As	R_p	151	263	368	471	574	677	781	885	991	1097
	(ΔR_p)	34	59	81	101	122	141	161	180	198	217
In	R_p	133	223	304	381	456	529	601	673	744	815
	(ΔR_p)		23	38	51	63	75	86	97	108	119
Sb	R_p	132	221	300	376	448	519	590	659	728	797
	(ΔR_p)	22	36	49	60	71	82	92	102	112	122

and N_D is the number of implanted atoms per square centimeter. The ion concentration $N(z)$ will be normally distributed about R_p; that is:

$$N(z) = N_{max} e^{\frac{-(z - R_p)^2}{2\Delta R_p^2}}, \tag{8.61}$$

dropping by one decade at:

$$z = R_p \pm 2\Delta R_p, \qquad N(z) = \frac{1}{10} N_{max},$$

two decades at:

$$z = R_p \pm 3\Delta R_p, \qquad N(z) = \frac{1}{100} N_{max},$$

and five decades at:

$$z = R_p \pm 4.8\Delta R_p, \qquad N(z) = 10^{-5} N_{max}.$$

In practice, R_p and ΔR_p are tabulated for most common impurities as a function of the ion energy.

Thus to find an implantation profile one needs to do the following:

- Look up R_p and ΔR_p in Table 8.2.
- Assume that the distribution is Gaussian.
- Calculate N_{max} from:

$$N_{max} = \frac{N_D}{2.5\Delta R_p}.$$

The profile is given by Eq. (8.61).

The doping profile is usually characterized by the projected range R_p and the standard deviation in the projected range, ΔR_p. Figure 8.40 shows $R_p, \Delta R_p$ and the transverse distribution ΔR_T for boron, and phosphorus and arsenic ions implanted into silicon [70]. The penetration depth is roughly linear with energy and is shallower the more massive the ion.

8.5.1.5 Masking

Spatially selective doping is achieved by means of masking. Because it is basically a mechanical process, ion implantation only requires some physical barrier of a sufficient thickness in order to be masked. There is actually a considerably wider choice of masking materials that can be used with ion implantation with diffusion. Figure 8.41 shows some of these [71] and offers a comparison of different masking layer thicknesses required to prevent all but about 0.1 % of an implant from penetrating. SiO_2 mask a good mask, Si_3N_4 is even better, Al is good, and the photoresistance in thicknesses commonly applied is adequate for all but high energy boron implants.

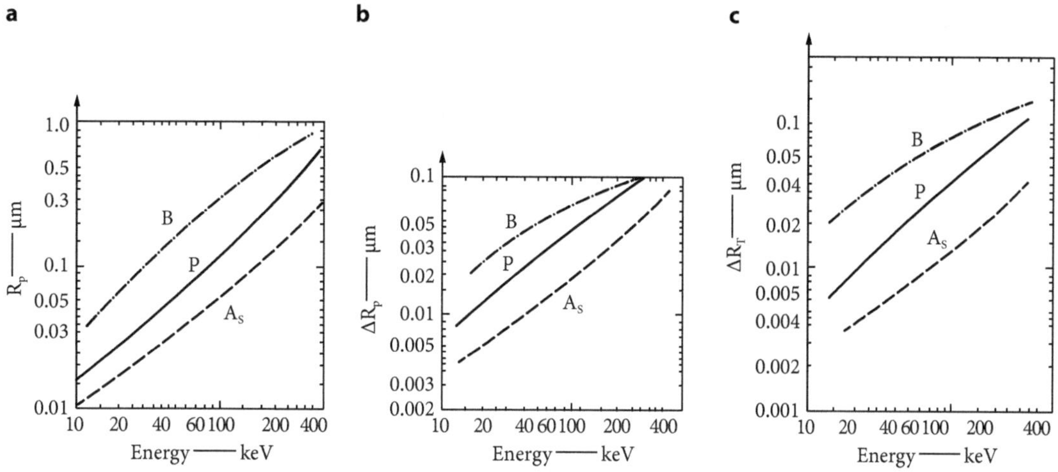

Fig. 8.40 **a** Projected range (R_p), **b** standard deviation (ΔR_p) and **c** transverse distribution (ΔR_T) for boron, phosphorus, and arsenic ions implanted into silicon

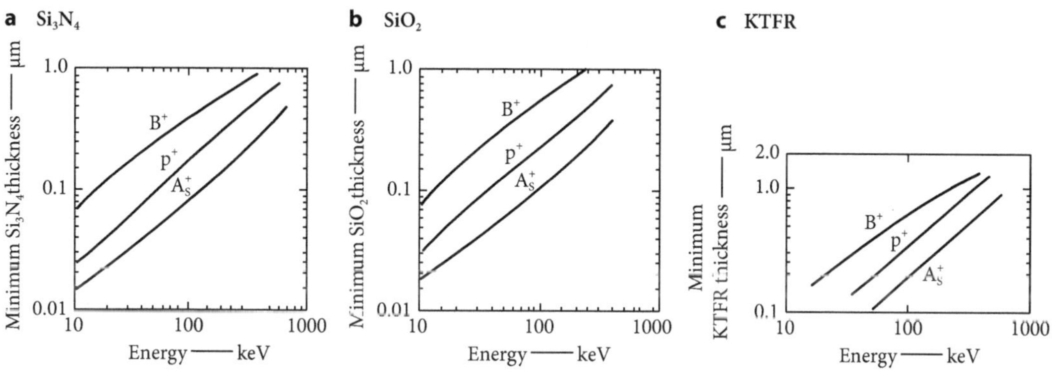

Fig. 8.41 Thickness required to mask implants

8.5.1.6 Channeling

The range distribution in single-crystal targets are different if the ion beam is aligned along the crystal axis [72]. In this case, because projectiles can exist along open direction, a deeper penetration can occur. The channeling angle Ψ is given as [73] (see Fig. 8.42):

$$\Psi \cong \left(\frac{a}{d}\psi_1\right)^{1/2} ,$$

where $\psi_1 = [(2Z_1Z_2q^2)/Ed]^{1/2}$, $a \sim 0.5\,\text{Å}$, and d is the atomic spacing (2.5 Å for Si). Critical angles as a function of energy for several different projectiles are also given in Fig. 8.42 for the three major orientations of a Si target.

The channeling effect can result in any of the several types of profiles shown in Fig. 8.42. The amorphous peak A or A' is generated by ions A' entering at random directions to the lattice direction, or by ions A in well-aligned trajectories but that impact substrate atoms at the surface end of a lattice rows. This amorphous distribution exhibits a Gaussian shape characterized by a mean projected range R_p and a standard deviation ΔR_p. The channeling peak C comprises ions whose trajectories are well within the critical angle of the crystal direction. These ions lose energy by electronic collisions and channel to approximately the maximum range. Ions with higher values of electronic stopping produce more-pronounced (higher) channeling

a

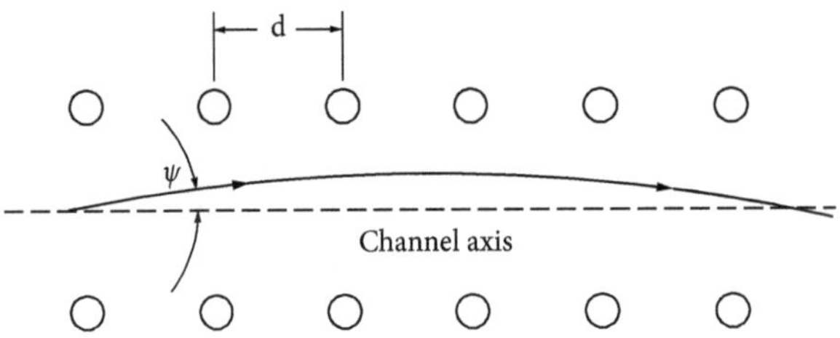

Ion	Energy (keV)	Channel direction		
		<10>	<111>	<100>
Boron	30	4.2°	3.5°	3.3°
	50	3.7°	3.2°	2.9°
Nitrogen	30	4.5°	4.8°	3.5°
	50	4.0°	3.4°	3.0°
Phosphorus	30	5.2°	4.3°	4.0°
	50	4.5°	3.8°	3.5°
	30	5.9°	5.0°	4.5°
Arsenic	30	5.2°	4.4°	4.0°

b

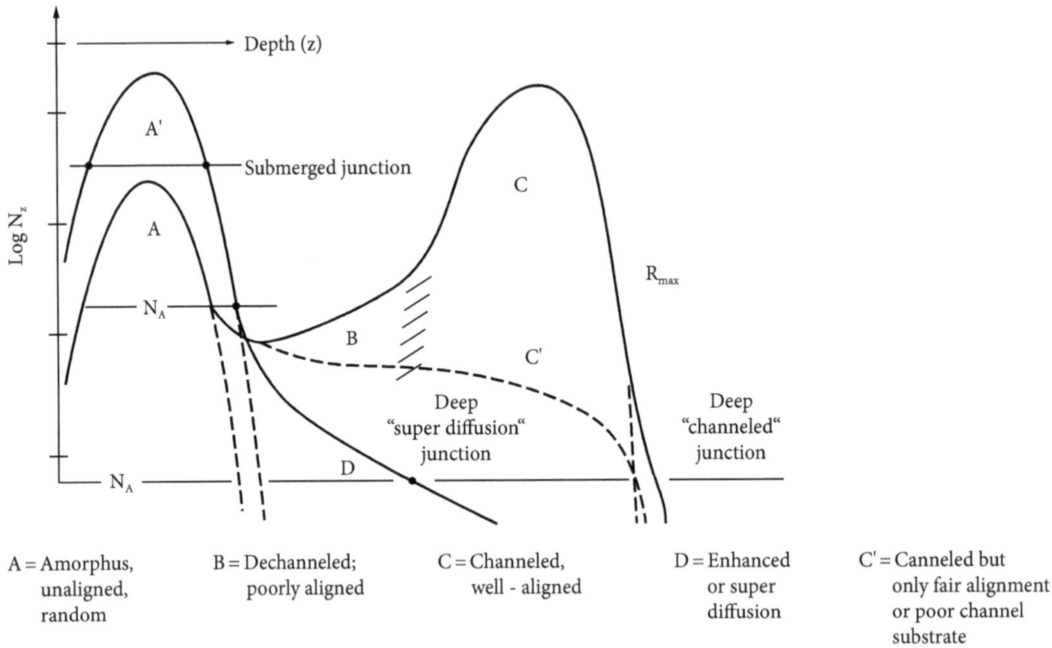

Fig. 8.42 Thickness required to mask implants. **a** Schematic trajectory of a channeled particle, **b** critical angles for channeling of selected ions in silicon

peaks. A channeling peak is not exhibited for all ions incident at room temperature. Channeling is purposely avoided in most devices because it is difficult to control accurately. Usually, implantation is done with the wafer tilted to obtain a random, "amorphous" profile. For example, a 7° tilt off the ⟨100⟩ axis results in more than 99 % of the ions being stopped, as if the silicon were amorphous.

8.5.1.7 Implantation

To summarize the previous results:

- For light ions (B) and high energies the electronic stopping is the major energy-loss mechanism.
- For heavy ions (As, Sb) and low energies the dominant energy loss is due to nuclear collisions.

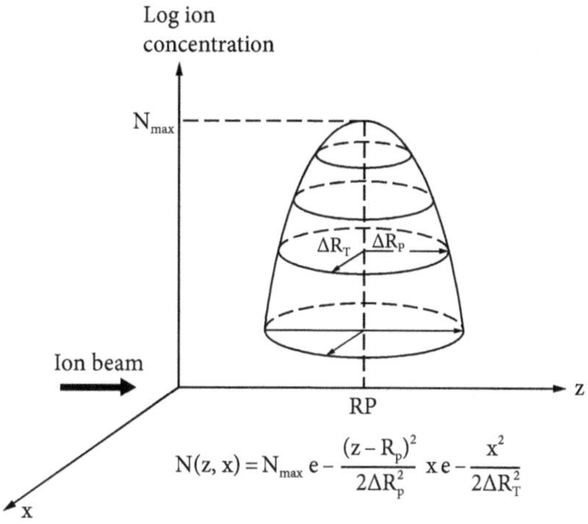

Fig. 8.43 Implanted longitudinal (z) and transversal (x) ion distribution

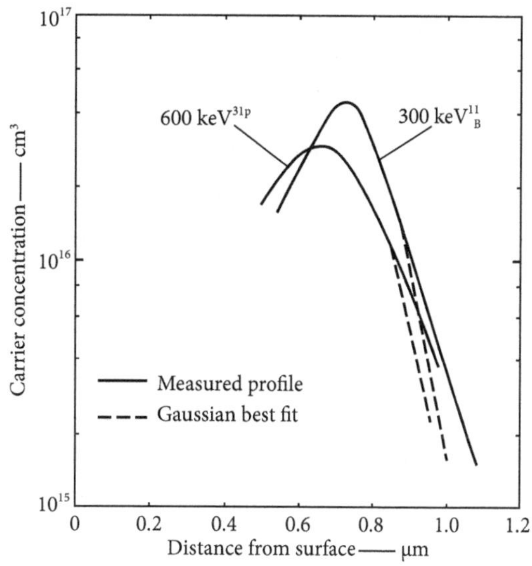

Fig. 8.44 Measured profiles and Gaussian fit

The ion distribution in the target is ***Gaussian*** in both the z and x directions (see Fig. 8.43), or:

$$N(z,x) = N_{\max} \exp\left(\frac{-(z-R_\mathrm{p})^2}{2\Delta R_\mathrm{p}^2}\right) \times \exp\left(\frac{-x^2}{2\Delta R_\mathrm{T}^2}\right) . \quad (8.62)$$

LSS (Lindhard, Scharff, Schiott) theory allows the prediction of R_p, ΔR_p, and ΔR_T for an arbitrary ion and substrate [74]. Channeling can be a problem if the beam is aligned along a crystal axis, but this is normally avoided in devices by tilting the wafer with respect to the beam. Deviations from the Gaussian shape are usually caused by an unwanted channeling effect or by enhanced diffusion during annealing as a result of damage. Figure 8.44 shows two actually measured profiles with a Gaussian fit [75].

8.5.1.8 Damage

For range determination both electronic and nuclear interactions are important. However, concerning radiation damage, only the atomic or nuclear collisions should be considered because only these interactions can transfer enough energy to the target atoms (see Sect. 8.2) [69].

For crystalline substrates, atoms are bound together with energy E_d. The necessary condition to cause damage is that $E_\mathrm{trans} \geq E_\mathrm{d}$, ($E_\mathrm{trans}$ = energy of the implanted ions). Heavy ions (Sb, As) are stopped primarily by nuclear collisions; therefore they cause more damage than B or P, which are stopped mainly by electronic interactions. Nuclear stopping dominates at low energies near E_d; therefore, one should expect that most of the damage for B occurs near where the ions stop.

8.5.1.8.1 Heavy Ions – Sb, As

Heavy ions are stopped primarily due to nuclear collisions ($E_\mathrm{c} > 500\,\mathrm{keV}$) for As and Sb, and $E_\mathrm{c} \approx 200\,\mathrm{keV}$ for P); therefore one can expect a great deal of damage.

The typical situation is understood by observing that the nuclear stopping power S_n of a 100-keV Sb projectile in a silicon target is about $0.2\,\mathrm{keV/Å}$ over the entire trajectory [75]. Since the spacing between the lattice planes in silicon is about 2.5 Å, this means that the Sb projectile loses about 500 eV per lattice plane on the average. One can expect the majority of this energy to be given to one primary silicon recoil atom, as the average recoil will have an energy approaching 500 eV; it would then have a range of roughly 25 Å. For comparison, the range of the 100 keV Sb ion is 500 Å, which is not substantially larger

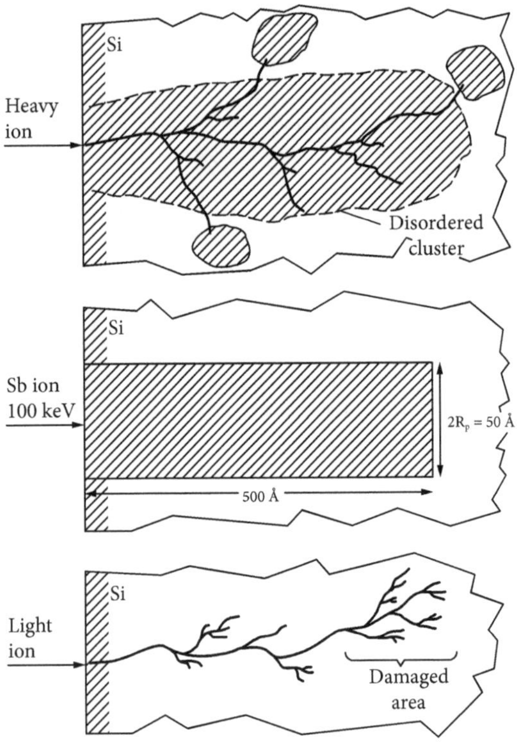

Fig. 8.45 Damage profiles

than the possible range of the average primary recoil atoms it creates.

The collision cascade of each primary recoil will contain roughly 15 displaced target atoms, since the number of recoils $\approx E/2E_d$ for heavy ions, where E_d is the bonding energy of the target atom, $\approx 15\,\text{eV}$ for Si. The damage volume $V_D \approx \pi(25\,\text{Å})^2\,(500\,\text{Å})$ (see Fig. 8.45).

Within this volume there are roughly 15 displacements per lattice plane, or 3000 total displacements. The average vacancy density is about $3000/V_D \sim 10^{22}/\text{cm}^3$, or about 20 % of the total number of atoms in V_D. As a consequence of the implantation, the material will be essentially amorphous.

8.5.1.8.2 Light Ions – B

Much of the energy loss is a result of electronic interactions, which do not cause damage. There is more damage near the end where nuclear stopping dominates. For boron $E_c \approx 10\,\text{keV}$. As a consequence, most of the damage is near the final ion position. For example, at 100 keV, B has $R_p \approx 3400\,\text{Å}$, which equals 500 recoils or vacancies created in the last 1700 Å, or about 200 recoils over the first 1700 Å. In the damage volume ($V_D \sim 1.6 \times 10^{-18}\,\text{cm}^3$) there are 500 displacements, and with this the average vacancy density is $500/V_D \sim 3 \times 10^{20}/\text{cm}^3$, which is less than 1 % of the atoms; higher doses are needed to create amorphous material with light ions.

8.5.1.8.3 Creation of Amorphous Material

To estimate the dose for converting a crystalline material to an amorphous form by heavy ion bombardment, consider the $E_c = 100\,\text{keV}$ Sb ion that has a range $R_p \sim 500\,\text{Å}$. For this ion beam the dose to make amorphous Si is:

$$D = \frac{(10^{21}\,\text{keV/cm}^3)\,R_p}{E} = 5 \times 10^{13}/\text{cm}^2\,.$$

For $E_c \approx 100\,\text{keV}$, for B, it is $R_p \approx 3300\,\text{Å}$. The dose to make amorphous Si is:

$$D \approx 10^{21}(3.3 \times 10^{-5})/100 = 3.3 \times 10^{14}/\text{cm}^2\,.$$

In practice, higher doses (D) are required for boron because the damage is not uniformly distributed along the path of the ion.

8.5.1.8.4 Damage Profiles

To the first order the damage profile may be assumed to be Gaussian. It is generally shallower than the ion profile (Fig. 8.46). Based on ***LSS theory***, it is possible to calculate the range values for the damage profile. For example, for boron:

$$\frac{M_{\text{Si}}}{M_{\text{B}}} = \frac{12}{5} = 2.4\,.$$

Using the ***Sigmund–Sanders*** tables [76], one can estimate the damage distribution by using:

$$\langle Z \rangle_D = R_D$$

and

$$\langle \Delta z \rangle_D = \Delta R_D\,.$$

The damage range can be calculated by taking ratios from the second column of the tables in

a Damage profile

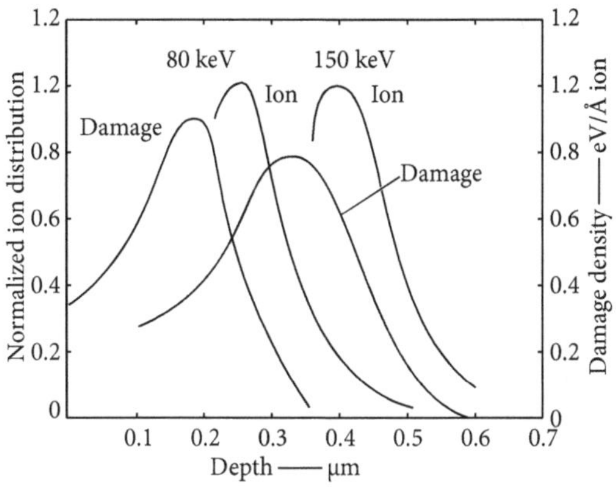

b Range distribution parameters

M_{target}/M_{ion}	$\langle z\rangle / (E/NC_1)$	$\langle\Delta z^2\rangle/\langle z\rangle^2$	$\langle x^2\rangle/\langle z\rangle^2$	$\langle zx^2\rangle/(\langle z\rangle\langle x^2\rangle)$	$\langle\Delta z^3\rangle/\langle z\rangle$
1/10	0.842	0.058	0.018	1.07	0.007
1/4	0.577	0.125	0.044	1.16	0.021
1/2	0.453	0.195	0.089	1.20	0.043
1	0.369	0.275	0.176	1.20	0.079
2	0.297	0.409	0.343	1.16	0.135
4	0.229	0.710	0.674	1.12	0.221
10	0.153	1.684	1.671	1.07	0.345

c Damage distribution parameters

M_{target}/M_{ion}	$\langle z\rangle / (E/NC_1)$	$\langle\Delta z^2\rangle/\langle z\rangle_D^2$	$\langle x^2\rangle/\langle z\rangle_D^2$	$\langle zx^2\rangle_D/(\langle z\rangle_D\langle x^2\rangle_D)$	$\langle\Delta z^3\rangle_D/\langle z\rangle_D$
1/10	0.692	0.434	0.192	1.80	0.580
1/4	0.489	0.437	0.181	1.71	0.433
1/2	0.376	0.386	0.152	1.47	0.218
1	0.295	0.380	0.157	1.41	0.172
2	0.241	0.457	0.257	1.42	0.272
4	0.198	0.623	0.485	1.31	0.391
10	0.143	1.215	1.153	1.16	0.567

Fig. 8.46 Damage profile and associated range and distribution parameters. **a** Damage profile, **b** range distribution parameters, **c** damage distribution parameters

Fig. 8.46. Thus

$$\langle z\rangle_D = \frac{0.22}{0.26} R_p \approx 0.8 R_p\,. \tag{8.63}$$

The standard deviation in the damage distribution is given from ratios in the third column of the table as

$$\langle \Delta z\rangle_D = 0.75(\Delta R_p)\,, \tag{8.64}$$

which can be used together with Eq. (8.64) to construct an actual damage distribution.

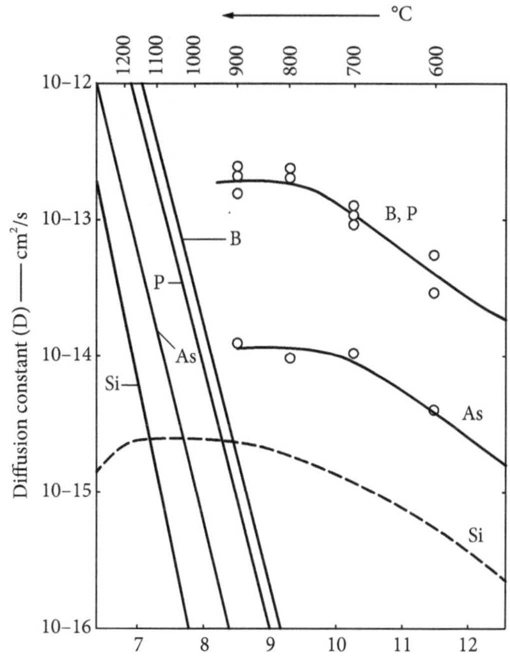

Fig. 8.47 Temperature dependence of enhanced diffusion (5×10^{11} proton/$cm^2\,s^{-1}$)

8.5.1.8.5 Location of Implanted Ions

For low doses many of the ions end up on lattice sites; they are substitutional and also electrically active. For high doses most ions end up as interstitials and are not electrically active. In general B shows a higher percentage of ions in interstitial or nonelectrical active sites.

8.5.1.9 Radiation-Enhanced Diffusion

All the impurities used in fabricating silicon devices diffuse either by a substitutional or an interstitial mechanism. As a consequence, the diffusion process is sensitive to the vacancy concentration and can be readily influenced by the supply of excess vacancies. One way of generating vacancies is by means of displacement reactions caused by nuclear collisions in the ion implantation process. This scheme is known as ***radiation-enhanced*** diffusion [77]. Figure 8.47 shows the experimental results for B, P, As, and Si under the intrinsic condition $N_A < n_i$.

From the theory at low temperatures and high fluxes the enhanced self-diffusion coefficient in

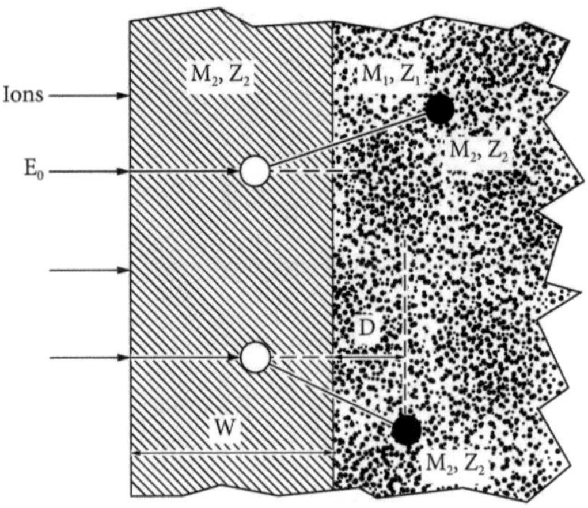

Fig. 8.48 Diagram of recoil collisions

Si depends on the flux by a square-root law:

$$D_S \sim k(\text{flux})^{1/2}$$

and at low fluxes $D_S \sim k'$ (flux) depends linearly on the flux.

8.5.1.10 Recoil Phenomena

Nuclear collisions are responsible not only for radiation damage in the target materials, which leads to radiation-enhanced diffusion of impurity profiles, but also for the modification of impurity distributions by the introduction of additional impurity species by recoil from surface layers (see Fig. 8.48).

This effect occurs when dielectric films are applied to the surface of a silicon wafer for selective doping. The total number of recoil atoms from the surface film reaching the silicon substrate depends on the type of bombarding ions, the energy (E_0), the type of atoms in the film, and the film thickness (W).

The depth distribution for the dielectric-silicon interface is shown in Fig. 8.49 and is given by [78]

$$N(D) = BN\left[\left(\frac{E_0}{W}\right)^{1/3} - \left(\frac{E_0}{D}\right)^{1/3}\right] \times \left[\left(1 + \frac{W}{W_0}\right)^{2/3}\right] \text{ atom/volume},$$

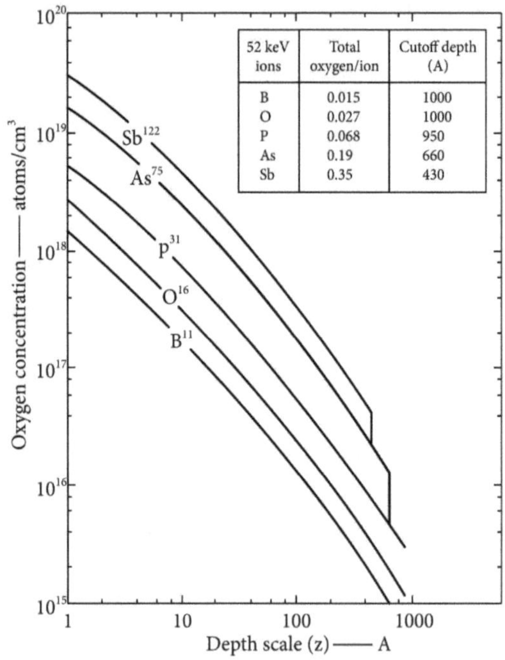

Fig. 8.49 Oxygen profile obtained from 50 keV recoil collision at 10^{13} ion/cm^2

where B is a coefficient that contains the atomic parameter of the film substrate and W_0 is an empirical factor that characterizes the formation of the collision cascade. Figure 8.49 shows the oxygen profile obtained from the 50 keV recoil collisions of 10^{13} ions/cm^2 traversing an oxide layer of 10 Å.

Recoil, range, and damage distributions following ion implantation are of considerable importance in device design. Calculations based on the theory of ***Lindhard, Scharff***, and ***Shiott (LSS)*** have been developed for range and damage profiles for ions implanted in semiconductors, and they give a satisfactory account of the primary-ion range distributions in semi-infinite substrates. However, practical device processing involves implantation into targets with one or more thin films. In these cases, we are interested in the spatial distributions of not only the primary ions but also the recoil-implanted atoms in each layer.

It was shown recently that numerical integration of the extended ***Boltzmann*** equation can be used to obtain estimates of the primary ion and recoil range distribution, the energy deposition profiles, and the energy and angular distributions.

8.5.1.11 High Dose Implantation Limits

One of the major advantages of ion implantation is the external control of the number of implanted ions. In high dose implantation this advantage is lost due to target sputtering. The maximum achievable concentration of an implanted species is determined in the limit by sputtering or erosion of the implanted surface. In the implantation process ion sputtering transfers energy from the incident ion to the target material, which results in the ejection of atoms from the surface.

For a dose of 10^{17} ion/cm^2 (about 100 monolayers) it is possible to remove 100–1000 layers of target material, which corresponds to a thickness change of 500–5000 Å. Then during implantation the surface profile is a result of erosion and sputtering of both the target and the implanted ions. The simplest estimate gives the concentration of implanted species to be proportional to r/S, where S is the sputtering yield of the target.

Experiments on the sputtering of compounds have indicated that preferential sputtering effects can influence the maximum achievable ion concentration. The influence of preferential sputtering (r) is generally to reduce the concentration of higher-mass atoms at the surface of the target. According to this model, the maximum concentration is proportional to r/S, where S is the total sputtering yield and r the preferential sputtering factor ($0.5 < r < 2$). Since lower mass elements tend to be preferentially sputtered, one can achieve a higher concentration of heavy elements than of lighter elements in the substrate.

8.5.1.12 Low Energy Implantation

Bombardment of solid surfaces with particle energies below 10 keV is always directed to more or less strong sputtering of the surface. By using reactive ion beams or also metal ions the bombardment particles are incorporated at the same time because of the chemical reactions taking place or during alloy formation at the surface. Above certain saturation fluences of the bombardment ions, due to the implantation of relatively low-energy beam particles it comes to the formation of thin and very compact films. In these layers the concentration conditions are determined by the saturation stoichiometry of the compound or alloy formed from the reactive bombardment particles or the metal ions and the substrate material used. The thickness of such films and the concentration process in the transient area to the undisturbed substrate material are specified mostly by the implantation profile for the bombardment particles used.

It is of great practical importance that dense layers can be manufactured at in principle arbitrarily low substrate temperatures also by low energy implantation. In this way, very dense **Si_3N_4*-layers*** can be produced by bombardment of ***Si-wafers*** with pure nitrogen ion beams [79]. At a beam energy of 5 keV and for fluences above 10^{18} ions/cm^2 a layer thickness from 120 Å develops.

8.5.2 Devices for Ion Implantation

For doping semiconductors one needs an ion accelerator that consists of an ion source, a beam guiding system, a magnet for beam characterization, and a mass separation and process chamber (Fig. 8.50).

For the calculation and optimization of the beam guiding system numeric computational programs are used. The ion beam is characterized by a number of ion trajectories, which are from the plasma meniscus or are extracted from the emitter surface. The envelope of these ion trajectories describes the form of the ion beam and concomitantly its parameters in the focal spot. The ion trajectories must be calculated thus for extraction geometry, which can be given in each case.

A fine focus source for implantation equipment should supply as high a current density with small energy uncertainty from as small an emission area as possible, i.e., ion guns must possess a high directed beam value. Important is not the total ion current, which the source supplies, but only the focusable part. The capability to fo-

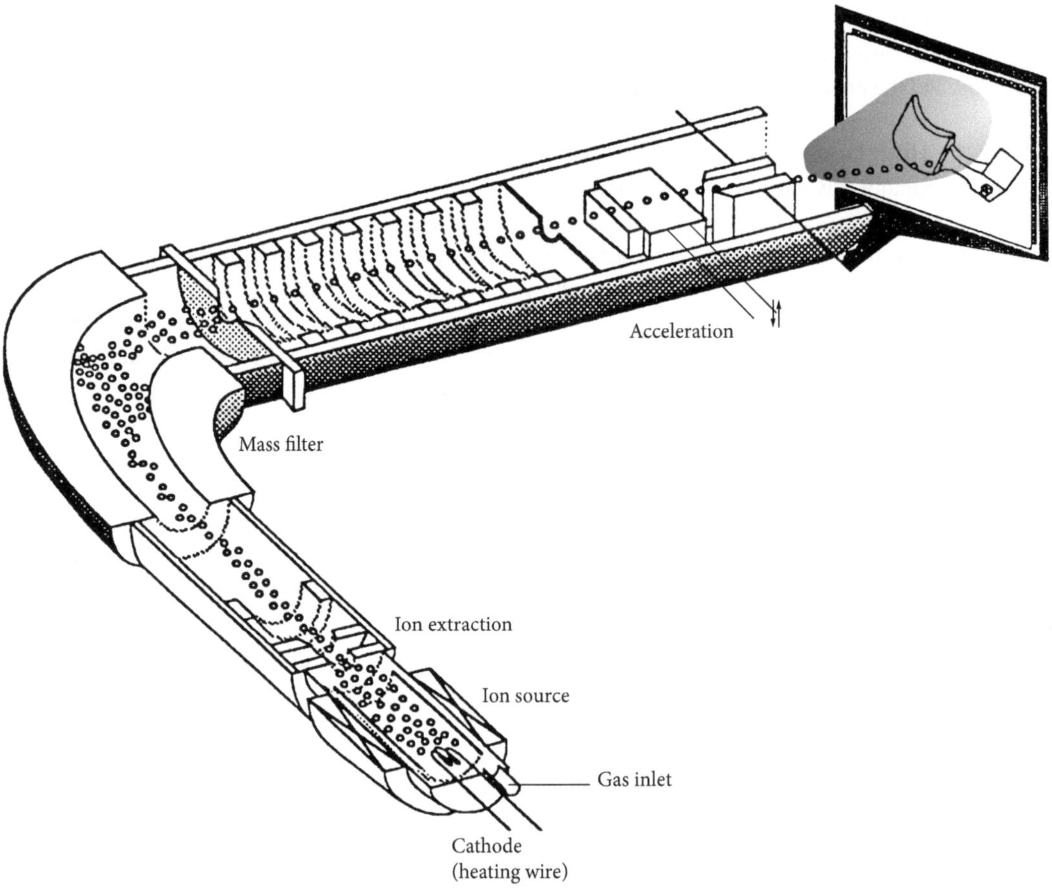

Fig. 8.50 Schematic representation of a beam guiding system with ion source, mass filter, acceleration line, and process chamber

cus depends on the size of the emission area, the solid angles, into which the source emits, and the energy uncertainty in the ion beam. The effective source size, i.e., the lateral expansion of the emission area, determines the smallest possible radiation diameter apart from the lens aberrations and the energy uncertainty. The energy uncertainty of the ion beam is a function of the potential fluctuations in this volume. The smallest energy uncertainties have sources that the neutral particles at a metal surface with exactly defined potential ionize and extract.

Besides simply positively charged ions an ion source emits simply or manifold charged molecules, partially in addition also simply or manifold charged ions of impurities, which come from the walls of the discharge chamber or from sputtering effects at the ion guiding system. Since these variances induce, for example, the physical characteristics desired by implantation in a semiconductor, an ion-optical system of ion implantation equipment must contain a mass filter for the separation of the unwanted isotopes (Fig. 8.51). For the separators we have a choice between ($\boldsymbol{E} \times \boldsymbol{B}$)-filters and a magnet. The ($\boldsymbol{E} \times \boldsymbol{B}$)-filter uses a crossed, constant electrical field E and constant magnetic field B. At a charged particle (charge Ze) the forces ZeE and/or $ZevB$ affect. The particle passes the filter unimpeded if its velocity:

$$v = \vec{E} / \vec{B} .$$

The ($\boldsymbol{E} \times \boldsymbol{B}$)-filter is a velocity filter. If all ions have the same energy, the ($\boldsymbol{E} \times \boldsymbol{B}$)-filter operated

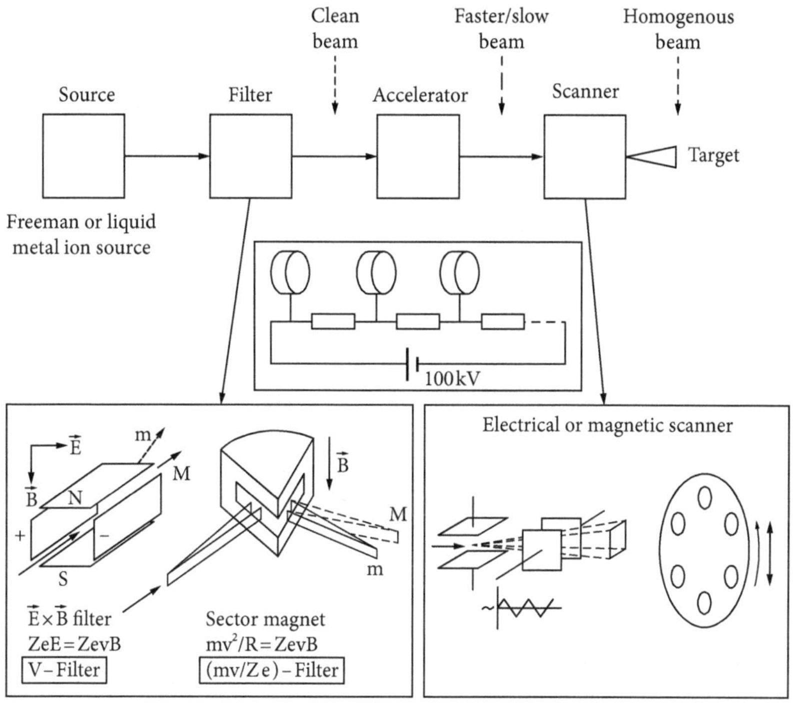

Fig. 8.51 Structure of ion implantation equipment

as mass filter

$$m = \frac{2E}{v^2},$$

since ions with the same velocity have the same masses. Ions flow through a magnet in such a way that the centrifugal force is equal to the Lorentz force, i.e.,

$$\frac{mv^2}{R} = ZevB$$

$$\frac{mv}{Ze} = RB,$$

where R as turning radius is an impulse filter, i.e., it focuses ions, for which mv/Ze is equal to a certain constant, on the exit slit. With constant charge e and same energy E of the ions it is thereby also a mass filter:

$$\frac{mv}{Ze} = \frac{2E}{Ze}m.$$

For detailed calculations of the $E \times B$ filter see [80].

There are two types of scan systems (Fig. 8.51):

- The electrical scan system, at which the ion beam is moved over the resting substrate.
- The mechanical scan system with moved substrate and immovable ion beam.

The electrical scan system consists of two capacitors arranged perpendicularly to each other. If one puts an alternating voltage to these, then the ion beam in the pulse of this voltage is diverted in the x and/or y-direction. The advantage of the electrical scanning is its high speed. It cannot be used for large surfaces which can be illuminated and large ion currents because of the change of the angle of incidence and/or the radiation of the beam focus. Mechanical scanning does not have these disadvantages; it is, however, slower and can produce particle abrasions. With mechanical scanning one usually uses a rotary disk, the axle of which is additionally moved. This movement must be equal so that the surface allocation density of the ions is constant everywhere.

8.5.3 Pulse Implantation

Pulse implantation is a method between plasma diffusion and ion beam implantation [81]. The interaction effects of the ions with the substrate

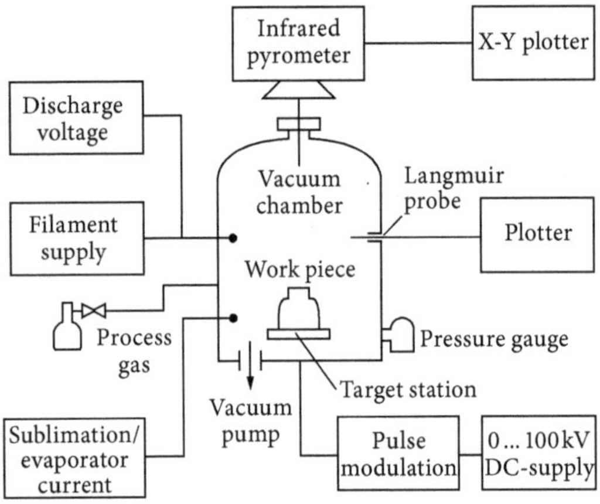

Fig. 8.52 Schematic structure of a pulse implantation plant

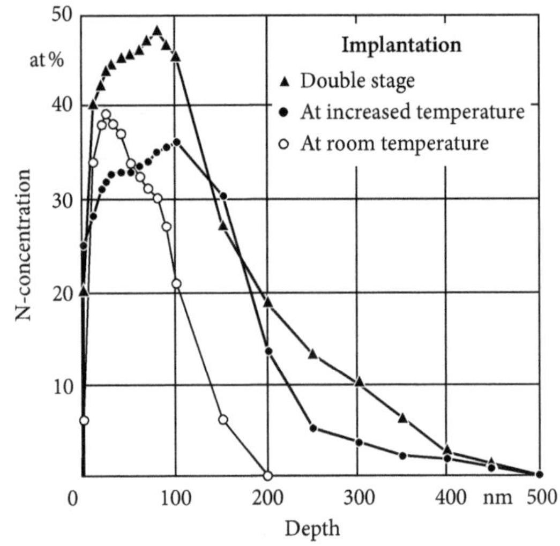

Fig. 8.53 Nitrogen depth profiles after nitrogen pulse implantation of Ti-6Al-4V

surface are the same as with the ion implantation. Because ion implantation works with adjusted ion beams, it is difficult to implant complex formed workpieces evenly; pulse implantation, however, succeeds to do this due to the all-around radiation. The ion energy is, however, limited to the range < 100 keV. Figure 8.52 shows the structure of a pulse implantation plant [82]. In a recipient there is a target station where the workpiece to be treated lies.

The target station is pulsed up to -100 kV. The process gas is ionized by means of a gas discharge. The ions are accelerated to the workpiece. At -100 kV the maximum energy for N^+ is 100 keV, for N_2^+ is 50 keV.

An example of the pulse implantation of nitrogen is the titanium alloy ***Ti-6Al-4V***, which is used in aircraft construction and for medical implants [83]. The implants were treated both at ambient temperature and at 600 °C. A combination of both was likewise accomplished. The substrate pre-voltage amounted to 50 kV and the operating pressure was 4.5×10^{-4} mbar. At one implantation duration of 3 h the total dose amounted to some 10^{18} atoms/cm^2. The nitrogen profiles are shown in Fig. 8.53. The profiles were determined by means of Auger spectroscopy.

The nitrogen distribution extends nearly to the surface and amounts to maximum 30–50 at.%. At higher temperatures it comes to an additional diffusion up to $\sim 0.5\,\mu$m. The highest peaks result with combined treatment. Measurements of the microhardness and the abrasion correspond to an ion-nitrided comparison sample.

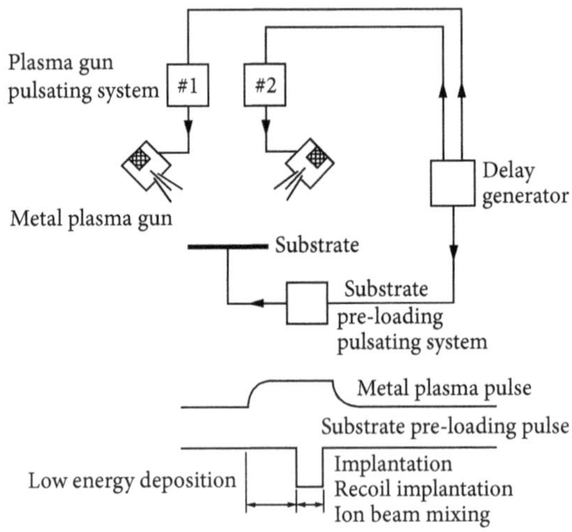

Fig. 8.54 Pattern of a pulse implantation plant for metal ions

Apart from the pulse implantation of gas ions, systems for the pulse implantation of metal ions have also been developed [84]. As is schematically represented in Fig. 8.54, a metal plasma current produced by plasma sources by means of one or several arc discharges is directed toward the substrate.

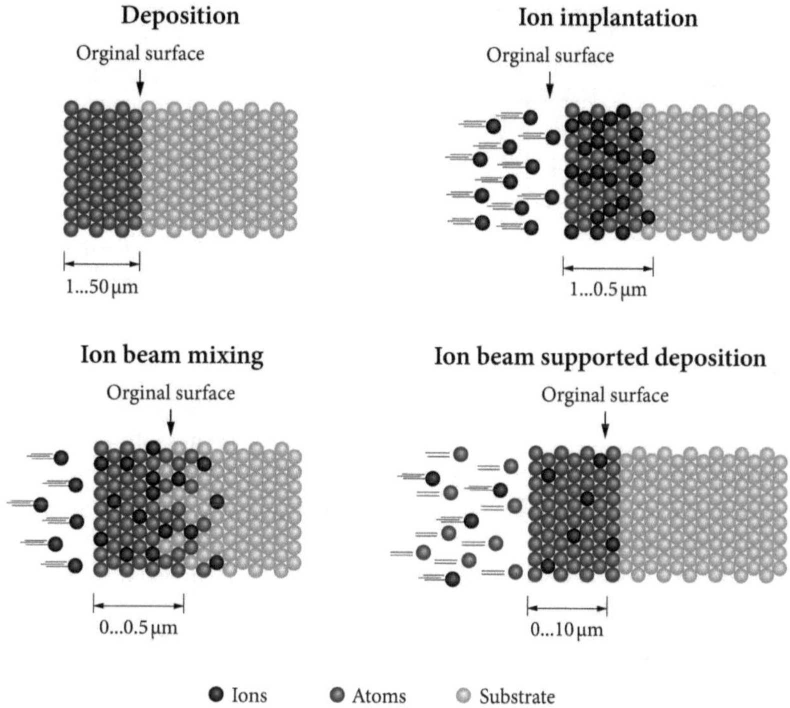

Fig. 8.55 Representation of ion beam-assisted deposition

8.5.4 Ion Beam-Assisted Deposition

8.5.4.1 Basic Principles

Particle beam-supported coating, such as ***IBAD*** (*ion beam-assisted deposition*), ***IBED*** (*ion beam-enhanced deposition*), ***IAD*** (*ion-assisted deposition*), ***IVD*** (*ion vapor deposition*), *dynamic recoil mixing or ion beam **PVD***, can be derived from both ion implantation and ***PVD*** procedures. In the first case, an ion accelerator is added to a vacuum coating chamber at the target end, and in the second case, an ion source is inserted into or flanged onto the ***PVD*** equipment.

The main difference between ion beam and plasma procedures is that the ions that are accelerated and strike the increasing film, with the ion beam procedure come from a source with defined opening; in the second case they come from diffuse plasma. With ion beam bombardment energy, direction, and ion species are clearly defined and are variable within wide limits. With plasma bombardment this is only possible in small limits. In Fig. 8.55 the basic principle of ***IBAD*** is shown. Energy is supplied by the ion beam to the increasing film; either in the form of electron excitation (electronic stopping) or atomic collisions (nuclear stopping). With substrates and films, which are present as chemical compounds, electron excitation leads to chemical modifications. The ion beam causes reactions that strongly affect the characteristics of the film/substrate system.

Before coating it is important to:

- Clean the substrate
- Mix and enlarge the threshold range layer/substrate.

During growth of the layer there is:

- Surface diffusion
- Compression of the layer
- Sometimes development of special morphologies or textures
- Sometimes phase formation or glass formation.

The results of the interaction processes depend on the ion dose and/or fluence (ion/cm^2) and the ion current density ($\text{ion/cm}^2\,\text{s}$). The relative parts of the electronic and collision interactions depend on the ion mass and ion energy. Thus, e.g., in Ni a 100 keV He^+ ion 99 % of its energy is already released by electronic interactions, a

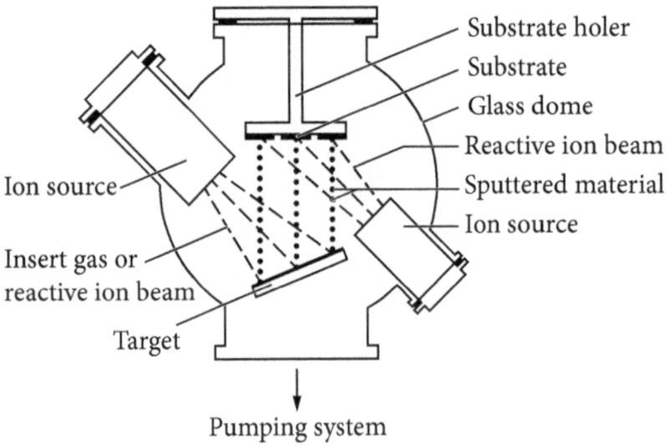

Fig. 8.56 Dual beam arrangement

10 keV Ni^+ ion already 90 % by collision. A typical arrangement of equipment with two sources is shown in Fig. 8.56.

An ion source is used for sputtering the target, another ion source for the bombardment of the growing layer.

References

1. McCracken GM (1975) Rep Prog Phys 38(2):241–372
2. Rutherford E (1911) The scattering of α and β particles by matter and the structure of the atom. Philosophical Magazine 21(6):669–688
3. Dearnaley G (1969) Ion bombardement and implantatio. Rep Prog Phys 32(4):405–492
4. MacDonald RJ (1970) The ejection of atomic particles from ion bombarded solids. Adv Phys 19(80):457–524
5. Biersack JP, Ziegler JF (1982) Nucl Instrum Meth 194:93
6. Lindhard J, Nielsen V, Scharff M (1968) K Dan Vidensk SelskMatFysMedd 16:10
7. Ziegler JF, Biersack JP, Littmark U (1985) Stopping power and ranges of ions in matter. In: Ziegler JF (ed) The Stopping and Range of Ions in Solids, vol 1. Pergamon Press, New York
8. Matsunami N, Yamamura Y, Hikawa Y, Itoh N, Kazumata Y, Miyagawa S, Morita K, Shimizu R, Tawara H (1983) Report IPPJ-Am-32. Insitute of Plasma Physics, Nagoya University
9. Fano U (1963) Ann Rev Nucl Sci 13:1
10. Lindhard J, Scharff M (1961) Phys Rev 124:128
11. Andersen HH, Ziegler JF (1977) Hydrogen stopping power and ranges in all elements. In: Ziegler JF (ed) The Stopping and Range of Ions in Solids. Series, vol 3. Pergamon Press
12. Ziegler JF (1977) Helium, stopping power and ranges in all elements. In: Ziegler JF (ed) The Stopping and Range of Ions in Solids, vol 3. Pergamon Press, New York
13. Littmark U, Ziegler JF (1980) Range distributions for energetic ions in all elements. In: Ziegler JF (ed) The Stopping and Range of Ions in Solids, vol 6. Pergamon Press, New York
14. Chu WK, Moruzzi VL, Ziegler JF (1976) JAppl Phys 46:2817
15. Boutard D, Müller W, Scherzer BMU (1988) Phys Rev B 38:2988
16. Bauer P, Rössler W, Mertens P (1992) Nucl Instrum Meth B 69:46
17. Lindhard J, Scharff M, Schiott H (1963) K Dan Vidensk SelskMatFysMedd 33:14
18. Schiott HE (1966) K Dan Vidensk SelskMatFysMedd 25:9
19. Merkle KL (1976) In: Petersen NL, Harkness SD (eds) Radiation Damage in Matals. Am Soc For Metals, Metals Park, Ohio
20. Lucasson P (1975) In: Robinson MT, Young FW (eds) Fundamental Aspects of Radiation Damage, (ERDA Conf-751006-p1, 1975)
21. Schilling W (1978) J Nucl Mater 72:1
22. Robinson MT, Oen OS (1982) J Nucl Mater 110:147
23. Robinson MT (1970) Nuclear Fusion Reactors. British Nuclear Energy Soc., London
24. Seitz F, Köhler JS (1956) In: Seitz F, Turnbull D (eds) Solid State Physics, vol 2.
25. Vineyard GH (1976) Rad Eff 29:245
26. Sigmund P (1974) Appl Phys Lett 25:169
27. Itho N, Tanimura K (1986) Rad Eff 98:269
28. Gibbons JF (1968) Ion implantation in semiconcuctors, Part I, Range distribution theory and experiments. Proc IEEE 56(3):295–319
29. Sigmund P, Gras-Marti A (1981) Nucl Instrum Meth 182/183:25
30. Desimoni J, Traverse A (1993) Phys Rev B 48:13266
31. Lam NQ, Wiedersich H (1987) Nucl Instrum Meth B 18:471
32. Müller KH (1987) Phys Rev B 35:7906
33. Gilmore CM, Sprague J (1992) J Vac Sci Technol A 10:1597
34. Choi C-H, Hultman L, Barnett SA (1993) J Vac Sci Technol B 11:1
35. Kitabatake M, Green JE (1993) J Appl Phys 73:3138

36. Haberland H, Insepov Z, Karrais M, Mall M, Moseler M (1993) Nucl Instrum Meth B 80/81:1320
37. Haberland H, Insepov Z, Moseler M (1993) Z Phys D 26:229
38. Windischman H (1991) J Vac Sci Technol A 9:2341
39. Robertson J (1993) Diam Rel Mater 2:984
40. Wilson RG, Brewer GR (1973) Ion Beams: With Applications to Ion Implantations. Wiley & Sons, New York
41. Valyi L (1978) Atom and Ion Sources. Wiley & Sons, New York
42. Smullin LD, Haus HA (eds) (1959) Noise in Electron Devices. MIT Press and Wiley & Sons, New York
43. Loeb LB (1960) Basic Prozesses of Gaseous Electronics. University of California Press, Berkely, California
44. Langmuir I (1924) Currents limited by space charge between concentric spheres. Phys Rev 24:49
45. Wilson RG, Brewer GR (1973) Ion Beams with Applications to Ion Implantation. R. E. Krieger Publishing Company, Amsterdam, Oxford, New York, Tokyo
46. Kaufman HR, Haper JMF, Cuomo JJ (1979) Focused ion beam designs for sputter depositions. J Vac Sci Technol 16:899
47. Kaufman HR, Robinson RS (1980) Gridded Brod Beam Ion Sources. In: Cuomo JJ, Rossnagel SM, Kaufman HR (eds) Handbook of Ion Beam Processing Technology. Noyes Publications, Park Ridge, New Jersey
48. Keller F (1983) A High-Bridhtness Duoplasmatron Ion Source. In: Ryssel H, Glawischnig H (eds) Springer Series in Electrophysics. (Eds, vol 11. Springer Verlag, Berlin, p 83
49. Freeman JH (1973) In: Dearnaley G (ed) Ion Implantation. North Holl. Publ. Comp., Amsterdam, p 292
50. Baumann H, Bethge K (1981) The Franfurt PIG Ion Source. Nucl Instr and Meth 189:107
51. Oechsner H (1974) Electron Cyclotron Wave Resonances and Power Adsorption Effects in Electrodeless Low Pressure H.F. Plasmas with a Superimposed Static Magnetic Field. Plasma Physics 16:835
52. Tahagi T, Ishikawa J (1990) Current status of development of ion sources in Japan from a standpoint of materials science. Rev Sci Instrum 61:517
53. Mahoney JF, Takiku AY, Daley HL, Moore RD, Perel J (1979) J Appl Phys 40:51001
54. Hanson GR, Siegel BM (1979) J Vac Sci Technol 16:1875
55. Swanson LW et al (1979) Emission characteristics of gallium and bismuth liquid metal field-ion sources. Scanning Electron Microsc. 1979/I, 45–51
56. Taylor GI (1964) Disintegration of water drops in an electric field. Proc R Soc London A 280:384
57. Swanson LW, Schwind GA, Bell AE, Brady JE (1979) Emission characteristics of a liquid gallium in source. J Vac Sci Technol 16(6):1864
58. Powell RA, Downey DF (1984) Dry Etching for Microelectronics. In: Powell RA (ed) Reactive Ion Beam Etching. North Holland Physics Publishing, Amsterdam
59. Anatech LTD, Alexandria VA, USA, Firmenschrift 1990
60. Kessler J, Tomcik J, Waldorf J, Oechsner H (1991) Plasma beam deposition of diamand like films. Vacuum 42:273
61. Keller F (1983) A High-Bridhtness Duoplasmatron Ion Source. In: Ryssel H, Glawischnig H (eds) Springer Series in Electrophysics. (Eds, vol 11. Springer Verlag, Berlin, p 83
62. Jaeger H, Schulze J, Frey H (1990) High-Temperature Superconducting Films Prepared by Low-Pressure RF-Plasma Deposition Conference on the Science and Technology of Thin-Film Superconductors, April 30-May 4, 1990
63. Sun SS (1986) Internal stress of ion beam sputtered molybdenium films. J Vac Sci Technol A 4:572
64. Giannuzzi LA, Stevie FA (2005) Introduction to Focused Ion Beams: Instrumentation, Theory, Techniques, and Practice. Springer
65. Kretschmer KH, Lorenz G, Casttrischer G, Kessler I, Baumann P (1990) LH Electron Cyclotron Resonance ((ECR) Plasma Source SPIE, Advanced Techniques for Integrated Circuit Processing, vol. 1392.
66. Marshall A, Natarajan S (2002) SOI Design: Analog, Memory and Digital Techniques. Springer
67. Williams JS, Poate JM (1984) Ion Implantation an d Beam Processing. Academic Press, New York
68. Gibbson JF (1968) Ion implantation in semiconductors – Part I: Range distribution theory and experiments. Proc IEEE 56(3):295
69. Gibbson JF (1972) Ion implantation in semiconductors – Part II: Damage production and annealing. Proc IEEE 60(9):1062
70. Sansbury J (1976) Applications of ion implantation in semiconductor processing. Solid state Technol 19(11):31
71. Meindl JD, Saraswqat KC, Dutton RW, Gibbons JF, Tiller W, Plummer JD, Deal BE, Kamins TI (1978) Computer aided engineering of semiconductor integrated circuits, Stanford University Integrated Circuit Laboratory, Report TR 4969-3, SEL-78.011. Stanford University, Stanford, California
72. Erramli H, Blondiaux G (1995) Ion chanelling. Applied Radiation and Isotopes 46(6–7):413–418
73. Mayer JW, Rimini E (1977) Ion Beam Handbook for Materials Analysis. Academic Press Inc, New York

74. Lindhard J, Scharff M, Schioett H (1963) Atomic collisions II. Range concepts and heavy ion ranges. K Dan Vidensk Selsk Mat Fys Medd 33(14):1

75. Sansbury J (1976) Applications of ion implantation in semiconductor processing. Solid State Technol 19(11):31

76. Sigmund P, Saunders JB (1967) Spatial distribution of energy deposited by ionic bombardment. In: Glotin P (ed) Proceedings of the International Conference on Applications of Ion Beams to Semiconductor Technology. Editions Orphrys, Grenoble, France, p 215

77. Rupprecht HS (1978) New advances in semiconductor implantation. J Vac Sci Technol 15(5):1669

78. Christel LA, Gibbons JF, Mylroie S (1980) An application of the Boltzmann transport equation to ion range and damage distribution in multi-layered targets. J Appl Phys 51(12):6179

79. Füßer HJ, Oechsner H (1991) High dose low energy implantation of nitrogen in silicon, niobbium and aluminium. Suf Coat Technol 48:97

80. Pailthorpe BA, Mahon P (1990) Thin Solid Films 193/194:34

81. Anders A (2000) Handbook of Plasma Immersion Ion Implantation and Deposition. Wiley, New York

82. Conrad JR (1989) Plasma Source Ion Implantation: A new Approach to Ion Beam Modification of Materials. Mater Sci Engin A 116:197

83. Chen A, Sridharan K, Conrad JJ, Fetherston RP (1991) Surface Modification of Ti-6Al-4V Surgial Alloy by Plasma Source Ion Implantation. Surface and Coatings Techn 50:1

84. Brown IG, Godechot X, Yu KM (1991) Plasma Immersion Surface Modification with Metal Ion Plasma. Report LBL-29914, Lawrence Berkely Laboratory, April 1991

Chemical Vapor Deposition (CVD)

9

H. Frey

9.1 Introduction

In chemical vapor deposition (CVD) the compounds of a vapor phase, often diluted with an inert carrier gas, react at a hot surface to deposit a solid film [1, 2]. The importance of CVD is due to the versatility for depositing a large variety of elements and compounds at relatively low temperatures and at atmospheric pressure. Amorphous, polycrystalline, epitaxial, and uniaxially oriented polycrystalline films can be deposited with a high degree of purity. Aspects of CVD include the chemical reactions involved, the thermodynamics and kinetics of the reactors, and the transport of material and energy to and from the reaction site.

The following is a list of examples of some of the common types of chemical reaction used in CVD.

Pyrolysis The simplest CVD process is pyrolysis, in which a gaseous compound decomposes on a hot surface to deposit a stable residue. Examples are the following: deposition of pyrolytic graphite from methane (CH_4), which takes place at a substrate temperature of 2200 °C; deposition of silicon from monosilane (SiH_4), which takes place in the range 800–1350 °C; and deposition of nickel from carbonyl from [$Ni(CO)_4$], which takes place at about 100 °C [3].

Reduction Hydrogen is the most commonly used reducing agent. Examples are deposition of silicon by the hydrogen reduction of silicon tetrachloride, which takes place at about 1000 °C, and deposition of tungsten by the hydrogen reduction of tungsten hexafluoride, which takes place at about 800 °C. Hydrogen reduction is also used to accelerate the pyrolytic process by removal of unwanted byproducts as gaseous hydrogen compounds, for which less energy is required.

Oxidation Silicon dioxide films can be deposited by the reaction of silane with oxygen.

Nitridation Silicon nitride films can be deposited by reaction of silane with ammonia.

Carbidization Titanium carbide films can be deposited by reaction to titanium tetrachloride with methane at a substrate temperature of 1850 °C.

Chemical-transport reaction For these processes, the transport of the desired material from the source to the substrate on which it is to form a film depends on the difference in equilibrium constants between the reactant source and carrier phase, and the substrate and the carrier phase, when each are held at different temperatures. For example, the deposition of gallium arsenide by the chloride process depends on the reversible reaction

$$6\,GaAs_{(g)} + 6\,HCl_{(g)} \underset{T_2}{\overset{T_1}{\rightleftarrows}} As_{4(g)} + Als_{2(g)} + GaCl_{(g)} + 3\,H_{2(g)}\,,$$

where T_1 is the temperature of the solid GaAs source, T_2 is the temperature of the

© Springer-Verlag Berlin Heidelberg 2015

H. Frey, H. R. Khan, *Handbook of Thin-Film Technology*, DOI 10.1007/978-3-642-05430-3_9

solid GaAs substrate, and $T_1 > T_2$. This allows, in effect, indirect distillation of gallium arsenide from the hot source at temperature T_1 to the cooler substrate at temperature T_2 through an intermediate gas phase of different chemical composition.

Spray pyrolysis In this process the reagents are dissolved in a carrier liquid, which is sprayed onto a hot surface in the form of tiny droplets. On reaching the hot surface the solvent evaporates and the remaining components react, forming the desired material. An example is the formation of cadmium sulfide films by spray pyrolysis of cadmium chloride and thiourea dissolved in water with the substrate at about 300 °C [4].

Table 9.1 shows many applications of CVD-processes.

9.2 Physicochemical Bases

At the CVD process gaseous chemical compounds, often diluted with an inert carrier gas, react on hot surfaces in the reactor to form solid films, Fig. 9.1.

The CVD process can be subdivided as follows:

- Production of a reaction gas mixture
- Transportation phase
- Deposition and film formation.

The relatively high gas pressure is characteristic in the transportation phase. It can amount to between 10 mbar and 10 bar. The mean free path λ is small in relation to the dimensions of the recipient and/or the distance source – substrate d. CVD processes occur at small Knudsen numbers Kn

$$\mathrm{Kn} = \frac{\lambda}{d}\,. \tag{9.1}$$

The high process pressure distinguishes CVD from PVD. Because of the collisions in the gaseous phase of the CVD process, the molecules, which create the film structure, must be sluggish in reaction, so that no powder developed in the gaseous phase and no damaging secondary reactions take place. This means that as original substances not atoms, like in the PVD process, but molecules that are sluggish in reaction must be used. An ideal characteristic curve for a thermal CVD process is shown in Fig. 9.2.

At low temperatures (range I) the deposition process is determined by the reactions taking place at the surface. The deposition is described by ***Arrhenius law***:

$$\dot{s} = \dot{s}_0 \exp\left(\frac{-E}{RT}\right), \tag{9.2}$$

where the E the activation energy and R the universal gas constant.

At higher temperatures the reaction rates at the surface become so fast that the deposition is

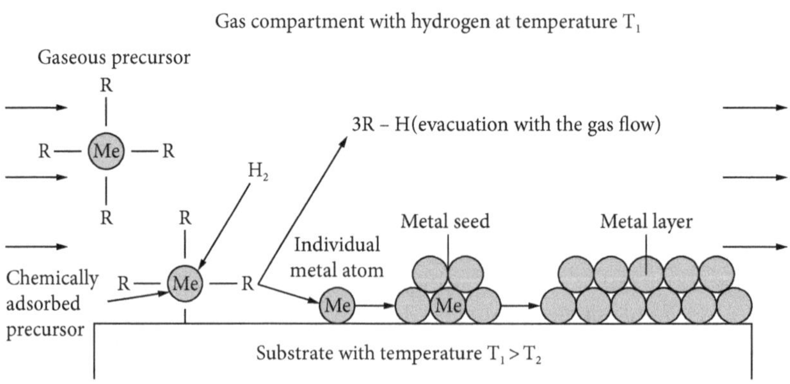

Fig. 9.1 With the chemical the vapor deposition, e.g., of metals, a precursor molecule at a temperature T_1 reacts with hydrogen (H_2) to a complex, which deposits itself to the coated surface. At the substantially higher temperature T_2 of the substrate the complex decomposes into an organic remainder of R and a metal atom Me, which forms the nucleus for layer growth with others

Table 9.1 Application of CVD processes

Thin film	Application
Epi-Si	Semiconductor material
a-Si	Solar cells
Ge	Tunnel diodes, solar cells
SiC	Protective against oxidation
BN	Diffusion source, crucible material
BP	Photovoltaics
AIN	Dielectric, piezoelectric material for acoustic wave beam
AIP	Electroluminescent diodes
AlAs	Electroluminescent diodes, solar cells
AlSb	Solar cells
GaN	Electroluminescent diodes
GaP	Electroluminescent diodes
ZnSe	Photoconductor, laser diodes
ZnTe	Electro-optical applications
CdS	Photoconductors, laser diodes, solar cells, detectors
$CdS_{1-x}Se_x$	Lasers
$Zn_{1-x}Cd_xS$	Lasers
$Cd_{1-x}Hg_xTe$	Photoconductors, laser, photo diodes
$Pb_{1-x}Sn_xTe$	Photoconductors, laser, photo diodes
$Pb_{1-x}Sn_xSe$	Lasers
PbS	Photoconductors, lasers
PbTe	Infrared detectors
$PbS_{1-x}Se_x$	Lasers
PbSe	Lasers
PbO	Photoconductors
SiO_2, PSG, BSG, ASG	Passivation, etching mask, gate dielectric in MOS structures
Si_3N_4	Diffusions masks, oxidation masks, anti-reflection coating in solar cells
GaAs	Semiconductor material
Gasb	Laser diodes
InP	Gun diodes, laser diodes
InAs	Laser diodes, Hall effect diodes
InSb	Photo electrodes
$GaAs_{1-x}P_x$	High temperature rectifiers, electroluminescent diodes, laser diodes
$InAs_{1-x}P_x$	Photo cathodes, electro-optical applications
$InAs_{1-x}Sb_x$	Laser diodes
$GaAs_{1-x}Sb_x$	Laser diodes, photo cathodes
$Ga_{1-x}In_xAs$	Laser diodes, photo cathodes
$Ga_{1-x}In_xP$	Luminescent diodes
$Ga_{1-x}Al_xAs$	Laser diodes, solar cells
$In_xAl_{1-x}P$	Electroluminescent diodes, laser diodes
W	Metallization, conductive path in integrated circuits
Mo	Schottky diodes, conductive path
Ta	Conductive path

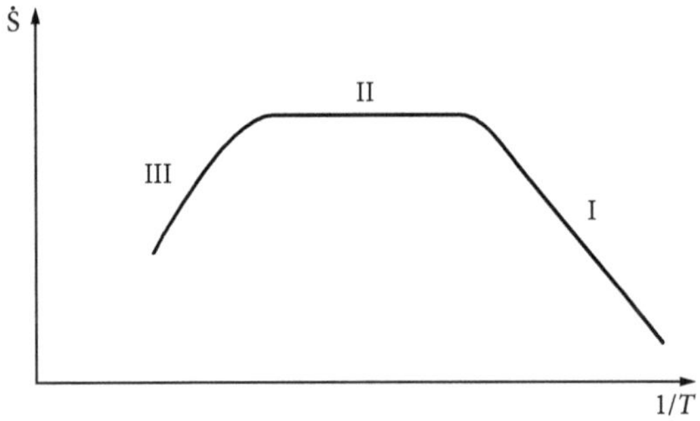

Fig. 9.2 Ideal CVD characteristics curve, *I* the deposition process is determined by the reactions taking place at the surface, *II* determined by the flow in the reactor and approximated temperature-independently, *III* homogeneous reactions in the gaseous phase

determined by the transport through the gaseous phase. In this range the deposition rate and the uniformity of the film is determined by the flow in the reactor and approximated temperature independently (range II). At still higher temperatures (range III) the deposition is controlled by homogeneous reactions in the gaseous phase, which can finally lead to powder formation. The deposition rate is thereby reduced. Figure 9.3 shows the different types of characteristic curves.

The selection of the operating point on the CVD characteristic curve depends on:

- The form of the workpieces
- The film structure.

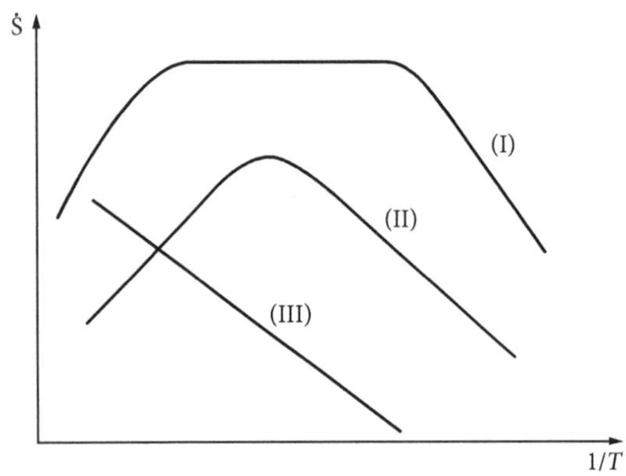

Fig. 9.3 Different types of CVD characteristics curves, *(1)* suitable for batch reactor, *(3)* for high depostion rates

If complicated forms or porous workpieces are to be coated on the inside, then the operating point is always selected within the kinetically controlled range, because of the low deposition rate at which no eluviation of the gaseous phase occurs. With simple forms, e.g., coating of plane surfaces, other aspects are of importance. Thus, for example, the desired film structure requires a defined operating point.

This is valid, for example, in the case of the ***Si epitaxy*** [5, 6]. At the silicon surface silane decomposes according to the chemical equation

$$\mathrm{SiHCl}_x\mathrm{H}_{4-x} + (x/2)\,\mathrm{H}_2 \rightarrow \mathrm{Si} + x\mathrm{HCl} + 2\,\mathrm{H}_2\,,$$

$T = 1100$–$1200\,°\mathrm{C}$.

Moreover, the silicon growth rate is proportional to the partial pressure of silane. For cases where the reaction is surface controlled, epitaxial growth proceeds according to the following steps:

- Mass transfer of the reactant molecules (SiH_4) by diffusion from the turbulent film across the boundary film to the silicon surface.
- Adsorption of the reactant atoms on the surface.
- The reaction of a series of reactions that occur on the surface.
- Desorption of the byproduct molecules.
- Mass transfer of the byproduct molecules by diffusion through the boundary film to the main gas stream.
- Lattice arrangement of the adsorbed silicon atoms.

The overall deposition rate is determined by the slowest process in the list above. Under steady-state conditions all steps occur at the same rate and the epitaxial film grows uniformly.

Depending upon deposition conditions different reactor types are used. For the kinetically controlled range (range I) batch reactors are suitable. The flux plays a subordinated role in this case only. Due to low eluviations larger numbers of workpieces can be coated at the same time. In the range II the deposition is strongly affected by the flux, however higher coating rates are to be expected.

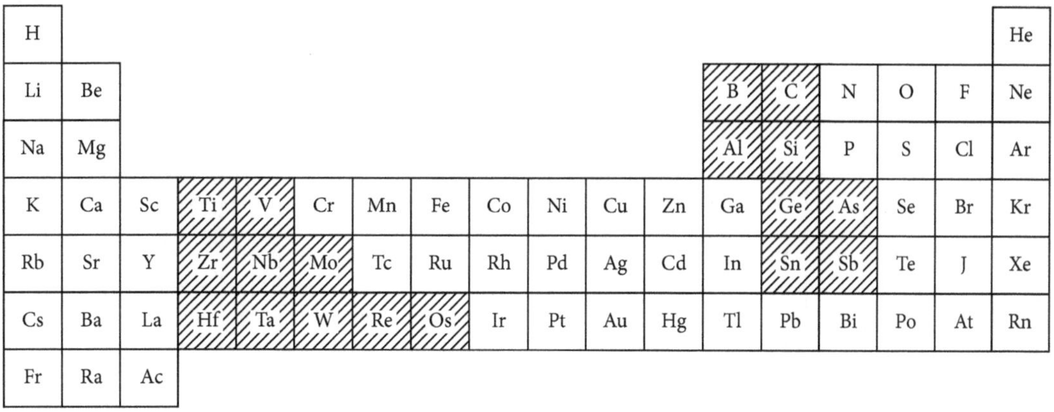

Fig. 9.4 Vaporable halides [7]

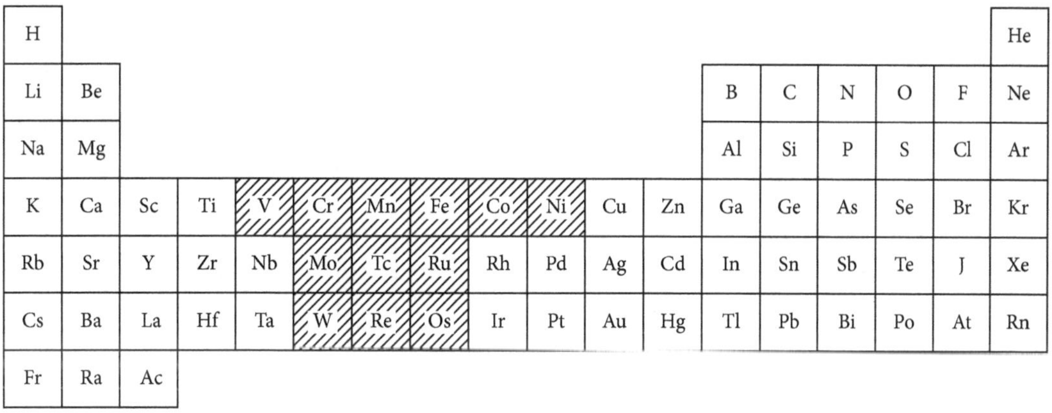

Fig. 9.5 Vaporable carbonyl [7]

9.2.1 Initiator Molecules

Initiator molecules for CVD processes should be stable without nucleation in the gaseous phase at partial pressure of $\sim\ p > 10$ mbar. For the process it is favorable, if the concentrations are attainable at temperatures around 300 °C. The most frequently used compound class are the halides. Figure 9.4 shows an overview of the metals whose halides are sufficiently volatile.

Halides are highly aggressive, particularly at high deposition temperatures (> 700 °C) when using chloride. Chloride is used for the deposition of TiC, TiN, and Al_2O_3 under addition of carbon (methane), nitrogen (N_2), or sources of oxygen (CO_2).

Carbonyls (Fig. 9.5) have a substantially lower aggressiveness. In former times these were used, among other things, for the production of pure nickel. Carbonyls possess a high evaporability at low coating temperatures ($\sim$ 300 °C). Because of their high toxicity, carbonyl applications are limited.

Hydrides are available for microelectronics on the Si basis (Fig. 9.6). In the last years the so-called ***MOCVD processes*** (metal organic CVD) have gained significance. By metal-organic compounds, we generally understand all compounds that contain organic components, thus not only with C-metallic bindings.

The following, for example, belong to these:

- Alkyls
- Alcoholates

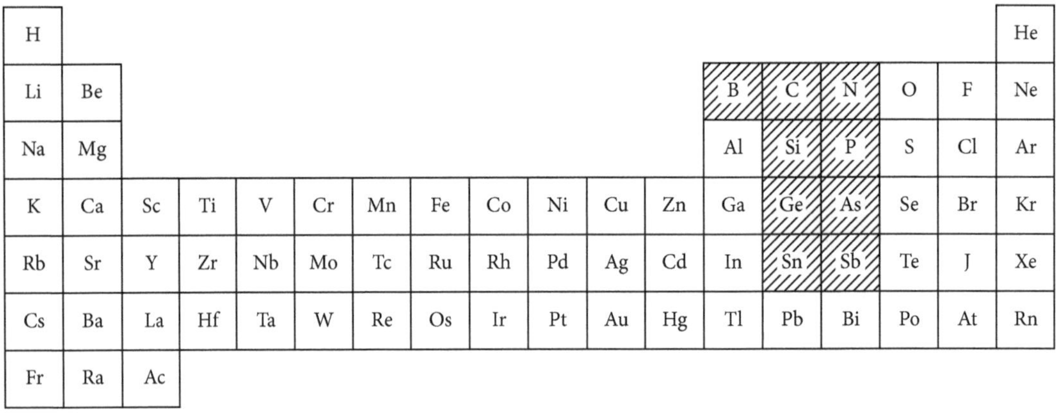

Fig. 9.6 Vaporable hydrides [7]

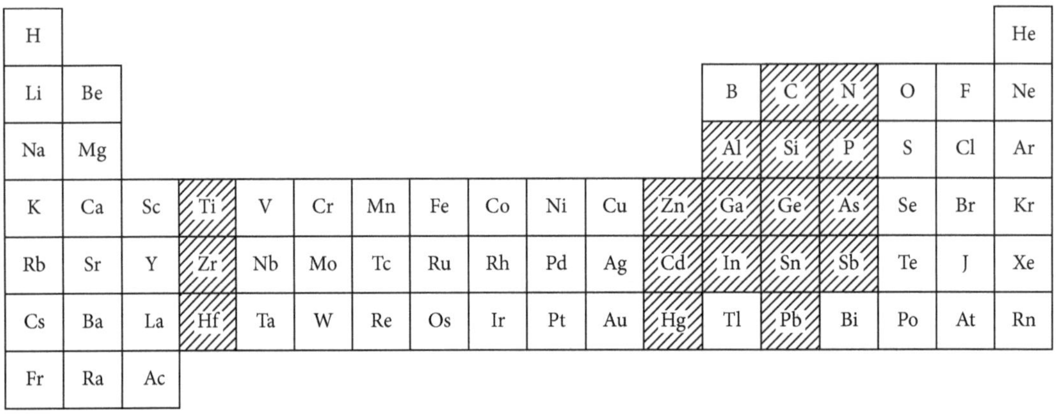

Fig. 9.7 Vaporable alkyls [7]

- Diketonates
- Cyclopentadienyl compounds
- Amido complexes
- PF_3-complexes.

The metal-organic initiator compounds are of great interest for the following reasons:

- Low coating temperatures are possible.
- New components can be brought into the gaseous phase.
- New film structures can be created.
- Availability of better output components (for example, not containing Cl, high vapor pressure).
- Suitable for photo CVD.

Alkyls [8] are suitable for the deposition of III–V connections (Fig. 9.7).

Triisobutylaluminium is used for aluminum deposition [9]. In contrast to vacuum evaporating processes for Al coating, with CVD high scattering can be used. A further group of initiator compounds is the group of alcoholates (Fig. 9.8), which are used for the production of oxides.

Figures 9.4, 9.5, 9.6, and 9.7 show that large parts of the periodic system are covered, not so however the elements of alkali and earth-alkali. Moreover, the noble metals Cu, Pd, Pt, Ag, Au, Co, Rh, and Ir are not depositable. Here two groups present themselves. The cyclopentadienyl compounds and the β-diketonate, which are formed by chelate compounds, are shown in Fig. 9.8. The diketonates [10] create easily volatile compounds (Figs. 9.9 and 9.10).

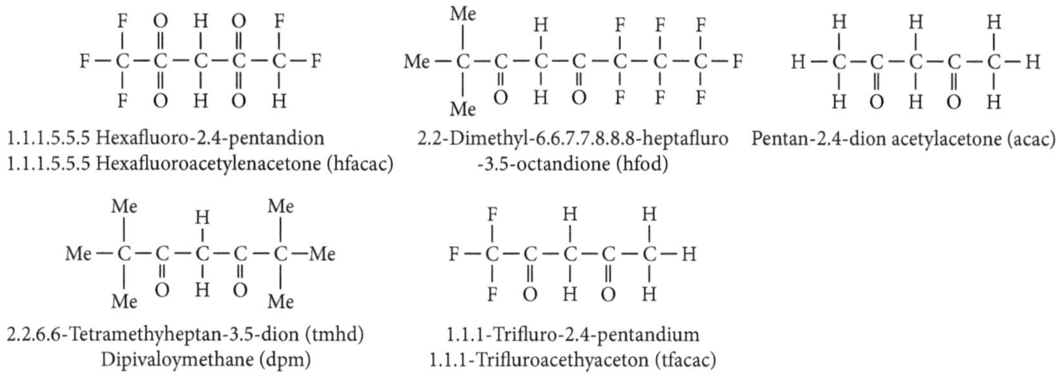

Fig. 9.8 Chelate compounds for the generation of films with CVD

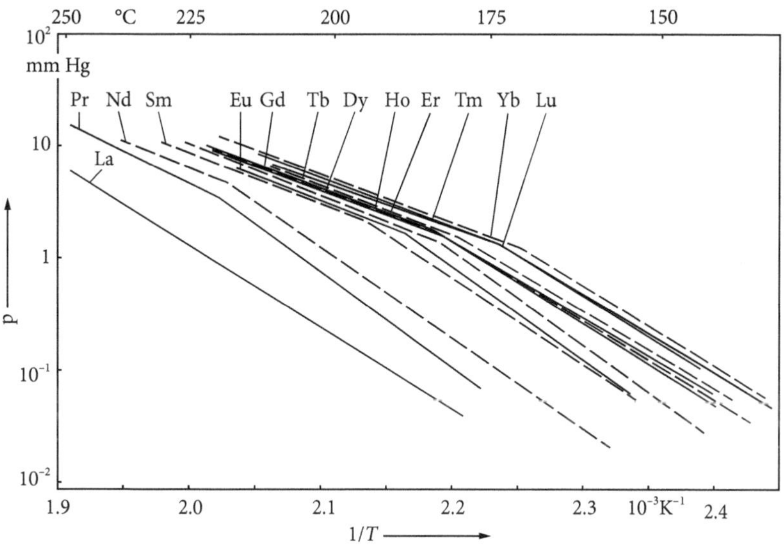

Fig. 9.9 Vapor pressure of thd (tetramethylheptanedione)complexes [10]

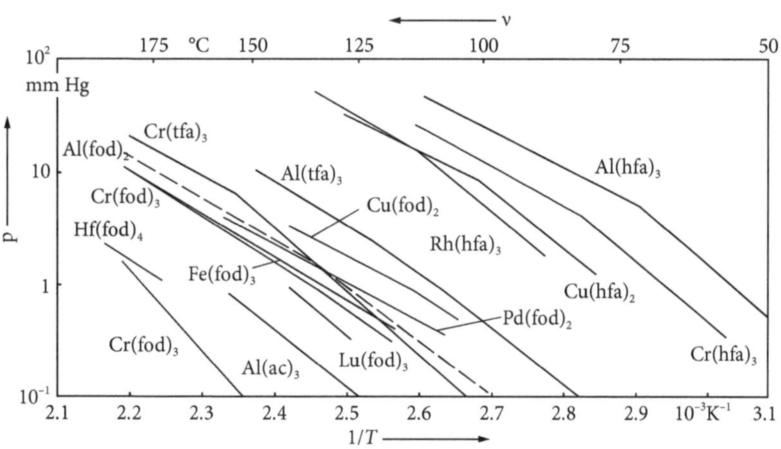

Fig. 9.10 Vapor pressure of β-diketonates [10], (Fig. 9.8, *hfa* = hfacac, *fod* = hfod, *tfa* = tfacac, *ac* = acac)

Fig. 9.11 Metal halides and complexes of metals with organic compounds for the production of metallic compounds

Connetion class	Example	Chemical formula	Layer material
Metal halogenide	Titan tetrachloride	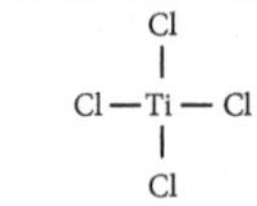	Titan nitride as hard material layer and diffusion barrier
Metal beta dicetonate	Tris (2,26,6-tetramethyl-3,5-heptandionato) yttrium	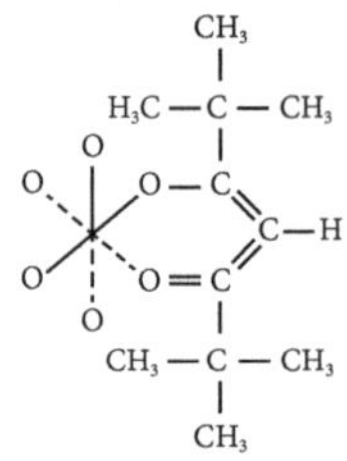	Yttrium barium cuprat as superconductor
Metal alcoholates	Tantalum (V)-ethylate	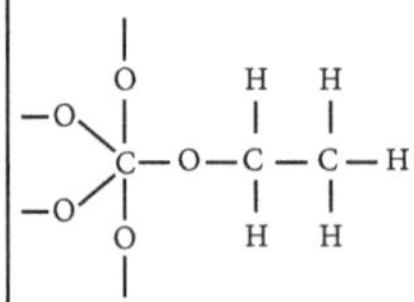	Tantalum oxide as dielectric
Metal carbonyles	Tungsten hexacarbonyl	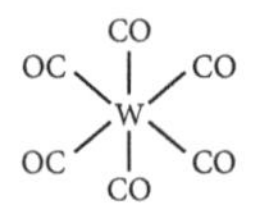	Tungsten layers for microelectronics
Metal alkyles	Trichlorethylene isobutyl aluminum	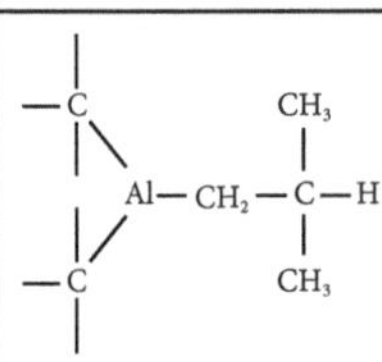	Aluminum layers as conductive strips
Metalamido compounds	Tetrakis-(dimethylamido)-titan	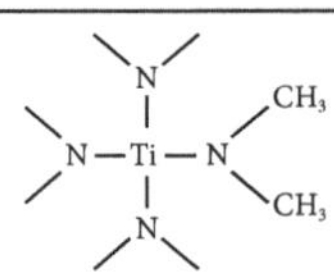	Titaniumnitride as diffusion barrier in microelectronics
Adduct compounds	Trimethylaminealan	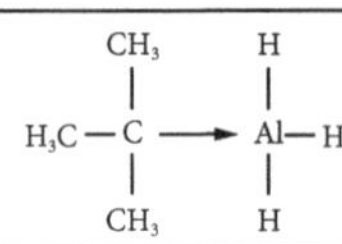	Aluminum layers as conductive strips

An overview of the compound classes for the deposition of industrially manufactured metallic compounds is given in Fig. 9.11.

To execute the deposition process the substances must be converted into he gaseous phase. For the execution of the coating an exact dosage of the reaction gases is necessary. To measure gas flows thermal flow meters are usually used. The principle is shown in Fig. 9.12. The gas is led across two different pipes; the measurement pipe and the laminar flow element. Depending upon quantity of the gas flow the flow sensor is adjusted. The temperatures T_1 and T_2 are measured over two temperature sensors, where warmth is supplied to the gas by way of winding heaters between the temperature sensors .

The temperature difference can be determined via the ***Reynolds number***

$$Re = \frac{\varrho v d}{\eta}, \qquad (9.3)$$

where ϱ is the gas density, η is the viscosity, v is the medium velocity, and d is the diameters of

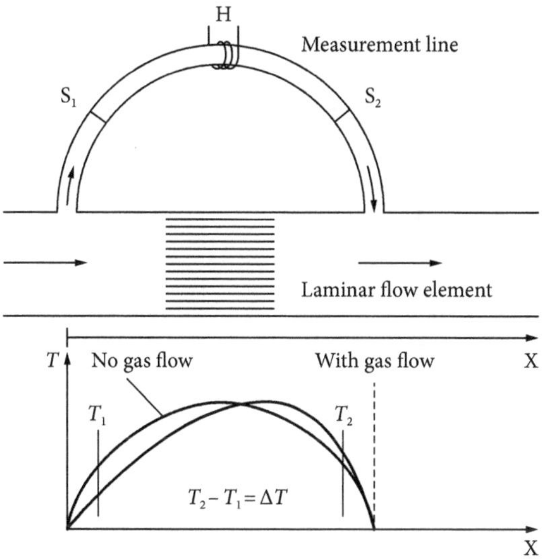

Fig. 9.12 Principle of a thermal flow meter (*S1*, *S2* temperature sensor, *H* heating)

the capillary and the ***Prandtl number***:

$$\mathrm{Pr} = \frac{\eta \varrho c}{\lambda}, \tag{9.4}$$

with the heat conductivity λ and the specific heat c at constant pressure, we can determine:

$$\Delta T = f(\mathrm{Pr}, Re). \tag{9.5}$$

The exact process is over the calibration determined. From Eq. (9.5) one can recognize that ΔT does not depend on the pressure of the flowing gas, since the Reynolds number does not depend on the pressure, but only on the mass flow, and the Prandtl number is likewise pressure independent.

For liquid substances bubblers are used. With these the feed gas is directly led through the liquid. Figure 9.13 shows such a bubbler. If the droplet size is too large pulsations occur [11].

The principle of a tube evaporator is shown in Fig. 9.14. A gas is led over the substance, which is evaporated and then enriched in the reaction gas. At higher gas velocities (hexafluoroacetylacetone) the mass content is determined

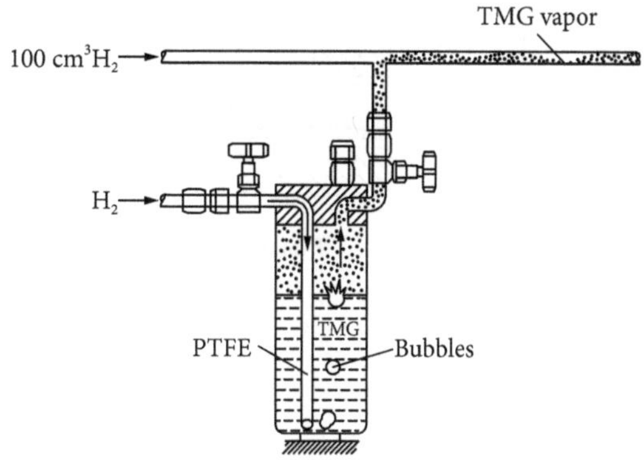

Fig. 9.13 Principle of bubblers (example, trimethylgallium TMG)

by $Cu(hfacac)_2$ by the evaporation velocity. At low velocities the mass content is determined by the equilibrium pressure.

A possible difficulty with bringing evaporable substances into the gaseous phase is the creation of fumigation or aerosol (nebulizer). Figure 9.15 shows different stages – depending on the coating conditions. For a CVD process stage c is optimal. As nebulizer, air nebulizers or ultrasonic nebulizers are used. Ultrasonic nebulizers have a closer particle spectrum than air nebulizers.

Ultrasonic waves produce capillary oscillations on the liquid surface (Fig. 9.16), with the wavelength [12]:

$$\lambda_\mathrm{K} = 2\pi \sqrt[3]{\frac{\sigma}{\varrho_0 \omega_\mathrm{p}^2}}, \tag{9.6}$$

where

σ is the surface stress,
ϱ_0 the density of the liquid, and
ω_p the angular frequency of the ultrasonic wave.

With increasing sound intensity the amplitude of the capillary wave increases, which form deviates due to increasing nonlinearity from the sine wave. Drops finally develop at high amplitudes; their diameter D lies in the order of magnitude of 1/4 of the wavelength. For ***water***, for example, we have ($\sigma = 0.073\,\mathrm{N/m}$)

$$D/\mu\mathrm{m} = (\nu/\mathrm{MHz})^{-2/3}.$$

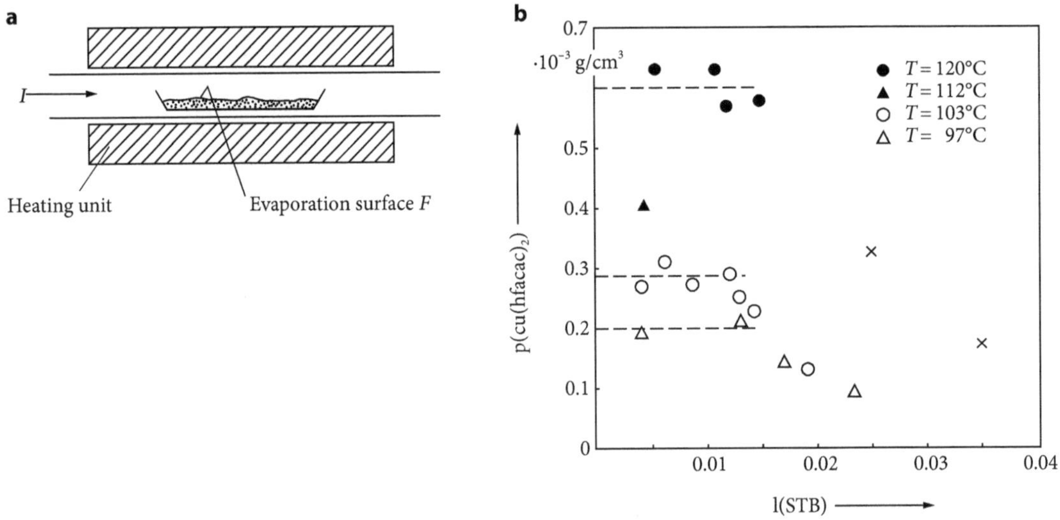

Fig. 9.14 Mass enrichment of Cu (hfacac) in a tube evaporator **a** against the gas flow

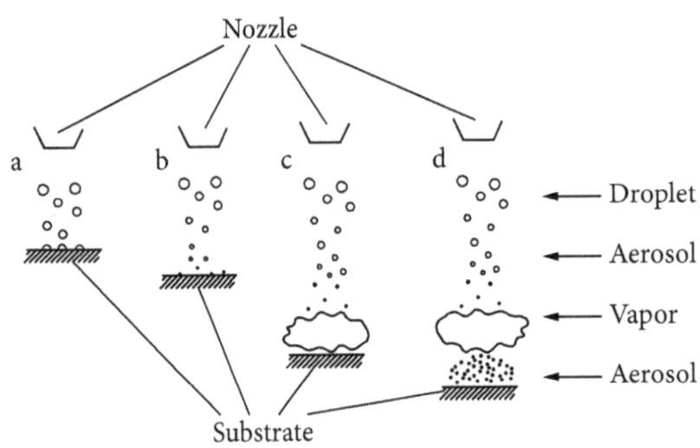

Fig. 9.15 Possibilities of the coating with droplet evaporation with different positions of the substrate

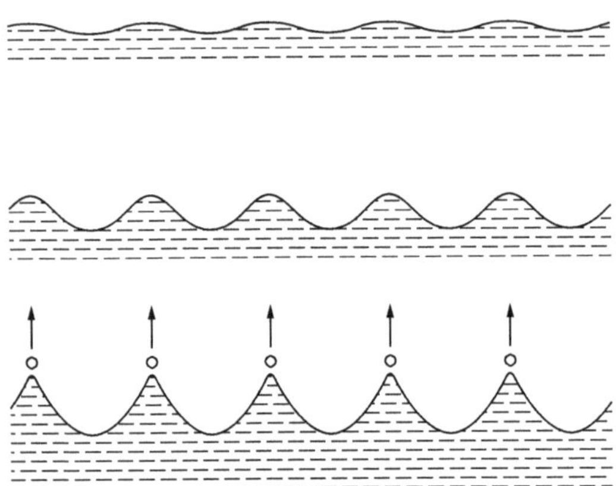

Fig. 9.16 Capillary waves with increasing amplitude (from *top* to *bottom*) on a liquid surface

9.2.2 Transportation Phase

9.2.2.1 General Function

The transportation phase mainly has the function to transport the reactive components to the surface to be coated. The gaseous phase is frequently still modified in order to optimize the coating conditions. The gaseous phase can be modified by a gas discharge, Fig. 9.17.

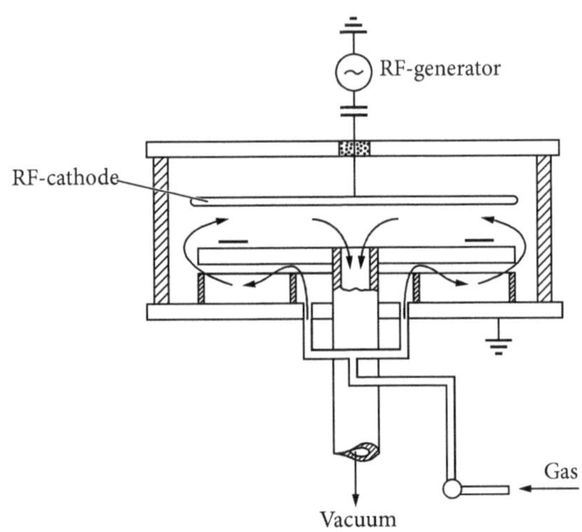

Fig. 9.17 Parallel disk reactor

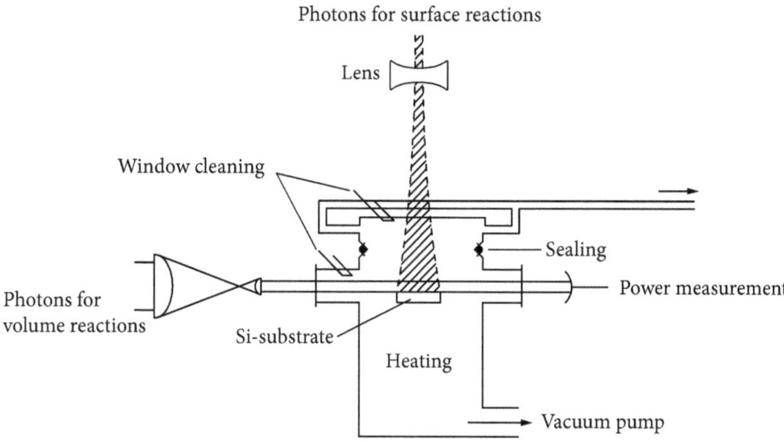

Fig. 9.18 Photo CVD reactor with modification of the gaseous phase by a photon beam

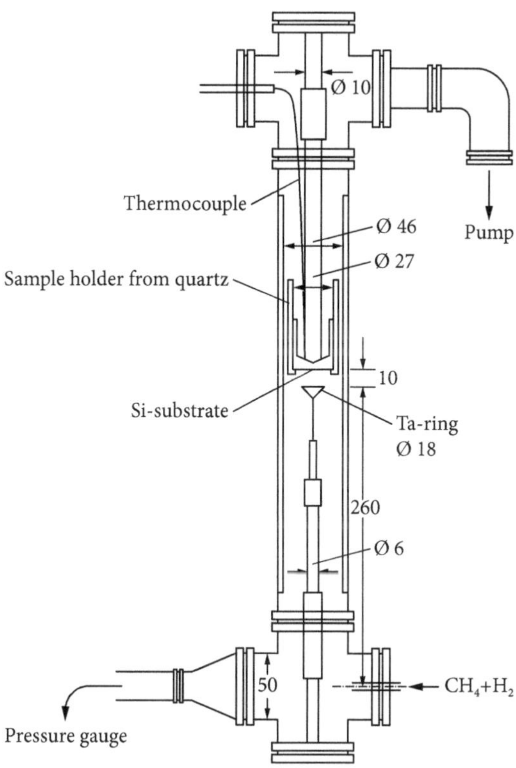

Fig. 9.19 Hot wire CVD reactor for diamond production

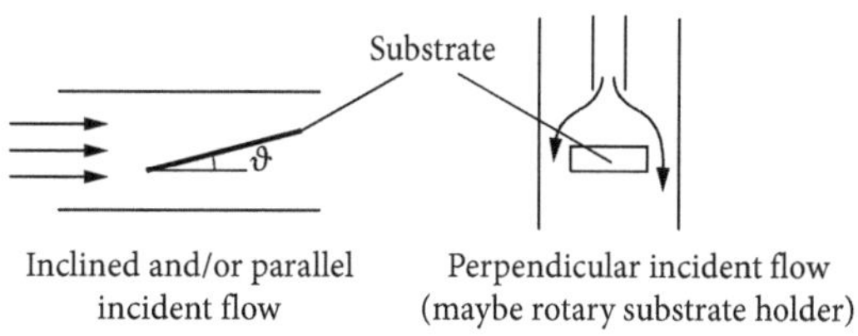

Fig. 9.20 Holdup flow reactor types

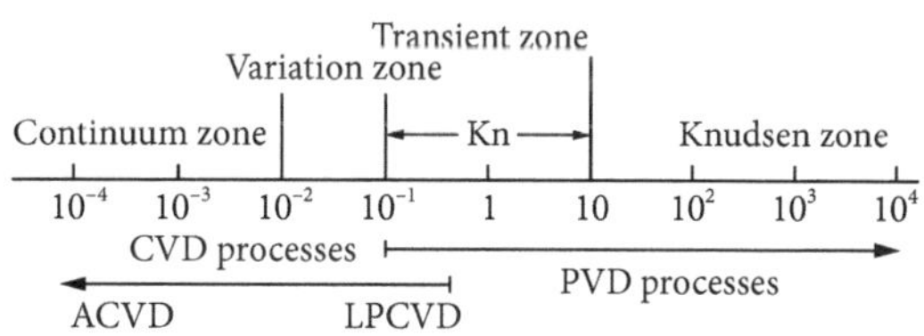

Fig. 9.21 Flow zones (ACVD = atmospheric CVD, LPCVD = low pressure CVD)

A further alternative for modification is the irradiation with photons (Fig. 9.18). A further alternative is shown in Fig. 9.19. The gas stream is preheated to high temperatures by means of a hot Ta wire and then brought into contact with the surface to be coated. This method can be used for the manufacture of diamond films from a mixture of hydrocarbon and hydrogen [13, 14]. To scale up CVD equipment a calculation of the flow distribution is necessary. In the following, the deposition conditions are analyzed, which describe the coating requirements within the diffusion-controlled range in a holdup flow reactor (Fig. 9.20). The characteristic index number for the description of the flow is the ***Knudsen number*** **Kn**. According to Fig. 9.21 different computer models are applicable, depending on the zone of the Knudsen number [15].

For Kn > 10 the kinetic gas theory is valid; for Kn < 0.01 the continuum equations with the velocity of $v = 0$ at the walls are valid. CVD processes predominantly take place within this flow zone. The transport of molecules to the substrate surface is a coupled process, which consists of convection and diffusion. Diffusion can take place in the concentration gradient, (***Fick's diffusion***) and in the temperature gradient (thermodiffusion). The transportation characteristics are calculated by simultaneous solution of the appropriate differential equations for the impulse transport, the heat transport, and the particle transport.

The most important boundary condition for the calculation is the reactor geometry. In the following analysis the holdup flow reactor will be considered (right-hand side of Fig. 9.20).

For the coating process it is simply accepted that a component i at the surface is deposited. Furthermore, at the substrate surface a definite mass break c_{is} is accepted, conditioned through the vapor pressure of the component i is adjusted. The entrance (feed) gas should possess the temperature T_0, and the substrate temperature T_S. The Knudsen number Kn should be $\ll 1$. The flow and the particle transport at small Knudsen numbers are then calculated by the following equations:

1. Continuum equation:

$$\frac{\partial \varrho u}{\partial x} + \frac{\partial \varrho v}{\partial y} + \frac{\partial \varrho w}{\partial z} = 0 \,. \qquad (9.7)$$

2. ***Navier–Stokes equations*** (here indicated for the x-component u of the speed):

$$\begin{aligned}\left(u\frac{\partial u}{\partial x} + v\frac{\partial u}{\partial y} + w\frac{\partial u}{\partial z}\right) &= -\frac{\partial p}{\partial x} + g_x(\varrho - \varrho_0) \\ &+ \frac{\partial}{\partial x}\left[\eta\left(2\frac{\partial u}{\partial x} - \frac{2}{3}\text{div } v\right)\right] \\ &+ \frac{\partial}{\partial y}\left[\eta\left(\frac{\partial u}{\partial y} + \frac{\partial v}{\partial x}\right)\right. \\ &\left.+ \frac{\partial}{\partial z}\left[\eta\left(\frac{\partial w}{\partial x} + \frac{\partial u}{\partial z}\right)\right]\right] \qquad (9.8)\end{aligned}$$

(g_x, x-component of acceleration due to gravity).

Similar equations are valid for the velocities of v and w in y and z-directions.

3. The thermal conduction equation:

$$\begin{aligned}\varrho c_p\left(u\frac{\partial T}{\partial x} + v\frac{\partial T}{\partial y} + w\frac{\partial T}{\partial z}\right) &= \frac{\partial}{\partial x}\left(\lambda\frac{\partial T}{\partial x}\right) \\ &+ \frac{\partial}{\partial y}\left(\lambda\frac{\partial T}{\partial y}\right) + \frac{\partial}{\partial z}\left(\lambda\frac{\partial T}{\partial z}\right). \qquad (9.9)\end{aligned}$$

4. Diffusion equation for the component i:

$$\begin{aligned}\varrho\left(u\frac{\partial c_i}{\partial x} + v\frac{\partial c_i}{\partial y} + w\frac{\partial c_i}{\partial z}\right) &= \frac{\partial}{\partial x}\left(\varrho D_i\left(\frac{\partial c_i}{\partial x}\alpha_i c_i\frac{\partial \ln T}{\partial x}\right)\right) \\ &+ \frac{\partial}{\partial y}\left(\varrho D_i\left(\frac{\partial c_i}{\partial y}\alpha_i c_i\frac{\partial \ln T}{\partial y}\right)\right) \\ &+ \frac{\partial}{\partial z}\left(\varrho D_i\left(\frac{\partial c_i}{\partial z}\alpha_i c_i\frac{\partial \ln T}{\partial z}\right)\right). \qquad (9.10)\end{aligned}$$

5. Ideal gas law

$$\frac{\varrho_0}{\varrho} = \frac{T}{T_0}\,.$$

The following boundary conditions must be fulfilled.

Substrate surface:

mass fraction of the component i	$c_i = c_{is}$,
velocity	$v = 0$,
surface temperature	$T = T_s$.

Walls:

gas velocity	$v = 0$,
wall temperature	$T = T_0$,
no deposition	$\text{grad}_\perp c_i = 0$.

Gas entrance:

density	ϱ_0,
mass fraction	$c_i = c_{i0}$,
gas velocity	$v_{\text{ein}} = w_0$,
gas entrance temperature	$T = T_0$.

Gas exit:

$$\frac{\mathrm{d}T}{\mathrm{d}z} = \frac{\mathrm{d}w}{\mathrm{d}z} = \frac{\mathrm{d}c_i}{\mathrm{d}z} = 0\,.$$

The mass separation rate is given by the expression

$$\dot{m}_i = \varrho D_i (\mathrm{grad}_\perp c_i + c_i \alpha_i \mathrm{grad}_\perp \ln T)\,.$$

For mathematical analysis the gas is regarded as feed gas (i.e., $c_i \ll 1$), η, c_p, and λ and the density ρ does not depend on the reactive component i. The higher mass fractions λ, c_p and η are dependent on the component i, and for diffusion transport the Stefan–Maxwell equations will be used [16].

Equations (9.7)–(9.10) are simplified if all sizes related to reference sizes:

$$x^* = \frac{x}{d}\,, \quad y^* = \frac{y}{d}\,, \quad z^* = \frac{z}{d}\,,$$

$$u^* = \frac{u}{w_0}\,, \quad v^* = \frac{v}{w_0}\,, \quad w^* = \frac{w}{w_0}\,,$$

$$T^* = \frac{T}{T_0}\,, \quad c_i^* = \frac{c_i - C_{0i}}{c_{is} - c_{0i}}\,, \quad \eta^* = \frac{\eta}{\eta_0}\,,$$

$$\lambda^* = \frac{\lambda}{\lambda_0}\,, \quad c_\mathrm{p}^* = \frac{c_\mathrm{p}}{c_{\mathrm{p}0}}\,, \quad D_i^* = \frac{D_i}{D_{0i}}\,,$$

$$\varrho^* = \frac{\varrho}{\varrho_0}\,, \quad p^* = \frac{p}{\varrho_0 v_0^2}\,,$$

$$g_x^* = \frac{g_x}{g}\,, \quad g_y^* = \frac{g_y}{g}\,, \quad g_z^* = \frac{g_z}{g}\,,$$

where the index 0 stands for the gas entrance and the index s for the values obtained on the substrate surface; g_x, g_y, and g_z are the components of gravity acceleration.

With these sizes we obtain for the differential equations:

$$\frac{\partial \varrho^* u^*}{\partial x^*} + \frac{\partial \varrho^* v^*}{\partial y^*} + \frac{\partial \varrho^* w^*}{\partial z^*} = 0\,, \tag{9.11}$$

$$\varrho^* \left(u^* \frac{\partial u^*}{\partial x^*} + v^* \frac{\partial u^*}{\partial y^*} + w^* \frac{\partial u^*}{\partial z^*} \right) = -\frac{\partial p^*}{\partial x^*} + \frac{d g g_x{}^* (\varrho^* - 1)}{w_0^2} + \frac{\eta_0}{\varrho_0 w_0 d} \frac{\partial}{\partial x^*} [\ldots] \ldots$$

$$\varrho^* c_p^* \left(u^* \frac{\partial T^*}{\partial x^*} + v^* \frac{\partial T^*}{\partial y^*} + w^* \frac{\partial T^*}{\partial z^*} \right) = \frac{\lambda_0}{\varrho_0 c_{p0}} w_0 d \left(\frac{\partial}{\partial x^*} \left(\lambda^* \frac{\partial T^*}{\partial x^*} \right) + \ldots \right.$$

$$\varrho^* \left(u^* \frac{\partial c^*}{\partial x^*} + v^* \frac{\partial c^*}{\partial y^*} + w^* \frac{\partial c^*}{\partial z^*} \right) = \frac{D_{0i}}{d w_0} \left(\partial^* D_i^* \frac{\partial c^*}{\partial x^*} \right) + \ldots \Bigg)\,. \tag{9.12}$$

The boundary conditions are given below.

Gas inlet:

density $\varrho^* = 1$,
velocity $w^* = 1$,
temperature $T^* = 1$,
concentration $c_i^* = 0$.

Substrate surface:
$v^* = 0$,
$T = T_\mathrm{S}^*$,
$c_i^* = 1$.

Walls:
$v^* = 0$,
$T^* = 1$,
$\mathrm{grad}_\perp c_i^* = 0$.

Gas exit: All gradients in z-direction = 0.

The deposition rate is given through:

$$\begin{aligned} \mathrm{Sh}_i &= \frac{\dot{m}_i d}{(c_{si} - c_{0i}) D_{0i} \varrho_0} \\ &= \varrho^* D_i^* \Bigg(\mathrm{grad}_\perp c_i^* \\ &\quad + \alpha_i \left(c_i^* + \frac{c_{0i}}{c_{si} - c_{0i}} \right) \mathrm{grad}_\perp \ln T^* \Bigg)\,, \end{aligned} \tag{9.13}$$

where Sh is the Sherwood number. With the Sherwood number the mass deposition rate $\dot{m}_i$ can be calculated,

$$\dot{m}_i = (c^*_{si} - c^*_{0i})\frac{\varrho_0 D_{0i}}{d}\,\mathrm{Sh}_i\,. \qquad (9.14)$$

We receive the molecular deposition rate under the condition $c \ll 1$:

$$J_i = (n^*_{si} - n^*_{0i})\frac{n_0 D_{0i}}{d}\,\mathrm{Sh}_i\,, \qquad (9.15)$$

where n^*_i is the mol fraction of the component i and n_0 the molecular density of the entrance gas.

Equations (9.12)–(11.14) depend on the following constants:

$$\frac{dg}{w_0^2}\,,\quad \frac{\lambda_0}{\varrho_0 c_{p0} d w_0}\,,\quad \frac{D_{0i}}{d w_0}\,,\quad \frac{\eta_0}{\varrho_0 w_0 d}\,,$$
$$T^*_{\mathrm{s}}\,,\ \eta^*\,,\ \lambda^*\,,\ c^*_p\,,\ D^*_i\,,\ \frac{c_{0i}}{c_{si} - c_{0i}}\,,\ \alpha_i\,.$$

The first four groups can be transformed into:

$$\frac{dg}{w_0^2} = \frac{1}{\mathrm{Fr}} \qquad \frac{\lambda_0}{\varrho_0 c_{p0} d w_0} = \frac{1}{\mathrm{Pr}\,Re}$$
$$\frac{D_{0i}}{d w_0} = \frac{1}{\mathrm{Sc}_i Re} \qquad \frac{\eta_0}{\varrho_0 w_0 d} = \frac{1}{Re}\,. \qquad (9.16)$$

The ***Reynolds number Re***, the ***Schmidt number* Sc**$_i$, the ***Prandtl number* Pr**, and the ***Froude number* Fr** are defined through:

$$Re = \frac{w_0 d \varrho_0}{\eta}$$
$$\mathrm{Pr} = \frac{\eta \varrho_0 c_p}{\lambda_0}$$
$$\mathrm{Sc}_i = \frac{\eta_0}{D_i \varrho_0}$$
$$\mathrm{Fr} = \frac{w_0^2}{gd}\,. \qquad (9.17)$$

The ***Sherwood number*** depends on these sizes:

$$\mathrm{Sh}_i = f\left(x^*_{\mathrm{s}}, y^*_{\mathrm{s}}, z^*_{\mathrm{s}}, Re, \mathrm{Sc}_i, \mathrm{Pr}, \mathrm{Fr}, \frac{c_0}{c_{is} - c_{i0}},\right.$$
$$\left. T^*_{\mathrm{s}}, \alpha_i, \eta^*, \lambda^*, c^*_p, D^*_i\right)\,. \qquad (9.18)$$

For the isotherm case the expression reduces to:

$$\mathrm{Sh}_i = f(x^*_{\mathrm{s}}, y^*_{\mathrm{s}}, z^*_{\mathrm{s}}, Re, \mathrm{Sc}_i)\,. \qquad (9.19)$$

The goal of the mathematical analysis and calculations is the determination of the ***Sherwood number***.

In addition the properties of the gases must be known. With many CVD processes a gas component (component 2) is available in excess (e.g., H_2, N_2, O_2). For this gas component λ, c_p, and η can be used directly.

The binary diffusion coefficients D_1 for the diffusion of the component 1 in the gas 2 are almost never admitted for the reactive gas components used. These sizes must, therefore, be calculated by means of the kinetic gas theory:

$$D_1 = 0.002628\frac{\sqrt{T^3\dfrac{(M_1 + M_2)}{2M_1M_2}}}{p\sigma_{12}^2 \Omega_{12}^{(1.1)*}}\,, \qquad (9.20)$$

with D_1 in $\mathrm{cm^2\,s^{-1}}$, p in atm T in K, M_1, M_2 in $\mathrm{g\,mol^{-1}}$, σ_{12} in Å.

The reduced impulse functions $\Omega_{12}^{(1.1)*}$ can be found in tabulated form in [17] and are approximated by analytic expressions in [18].

The impulse functions depend on the quantity ε_{12}/kT, whereby ε_{12} is the ***Lennard-Jones*** interaction energy, which in the Lennard-Jones interaction potential V_{12} as well as the Lennard-Jones length σ_{12} determines the dependence on the distance r:

$$V_{12} = 4\varepsilon_{12}\left[\left(\frac{\sigma_{12}}{r}\right)^{12} - \left(\frac{\sigma_{12}}{r}\right)^{6}\right]\,. \qquad (9.21)$$

Only a few interaction parameters are known [17], so they must usually be determined from empirical rules. To this end, first the interaction parameters between the same molecules ii are estimated, for example, from the boiling temperatures T_{S} and the molecular volumes V_{S} at the boiling point ($i = 1.2$) [16, 18]:

$$\frac{\varepsilon_{ii}}{k} = 1.15\,T_{si}$$
$$\frac{\sigma_i}{\text{Å}} = 1.18\,(V_{si}/\mathrm{cm^3\,mol^{-1}})^{1/3}\,. \qquad (9.22)$$

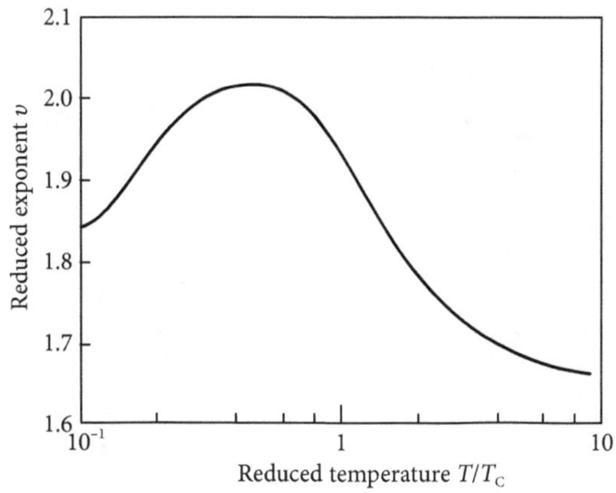

Fig. 9.22 Exponent ν as function of the reduced temperature T/T_c

If V_{si} is not known it must be numerically calculated. By means of arithmetic and geometrical averaging the interaction energy and interaction length can be calculated:

$$\sigma_{12} = \frac{\sigma_1 + \sigma_2}{2} \quad \text{and} \quad \varepsilon_{12} = \sqrt{\varepsilon_{11}\varepsilon_{22}}\,. \tag{9.23}$$

The temperature dependence of the diffusion coefficients can be determined through

$$D_1 = D_{10} \times T^{\nu}\,, \tag{9.24}$$

where ν depends on ε_{12}/kT.

With Eq. (9.20), in Fig. 9.22 the connection is shown, where ν is plotted as a function of the temperatures T/T_c. As a reference temperature $T_c = (4/3)(\varepsilon_{12}/k)$ has been selected [18].

For the deposition in non-isotherm reactors the thermodiffusion can play an important role. The thermodiffusion factor α_{12} depends substantially more sensitively on the concrete form of the interaction potential. Therefore, this value can be calculated only approximately from the kinetic gas theory:

$$\alpha_{12} = -\left[\frac{6C_{12}^* - 5}{6}\right]\left[\frac{\lambda_2}{\lambda_{12}}\right]\left[\frac{S^{(2)}}{1 + U^{(2)}}\right]$$

$$U^{(2)} = \frac{4}{15}A_{12}^* - \frac{1}{12}\left[\frac{12}{5}B_{12}^* + 1\right]\frac{M_2}{M_1} + \frac{(M_2 - M_1)^2}{2M_1M_2}$$

$$S^{(2)} = \left[\frac{(M_1 + M_2)}{2M_1}\right]\frac{\lambda_{12}}{\lambda_2} - \frac{15}{4A_{12}^*}\left[\frac{M_1 - M_2}{2M_2}\right] - 1$$

$$C_{12}^* = \Omega_{12}^{(1.2)*}/\Omega_{12}^{(1.1)*}$$

$$A_{12}^* = \Omega_{12}^{(2.2)*}/\Omega_{12}^{(2.2)*}$$

$$B_{12}^* = \left(5\Omega_{12}^{(1.2)*} - 4\Omega_{12}^{(1.3)*}\right)/\Omega_{12}(1.1)^*$$

$$\frac{\lambda_2}{\lambda_{12}} = \frac{\sigma_{12}^2\,\Omega_{12}^{(2.2)*}}{\sigma_2^2\,\Omega_2^{(2.2)*}}\left[\frac{2M_1}{M_1 + M_2}\right]^{1/2}\,. \tag{9.25}$$

The reduced collision integrals Ω are also registered in [18]. The thermal diffusion factors are located between 0 and 1.

With the discussed set of equations and the calculated gas characteristics the flow can be calculated (Fig. 9.23); on the basis of these results we obtain the Sherwood numbers [16].

The theoretical mathematical analysis shows the dependence on the Reynolds number and the Schmidt number. The Sherwood number reads:

$$\mathrm{Sh} = \beta\sqrt{Re}\,\sqrt[3]{\mathrm{Sc}}\,. \tag{9.26}$$

9.2.3 Deposition and Layer Generation

The process steps leading to the development of the film are represented in Fig. 9.24.

If all process steps interact the film structure is developed, while for the growth rate usually only one of the process steps is important. If only the mass transport in the gaseous phase determines the deposition, then the considerations discussed in the previous section can be used directly.

If steps 2 to 4 are determinant for the deposition, then the methods of the catalysis for the calculations of the deposition are used. An example is the SiO_2 deposition from SiH_4, which can be described by the ***Langmuir–Hinshelwood mechanism***.

SiH_4 and O_2 are adsorbed on the surface and react with one another to from SiO_2. The deposition can be reduced with constant SiH_4 concentration if too much O_2 is in the gaseous phase, since O_2 is more strongly adsorbed than

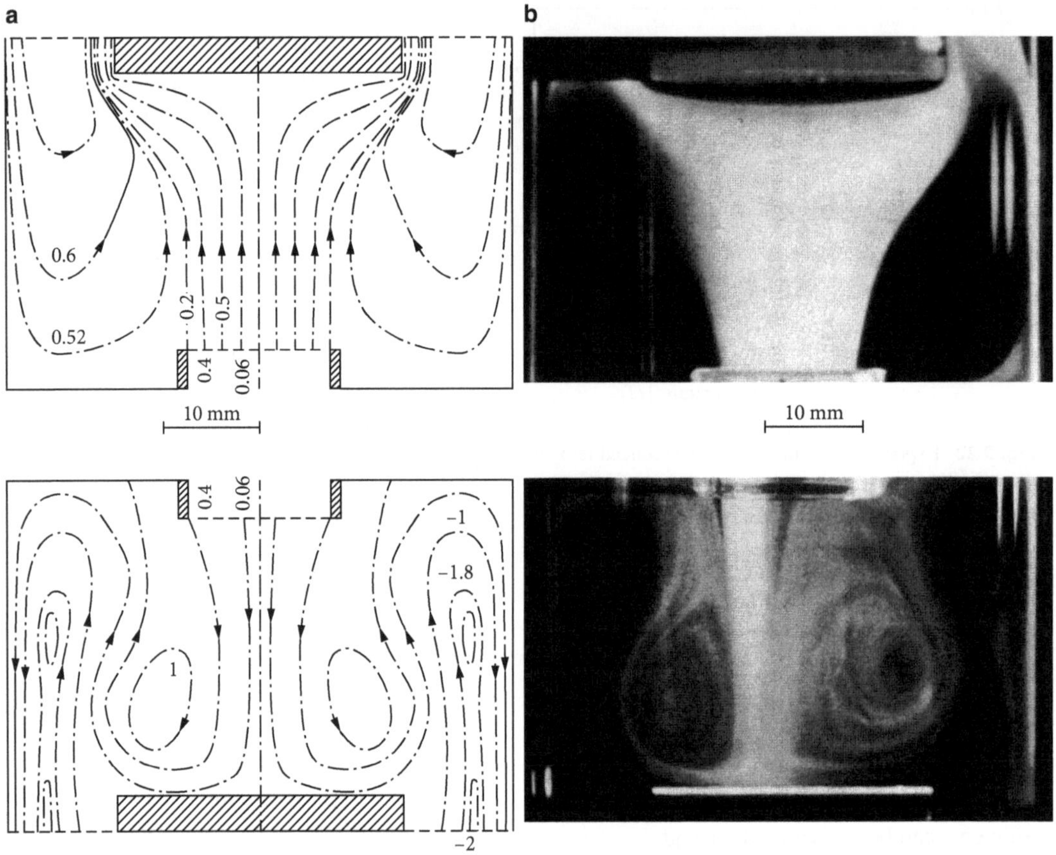

Fig. 9.23 Flow lines in a holdup flow reactor (*top* calculated, *bottom* made visible by TiO_2)

SiH_4. The deposition curves for SiO_2, which were measured in a holdup flow reactor, are represented in Fig. 9.25 [19].

On this basis, a simple model of nucleation is analyzed and integrated into the growth process of the developing film. During the nucleation the following are to be differentiated: the homogeneous nucleation within the condensing and crystallizing phase, for example film and substrate from same material, and the heterogeneous nucleation on a substrate surface from different material. Theories for nucleation are described in [20–22]. The heterogeneous nucleation is to be analyzed somewhat more exactly. According to Fig. 9.26, two molecular species A and B are to be adsorbed on the surface and reacted to C.

The collision rate of the molecules on the surface is obtained by the ***Hertz–Knudsen*** formula

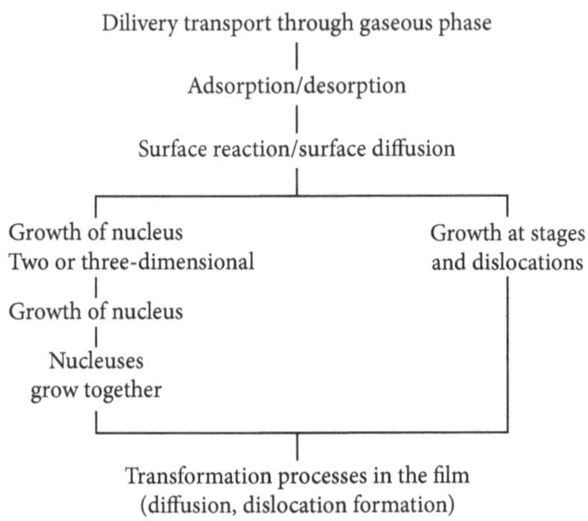

Fig. 9.24 Fundamental processes of film growth

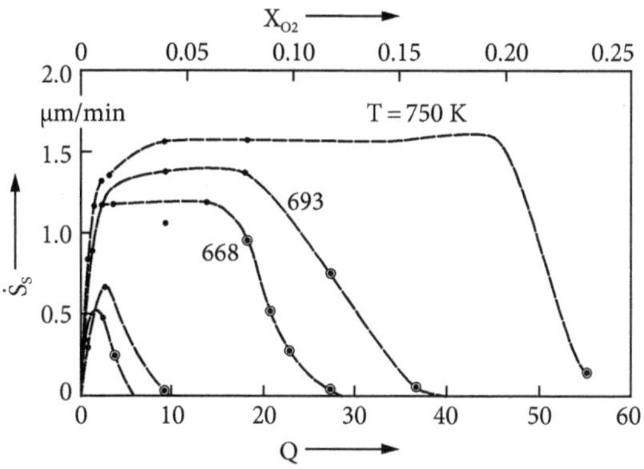

Fig. 9.25 SiO_2 deposition rate against the quotient Q from the gas flow SiH_4 and O_2. SiH_4 is constant

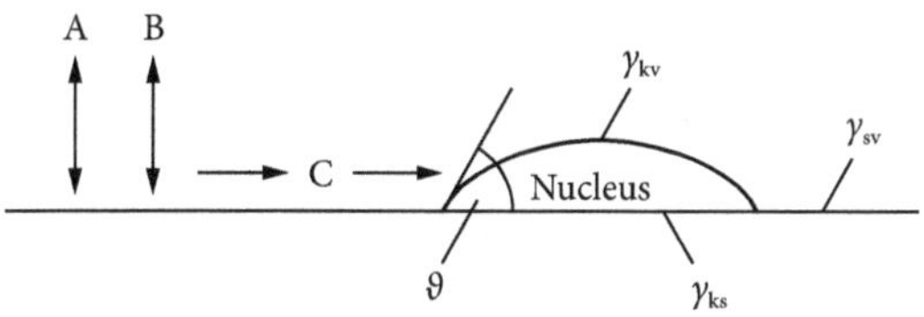

Fig. 9.26 CVD model including nucleation

$(i = \mathrm{A,B})$

$$J_i = \frac{p_i}{(2\pi m_i kT)^{1/2}}, \qquad (9.27)$$

where

p_i is partial pressure of the component i;

T the temperature in the proximity of the surface (for instance, in the distance of the mean free path); and

m_i the molecular mass.

A part of the molecules is chemisorbed, which is described by the condensation coefficient α_i: α_i = adsorption rate/impact rate so that the adsorption rate is given by: $J_{i\mathrm{ad}} = \alpha J_i$.

With metal on metal the condensation coefficient is $\alpha_i \sim 1$, as long as the substrate temperature is far from the melting point. At the adsorption of external gas molecules, as is usual with CVD processes, α_i can deviate strongly from 1.

The adsorption coefficient is then independent of 1; in the simplest case we obtained:

$$\alpha = (1 - \Theta), \qquad (9.28)$$

where Θ is the total degree of coverage of the surface.

Here it is accepted that the surface consists of similar types of adsorption locations, which can be occupied by different molecules:

$$\Theta = \sum_i \Theta_i, \qquad \Theta_i = \frac{\sigma_i}{\sigma_0}, \qquad (9.29)$$

where σ_i is the surface density of the sort of particle I and Θ_i is the degree of coverage of the particle species i.

The degree of coverage results from the total balance. It is accepted that the species of particle C is formed by a binary surface reaction, so that the reaction rate referred to the surface is given through:

$$\dot{\sigma}_\mathrm{C} = k_\mathrm{AB}\sigma_\mathrm{A}\sigma_\mathrm{B}, \qquad (9.30)$$

where σ_A and σ_B are the surface densities of the particles A and B. The reaction

$$\mathrm{C} \rightarrow \mathrm{A} + \mathrm{B} \qquad (9.31)$$

is given by a similar equation:

$$\dot{\sigma}_\mathrm{A} = \dot{\sigma}_\mathrm{B} = k_\mathrm{C}\sigma_\mathrm{C}. \qquad (9.32)$$

The concentrations σ_A, σ_B, and σ_C, and the degrees of coverage Θ_A, Θ_B, and Θ_C can now be calculated from the following balances.

Particle species A:

$$(1 - \Theta)J_\mathrm{A} = J_\mathrm{Ads} + k_\mathrm{AB}\sigma_\mathrm{A}\sigma_\mathrm{B} - k_\mathrm{C}\sigma_\mathrm{C}.$$

Particle species B:

$$(1 - \Theta)J_\mathrm{B} = J_\mathrm{Bds} + k_\mathrm{AB}\sigma_\mathrm{A}\sigma_\mathrm{B} - k_\mathrm{C}\sigma_\mathrm{C}.$$

Particle species C:

$$k_\mathrm{AB}\sigma_\mathrm{A}\sigma_\mathrm{C} = k_\mathrm{C}\sigma_\mathrm{C}. \qquad (9.33)$$

It is accepted that the balances are not affected by the nucleation.

$J_{i\mathrm{ds}}$ $(i = \mathrm{A,B})$ are the desorption flow densities and are described as equations of the form

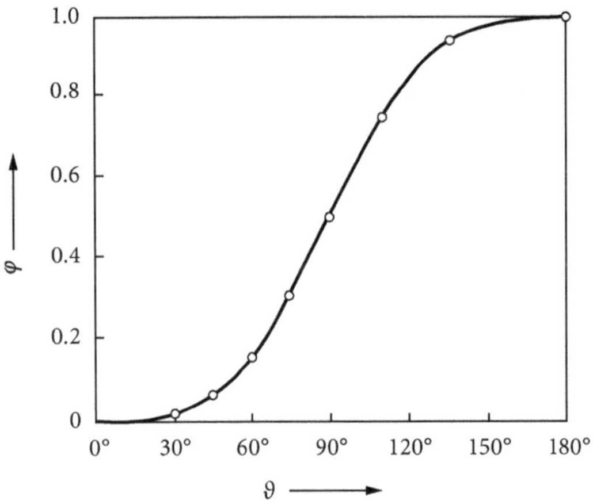

Fig. 9.27 The correcting function φ against the angle of contact δ

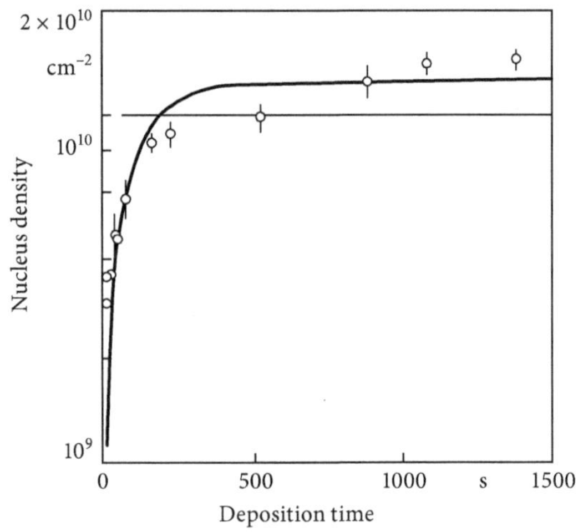

Fig. 9.28 Nucleus density; depositing of Au on NaCl

$J_{i\text{ds}} = \sigma_i \nu_i \exp(-E_{i\text{ds}}/kT)$, (where $E_{i\text{ds}}$ is the activation energy of the desorption and ν_i = the frequency factor).

For the calculation of the nucleuses either the molecular theory by ***Walton*** [23] can be used or the classical nucleation theory, which goes back to ***Becker and Volmer*** [24].

According to the classical nucleation theory the nucleation density is given by

$$\sigma_K = \sigma_C \exp\left(-\frac{\Delta G^* \times \varphi(\vartheta)}{RT}\right), \quad (9.34)$$

where ΔG^* is the free enthalpy for the formation of the critical spherical nucleus, and $\varphi(\delta)$ the formation of calottes at the surface.

The function $\varphi(\delta)$ as given reads,

$$\varphi(\vartheta) = \tfrac{1}{4}(2 + \cos\vartheta)(1 - \cos\vartheta)^2, \quad (9.35)$$

see Fig. 9.27. The angle of contact δ is described by the ***Dupres*** relation

$$\gamma_{\text{sv}} - \gamma_{\text{ks}} = \gamma_{\text{kv}} \cos\vartheta . \quad (9.36)$$

We can calculate the angle of contact δ from the surface stresses γ_{sv}, γ_{kv}, and the interface stress γ_{kv} (Fig. 9.28).

For $\delta = 0$ a complete wettability results and from Eq. (9.34) it follows that

$$\sigma_K = \sigma_C , \quad (9.37)$$

which means that all particles C are nucleuses at the same time.

For $\delta = 180°$ a homogeneous nucleation takes place, but no calotte develops on the surface. The nucleation rate is given by the rate with which the critical nucleuses, whose density is given by Eq. (9.34), grow further, i.e., becomes supercritical. This rate equals

$$J_K = \omega\sigma_K , \quad (9.38)$$

where ω is the frequency, by which particles at the nucleuses are adsorbed. This frequency can be estimate by:

$$\omega = \sigma_C \times U \times a\nu e^{-\frac{E_d}{kT}} , \quad (9.39)$$

where

U is the contour of the nucleuses,
a the jump width of the particles,
E_d the activation energy for surface diffusion,
ν the pre-exponential factor, and
σ_C the molecular surface density in Eq. (9.39) (particle/unit area).

The number of supercritical nucleuses grows temporally; this was found also experimentally, as shown in Fig. 9.28 [25, 26] for the condensation of Au on NaCl.

The number of nucleuses becomes constant because all formed particles bond on the surface

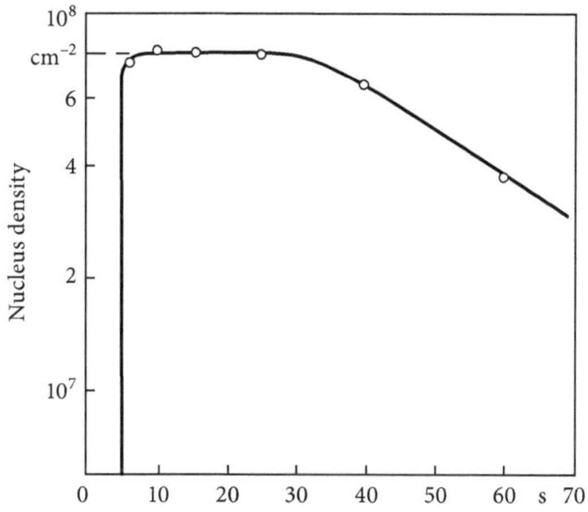

Fig. 9.29 Nucleus density by the deposition of SiO_2 from SiH_4 (0.1 Vol.%) and HCl (0.5 Vol.%) in H_2 at 10^3 mbar and 1000 °C

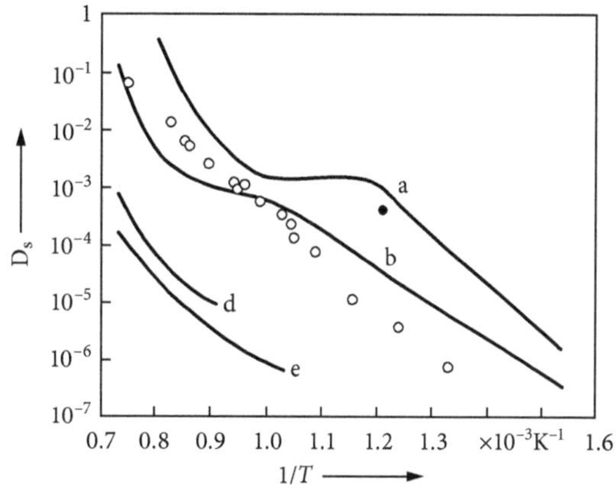

Fig. 9.30 Changes of the surface diffusion constants of Cu on Cu with simultaneous adsorption of halides for the systems Cu–Cl (*a*), Cu–Br (*b*), Cu–F (*c*), *d* and *e* pure copper

already on the existing nucleuses. This is shown in Fig. 9.28, as well as in Fig. 9.29, where the number of nucleuses is illustrated against time with the generation of Si on SiO_2 according to the CVD process [27].

The nucleuses formed in this way grow together with surface diffusion processes, and the number of nucleuses decreases again, as is shown in Fig. 9.31. The time constant for spherical nucleuses growing together is given by

$$t_c = 0.56 \frac{RT}{\Omega \gamma \Delta_s D_s} \times a^4 , \qquad (9.40)$$

where
D_s is the surface self-diffusion constant,
R is the general gas constant,
Ω is the molecular volume of the molecules C,
γ is the surface stress,
Δ_s is the thickness of the diffusing film,
T is the temperature, and
A the surface stress.

The surface diffusion constant D_i for any component i is defined through:

$$J_i = \frac{\sigma_i D_i}{RT} \nabla \mu_i , \qquad (9.41)$$

where
∇ is the nabla operator relating to the surface coordinates,
σ_i is the molecular surface density of the diffusing particles,
J_i is the molecular molecule flow density at the surface,
R is the general gas constant,
T is the temperature, and
μ_i is the chemical potential at the surface.

D_i depends on impurities and can, therefore, vary by an order of magnitude of 10, as is shown in Fig. 9.30 with the example of the self-diffusion of Cu on Cu [28].

The temperature dependence of the surface diffusion constant is described by

$$D_s = D_{s0} e^{-E/RT} , \qquad (9.42)$$

where E is additionally temperature dependent. Equation (9.42) shows that D_S increases with temperature increase.

A correlation between the deposition parameters and the film structure is difficult, because during the coating the entire film can change. For example, a film is to grow epitaxially, which has a mismatching f of the lattice parameters a_s, a_0 of the substrate and the film

$$f = (a_s - a_0)/a_0 .$$

This film grows first, as is shown in Fig. 9.31. Only at a critical thickness δ_k does the film relax partially and it develops dislocations at the

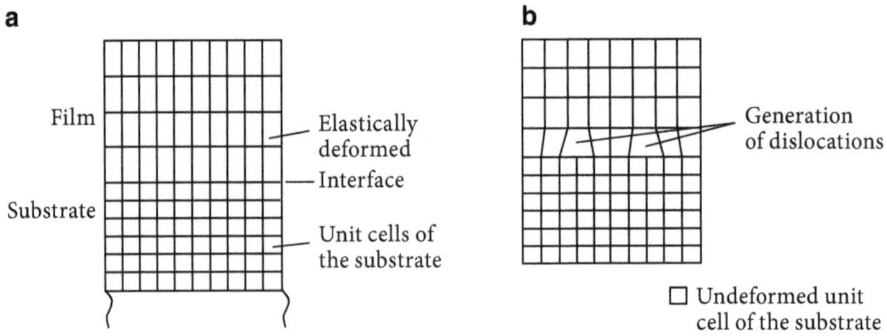

Fig. 9.31 Epitaxial deposition of films with mismatching for **a** deposition of thin films, and **b** partially relaxed films, which develop during the deposition of thicker films

boundary surface. This critical thickness lies within the range of 50 nm ($f = 1\,\%$) up to 1 nm ($f = 6\,\%$). A modeling of the deposition did not succeed for the nucleation because the number of parameters in particular is too large.

Numeric calculations, in good agreement with experimental results, lead qualitatively to the following statements:

- The nucleus size develops a distribution in the form of an asymmetrical bell-shaped curve, the maximum shifts with progressive condensation to larger nucleuses.
- In the neighborhood of nucleuses the surface of adsorbent particles becomes impoverished, and therefore also the probability is small that in the environment of a nucleus, new nucleuses will develop.

The entire nucleus density increases first, and afterward it finally reaches a maximum, since practically all condensed particles are again attached to the existing nucleuses, which is followed by a decay, because with further growth the nucleuses merge with into a closed film.

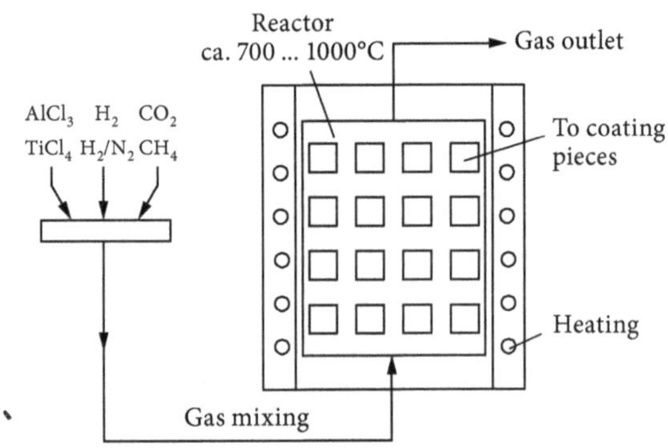

Fig. 9.32 Schema of a batch reactor for coating with Al_2O_3, TiC, and TiN

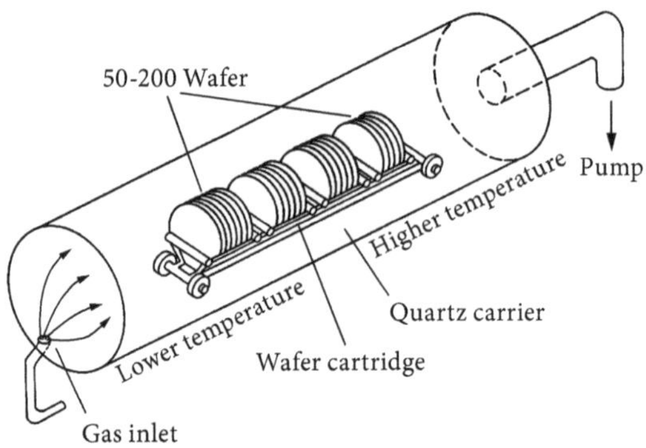

Fig. 9.33 Tubular reactor

9.3 Equipment for CVD Processes

9.3.1 Batch Reactors with Operating Points Within the Kinetically Controlled Range

With CVD, which operates within the kinetically controlled range, workpieces can be coated three-dimensionally. This is used mainly for the deposition of hard material films. Figure 9.32 shows the schema of a batch equipment for deposition hard material films.

In the semiconductor industry, tubular reactors (Fig. 9.33), for polycrystalline silicon, SiO_2 and Si_3N_4 from silane coating are used [29]. Table 9.2 shows some processes that are accomplished with tubular reactors.

Table 9.2 CVD processes in tubular reactors

Process	Name	Reactants	Temperature
SiO_2	HTO (high temperature oxide)	$SiH_2Cl_2 + N_2O$	850–900 °C
	TEOS (tetra ethyl ortho silicate)	$Si(OC_2H_5)_4$	650–750 °C
	LTO (low temperature oxide)	$SiH_4 + O_2$	380–430 °C
	(B)PSG (B, P-doped LTO)	$LTO + PH_3 + B_2H_6$	380–430 °C
Si_3N_4	Nitrides	$SiH_2Cl_2 + NH_3$	700–800 °C
$Si_xO_yN_z$	Oxynitrides	$SiH_2Cl + NH_3 + N_2O$	700–800 °C
Poly Si	Polysilicon	SiH_4	580–630 °C
	Doped polysilicon	$SiH_4 + PH_3$	580–630 °C
W	Tungsten	$WF_6 + H_2$	250–400 °C
Al	Aluminum	$Al_2(C_4H_6)_3$	250–400 °C

Table 9.3 Films manufactured in holdup flow reactors with atmospheric pressure

SiO_2 :	$O_2 + SiH_4 \rightarrow SiO_2 + H_4$	$T = 650$–800 K
SiO_2 :	$O_3 + TEOS \rightarrow SiO_2$	550–800 K
BSG:	$O_2 + SiH_4 + B_2H_4 \rightarrow BSG$	650–800 K
PSG:	$O_2 + SiH_4 + PH_3 + B_3H_4 \rightarrow PSG$	650–800 K
BPSG:	$O_2 + SiH_4 + PH_3 + B_2H_4 \rightarrow BPSG$	650–800 K
$S_nO_2 : F$:	$O_2 + Sn(CH_3)_4 + CF_3Br \rightarrow SnO_2 : F$	750–800 K
$S_nO_2 : Sb$:	$H_2 + CH_3OH + SnCl_4 + SbCl_5 \rightarrow SnO_2 : Sb$	750–800 K
$S_nO_2 : F$:	$SnCl_4 + H_2O + CH_3OH + HF \rightarrow SnO_2 : F$	750–800 K
TiO_2 :	$Ti(OC_3H_7)_4 + H_2 \rightarrow TiO_2$	450–650 K

9.3.2 Reactors Operating Within the Mass-Transport Controlled Range

For Si and/or III–V semiconductors epitaxy barrel reactors are used (Fig. 9.34), where the angle between the incident flow and the substrate surface holder is adjusted so that the coating on the wafer is constant.

A further variant of a holdup flow reactor is shown in Fig. 9.37. A continuous line spray coating system for the production of the films listed in Table 9.3 is illustrated schematically in Fig. 9.38.

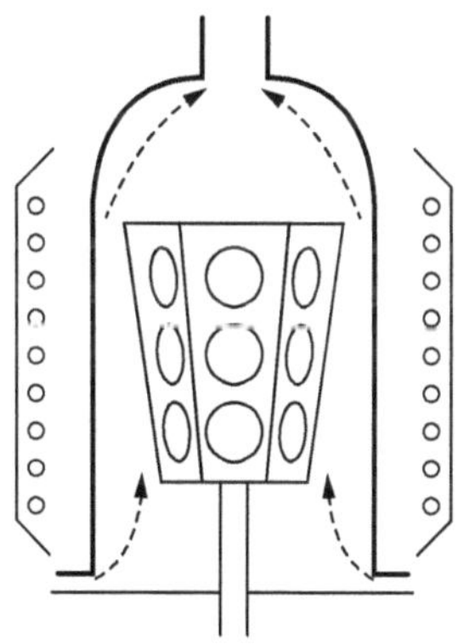

Fig. 9.34 Schema of a barrel reactor

A further reactor type, with parallel flow, is the rotation reactor (Fig. 9.35). The gas mixtures used feed over two gas inlets. The wafer and wafer bracket are being floated with the foarming gas and move in rotation. Figure 9.36 shows a holdup flow reactor for the deposition of SiO_2, phosphorus-glass (PSG), and boron-phosphorus-glass (BPSG).

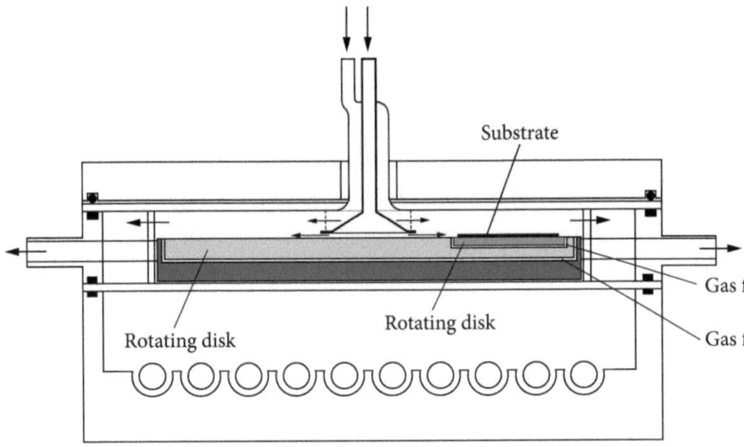

Fig. 9.35 Rotation reactor

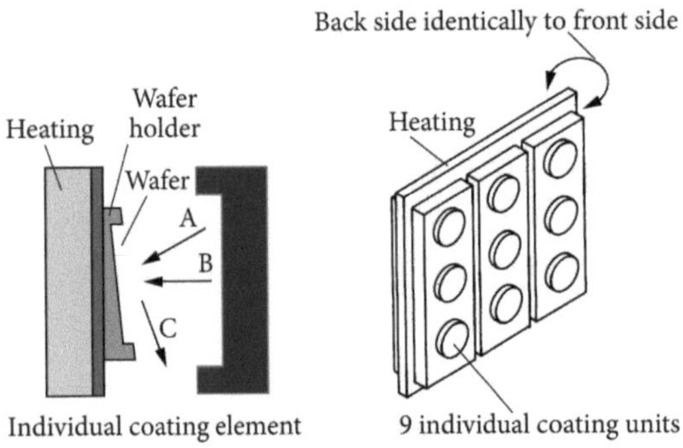

Fig. 9.36 Reactor configurations for the production of SiO_2, PSG, and BPSG

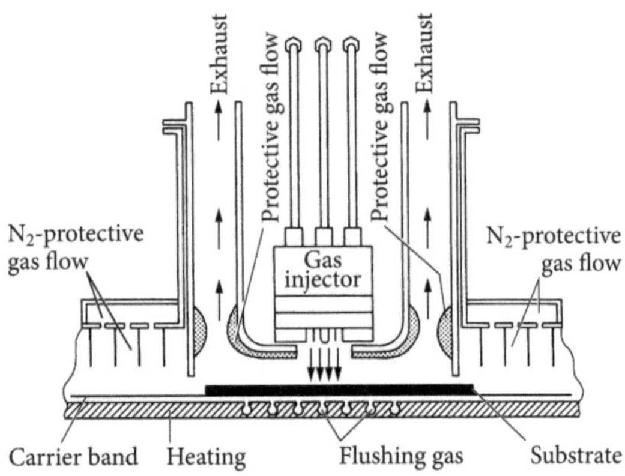

Fig. 9.37 Spray coating system [30]

9.4 Plasma-Enhanced Chemical Deposition from the Gaseous Phase (PECVD)

In the case of the plasma-supported chemical deposition from the gaseous phase (PECVD) the non-thermal, i.e., kinetic electron collision dissociation of the background gas or gas mixture becomes displaced on thermal activation [31].

This is possible where the electron energy typically occurs between $T_e \approx$ 1–5 eV. Since the electron temperature is very much higher than the substrate temperature and the heavy particle temperature (due to the thermal uncoupling of the atoms and ions of the electrons), the deposition occurs during a clearly smaller thermal load of the substrate instead as compared with ***CVD***.

Since, however, chemical reactions between heavy particles from the gaseous phase are necessary with ***PECVD***, the discharge pressure range with approx. 0.05–10 mbar lies clearly over the pressure range of the plasma etching (ion-supported etching and reactive ion etching: 0.01–0.10 mbar). For this reason, the mean free length of paths lies within the range of approximately 3–300 µm, and is clearly smaller than sputtering or etching.

The plasma density typically lies between $n_e = n_i = 10^9$ and $5 \times 10^{11}\,cm^{-3}$, and the ionization degree ε between 10^{-6} to 10^{-3}.

The surface activation energies for ***PECVD*** are usually small. Therefore, the deposition rate is only marginally dependent on the substrate temperature. However, many coating properties that determine the morphology, the composition,

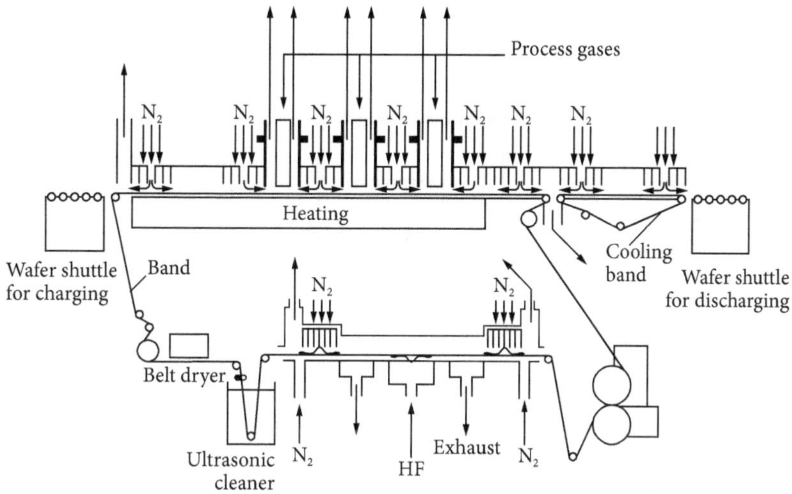

Fig. 9.38 Schema of a continuous line CVD plant; *outward pointing arrows* show that gas exits

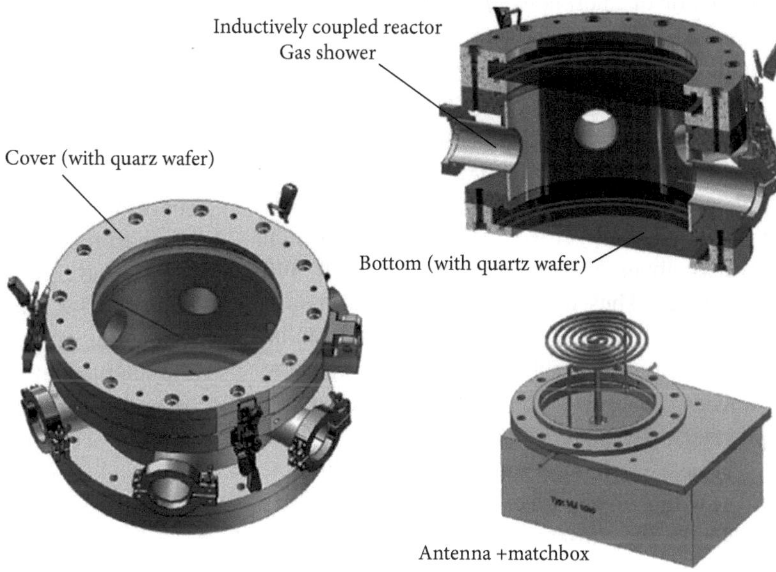

Fig. 9.39 Biodecon reactor (with quartz disk), ground (with quartz disk) inductive coupled reactor (ICP), antenna and match box

the internal stresses, often impact the substrate temperature T_{sub} very strongly. For this reason, T_{sub} must be optimized and regulated with respect to the required coating properties.

Because of the relatively high pressures and, therefore, small free length of paths and the temporally high reaction rates, film uniformity with the ***PECVD*** technique is a critical point. This is intensified by the fact that high sticking probabilities with large radical densities are combined, in particular for radicals. Due to this unfavorable combination of factors the uniformity of the gaseous phase dynamics, i.e., the transport of the gas particles of the inlet must be laid out over the gas shower up to the pumping system with great accuracy (Fig. 9.39).

Further, the power density (on the film surface) must be controlled and distributed as evenly as possible. For this reason, the ***CCP*** arrangement is still important. However, HD plasmas are also used in some cases (Fig. 9.40).

9.4.1 Deposition of Amorphous Silicon (a-Si : H)

The deposition of amorphous silicon is needed (a-Si : H) for many technical applications. Thus,

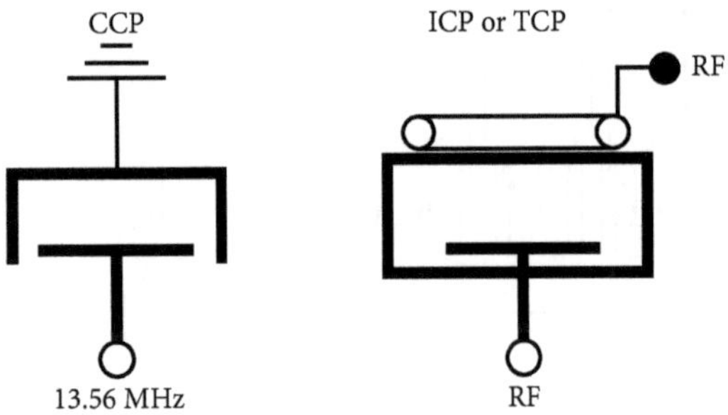

Fig. 9.40 Principles of PECVD systems

for the production of thin film transistors several process steps are needed, in which a-Si : H is to be deposited. There are other needs for the production of amorphous solar cells a-Si : H coatings. Also flat screens and displays are based on the deposition of a-Si : H films. Contrary to the epitaxial growth, i.e., crystalline silicon with a density of $2.33\,\mathrm{g/cm^3}$, amorphous Si has a smaller density of approximately $2.2\,\mathrm{g/cm^3}$. For the deposition of amorphous silicon, SiH_4 (silane) is usually used. With the deposition process about 5–20 % H-atoms are built into the lattice. Thus we obtain amorphous, hydrogen doped silicon: a-Si : H. The material can be manufactured favorably and deposited on large surfaces. As substrate materials glass, metals, polymers, and ceramic(s) are used.

Typical gas mixtures for the production of a-Si : H with ***PECVD*** are mixtures of $Ar : H_2 : SiH_4$ within the pressure range of 0.25–1.20 mbar. For doping typically B_2H_6 (diborane, p-doping) and PH_3 (phosphine, n-doping) are used. In capacitive coupled discharges (***CCP***) at power densities between $10–100\,\mathrm{mW/cm^2}$, we obtain deposition rates of 5–50 nm/min. The substrate temperatures are typically between 25–400 °C. The molecule forms tetrahedrons with the Si atom in the center. The energy of formation amounts to 24.3 kJ/mol. The two radicals SiH_2 and SiH_3 have a positive electron affinity, i.e., they attract free electrons and thereby become negatively charged ions. Binding energies of the radicals SiH_2–H, SiH–H, and Si–H amount to 3.0, 3.4, and 3.0 eV.

Table 9.4 lists some important rate constants. These are usually two-body collisions. As far as possible, a complete reaction mechanism has been specified by ***M. Kushner*** [32]. This model contains more than 35 electron collision reactions, 90 neutral particle–neutral particle reactions, and 80 ions (positive)–neutral particle reactions, as well as all recombination reactions.

9.4.1.1 Atomic Coating Mechanism in Four Steps

1. Activation and passivation of the (Si) surface
   A passive or passivated surface (coverage θ_p) consists of Si (bulk material) or e.g. H atoms. An active surface (coverage θa) from free bonding arms, the so-called ***dangling bond*** (Fig. 9.41). Dangling bonds (***DBs***) are produced by ion bombardment with primarily SiH_3^+, which can also remove H atoms from the surface.
2. The SiH_2 radicals can be built by collision into the solid lattice. This can happen on active and passive surfaces. In both cases, it concerns physisorption.
3. The adsorbed SiH_3 radicals diffuse along the surface and contribute to growth only at active ranges. This leads to equalizing of surface roughness and contributes thus to growth of smooth films of high quality.
4. The adsorbed SiH_4 molecule can lose the H atom at active ranges and thereby passivate the dangling bond, Fig. 9.42:

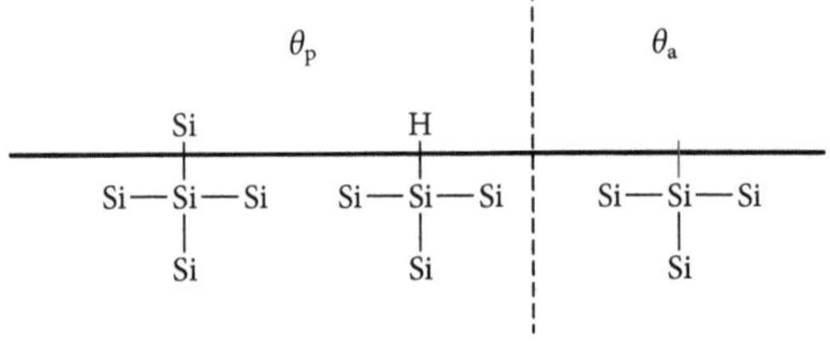

Fig. 9.41 Formation of amorphus silicon layers

Fig. 9.42 Saturation of the dangling bonds with hydrogen

Table 9.4 Simplified silane reaction mechanism [32]

Serial no.	Reaction	Rate constant (cm^3/s)
1	$e + SiH_4 \rightarrow SiH_3 + H + e$	$1.5 \times 10^{-8} \exp(-10/T_e)$
2	$e + SiH_4 \rightarrow SiH_2 + 2H + e$	$1.8 \times 10^{-9} \exp(-10/T_e)$
3	$e + SiH_4 \rightarrow SiH_3^- + 2H$	$1.5 \times 10^{-11} \exp(-9/T_e)$
4	$e + SiH_4 \rightarrow SiH_2^- + H_2$	$9 \times 10^{-12} \exp(-9/T_e)$
5	$e + SiH_4 \rightarrow SiH_3^+ + H + 2e$	$3.3 \times 10^{-9} \exp(-12/T_e)$
6	$e + SiH_4 \rightarrow SiH_2^+ + H_2 + 2e$	$4.7 \times 10^{-9} \exp(-12/T_e)$
7	$e + SiH_n^+ \rightarrow SiH_{n-1} + H$	$2.5 \times 10^{-7} T_e^{-1/2}$
8	$SiH_4 + H \rightarrow SiH_3 + H_2$	2.7×10^{-12}
9	$SiH_4 + SiH_2 \rightarrow Si_2H_6^+$	1.0×10^{-11}
10	$Si_2H_6^+ \rightarrow Si_2H_4 + H_2$	$5.0 \times 10^{+6} (1/s)$
11	$Si_2H_6^+ + SiH_4 \rightarrow Si_2H_6 + SiH_4$	1.8×10^{-10}
12	$SiH_4 + SiH_3 \rightarrow Si_2H_5 + H_2$	1.8×10^{-15}
13	$SiH_3 + H \rightarrow SiH_2 + H_2$	1.0×10^{-10}
14	$SiH_3 + SiH_3 \rightarrow SiH_2 + SiH_4$	7.0×10^{-12}
15	$SiH_m^- + SiH_n^+ \rightarrow SiH_m + SiH_n$	5.0×10^{-7}

Under the condition that the complete surface consists of active and passive or passivated surfaces, i.e., $\theta_p + \theta_a = 1$ is valid.

With 1:

$$SiH_3^+ + \Theta_p \overset{K_{ion}}{\rightarrow} \Theta_a + Y_{ion}H(g)\,.$$

where Y_{ion} describes the yield of H atoms per incident ion.

With 2:

$$SiH_2 + \Theta_a \overset{K_{2a}}{\rightarrow} \Theta_a\,,$$

$$SiH_2 + \Theta_p \overset{K_{2p}}{\rightarrow} \Theta_p\,,$$

With 3:

$$SiH_3 + \Theta_a \overset{K_3}{\rightarrow} \Theta_p\,.$$

With 4:

$$SiH_4 + \Theta_a \overset{K_4}{\rightarrow} \Theta_p + SiH_3(g)\,.$$

With this set of equations the atomic processes are described. However, this represents a simplification of the actual conditions. In the next step, the rate constant (of the particle densities involved independently) must be determined:

$$K_i \approx \frac{v_{Bohm}}{n'_0} \quad \left[\frac{m^3}{s}\right] \tag{9.43}$$

$$K_{2a} \approx \frac{1}{4}\frac{s_{2a}\bar{v}_2}{n'_0} \qquad K_{2p} \approx \frac{1}{4}\frac{s_{2p}\bar{v}_2}{n'_0} \tag{9.44}$$

$$K_3 \approx \frac{1}{4}\frac{M s_3 \bar{v}_3}{n'_0} \tag{9.45}$$

$$K_4 \approx \frac{1}{4}\frac{s_4 \bar{v}_4}{n'_0}\,, \tag{9.46}$$

where the ***sticking coefficient*** is shown by s_2, s_3 and s_4, and M defines the number of dangling bonds that pass the SiH_3 radicals on the average by diffusion before being desorbed (chemisorption).

Under stationary conditions the generation and elimination of dangling bonds are in the equilibrium: where the installation of the SiH_2 radicals leaves the surface coverage uninfluenced θ_a

$$\frac{d\Theta_a}{dt} \overset{!}{=} 0 = Y_{ion}K_{ion}n_{ion}(1 - \Theta_a) - K_3 n_{3S}\Theta_a - K_4 n_{4S}\Theta_a\,. \tag{9.47}$$

The deposition rate is based on the chemical relations:

$$D_{Si} = (K_{2a}n_{2S} + K_{2p}n_{2S} + K_3 n_{2S}\Theta_a)\frac{n_0'}{n_{Si}}\,, \tag{9.48}$$

with the solid density n_{Si}.

We summarize $K_{2a} + K_{2p} = K_2$ and consider that the SiH_4 molecule density is clearly larger than the ions and the SiH_2-radicals densities, i.e.,

$$K_4 n_{4S} \gg Y_{ion}K_{ion}n_{ion} + K_3 n_{3S}\,, \tag{9.49}$$

then with Eq. (9.47) we obtain:

$$\Theta_a = \frac{Y_{ion}K_{ion}n_{ion}}{Y_{ion}K_{ion}n_{ion} + K_3 n_{3S} + K_4 n_{4S}} \approx \frac{Y_{ion}K_{ion}n_{ion}}{K_4 n_{4S}}\,. \tag{9.50}$$

For typical discharges, $n_{ion}/n_{4S} \approx 10^{-4}$, $Y_{ion} \approx 10$ and K_{ion}/K_4 is valid:

$$\frac{K_{ion}}{K_4} = \frac{v_{Bohm}}{1/4 \times s_4\bar{v}_4} = \frac{\sqrt{kT_e/m_{ion}}}{1/4\sqrt{kT_g/m_{ion}}} \approx \sqrt{\frac{T_e}{T_g}} \approx \sqrt{\frac{10^4}{10^2}} = 10\,. \tag{9.51}$$

Thus we have

$$\Theta_a \approx \frac{10 \times 10}{10^4} = 10^{-2}\,, \tag{9.52}$$

with $M \approx 10$ (M is the number of dangling bonds) for SiH_3 radicals the effective sticking coefficient is

$$s_{3,eff} \approx s_3\overline{M}\Theta_a \approx 0.1\,. \tag{9.53}$$

For typical deposition conditions the density of the SiH_2 radical is clearly smaller than those of the SiH_3 radicals, i.e.,

$$S_{3,eff} \approx S_3\overline{M}\Theta_a \approx 0.1 \tag{9.54}$$

$$n_{2S} \approx 10^{-2}n_{3S}\,, \tag{9.55}$$

and thus the third term in Eq. (9.48), which is responsible for good film quality, is approximately ten time larger than the first two terms, which are responsible for bad film quality.

Thus with Eqs. (9.48) and (9.50) it becomes clear that a high ion flux is decisive for good film quality, just like a large n_{3S}/n_{2S} relation as well as a high surface diffusion of SiH_3.

As was already suggested, the mechanism is strongly simplified in the chemical formulas, since, e.g., the first chemical formula is valid for a whole class of ions (and thus of formulas), which can produce dangling bonds by ion bombardment of the surface, like, e.g., Ar^+ or SiH_n^+ ions. The same is valid also for occupation or free etching of dangling bonds by hydrogen atoms:

$$H + \theta_a \rightarrow \theta_p$$
$$H + \theta_p \rightarrow \theta_a + H_2(g)\,.$$

9.5 Photoinitiated Vapor Deposition

Photoinitiated chemical vapor deposition (***piCVD***) is an evolutionary ***CVD*** technique for depositing polymeric thin films in one step without using any solvents [33]. The technique requires no pre or post-treatment and uses a volatile photoinitiator to initiate free-radical polymerization of gaseous monomers under UV irradiation. Glycidyl methacrylate (***GMA***) was used as a test monomer for its ability to undergo free radical polymerization, and 2,2′-azobis (2-methylpropane) (***ABMP***) was used as the photoinitiator, as it is known to produce radicals when excited by photons. ***GMA*** and ***ABMP*** vapors were fed into a vacuum chamber in which film growth was observed on a substrate exposed to UV irradiation. The resulting poly(glycidyl methacrylate) (***PGMA***) thin films were comprised of linear chains and had high structural resemblance to conventionally polymerized ***PGMA***, as shown by the high solubility in tetrahydrofuran and the infrared and X-ray photoelectron spectroscopy measurements. The introduction of ***ABMP*** into the vacuum chamber significantly increased growth rates. The maximum growth rate achieved was ~ 140 nm/min, which represents a sevenfold enhancement over the case without ***ABMP***. The molecular weight was found to increase with increasing monomer-to-initiator (M/I) feed ratio, and the polydispersity indexes (***PDIs***) of the samples were between 1.8 and 2.2, which is lower than the values obtained in conventional batch polymerization but in agreement with the theoretical expressions developed for low-conversion solution-phase polymerization, which are applicable to continuous processes such as ***piCVD***.

Molecular-weight distributions can be narrowed by filtering out wavelengths shorter than 300 nm, which induce branching and/or cross-linking. The strong dependence of the molecular weight on the M/I ratio, the rate enhancement due to the use of a radical photoinitiator, the good agreement between the experimental, and the theoretical PDIs provide evidence of a free-radical mechanism in piCVD. The clear films obtained in this work had number-average molecular weights between 12,500 and 97,000 g/mol. The similarities in growth conditions, growth rates, and molecular weights between the initiated ***CVD***, a previously reported thermal process able to synthesize a wide range of polymers, and the ***piCVD*** of ***PGMA*** suggest that ***piCVD*** can also be used to produce those polymers and potentially others whose monomers undergo free-radical mechanisms. This chapter serves as an introduction to the technique by demonstrating the ability of ***piCVD*** to synthesize high molecular weight ***PGMA*** thin films with narrow molecular weight distributions from vapors of ***GMA*** and ***ABMP*** in a single, dry step under UV irradiation.

PICVD is based on a new principle of the formation of an organometallic vapor phase. A simple idea became the basis for this principle – the entire amount of the precursor mixtures must be stored cool, only their microdoses can be introduced into the evaporator. This idea allowed us to elaborate a simple and precise liquid delivery system for ***CVD*** sources based on the principle used for the control of fuel injection in the latest thermal motors. The injected microdose evaporates instantaneously (flash evaporation), thus there is ***no time for chemical changes of precursors***.

Volatile organometallic precursors or their mixture (e.g., Y, Ba, and Cu β-diketonates for ***YBCO*** deposition) are dissolved in the appropriate organic solvent. The precise microdoses of solution are injected into the evaporation zone by means of a computer-driven injector, where they instantly evaporate. The resulting mixture of solvent and precursor vapor is transported by carrier gas (Ar, O_2) into a high temperature zone where the decomposition reaction and film growth occur on a substrate. An injector is a high speed micro electro valve.

9.5.1 Advantageous Features of PICVD

A microdose (1–10 mg) of solution evaporates instantly (flash evaporation) in vacuum ($\sim$ 5 mbar), so a mixture of precursors is heated for only a short time, insufficient for chemical change, so instability of precursors becomes insignificant; the composition of the vapor phase corresponds to the composition of the precursor solution; consequently, it is easy to change and reproduce it.

A precise microdosing (within $\sim$ 3 %) of the precursor mixture, as well as the possibility to change the injection parameters easily, enables the precise control of the vapor pressure, the growth rate, and the improvement of the reproducibility of film properties.

It is possible to deposit films of various composition *in situ* in the same ***PICVD*** reactor by a simple change of solution composition and deposition conditions. The use of several injectors allows production of *in situ* multi-layered structures.

Another particularity of the process is that it works "digitally": the thickness may simply be controlled by the number of drops injected, which is very attractive for multi-films. For example, for the deposition of oxide superlattices, the solution concentration, opening time, and droplet size can be adjusted to obtain a mean thickness increase at each injection of less than 0.1 nm.

References

1. Powel CF, Oxley JH, Blocher JM Jr. (eds) (1966) Vapor Depositions. Wiley & Sons, New York
2. Wahl G, Davies PB, Bunshah RF, Joyce BA, Bain CD, Wegner G, Remmers M, Walsh FC, Hieber K, Sundgren J-E, Bachmann PK, Miyazawa S, Thelen A, Strathmann H (2000) "Thin Films" Ullmann's Encyclopedia of Industrial Chemistry. Wiley-VCH, Weinheim
3. Tavares J, Swanson EJ, Coulombe S (2008) Plasma Synthesis of Coated Metal Nanoparticles with Surface Properties Tailored for Dispersion. Plasma Processes and Polymers 5(8):759
4. Chamberlain RR, Skarman JS (1966) Chemical spray deposition process for inorgqanic films. J Electrochem Soc 113(1):88
5. Bloem J (1980) Nucleation and growth of silicon by CVD. J Cryst Groth 50, pp 581/604
6. Simmler W (2005) "Silicon Compounds, Inorganic", Ullmann's Encyclopedia of Industrial Chemistry. Wiley-VCH, Weinheim
7. Schleich DM (1989) CVD: a chemical approach to ceramic materials. Proc. VIIth Euro CVD Conf. Perigan 1989, Ducarroir M, Bernard C, Vandenbulcke L (eds). Publ. by Ed. De Physique, coll de Phys. Additif au Coll C5 suppl. no5 (1989), pp C5-961/C5-979
8. Scalilotti M, Horiuuchi L, Decobert J, Brasil MJ, Cardoso LP, Ossart P, Ganiere JD (1992) Growth and characterization of GaAlAs/GaAs and GaInAs/IP structures: The effect of a pulse metalorganic flow, J Appl Phys 71:179/186
9. Schmaderer F, Wahl G, Dietrich M, Dustmann CH (1984) Preparation of stabilized superconducting NbCl-y Ny Carbon fibres. In: Mc Robinson D, v d Brekel CHJ, Cullen GW, Blocher JM, Rai-Choudhury P (eds) Proc. IX Int. CVD Conf.. ESC, Pennington, pp 663–672
10. Sievers RE, Sadlowski JE (1978) Science 201:217
11. Sicre JE, Dubois JT, Eisentraut KJ, Sievers RE (1969) Volatile Lanthanide Chelates. II. Vapor Pressures, Heat of Vaporization and Heats of Sublimation. J Am Chem Soc 91:3476/3481
12. Kuttruff H (1988) Physik und Technik des Ultraschalls. Hirzel Verl., Stuttgart
13. Bachmann PK, Leers D, Lydtin H (1991) Towards a general concept of diamond CVD Diamond an related Materials, vol. 1., pp 1–12
14. Isberg J (2004) Single crystal diamond for electronic applications. Diamond and Related Materials 13(2):320–324
15. Kennard EH (1938) Kinetic theory of gases. McGraw-Hill, New York
16. Hirschfelder JO, Curtis CF, Bird RB (1954) Molecular Theory of Gases and Liquids. Wiley, New York
17. Reid RC, Prausnitz JM, Sherwood TK (1977) The Properties of Gases and Liquids. McGraw Hill, New York
18. Neufeld PD, Jansen AR, AziZ A (1972) Emirical Equation to Calculate 16 of Transport Collision Integrals $\Omega^{(1,s)}$ for the Lennard-Jones (12-6). Potential Transport. J Chem Phys 57:1100–1102
19. Wahl G (1975) Bocker JM, Hintermann HE, Hall LH (eds) CVD of SiO_2 and Boro- and Phospho-Silicate glasses by the reaction of SiH_4, B_2H_6 or PH_3 with O_2, Proc. Of V Int. Conf. on CVD. ECS, Princeton, pp 391–406
20. Volmer M (1939) Kinetik der Phasenbildung. Verlag. Th. Steinkopff, Dresden, Leipzig
21. Stranski IN (1943/44) 2. Krist. 105:91/123
22. Tammann G (1922) Aggregatzustände. Verlag von Leopold Voss, Leipzig
23. Walton D (1962) Nucleation of Vapor Deposits. J ChemPhys 37:2182–2188
24. Hirth JP, Pound GM (1963) Condensation and evaporation. Pergamon Press, Oxford
25. Velfe HD, Stenzel H Krohn M (1982) Nucleation Data for Gold on Pure NaCl Surfaces, Thin Sol Films 98:115/124
26. Jaeger RC (2002) "Film Deposition". Introduction to Microelectronic Fabrication, 2nd edn. Prentice Hall, Upper Saddle River
27. Bloem J (1989) Nucleation and growth of silicon by CVD. J Cryst Growth 50:581/604
28. Bonzel HP (1973) Surface Diffusion of Metals Structure and Properties of Metal Surface, vol. 1. Maruzen Company, Tokyo, pp 248–327
29. Sze SM (2008) Semiconductor devices: physics and technology. Wiley-India, ISBN, p 384
30. Watkins Johnson Company, Scotts Valley, CA. Internal Report
31. Ay, Aydinli (2004) Comparative investigation of hydrogen bonding in silicon based PECVD grown dielectrics for optical waveguides. Optical Materials 26(1):33–46
32. Kushner KJ (1988) J Appl Phys 63:2532
33. Felten F, Senateur JP, Weiss F, Madar R, Abrutis A (1995) Deposition of oxides layers by computer controlled injection-LPCVD. *J. Phys. IV France* 5(C5):1079–1086
34. Kelvin C, Gleason KK (2005) Photoinitiated Chemical Vapor Deposition of Polymeric Thin Films Using a Volatile Photoinitiator, Department of Chemical Engineering, Massachusetts Institute of Technology, Cambridge, Massachusetts 02139

Physical Basics of Modern Methods of Surface and Thin Film Analysis

10

H. Frey

10.1 Introduction

The characterization of surfaces contains the identification of the atoms existing at the surface, their classification in crystalline or amorphous structure, the chemical bonding, and the electronic structure of the surface. By means of this knowledge we can determine further characteristics, such as dynamic, electrical, thermal, and chemical properties, etc.

With the analysis methods discussed in this chapter, the atoms and molecules at the surface are identified due to their mass or their characteristic electronic energy states. The method of the diffraction of electrons (low energy electron diffraction LEED, reflection high energy electron diffraction RHEED) uses the periodic arrangement from surface atoms to the determination of the crystalline structure. The investigation of the electron work function also uses a collective characteristic, the potential jump at the surface.

For the investigation of surface structures new microscopic methods have been developed, which penetrate into the resolution range of atomic order of magnitude and also permit time-dependant investigations of dynamic modifications of surface structures or layer growth. The surface-analytic methods are based, in principle, on the excitation of the solid surface with photons, electrons, ions, or neutral particles and receive information by emitted ions, neutral particles, electrons, and photons or by backscattered electrons and ions of the solid surface (Fig. 10.1).

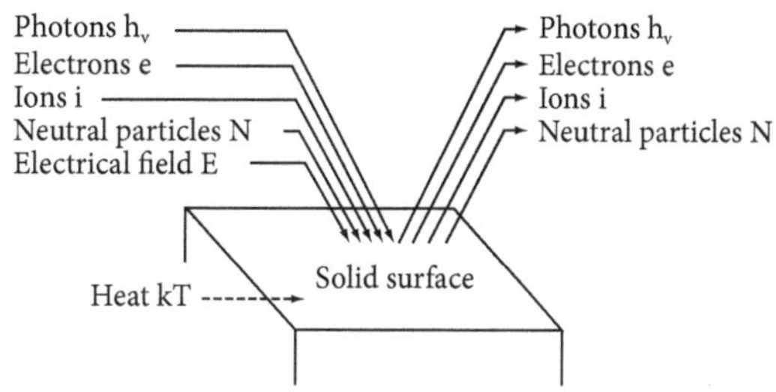

Fig. 10.1 Possible excitation processes and emissions

Thin film analysis methods have a finite resolution of depth. The measurable signal comes from a certain film depth of the material, and the appropriate surface composition over this depth is averaged. The depth of the analyzed film ranges from approximately one atomic layer with the most surface sensitive methods (e.g., ion surface scattering – ISS (ion scattering spectrometry), electron energy loss spectrum EELS) to some 100 nm (e.g., Particde Induced X-Ray Emission PIXE).

Methods with monofilm resolution are very much suitable for adsorption and desorption investigations. They are also used for the investigation of film growth. The question is, whether a thin film monolayer-to-monolayer grows, in three-dimensional islands or in a combination on the substrate [1] (Fig. 10.2). In order to be able to differentiate between these growth forms, atom-layer specific research methods are necessary. The depth of the information also has a connection in relationship with the sample purity and the

© Springer-Verlag Berlin Heidelberg 2015

H. Frey, H. R. Khan, *Handbook of Thin-Film Technology*, DOI 10.1007/978-3-642-05430-3_10

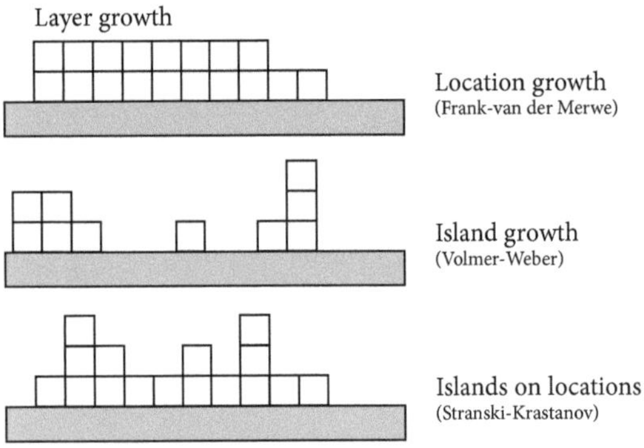

Fig. 10.2 Different modes of film growth caused by the specific free energy of the surface, that (from *top* to *bottom*) decreases due to the occupancy, increases or decreases only for the first monolayer

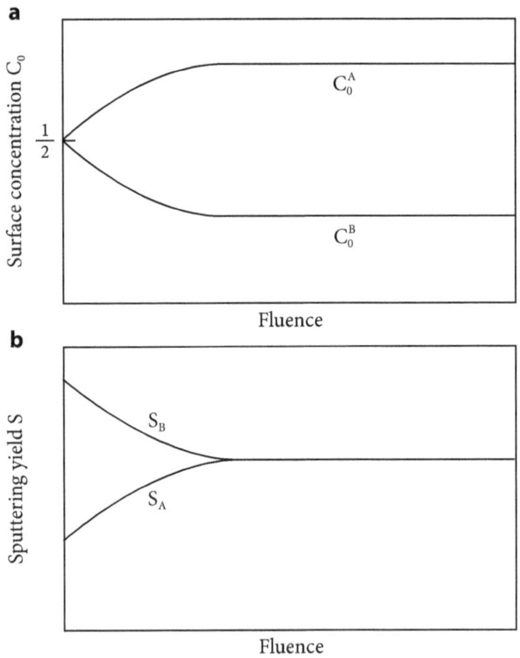

Fig. 10.3 Idealized process of sputtering a binary solid surface with the components A, B, and initial sputtering yields $S_b = 2S_a$. **a** Analysis of the surface, **b** Analysis of the flux of sputtered particles

vacuum requirements. Methods that are sensitive for the topmost atomic layer require appropriate sample preparation and ultra high vacuum technology.

Methods, such as (RBS) Rutherford backscattering [2–4] with a typical analysis depth of 10 nm, can also be operated in the high vacuum range (10^{-6} mbar), since the contribution of the topmost atomic layer to the total signal is sufficiently small.

Depth profiles of the sample compound are usually gained by means of gradual abrasion of the sample by sputtering. Generally, Ar- ions are used with energies of 1–2 keV. It is to be considered that different materials of compounds have different sputtering rates. The erosion rate can also change with the depth. The compound of the sputtered particles can likewise change [5]. Analysis methods that register the particle flux dropping out from the surface, can, therefore, show different results to methods that observe the respective surface compound in the inside depth [6]. Figure 10.3 illustrates this.

In the case of a two-component system (A and B) particle conservation in the stationary condition results for the concentrations C_V in the volume and C_0 at the surface in:

$$C_0^A/C_0^B = \left(C_V^A/C_V^B\right)(S_B/S_A) ,$$

where S is the concentration-dependant sputtering yield.

Analysis methods that measure the particle flux from the surface, e.g., SIMS (secondary ion mass spectrometry) [3] or SNMS (secondary neutral mass spectrometry) [7], register a particle mixture after a transient phase, which corresponds to the concentration of the bulk material.

10.2 Ion Spectroscopes

10.2.1 Ion Backscattering

The backscattering of ions is an important tool for identification of existing atoms at a surface and in films near the surface. From the energy of the projectile ion, after the scattering process, the surface atom is identified, and the concentration is identified from the scattering intensity (Fig. 10.4).

Chemical bonds or molecules cannot be identified as such. The elastic backscattering of

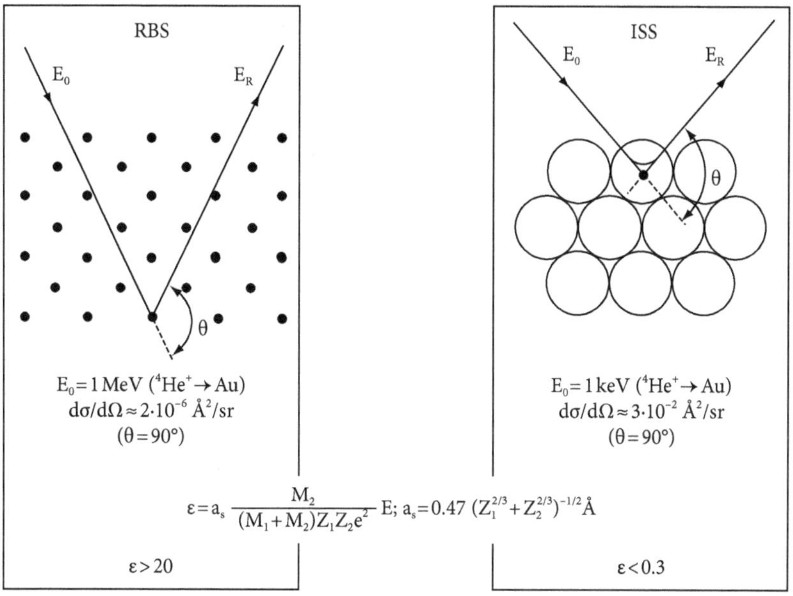

Fig. 10.4 Schematic representation of the ion scattering within different projectile ranges. RBS (high projectile energy), ISS (low projectile energy). The different differential scattering cross-sections $d\sigma/d\Omega$ are show by *circles* around the scattering centers. An exact classification possibility offers the reduced energy ε [8]. E is the projectile energy before the collision, M_1, M_2 mass numbers of the projectile and/or target atom, and Z_1, Z_2 the atomic number of the projectile and/or target

high-energy light ions, usually called Rutherford backscattering spectrometry (RBS) [2, 10], has also proved to be a very useful tool in the study of implantation damage and annealing, the location of impurity atoms in crystal, and the investigation of surfaces and thin films.

A well-collimated beam of singly ionized 1–3 MeV helium ions or hydrogen ions is used. The specimen to be analyzed is placed on a goniometer and exposed to the beam in a target chamber at a pressure of 10^{-6} mbar. The interaction between the projectiles and the surface atoms is determined by the Coulomb repulsion of the atomic nuclei. This method provides absolute quantitative results. Information about the specimen is contained in the energy spectrum of the ions backscattered through an angle of 140°. These ions are detected by a silicon surface barrier detector. The signal is amplified and analyzed, stored, displayed, and recorded.

A silicon sample, for example, exposed to a beam of energy E_0 gives rise to a spectrum consisting of a sharp edge at 0.6 E_0, corresponding to helium ions elastically scattered by surface silicon atoms and deflected into the detector. The sharp edge is followed by a smoothly varying yield at lower energies. Helium ions not scattered at the surface penetrate into the target, losing energy through inelastic collisions and being scattered elastically in the bulk. They eventually leave the target, to be detected after losing more energy inelastically on the way out. If the stopping power and the geometry of the system are known, one can convert the energy scale into a depth scale. If the ion beam is incident in a low index direction of a high-quality crystal specimen channeling develops (see Sect. 6.2.2).

Channeling minimizes collisions of incident ions with target atoms and reduces the observed backscattered yield for the entire depth in the crystal. A small surface-damage peak observed only in the channeled spectrum is due to localized surface disorder. Even high-quality crystal samples show this surface-disorder peak. If the crystal has been damaged, for example, by ion implantation, and the displayed atoms occupy sites in the channel areas, the channeled ions suffer elastic collisions with these atoms, resulting

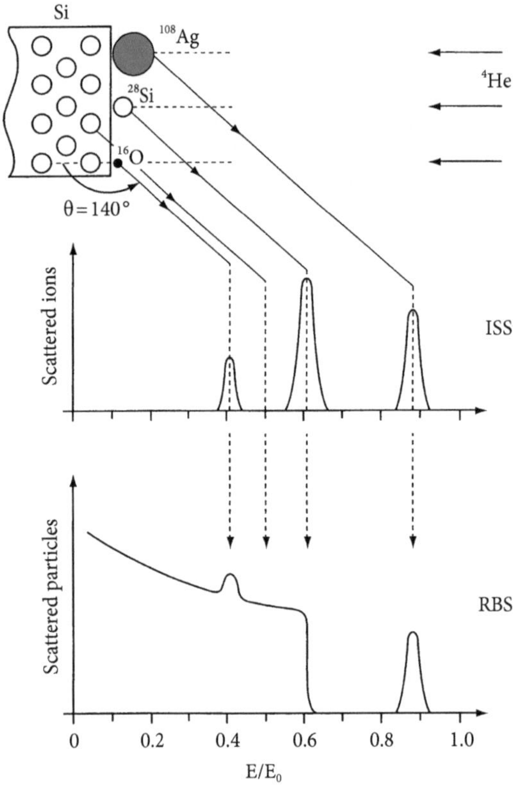

Fig. 10.5 Schematic representation of the energy spectra of helium ions backscattered under an angle of 140° of a Si substrate with Ag, Si, and O on the surface. *Upper spectrum* ISS ($E_0 \sim 1\,\text{keV}$), *lower spectrum* of RBS ($E_0 \sim 1\,\text{MeV}$)

in a higher scattered yield. The ratio of the yield in the channeling and nonchanneling directions is a measure of the quality of single-crystal specimens.

RBS, in conjunction with channeling, can also be used to determine the concentration and position of impurity atoms in the crystal; the technique has also been applied in a variety of surface an thin-film problems.

Depth profiles can be analyzed with RBS without damaging the surface. The resolution is not so exact that we can analyze mono-atomic films. With low energies around 1 keV the interaction is limited to a few atomic layers near the surface. This method is designated as ISS or LEIS (low energy ion scattering) [9].

In this energy range the shielding of the nuclear charge by the electron shell is effective. The cross-sections are not as exactly calculable as with RBS. The quantitative identification is possible only with calibration. The neutralization of the primary ions by the scattering is a further problem. Generally, He^+ or Ne^+ ions are used, and the backscattered parts of the ions are analyzed after their energy (Fig. 10.5).

Only ions that are scattered on the top atomic layer are not neutralized ($\sim 5\,\%$) and are thereby detectable. With penetration into deeper atomic layers less than 10^{-3} of the backscattered particles are ionized. Recombination is the main cause of the problem of quantification using ISS. At high energies this problem does not arise, since the number of ions is substantially larger. In addition, therefore, the particle used counters neutral particles and ions are countered equally.

The characteristics of both methods determined by the kinetics of the collision are treated together. Both are based on the same physical principles. An ion projectile with the mass M_1 has the energy E_0 before the collision and after the collision with a surface atom of the mass of M_2, the energy E_1, which can be calculated from energy conservation and momentum conservation

$$E_1 = E_0\, K\,(M_2/M_1, \theta)\,, \tag{10.1}$$

where the kinematic factor K is

$$K = \left[\frac{\cos\theta + \sqrt{\left(\frac{M_2}{M_1}\right)^2 - \sin^2\theta}}{1 + \frac{M_2}{M_1}}\right]^2 . \tag{10.2}$$

This means that at a given scattering angle Θ, defined through the analyzing system, the energy is determined after collision by the mass M_2 of the surface atom. The measured energy spectrum can be converted into a mass spectrum. Particularly for the scattering angles $\Theta = 90°$ and $\Theta = 180°$ we obtain

$$K(90°) = \frac{M_2 - M_1}{M_2 + M_1} \tag{10.3}$$

and

$$K(180°) = \left(\frac{M_2 - M_1}{M_2 + M_1}\right)^2 . \tag{10.4}$$

For the scattering angles $\Theta > 90°$ the ion projectile must have a lower mass than the surface atom.

From the analysis of the collision kinetics the resolution of mass $M_2/\Delta M_2$ arising in the energy spectrum also results. It is the larger, the closer the projectile atomic mass and the sample atomic mass are to each other. For $\Theta = 90°$, we have

$$M_2/\Delta M_2 = \frac{E_1}{\Delta E_1}\frac{2M_2/M_1}{(M_2/M_1)^2 - 1}. \quad (10.5)$$

By using an energy analyzer with constant relationship $E_1/\Delta E_1$ the resolution of energy can only be improved by a choice of large scattering angles and suitable ion masses. For analyzers with constant resolution ΔE one can increase the distance of appropriate maxima in the spectrum by increasing the primary energy.

The analyzer intensity from scattered ions in the solid angle $\Delta\Omega$ depends on the differential cross-section $d\sigma/d\Omega$. For Rutherford Backscattering it is accurately defined. In the laboratory system we obtain

$$\frac{d\sigma}{d\Omega} = \left(\frac{Z_1 Z_2 e^2}{2E_0 \sin^2\theta}\right)^2 \times \frac{\left[\left(1 - A^{-2} \times \sin^2\theta\right)^{1/2} + \cos\theta\right]^2}{\left(1 - A^{-2} \times \sin^2\theta\right)^{1/2}}, \quad (10.6)$$

where e is the elementary charge, Z_1, Z_2 is the nuclear charge numbers of projectile and sample atom, and $A = M_2/M_1$.

For the scattering intensity I from atomic species from the areal density N existing on the surface we obtain with a primary flux I_0:

$$I = N\frac{d\sigma}{d\Omega}\Delta\Omega I_0. \quad (10.7)$$

Such measurements of a surface distribution of the atomic values N gave an accuracy better than 10 %, depending on the sample conditions and can detect less than 10^{12} atoms/cm^2.

With RBS, for scattering from the volume, in addition to the elastic collision the inelastic loss of energy for each paths unit dE/dx must still be considered. Mainly, dE/dx is generated by interaction with electrons. For a certain atomic mass, one keeps from Eq. (10.1) an edge in the energy spectrum a continuous distribution to smaller energies. These reflect the distribution of the element concerned in the sample depth (Fig. 10.6). The resolution of depth is given by the collision slowing down power. Typical values dE/dx for metals are of an order of magnitude ~ 800 eV/nm. In the case of a typical resolution of an energy detector of $\Delta E = 6$ keV a depth resolution of approximately 7.5 nm results. The values for dE/dx are tabulated in [11].

An example of a film mixture analysis is shown in Fig. 10.6. It shows an RBS spectrum, which was taken up with 2.5 MeV He^{2+} ions at a film of a high temperature superconductor from $YBa_2Cu_3O_7$ on a substrate of $SrTiO_3$ [12]. The top of the figure shows the spectrum experimentally taken up (added noised curve) and a numeric fit. The continuous arrows indicate the edges that correspond to the scattering of the particular element at the surface, to Eq. (10.1), and the broken arrows correspond to the scattering of these elements of the substrate surface. The broken arrows contain the additional loss of energy of the ions by the passage through the film.

The lower part of the diagram shows the appropriate decomposition of the spectrum into the scattering contributions in the different elements of the film (Ba, Y, Cu). The stoichiometry of the sample can then be calculated from the respective signal level H_i of the component i.

One receives approach for conditions of the components i, k,

$$N_i/N_k = (H_i/H_k) \times (Z_k/Z_i)^2.$$

ISS is not so exactly calculable and Eq. (10.7) must be extended with the factor P, the so-called ions escape probability. The effective cross-sections are approximately higher around the factor 10^4, P is in the order of magnitude 10^{-2}, and for light atoms (special for oxygen) the detection limits also lie within the range of 10^{12}/cm^2;

$$I = N\frac{d\sigma}{d\Omega}\Delta\Omega\, P\, I_0. \quad (10.8)$$

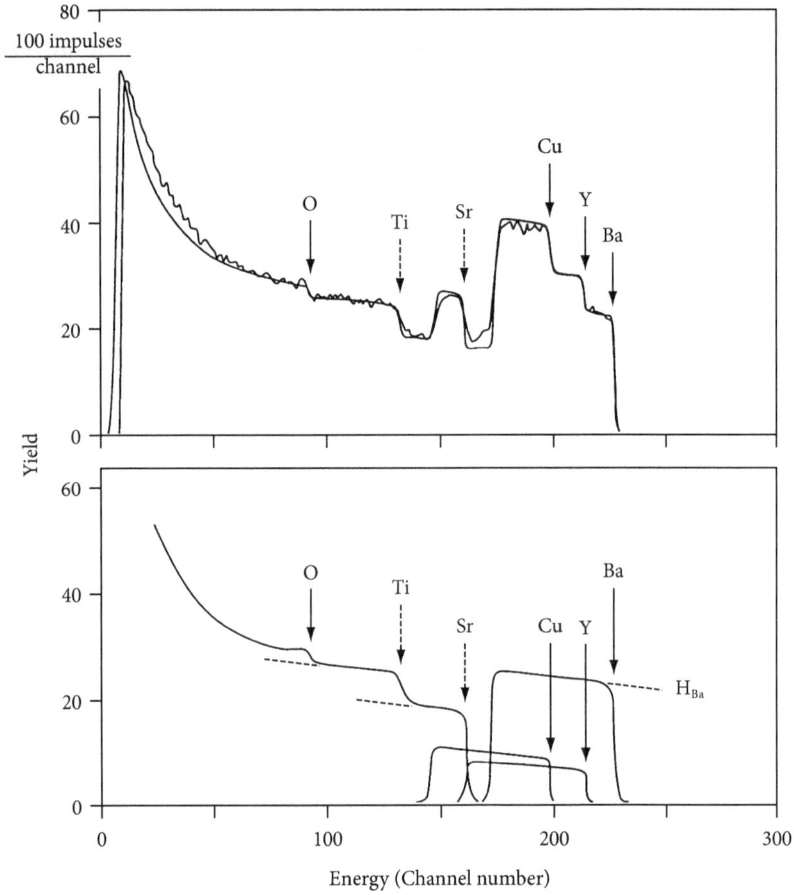

Fig. 10.6 RBS spectrum for the scattering of 2.5 MeV He^{2+} ions at a superconducting YBa_2Cu_3O layer. *Top*: experimental spectrum and numeric simulation (*smooth curve*). *Bottom*: decomposition into the scattering fractions of the different components. *Continuous arrows* refer to the surface of the layer, *broken arrows* to the boundary layer to the $SrTiO_3$ substrate

With the penetration into the sample volume, generally P degrades on values $P \leq 10^{-4}$, so that from the sample volume backscattered ions play practically no role. This is valid already for depths of more than two atomic layers.

10.2.2 Secondary Ion Mass Spectrometry (SIMS)

In secondary ion mass spectrometry (SIMS) the ions sputtered from the surface by a primary ion beam are monitored in a mass spectrometer. Typically, quadrupole mass spectrometers are used at resolutions of 1 amu. The imaging of surface inhomogeneities can be effected by projection images of the secondary ions or by raster scanning. Resolutions of the order of 1 μm have been reported in both cases. The primary beam is also used to remove surface layers for depth profiling, and depth resolutions to 50 Å have been reported. Figure 10.7 shows the schematic of a SIMS in which secondary ions are extracted from a sample by an electric field and analyzed according to their mass-to-charge ratio by a mass spectrometer.

For the analysis not only backscattered primary particles can be used, but also ion bombardment of the surface replaced secondary particles. When the sample of stripped particles is sputtered with an ion beam (or also by replacement with a laser beam), some of the particles leave the surface positive or negatively ionized (q-fold loaded). These secondary ions can be separated and detected directly in electromagnetic fields according to their specific mass M/q (Fig. 10.8).

Also time-of-flight methods are used, where the ion characteristic is used for the energy de-

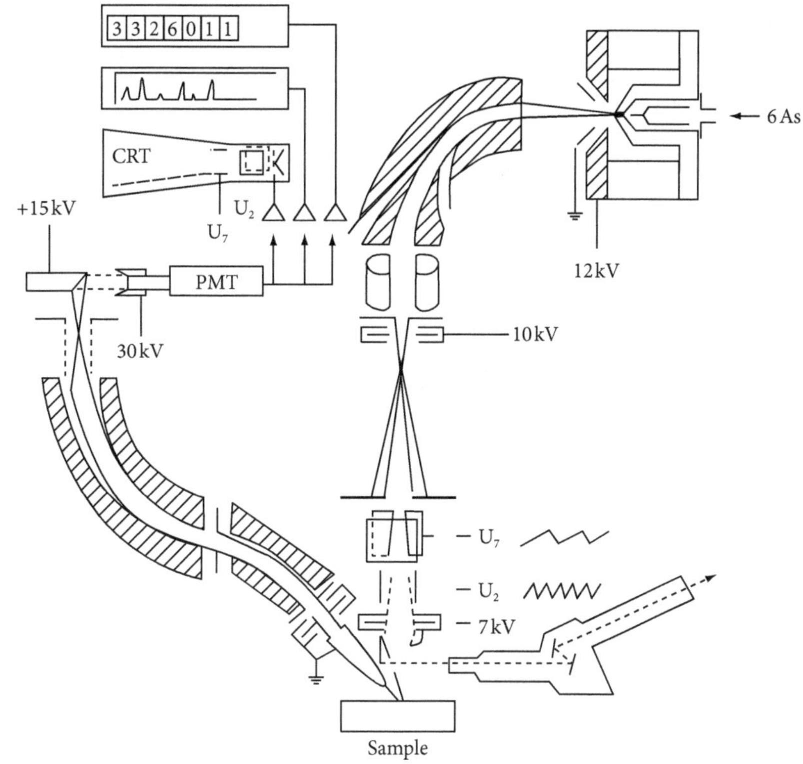

Fig. 10.7 Schematic diagram of a SIMS system

termination and for the proof. SIMS is based in principle on the sputtering of surface layers. In addition, SIMS can be used for the analysis of the surface composition. The abrasion of 10^{10} atoms is sufficient to receive an effectual signal; one speaks of static SIMS [13] (Fig. 10.9).

Figure 10.9 connects the ablated volume with the particle number n necessary for the detection (with the statistic error $\pm\sqrt{n}$), and the diameter of the ion beam used for the sputtering. γ is the ionization-factor and η the total transmission. Figure 10.9 shows the following correlations. In

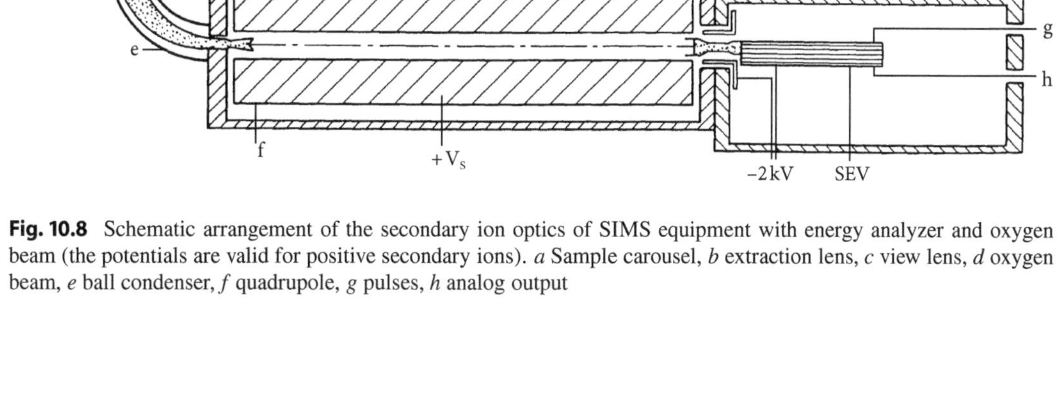

Fig. 10.8 Schematic arrangement of the secondary ion optics of SIMS equipment with energy analyzer and oxygen beam (the potentials are valid for positive secondary ions). *a* Sample carousel, *b* extraction lens, *c* view lens, *d* oxygen beam, *e* ball condenser, *f* quadrupole, *g* pulses, *h* analog output

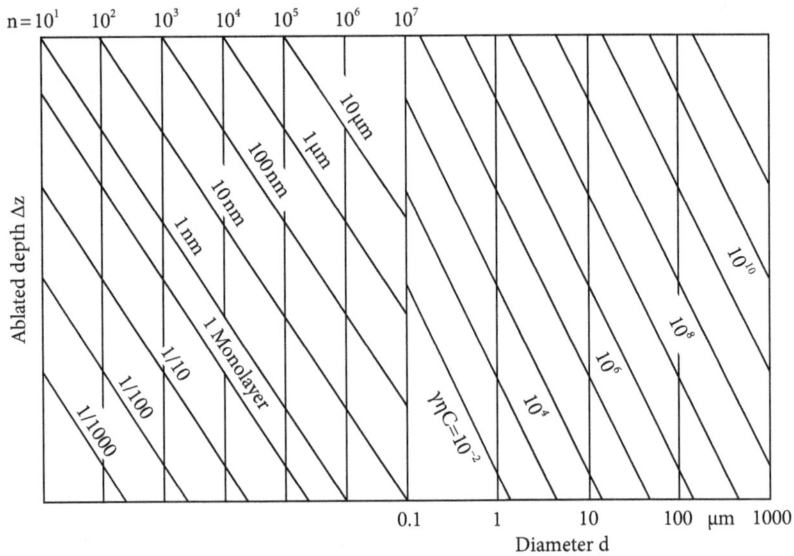

Fig. 10.9 Correlation between trace element analysis (atomic concentration C and density N) and microanalysis, γ is the ionization factor, η the total transmission, $V = \Delta z d 2\pi/4$ volume, $N = \eta\gamma CNV$ the number of proven ions

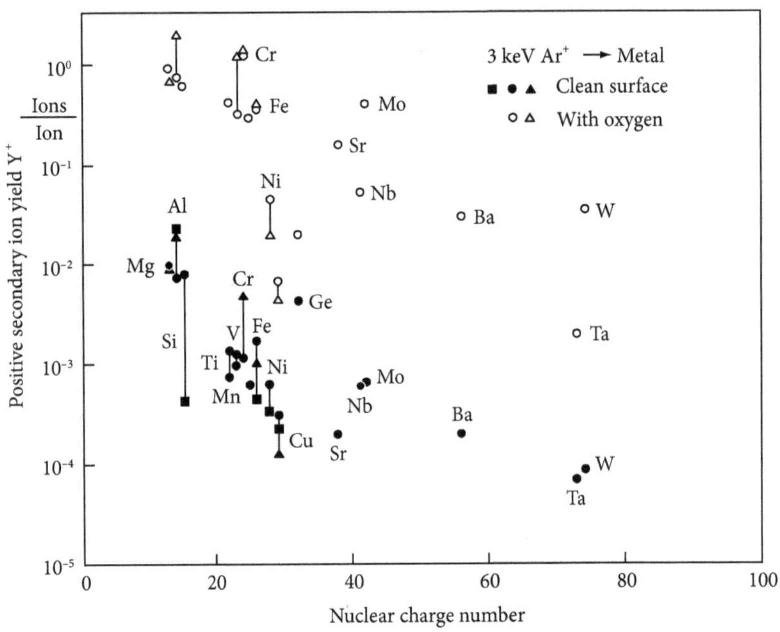

Fig. 10.10 Positive secondary ion yield of clean and oxygen-covered metal surfaces with bombardment by 3 keV Ar^+ ions

order to generate a picture with a lateral resolution of $d = 1\,\mu m$, and thereby detect a surface component, which has a concentration of 1 % ($C = 10^{-2}$), we must operate with an ionization yield of $\gamma = 0.1$ and 10 % transmission only a quantity from 1/100 atomic layers according to the ablated atomic number. One presupposes that in the analysis system with 10 ions can be already produced a sufficient bright picture element. A central problem with SIMS is the probability of ionization; more generally, the probability that a particle leaves the surface in the electronic initial state, in an excited state or in a positive or negative ionization state. The probability of the formation an ion can vary by orders of magnitude, dependant on the chemical status of the surface [14]. Oxygen on the surface increases the yield to positive ions (Fig. 10.10),

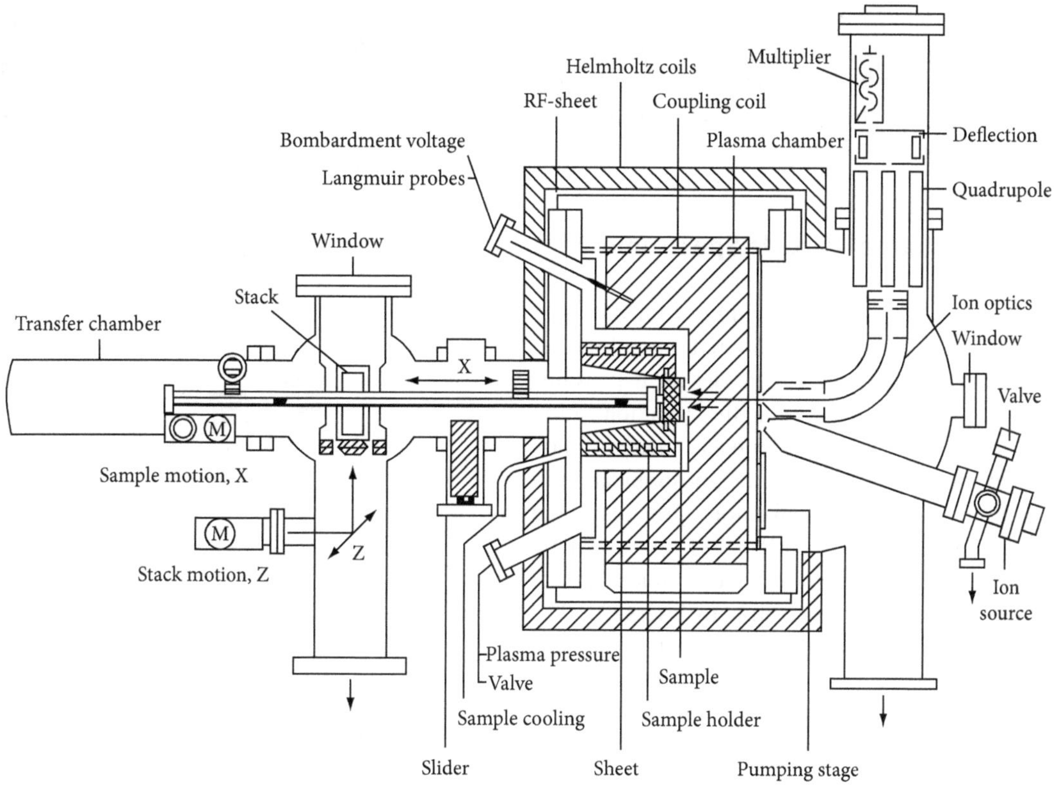

Fig. 10.11 Schema of SNMS equipment

and alkali atoms on the surface increase the yield at negative ions.

In order to achieve as constant and reproducible conditions as possible, it is favorable, for *positive SIMS*, to saturate the surface with oxygen [15]. For the generation of negative ions an alkali ion beam [16] is suitable. The uncertainty in the ion yield is the main problem with quantitative SIMS analyses.

The SIMS instrument can operate in three modes:

- An ion microprobe may be used, in which local analysis is performed by microfocusing of the primary ion beam.
- A true image may be formed by geometrical extraction of the secondary ions and aperturing within this image.
- Broad-beam instruments may be used, in which analysis over areas greater than 100 μm is performed instead of microanalysis.

SIMS has the following capabilities:

- All chemical elements can be observed, including hydrogen.
- Separate isotopes of an element can be measured, offering the possibility of determining self-diffusion coefficients, isotopic dating, and isotopic labeling.
- The area of analysis can be as small as 1 μm in diameter.
- The secondary ions are emitted close to the surface of the sample.
- The sensitivity of this technique exceeds that of AES or electron probe microanalysis in many but not all situations, with detection limits in the 10 ppm range under favorable conditions.
- The microanalytical part of this technique can be used to prepare ion images (ion microscope) or ion area scans (ion microprobe),

which can give pictorial information about the distribution of the species of interest.

- Material can be eroded away in a regular and carefully controlled fashion, and a profile can be prepared in depth.

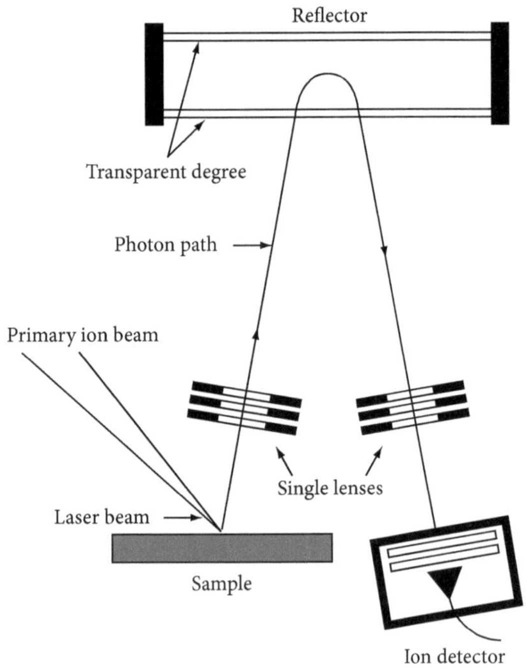

Fig. 10.12 Pattern of the SALI method for the surface analysis by means of a flying time mass spectrometer

10.2.3 Neutral Particle Spectrometry

In order to avoid the difficulties with the yield of ionized or excited sputtered particles, the postionization of these particles is a possibility. Methods for this are: the after ionization by electron collision (secondary neutral mass spectroscopy, SNMS) [17, 18], postionization in a plasma by Penning ionization (glow discharge mass spectroscopy, GDMS) [19], excitation by tuned laser radiation (laser induced fluorescence, LIF) [20], and photoionization by lasers [21, 22].

With the direct bombardment method the sample is in a plasma (Fig. 10.11). Between the RF-plasma and the sample a potential difference is applied for the bombardment (for the excitation of neutral particle emission).

This method can be used for conducting and semiconducting samples. The ablated area is relatively large (of some mm^2 up to some cm^2) and can take place at low energies with high lateral homogeneity.

With isolators instead of a DC voltage potential, a high RF-voltage with suitable frequency and form is used on the back of the sample. Electrostatic charges are thus avoided and in periodically returning time intervals ($f = 10$ kHz–1 MHz) the sample is bombarded with plasma ions of constant energy (high frequency method, HFM).

With photoionization three methods can be differentiated: resonant multi-photon ionization (REMPI), nonresonant multi-photon ionization (NRMPI), and single-photon ionization (SPI). For most atoms and usual lasers the ionization energy is larger than the energy of the photon, therefore a multi-photon process is necessary for the ionization.

The electrons in the first step are lifted either into a real (lifetime $> 10^{-9}$ s) or a virtual (lifetime $\sim 10^{-15}$ s) intermediate state. The typical power for a pulsed laser is at resonant excitation by $10^5\,\mathrm{W/cm^2}$ and by not resonant excitation $10^9\,\mathrm{W/cm^2}$. The non-resonant excitation has the advantage that one can ionize a broad spectrum of atomic species at the same time. With REMPI one can reach extreme high detection sensitivities under the ppb (parts per billion) range [23].

Figure 10.12 schematically shows the arrangement of a surface analysis by means of laser ionization (surface analysis by laser ionization, SALI).

The sample atoms sputtered by an ion beam are ionized by the laser beam and flying and through a reflecting flying time mass spectrometer with a secondary electron multiplier (microchannel plate). The GDMS arrangement is shown in Fig. 10.13 [19].

The sample surface is sputtered by ions from the plasma; the ions obtain energy in the cathode fall of approximately 200 eV. Neutral particles from the surface penetrate into the plasma and are ionized by Penning ionization of metastable

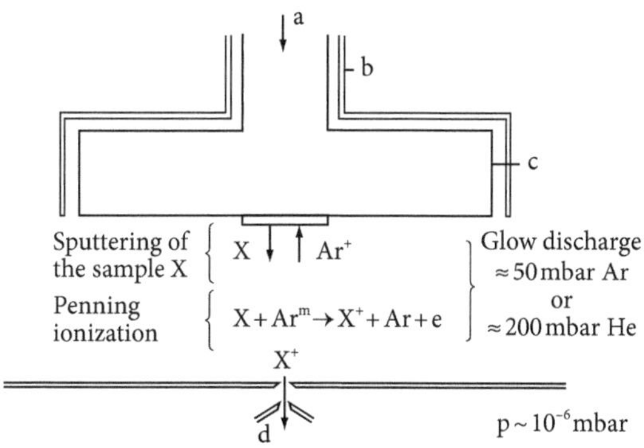

Fig. 10.13 Principle and schematic structure of the glow discharge spectrometry (GMDS). *a* Hf and cooling water, *b* grounded screen, *c* sample holder, *d* to the mass spectrometer

He or Ar atoms:

$$X + \mathrm{Ar}^* \rightarrow X^+ + \mathrm{Ar} + e^- .$$

The ionization probability varies around less than an order of magnitude over the periodic system; all elements except helium and neon can be proven. The analysis is independent of the electrical conductivity of the sample; trace element analysis up to less than 1 ppm is easily possible. A lateral resolution is not possible. The sample is ablated and destroyed by the analysis.

SNMS used a plasma with hot electrodes or an ion beam [24]. The method functions in principle with an ion beam or with ions from the plasma (Fig. 10.14). With SNMS extraordinarily sharp depth profiles can be obtained, because bombardment energies down to 10 eV are possible, and the plasma boundary at the sample surface can be adjusted. Sensitivity is comparable with by SIMS and quantitative analyses are possible.

Table 10.1 Absolute ion and photon yield for unsaturated Si and Ni surfaces

Species	Ions or photons/ sputtered atom
Si^+	1.0×10^{-2}
Si^{2+}	1.0×10^{-6}
Si I 2882 Å	1.6×10^{-3}
Si I 2516 Å	2.2×10^{-4}
Si II 3856 Å	1.4×10^{-5}
Si II 6347 Å	1.0×10^{-5}
Ni^+	6×10^{-4}
Ni I 3415 Å	1.0×10^{-4}
Ni I 2325 Å	3.1×10^{-7}

10.2.4 IPP (Ion-Induced Photo Production)

With bombardment of solids, also secondary particles in the electronic initial state (neutral particle and ions) and electronically excited secondary particles are observed [25]. This can be used also for surface analysis. This method is called SCANIIF (surface chemical analysis by neutral or ion-induced radiation) [26].

The instrumentation is comparable to SIMS. The mass spectrometer is replaced by equipment for optical spectrographic analysis (Fig. 10.15). The spectrographic analysis corresponds to chemical analysis by means of spectroscopy. Table 10.1 shows some estimations to sensitivity compared with SIMS [27].

Fig. 10.14 Schematic arrangement of SNMS. **a** Sputtering with external ion gun, **b** direct ion bombardment from the plasma. *a* Ion gun, *b* RF-plasma, *d* lens system, *e* quadrupole mass spectrometer, *f* Faraday cup, *g* SEM (secondary electron multiplier)

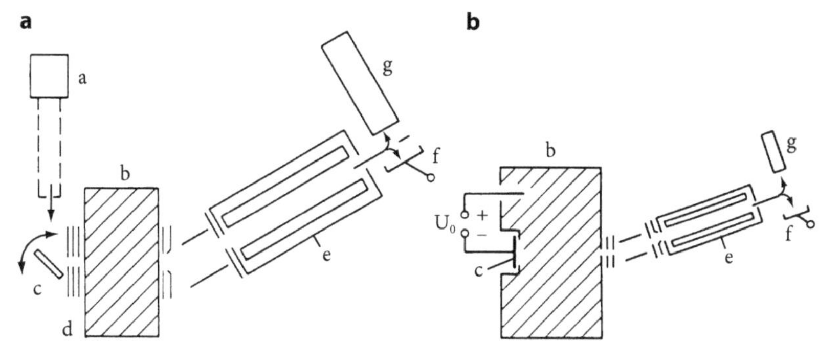

10.2.5 PIXE (Particle-Induced X-Ray Emission)

With PIXE similarly as with electron-induced X-ray spectroscopy (XES), the characteristic X-rays of the sample elements emitted from the surface by particle bombardment are measured. For the generation ions used (frequently H^+ or He^{2+}) in the energy range of some 100 keV to some MeV. The advantage of the ions is that the underground at bremsstrahlung in the X-ray spectrum is substantially smaller and thus the sensitivity can be increased. The radiating power of the bremsstrahlung is reversely proportionally to the square of the particle mass; around the factor 3×10^6 a proton thus creates less bremsstrahlung than an electron of the same energy.

Protons and He ion beams are preferred if RBS analyses are accomplished at the same time. For the use of PIXE the following is valid: the effective cross-section with proton-induced X-ray emission depend only a little on the atomic number Z_2 of the sample, which means the sensitivity varies only within an order of magnitude [28].

The basic effect of PIXE is, as in the case of AES (Auger electron spectroscopy)and PES (Photoemission spectroscopy), the ionization – prefers in the K-shell – by a high-energy particle with the ionization cross-section σ_1. Over the fluorescence yield w the effective cross-section σ_x for X-ray emission is connected with the ionization cross-section $\sigma_x = (1/w)\sigma_1$. w is the mean fluorescence yield of the accordingly more highly lying shell. The activation cross-section for the X-ray emission then results with the experimental photon yield

$$\sigma_x = \frac{1}{N}\frac{\mathrm{d}I}{\mathrm{d}E}\frac{\mathrm{d}E}{\mathrm{d}x} + \frac{1}{N}\varkappa I\,, \tag{10.9}$$

where

N is the number of the atoms/weight unit,

$\mathrm{d}I/\mathrm{d}E$ the experimental dependence of the intensity from the energy (keV) (photons/particle),

$\mathrm{d}E/\mathrm{d}x$ the stopping power (keV cm^2/g), and

$\varkappa$ the self-absorption coefficient of the measured X-ray in the sample [29].

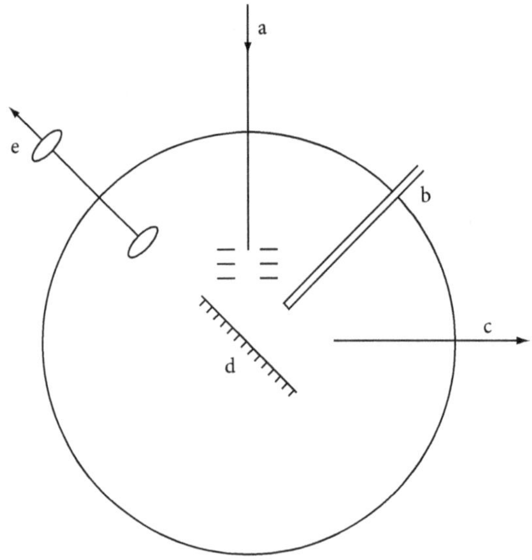

Fig. 10.15 Pattern of IPP equipment. *a* Ion gun, *b* CO beam, *c* to the ISS analyzer, *d* sample, *e* to the monochromator

The data in Fig. 10.15 [30] are average values about experimental data from 81 publications.

Q is the averaged normalized effective cross-section $Q = \sigma_1/\sigma_{\mathrm{theor}}$, where σ_1 are the averaged experimental K shell ionization cross-sections and σ_{theor} calculated values [31]. $\xi = 2\sqrt{\eta}/E_i$ with the velocity relationship $\sqrt{\eta} = v_1/v_{2\mathrm{k}}$, where $v_{2\mathrm{k}} = (Z_2 - 0.3)e^2/h$ is the velocity of the electron in the K-shell, and Z_2 is the atomic number of the bombarded element. E_i is the ionization energy of the element concerned in units of the hydrogen ionization energy, i.e., $(Z_2 - 0.3)^2$ Rydberg.

On the basis of the data given in Fig. 10.16 it is possible to indicate analytic formulas for the effective cross-section with good agreement with experimental values.

Since the energy range relations for ions in solids are well known [11], with PIXE quantitative analyses with high sensitivity can be performed. In practice, relatively simple formulas can be used for the analysis for thin layers. One determines the effective cross-section as the sensitivity factor P for the layer (impulses $\mathrm{cm}^3/\mu\mathrm{Cg}$) with a correction factor F for the

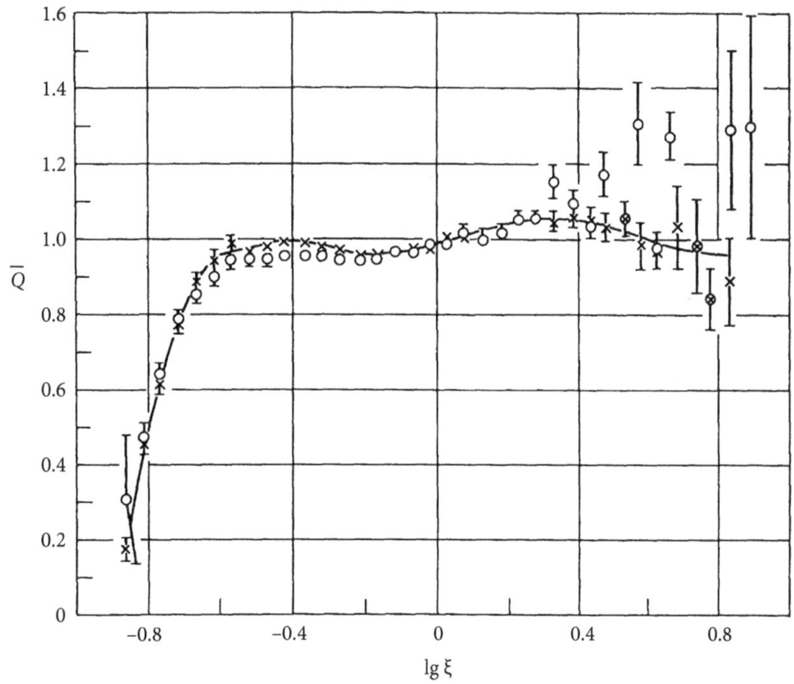

Fig. 10.16 Experimental average values $\overline{s}$ for the ionization effective cross-section of the K-shell for the elements of the ordinal numbers $4 \leq Z_2 \leq 92$ with proton bombardment as function of the scaled sample velocity ξ in the comparison to the theory (line = 1) and to an analytic approximation formula (*solid line*)

losses of energy of the protons and the X-ray absorption losses in the layer. Then the linear mass density of the element i is given by

$$\varrho_i \approx \frac{I_i}{\dot{N} P F}, \qquad (10.10)$$

where

I_i are the impulses in the X-ray peak of the element i, and

$\dot{N}$ in μC (cm^2 s) is the particle current density.

PIXE is non-destructive like RBS and can supplement RBS in the case of neighboring masses with higher resolution.

10.3 Electron-Spectroscopic Methods: Auger Electron Spectroscopy (AES) and X-Ray Photoelectron Spectroscopy (XPS)

Information carriers in the two most important thin film analysis methods are electrons (Auger and/or photoelectrons), which leave the solid body with an element-specific kinetic energy and are analyzed in an electron spectrometer. A schematic spectrum of electrons emitted from a thin layer under electron bombardment is shown in Fig. 10.17 [32].

Electrons of different origin are observed in the four energy ranges (I–IV) in Fig. 10.17. The relative intensity of the various contributions of the spectrum depends on the primary energy E_0 (5–50 keV), the angle of incidence, and the angle of emission. An actual elastic energy spectrum obtained under particular experimental conditions may, therefore, differ significantly in shape from that of Fig. 10.17. The narrow peak at the incident energy E_0 is made up of elastically scattered electrons plus diffracted electrons if the thin layer is crystalline. Energy losses in range I due to photon excitations (see insert (a) in Fig. 10.17) can be resolved only by the sophisticated energy analyzers.

In metals and semiconductors most of the energy lost in range II (due to electronic excitations and ionization losses) is transferred to the conduction or valence electrons through individual or collective excitations. In metals, the free electrons

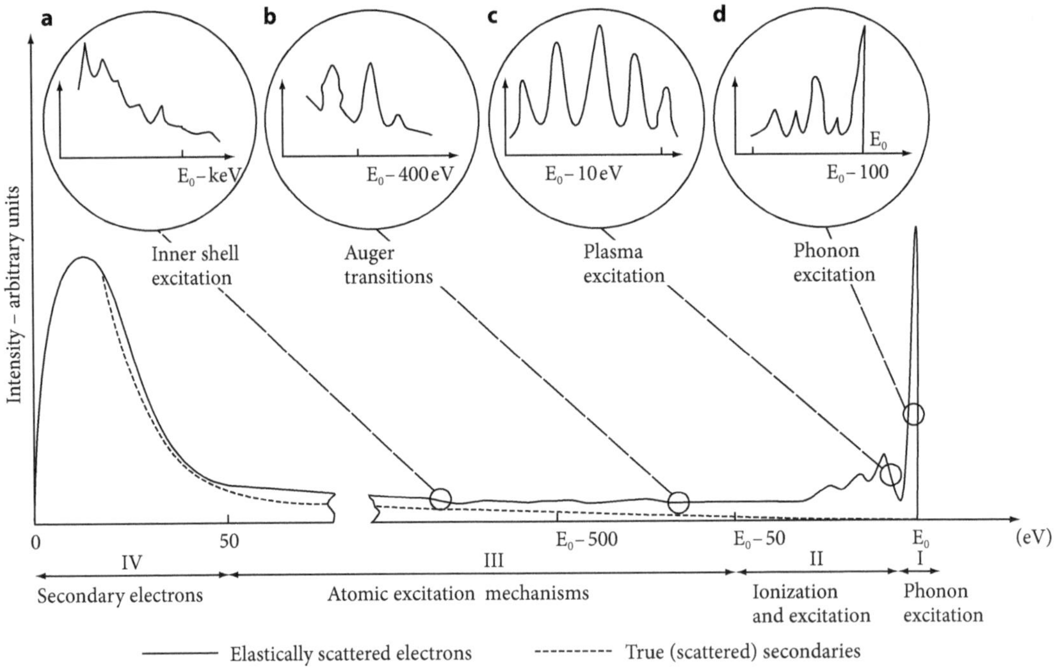

Fig. 10.17 Spectrum of inelastic electrons

can be excited in unison by the incident electron beam, resulting in a collective or plasma oscillation of the valence electrons. Typical energy losses for plasma excitation are around 10–20 eV (see insert (b) in Fig. 10.17). Above 50 eV energy losses (range III), the energy spectrum is shaped by various atomic excitation mechanisms.

The most striking is the appearance of sharp edges corresponding to the excitation of inner shell atomic electrons into the vacuum continuum (see insert (d) in Fig. 10.17). These edges correspond to the energy necessary to ionize the atoms, and the energy loss at which they occur is related to the atomic number. In these processes a core hole is generated by a primary or secondary electron and filled by another core or valence electron. In the Auger process an electron drops down to fill the hole and a second electron (Auger electron) is emitted (see insert (c) in Fig. 10.17). In the case of X-ray emission (X-ray fluorescence), instead of the second electron a photon is emitted. These processes (Auger, X-ray emission) are shown in Fig. 10.18.

Finally, range IV (0–50 eV) contains the true secondary electrons, which are the result of the cascade collision processes of the primary and target electrons. The secondary electron distribution (dashed line in Fig. 10.17) shows the most probable energy at 3–5 eV. One cannot find a clear upper energy limit for secondary electrons, but the use of 50 eV is the present convention.

For the energy analysis with AES and XPS hemisphere or cylinder mirror analyzers are usually used [33]. The mean inelastic free path of the electrons in the solid body is limited to a few atom layers and essentially depends on their kinetic energy. Figure 10.19 shows the mean electron emission depth of Auger and photoelectrons as a function of the kinetic energy [31]. With AES and XPS fractions of an individual atom layer can be analyzed. By the analysis, therefore, the sample surface may not be contaminated by rest gases. AES and XPS analyses must, therefore, be executed under ultra high vacuum conditions (10^{-9}–10^{-11} mbar).

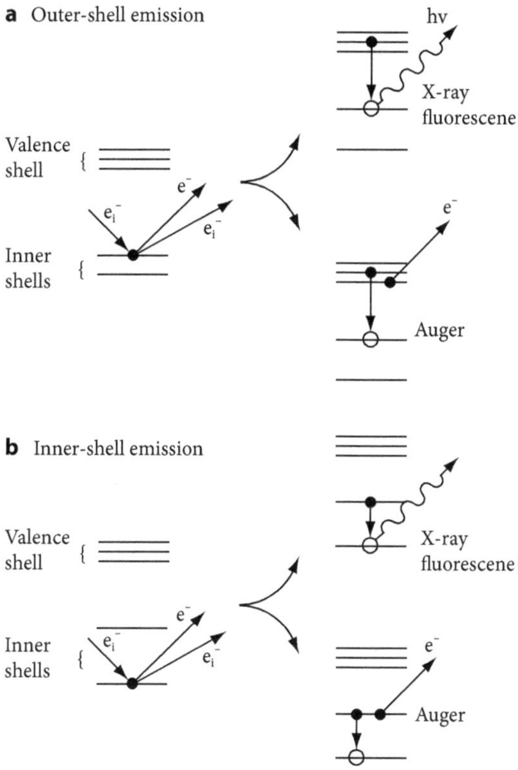

Fig. 10.18 Two-step Auger and X-ray fluorescence process. **a** Outer shell emission, **b** inner shell emission

10.3.1 Auger Electron Spectroscopy (AES)

With the Auger process by electron (energy 1–10 keV) excitation or photon excitation electrons are released from an internal shell of the sample atoms. After this process a hole remains in this shell, which can be filled by a transition from a higher shell. The energy difference between both shells can be emitted as radiation (characteristic X-ray) or to a further electron from a third shell, which is then emitted in the vacuum. It remains a doubly ionized atom. The kinetic energy of the emitted Auger electron is in first approximation

$$E_e = E_1 - E_2 - E_3 - \Phi_A .$$

At homogeneous samples the compound between the concentrations C_A of element A and the appropriate AES signal I_A is given by [35]:

$$C_A = \frac{\frac{I_A}{I_A^{\infty}}}{\sum_j F_{jA}^{A} \times \frac{I_j}{I_j^{\infty}}} ,$$

where I_j are the intensities of the elements j in the Auger spectrum of the examined sample and

Fig. 10.19 Middle emission depth of Auger and photoelectrons as a function of their kinetic energy

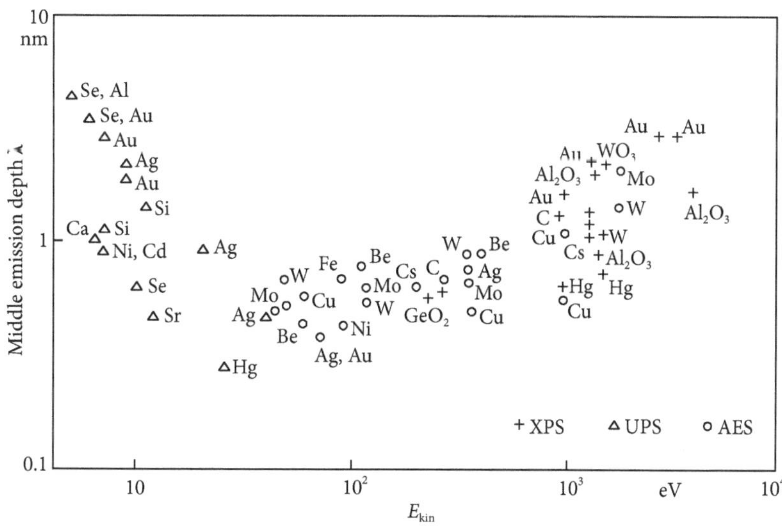

I_j^{∞} the intensity measured under same experimental conditions at an elementary standard. The Auger matrix correlation factor F_{jA}^{A} provides the energy and matrix dependence of backscattered primary electrons, the density of the target atoms, and in general, the matrix-dependant inelastic mean free path λ of the Auger electrons [36].

The line intensities of the differentiated spectrum can be used for the estimation of the surface concentrations (see Fig. 10.20) [37]. There the peak level in the not differentiated $n(E)$-spectrum of the number of exited atoms is proportional, if in the differentiated spectrum $\mathrm{d}N/\mathrm{d}E$, the range between the two peaks, which belong to a line, the concentration is proportional.

In the energy spectrum the Auger lines appear as weak structures on the underground of the remaining secondary electrons [37, 38]. In the energy spectrum it is also recognized that with the Auger electrons also different excitations can be connected. For example, plasmon excitations make a quantitative evaluation more difficult.

The surface sensitivity of AES depends on the emission depth of the electrons in the energy of the measured line. Generally, low energy lines from atoms near the sample surface are to be expected. This can be used for simple estimations of film thicknesses within the monatomic range. With adsorbate layers and with the investigation of isolators it should be considered that the primary electron beam can also cause dissociation and desorption. Figure 10.20b shows an Auger microanalysis of a Raney nickel sample in comparison with an X-ray microanalysis. The microanalysis shows that the copper doping is not evenly distributed over the catalyst surface.

Theoretically, the emitted Auger electron flux $I_{i,s}^{AES}$ can be described as follows:

$$I_{i,s}^{AES} = I_0 T d_{i,s} Q_{i,s} N_i r \,, \qquad (10.11)$$

where I_0 is the primary current, T the total detection sensitivity, $d_{i,s}$ the middle depth from which the electrons are emitted, $Q_{i,s}$ the cross-section for the ionization of the K-shell of the atomic species i, N_i the concentration (atoms cm^{-3}) of the atoms of the species i, and r the reflection factor. Eq. (10.11) is only approximately valid. An estimation of the activation cross-sections is sufficient with Eq. (10.11) to estimate relative concentrations at the substrate (constant r).

10.3.2 ESCA (Electron Spectroscopy for Chemical Analysis), X-Ray Photo Electron Spectroscopy (XPS)

A collimated beam of X-rays impinging on a surface is more penetrating than an electron beam of the same particle energy and can, therefore, provide more information concerning the constitution of the material at greater depths, although the depth is still limited by the range of the photoelectrons generated. X-rays also produce photoelectrons from the inner shells that carry information concerning the chemical bonding state of the excited atom. Thus, analysis of the energy distribution of the electrons emitted from a surface bombarded by X-rays is useful for subsurface chemical analysis. Instruments operating on these principles are known as ESCA (electron spectroscopy for chemical analysis) or XPS (X-ray photoelectron spectroscopy) instruments [39].

Depending upon the photon energy $h\nu$ one distinguishes between ultraviolet photo electron spectroscopy (UPS, uses UV-radiation, $h\nu$ within the range of 10–40 eV) and X-ray photo electron spectroscopy (XPS, uses X-rays, $h\nu = 10^2$–10^4 eV) [39].

According to Fig. 10.18, the kinetic energy of the photoelectrons is:

$$E_{\mathrm{kin}} = h\nu - E_{\mathrm{B}} - \Phi_{\mathrm{A}} \,,$$

where h is Plank's quantum of action, and $\Phi_{\mathrm{A}} =$ work the analyzer of emission.

At constant photon energy the kinetic energy of the emitted photoelectron depends only on the binding energy E_0 of one atomic level and can, therefore, be used for the identification of the element concerned directly.

XPS supplies information about the bonding status of atoms, since with changing distribution of the bonding electrons a shift of the bonding energy in internal electron shells can be observed of around several eV [40]. An XPS analysis system

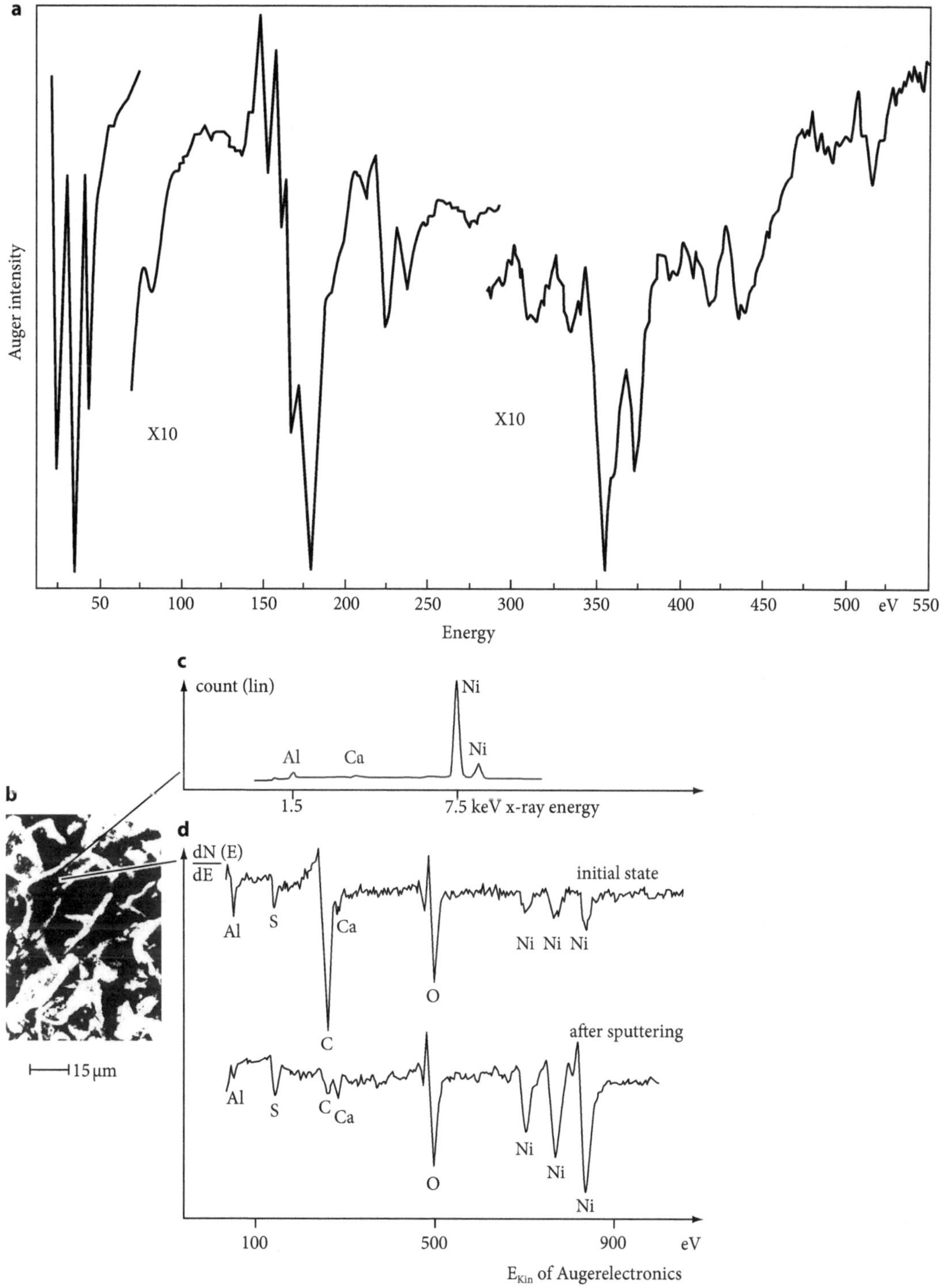

Fig. 10.20 X-ray microanalysis and Auger microanalysis of a Raney Ni sample [38]. **b** Scanning electron microscope picture, **c** X-ray microanalysis, **d** Auger microanalysis, uncleaned (*upper spectrum*) and cleaned by sputtering (*lower spectrum*)

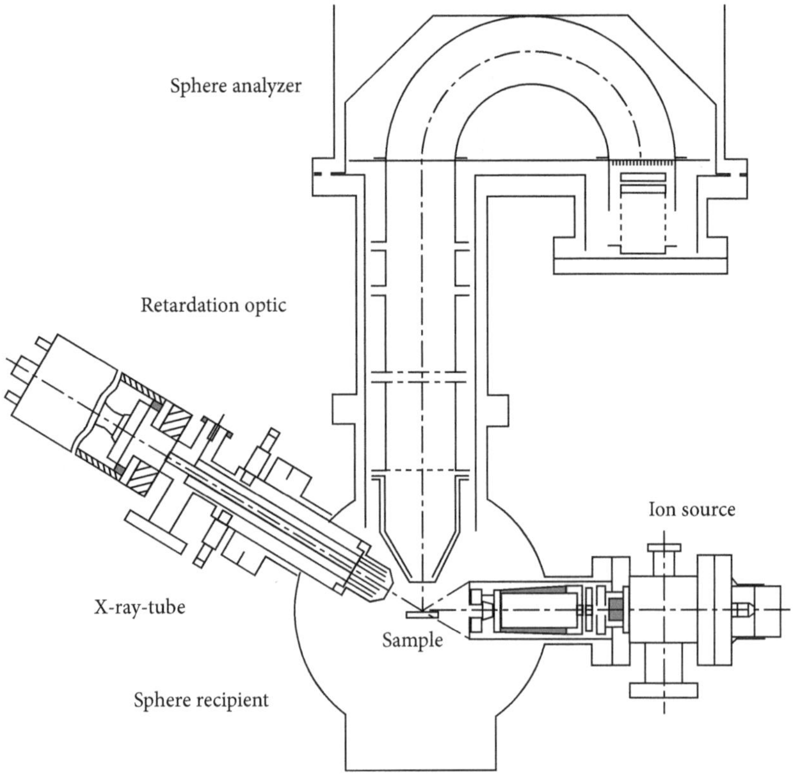

Fig. 10.21 Schematic representation of XPS equipment

consists of a radiation source and an energy analyzer with topped electron multiplier. For excitation (radiation source) frequently the Mg-$K_{\alpha 1.2}$ line with 1253.6 eV and a half width of 0.68 eV is used or the Al-$K_{\alpha 1.2}$ line 1486.6 eV with a half width of 0.83 eV. If the X-ray is by Bragg reflection at a monochromatized crystal, then the half width is reduced to approximately 0.1–0.2 eV; however, with reduction of the intensity [41].

In Fig. 10.21 the schema of XPS equipment with an X-ray tube, a sputtering gun for sample etching and/or for depth profile analysis, and the spherical sector field analyzer is represented [42, 43].

10.3.3 UPS (Ultraviolet Photoelectron Spectroscopy)

UPS is operated particularly with UV line radiation by He gas discharge lamps, i.e., with constant wavelength and/or photon energy. With synchrotron radiation it is possible to take up the entire energy sector from UPS to XPS continuously. Figure 10.22 schematically shows the differences of the diverse photoelectron spectroscopy methods.

Photon energy in the range between 10 to 100 eV is used substantially for the band structure analysis of solids [44]. Primarily with angle resolved measurements (ARUPS, angular resolved ultraviolet photoelectron spectroscopy) it is possible to determine both the occupied and the empty bands. Figure 10.23 shows the experimentally determined copper band structure [45] compared with theoretical results [46]. The theory is complicated and the calculation is complex. Exact theoretical models exist only for simple metals and semiconductors.

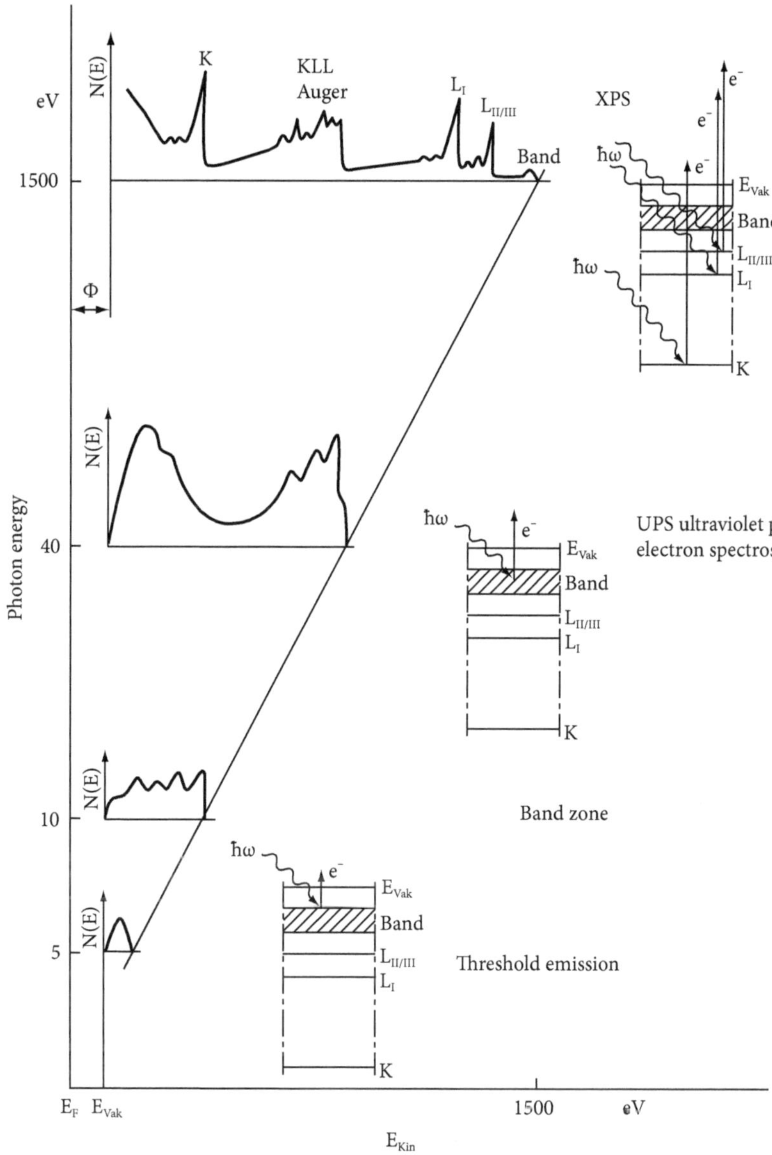

Fig. 10.22 Schematic representation of photoelectron spectra with (from *bottom* to *top*) increasing photon energy. Φ is the electron affinity of the spectrometer (strongly increasing, typically 4–5 eV); E_F is the Fermi energy and E_{Vak} is the vacuum energy

10.4 Inverse Photoemission (IPE)

A complementary method to photoelectron spectroscopy methods is inverse photoemission spectrometry (IPE), which in the last years has increased in importance for the spectroscopy of unoccupied electronic states at surfaces [47]. Monochromatic electrons of variable energy are bombarded to a sample and the UV-radiation with a constant energy $\hbar\omega$ that develops thereby is measured. Figure 10.24 shows the relations in the energy impulse (k), compared with photoemission spectroscopy.

The unoccupied electronic states between the Fermi level E_F and the vacuum level E_V are only

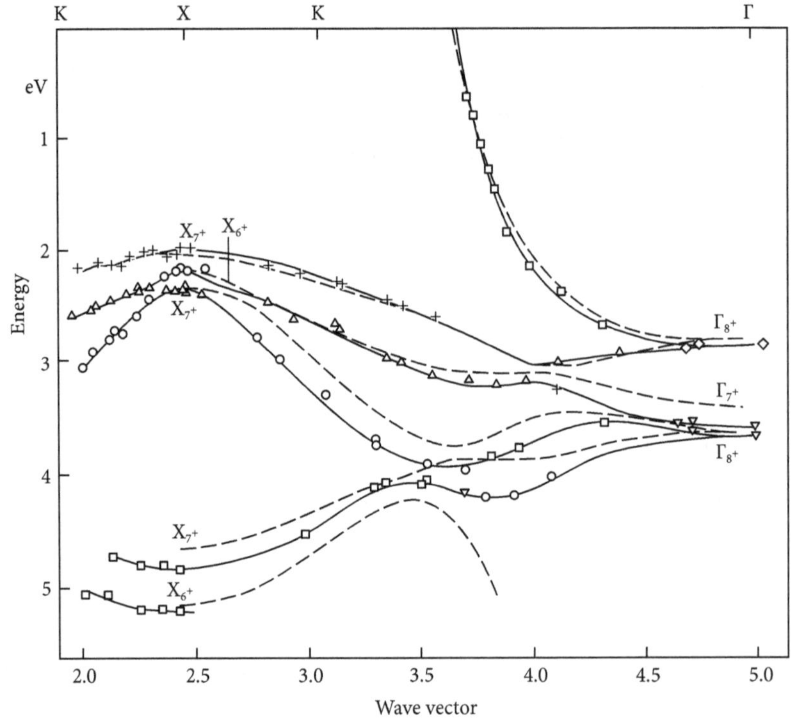

Fig. 10.23 Detail from a band structure identification, if electrons from valence levels are ionized or excited

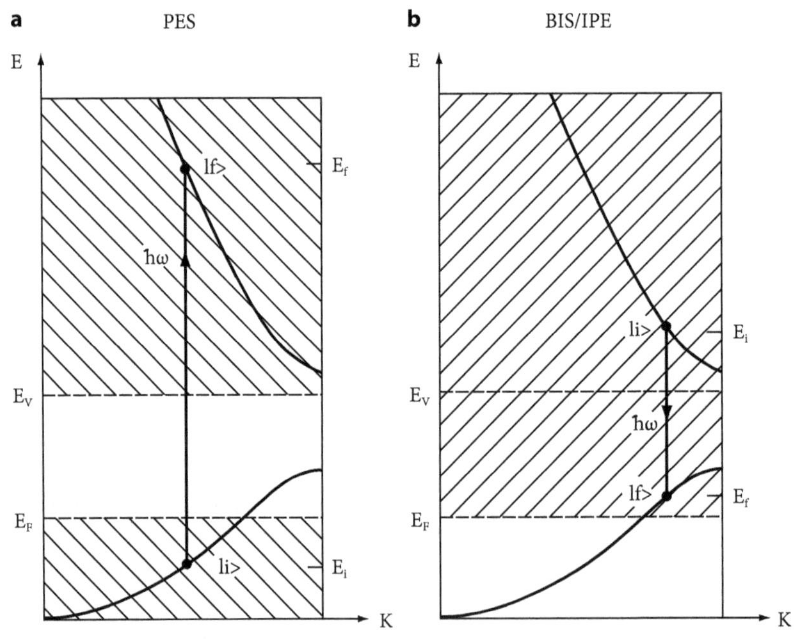

Fig. 10.24 Energy impulse diagram for **a** photoemission spectroscopy PES and **b** inverse photoemission IPE

available with the IPE method. The energy states above E_V are the final state $|f\rangle$ for PES and the initial states $|i\rangle$ for IPE. The IPE method permits us thus to measure the dispersion curves (energy vs. impulse) of the unoccupied states. This was successfully applied to volume states, surface states, and picture potential states. For thin Fe layers on Au (100), for example, also quantum well states were observed [48].

By use of polarized electrons there also exists the possibility to do a spectroscopy of the electronic states connected with the magnetism (e.g., spin fragmentation of the d bands at Ni) from surfaces and thin layers [49].

10.5 Electron Energy Loss Spectroscopy (EELS)

Contrary to the spectroscopy methods described above, by which an electron bound in the solid atom is spectroscopically analyzed, EELS (electron energy loss spectroscopy) is a method that is based on the scattering of electrons. For the scattering process energy and momentum conservation arc used:

$$E_1(k_1) = E_0(k_0) - \hbar\omega\,, \tag{10.12}$$

where E_1 (k_1) are the energy and the wave vector of the scattered electron, E_0 (k_0) for the primary electron, and $\hbar\omega$ the oscillation energy or the phonon energy.

For the momentum conservation the components are to be regarded parallel and perpendicularly to the surface, separately; for excitation for phonons, for example, only the parallel component is substantial:

$$\hbar k_1'' = \hbar\left(k_0'' - p'' \pm G''\right)\,, \tag{10.13}$$

where p is the phonon impulse and G a lattice factor.

Figure 10.25 shows the EELS spectra of C_2H_4 on platinum, where independently of the scattering angle different oscillations become excited. EELS show the local character of the available bonding very exactly.

10.6 Diffraction of Slow and Fast Electrons (LEED and RHEED)

The arrangements of the atoms in the surface can be detected by diffraction methods. Diffraction methods are analytic methods for the determination of the chemical composition of surfaces and thin films in such a way that they give information about the arrangement of surface atoms. A condition for this is that the samples which are to be analyzed possess sufficiently large areas with periodically arranged structures. We sustain the position of the diffraction maxima by the Ewald sphere construction (Fig. 10.26). In the case of perfect surfaces, points result in the diffraction pattern, which are frequently deformed to streaks through defects in the surface.

With RHEED the electrons are scattered over larger distances along the surface than with LEED. For RHEED lines are characteristically in the diffraction pattern (Kikuchi lines [50]), which are explained by the diffraction of inelastic scattered electrons at the bulk lattice. Defect sensitivity is also connected with the geometry of the experimental structure.

The position of the diffraction maxima is given by the Laue conditions

$$\begin{aligned} a_1\left(k' - k_0\right) &= 2\pi h\,,\\ a_2\left(k' - k_0\right) &= 2\pi k\,,\\ a_3\left(k' - k_0\right) &= 2\pi l\,, \end{aligned} \tag{10.14}$$

with the wave vectors k' and k_0 of the scattered and incident waves, the lattice constant a_1, a_2, a_3, and the integer numbers of $h, k, andl$. With RHEED the calculation of the intensities is somewhat more complicated than with LEED, because several atomic layers must also be included in the calculation [51].

RHEED has an experimental advantage compared to LEED since for the primary beam and scattered beam, because of the flat incidence and the small angle on dispersion, only a small solid angle in the equipment is coupled by an RHEED arrangement. It is possible to pursue the growth process of a film in situ. The RHEED intensity depends on the quality of the film at the surface. The diffraction intensity is max-

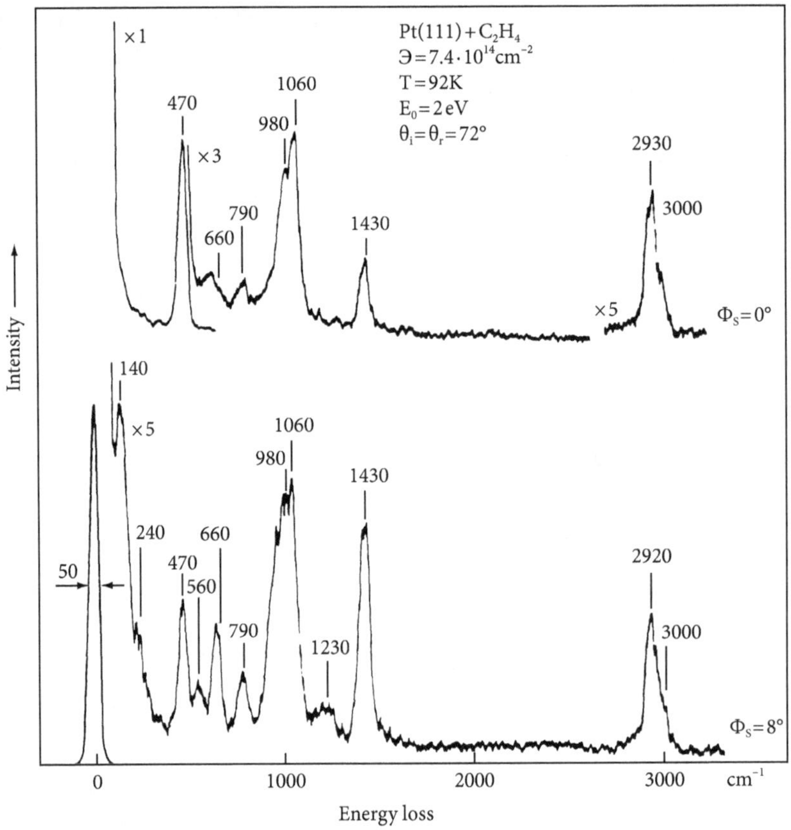

Fig. 10.25 EELS spectrum of a Pt (111) surface covered with C_2H_4 Pt (111) in mirror direction (*upper spectrum*), ($\theta_s = \theta_i - \theta_r = 0°$, $\theta_i = 72°$ and with $\theta_s = 80°$ (*lower spectrum*). The line identification is indicated with the lines. The coverage amounts to $7.4 \times 10^{14}\,cm^{-2}$)

imally perfect at the surface and goes through a minimum with maximum disorder, thus e.g., with an unordered coverage with a half monolayer. With more unordered multi-layer growth we achieve an exponentially damped oscillation of the RHEED intensity, as is represented in Fig. 10.27 for GaAs. RHEED permits us to observe the growth and the quality of epitaxial films in real time.

10.7 Image Methods

Image spectroscopy methods inform us about surface structures up to atomic dimensions and also permit dynamic procedures at surfaces with a time resolution from approximately 1 ms to seconds to observe.

In the photoemission microscope the local variations of the photo electron yield (due to a different electron affinity $e\Phi$) on the sample surface are used during photon irradiation.

In the SPM (scanning photoemission microscope) [52] a light beam ($h\nu \sim 6\,eV$) is focused on the sample and moved laterally with stepping motors. In the photo emission electron microscope (PEEM) the photoelectrons are imaged by an electrostatic lens system on a fluorescent screen (Fig. 10.28) [53].

With the LEEM (low energy electron microscope) a surface can be imaged both in the light of directly reflected electrons (local area) and in the light of a special LEED reflex.

Figure 10.29 shows an LEEM photo of Cu islands of the second monolayer on Mo (110), which recrystallizes spontaneously with this cov-

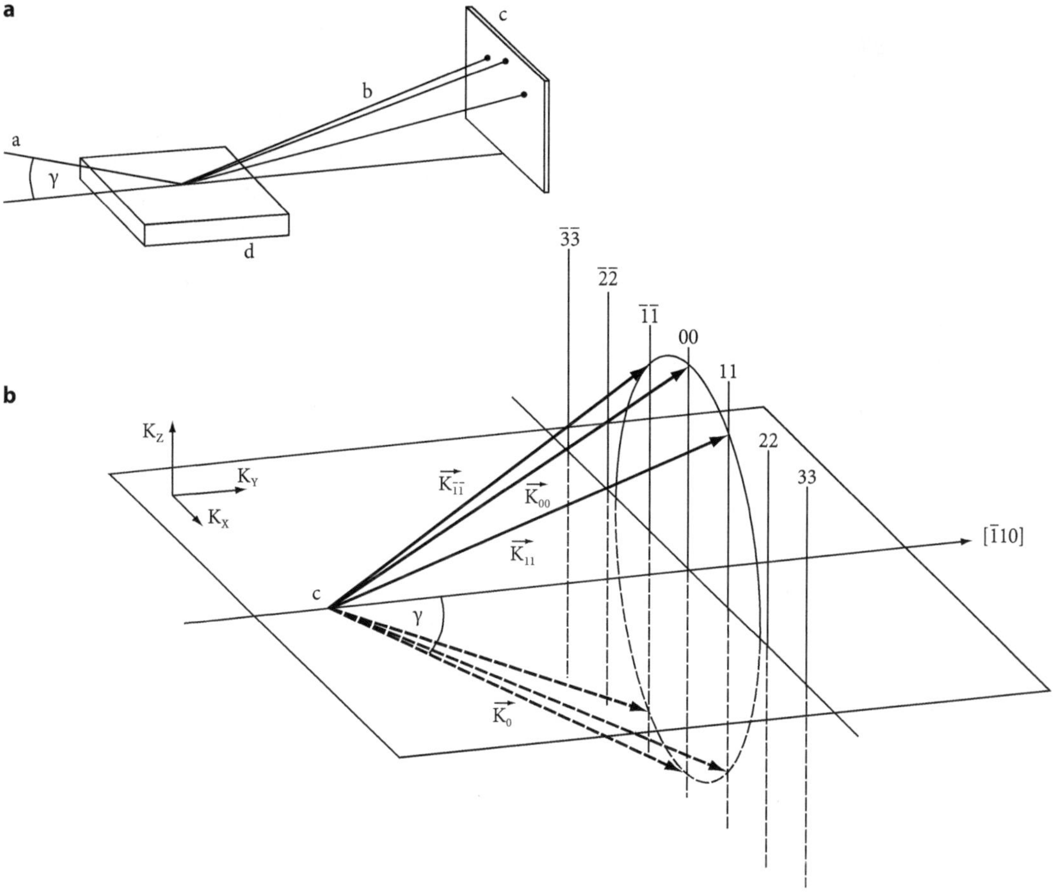

Fig. 10.26 **a** Pattern of the Ewald construction. **b** In the reciprocal area intensity is expected, where the Ewald sphere with the radius k_0 (wave vector of the incident electrons (19 keV)) reciprocal lattice rods of the zeros Laue zone cuts. That results in the reflexes $\overline{1}\,\overline{1}$, 00 and 11. The crystal is thereby in [$\overline{1}$10] – direction orients

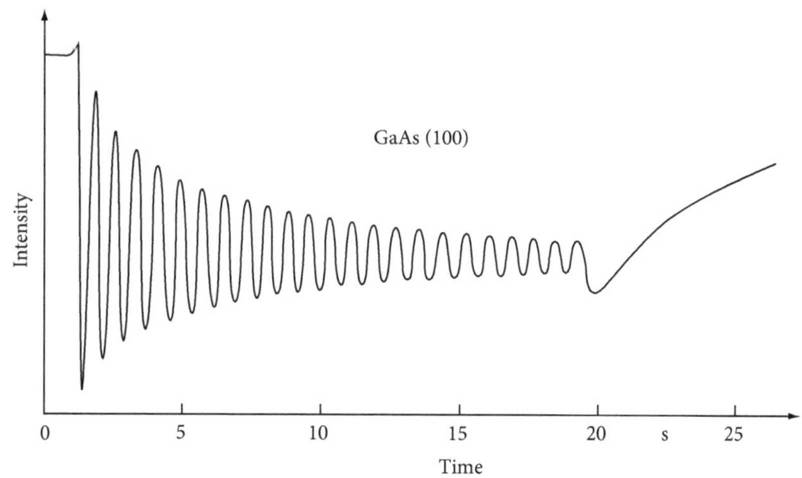

Fig. 10.27 RHEED oscillations by epitaxial growth of GaAs layers on GaAs (100). There are over 20 maxima observed, the attenuation refers to unordered multi-position growth

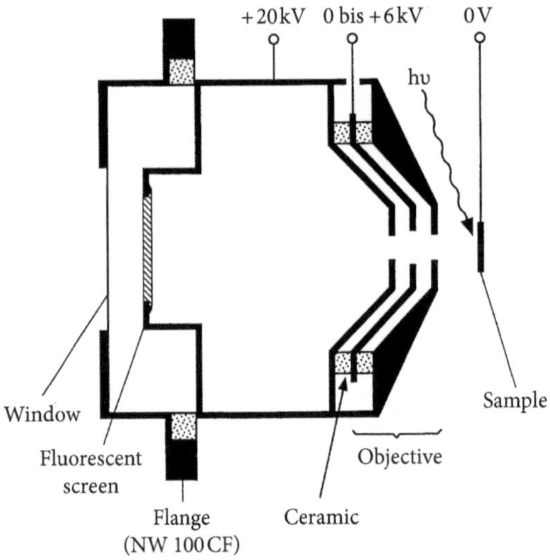

Fig. 10.28 Schematic cross-section by the photoemission electron microscope (PEEM)

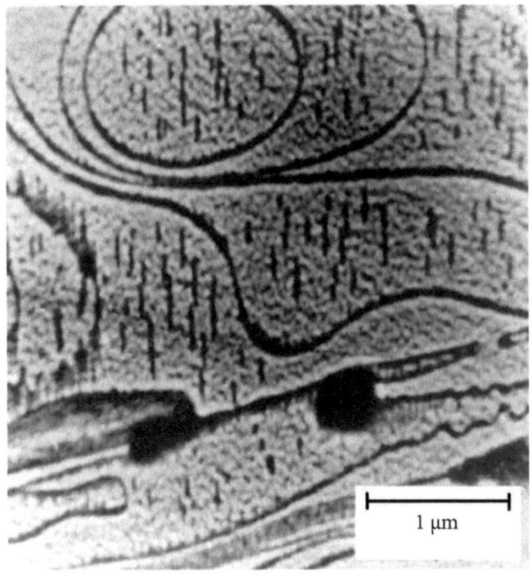

Fig. 10.29 LEEM photo of Cu islands of the second monolayer on Mo (110)

erage from the pseudomorphous first situation into a commensurable second situation [54]. A scheme of the arrangement used is shown in Fig. 10.30.

With the development of the scanning tunnel microscope [55] surface ranges can be successfully expanded with atomic resolution to imaging. A fine metal tip is brought so close to the sample surface (∼ 0.1 nm) that a tunneling current can flow. This current, which depends exponentially on the distance to tip sample, is usually kept constant during scanning by variation of the distance with a piezo holder. From the signal one receives a map of the topology of

Fig. 10.30 Scheme of the LEEM microscope

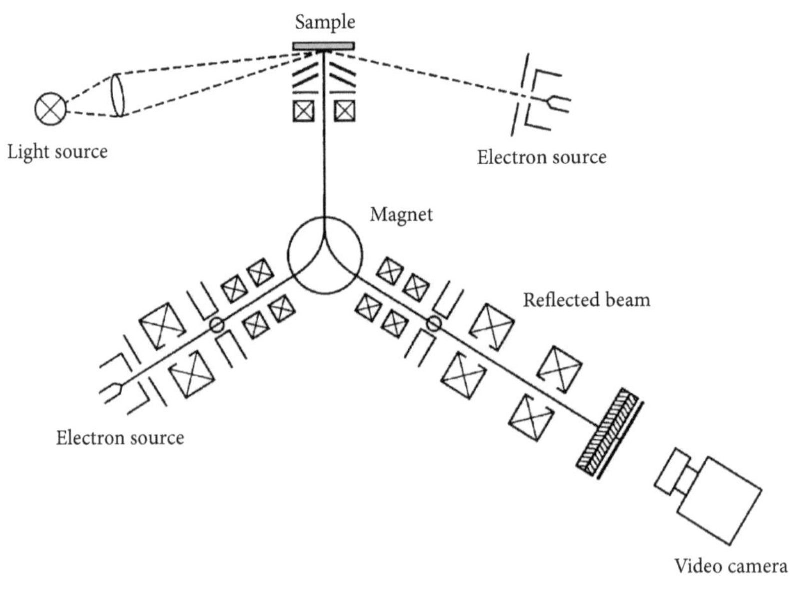

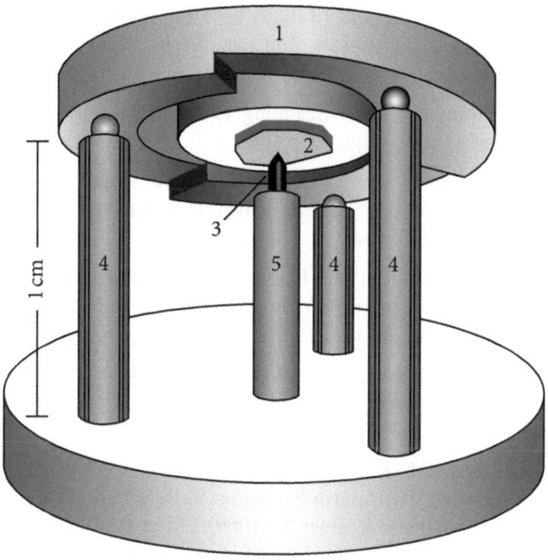

Fig. 10.30 Scheme raster tunnel microscope. *1* Sample holder (triple helix), *2* sample, *3* point, *4* Piezo tubes (leg), *5* Piezo tubes (feelers)

Fig. 10.31 Scanning tunnel microscopy image (36 nm × 50 nm)

the surface (more exactly, the electronic density of states).

According to their electronic properties, the observed corrugation of semiconductor surfaces (stronger localization) is higher than for metals by about a factor of 5. Figure 10.30 shows the structure of scanning tunnel microscope equipment [56] with the sample approach using a piezo controlled screw connection of the sample helix.

Figure 10.31 shows the image of a Cu surface with atomic resolution faceted by oxygen allocation.

References

1. Bauer E (1984) Z. Krist. 110 (1958), pp. 372, 395. Venables, J.A., G.D.T. Spiller & M. Hanbücken. Rep Progr Phys 47:399
2. Chu WK, Mayer JM, Nicolet MA (1978) Backscattering Spectrometry. Academic Press, New York
3. Feuerstein A, Grahmann H, Kalbitzer S, Oetzmann H (1976) In: Meyer O, Linker G, Kappeler F (eds) Ion Beam Surface Layer Analysis. Plenum Press, New York
4. Davis JA (1978) In: Thomas JF, Cachard A (eds) Material Characterization Using Ion Beams. Plenum Press, London
5. Taglauer E (1982) Appl Surf Sci 13:80
6. Coburn JW, Kay E (1974) Crit Rev Solid State Sci 4:63
7. Oechsner H (1995) Secondary Neutral Mass Spectrometry (SNMS) - Recent Methodical Progress and Applications to Fundamental Studies in Particle/Surface Interaction. Int J Mass Spectrom Ion Processes 143:271–282
8. Lindhard J, Scharff M, Schiøtt HE (1963) Kgl Dan Vid Selsk Mat Fys Med 33(14):1
9. Smith DP (1991) J. Appl. Phys. 38 (1967), p. 340. Niehus, H. W. Heiland & E. Taglauer: Surf. SCi. Rep. 17 (1993), p. 213, Taglauer. E. In: Ion Spectroscopies for Surface Analysis, Ed.: A. W. Czanderna & D.M. Hercules; Plenum Press, New York, p 363
10. Feldman LC (1978) In: Datz S (ed) Applied Atomic Collision Physics, vol 4. Academic Press, New York, p 261
11. Ziegler JF (1977) Helium Stopping Power and Ranges in All Elemental Matter vol. 4. Pergamon, New York
12. Berberich P, Dietsche W, Kinder H, Tate J, Thomson C, Scherzer B (1988) Proc. Int. Conf. High T Supercond. Mater Interlaken, CH.
13. Benninghoven A (1973) SurfSci 35:427
14. Wittmaak K (1977) In: Tolk NH, Tully JC, Heiland W, White CW (eds) Inelastic Particle Ion-Surface Collisions. Academic Press, New York, p 153
15. Magee CW, Harrington WL, Honig RF (1978) Rev Sci Instr 49:477
16. Liebl H, Harrison W (1976) Int J Mass Spectrom, Ion Phys 22:237
17. Kato S (ed) (1993) Contributions Third Int. Workshop on Postionization Technique in Surface Analysis Lake Kawaguchi, Japan.
18. Schoof, H: Untersuchung an anodischen Tantal- und Nioboxidschichten mit AES und SNMS, PhD , Dissertation, TU Clausthal 1981
19. Coburn JW, Taglauer E, Kav E (1974) J Appl Phys 45:1779

20. Schweer B, Bay HL (1982) Appl Phys A 29:53
21. Becker CH (1991) In: Czanderna AW, Hercules DM (eds) Ion Sopectroscopies for Surface Analysis. Plenum Press, New York, p 273
22. Wulff M, Wucher A (1998) In: Gillen G, Lareau R, Bennet J, Stevie F (eds) Quantitation of Single Photon Ionization Laser-SNMS, Secondary Ion Mass Spectrometry SIMS XI. Whiley, New York, pp 665–666
23. Pellin M, Young CE, Calaway WF, Gruen DM (1984) SurfSci 144:619
24. Hamagaki M, Kato S, Hara T, Hayashi S (1990) Vacuum 41:1730
25. Goldstein E (1889) Monatsberichte der Köngl Preuß Akademie der Wissenschaften zu Berlin 33:108
26. White CW, Simms DL, Tolk NH (1972) Science 117:481
27. Tsong IST (1981) In: Taglauer E, Heiland W (eds) Inelastic Particle-Surface Collisions. Springer Series in Chemical Physics, vol 17., p 258
28. Pierce TB (1974) In: Kane PF, Larabee GB (eds) Characterization of Solid Surfaces. Plenum Press, New York, p 419
29. Merzbacher E, Lewis HW (1958) In: Flügge S (ed) Handbuch d. Physik, vol 34. Springer-Verl., Berlin
30. Paul H (1984) Nucl Inst Meth B 3:5
31. Brundle CR (1974) The Application of Electron Spectroscopy to Surface Studies. J Vac Sci Technol 11:212–224
32. Hedberg CL (ed) (1995) Handbook of Auger Electron Spectroscopy, 3rd edn. Physical Electronics, Eden Prairie
33. Holm H, Storp S (1980) Methoden zur Untersuchung von Oberflächen. In: Ullmans Enzyklopädie der technischen Chemie, vol 5., pp 519–576
34. Brundle CR (1974) The Application of Electron Spectroscopy to Surface Studies. J Vac Sci Technol 11:212–224
35. Seah MP (1983) A Review of Quantitative Auger Electron Spectroscopy. Scanning Electron Microc II:521–536
36. Seah MP, Dench W (1979) Quantitative Electron Spectroscopy for Surfaces: A Standard Data Base for Electron Inelastic Mean Free Path in Solids. Surf Interf Anal 1(1):2–11
37. Ertl G, Küppers J (1985) Low Energy Electrons and Surface Chemistry, 2nd edn. Verlag Chemie, Weinheim
38. Holm R (1982) Scanning Electron Microscopy III. SEM Inc., Chicago, p 1043
39. Palmberg PW (1976) Quantitative Auger Electron Spectroscopy Using Elemental Sensivity Factors. J Vac Sci Technol 13:214–318
40. Siegbahn K (1970) Electron Spectroscopy for Chemical Analysis (ESCA) Instrumentation. Phil Trans Roy Soc Land A 268:33–57
41. Felner-Feldegg H et al (1974) New Developments in ESCA-Instrumentation. J Electron Spectrosc Rel Phenom 5:643–689
42. Siegbahn K, Nordling C, Johanson G, Hedman J, Heden PF, Hamrin K, Gelius V, Bergmark T, Werme LO, Manne R, Baer Y (1969) ESCA Applied to free molecules. North-Holland, Amsterdam
43. Cardona M, Ley L (1978) Photoemission in Solids vol. 1. Springer-Verl., Berlin
44. Spicer WE (1958) Phys Rev 112:225
45. Baalmann A (1984) Dissertation, Universität Osnabrück
46. Eckhardt H, Fritsche H, Noffke J (1983) J Phys F 14:97
47. Dose V (1989) Rendiconti S.I.F. – CVIII 108
48. Himpsel FJ (1991) Phys Rev B 44:5966
49. Donath M, Dose V, Ertl K, Kolac U (1990) Phys Rev B 41:5509
50. Kunz C (ed) (1979) Synchrotron Radiation, Techniques and Applications Topics in Current Physics. Springer, Berlin
51. Marten H, Meyer-Ehmsen G (1985) Surf Sci 151:570
52. Rotermund HH, Ertl G, Sesselmnn W (1989) Surf Sci 217:L 383
53. Rotermund HH, Engel W, Kordesch ME, Ertl G (1990) Nature 343:355
54. Bauer E, Mundschau M, Swiech W, Telieps W (1991) Vacuum 41:5
55. Binning G, Rohre H, Gerber C, Weibel E (1983) Phys Rev Lett 50:120
56. Besocke Delta Phi GmbH

11 Insitu Measurements

H. Frey

11.1 Determination of Film Thickness by Resistance Measurement

A contactless method to determine the film thickness is the eddy current method, which can be used also for in-situ analysis [1].

An alternating primary magnetic field induces electric currents, so-called eddy currents in an electrical conductor. Their magnetic field is converse to the primary field. Its intensity depends, among other things, on the electrical conductivity of the material, in which the eddy currents are induced. Over the connection of the electrical conductivity of the film with the attenuation of the primary magnetic field the film thickness can be determined.

Figure 11.1 shows the principle configuration of an eddy current system for film thickness measurement.

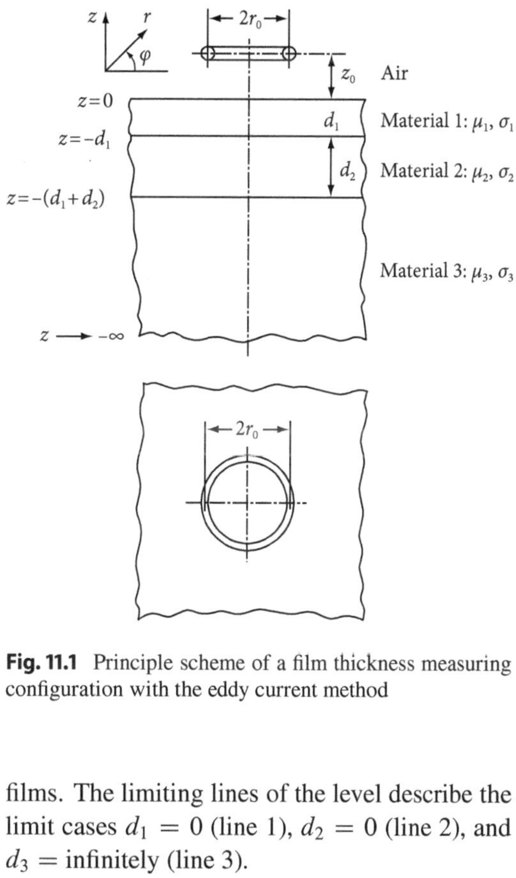

Fig. 11.1 Principle scheme of a film thickness measuring configuration with the eddy current method

The primary magnetic field is generated with a coil, which is arranged in the distance z_0 over the test sample. A high frequency current with frequencies between 100 kHz and some MHz is circulating in the coil.

The film and the substrate materials are characterizing through the film thickness d, the electrical conductivity σ, and the magnetic permeability μ [2].

Measuring instruments use the change of the impedance of the exciting coil by the eddy currents induced in the sample [3]. Figure 11.2 shows the impedance level for the example of Ni/Cu on Fe as function of the thicknesses of both films. The limiting lines of the level describe the limit cases $d_1 = 0$ (line 1), $d_2 = 0$ (line 2), and $d_3 =$ infinitely (line 3).

Figure 11.2 was taken up at a frequency by 1 MHz. Variation of the measuring frequency changes position and shape of the impedance level, so that the measuring accuracy of the method for film thickness ranges can be influenced above. In principle is valid: Higher measuring frequencies, better measuring accuracy.

© Springer-Verlag Berlin Heidelberg 2015

H. Frey, H. R. Khan, *Handbook of Thin-Film Technology*, DOI 10.1007/978-3-642-05430-3_11

11.2 Rate Measurement by Particle Ionization and Excitation

11.2.1 Introduction

For the measurement of the coating rate by ionization or excitation of the coating particles different methods were developed. By one method the coating particles are ionized with low energy electrons, then the ions passing through a measuring cell where the ions integral or specifically measured [4]. To it belong the ionization gauge, the periodic beam deflection over a measuring window and the mass spectrometer [5]. Instead of electrons also ions (ion probe) can be used.

Another method uses the emission radiation of excited coating particles induced by electron collision (EIES, electron impact emission spectroscopy) [6].

Commercially the method with ion probe, mass spectrometer, and EIES is used.

In Table 11.1 is listed some data for the comparison of these methods.

11.2.1.1 Ion Probes

For rate regulation, ion probes can be used in all cases of high vapor ionization [8]. The ionized fraction of the vapor can be exhausted with a biased probe and used as an analog rate signal.

By electron beam evaporators the possibility that the vapor ions can be turned back by the magnetic field in the direction to the cathode can be used as an ion probe (Fig. 11.3).

If one places a probe sheet near the cathode an ion current can be collected with a suction voltage from −50 to about 100 V, which is proportional over a wide range to the rate of deposition. By integral measurement also residual gas ions are measured. For reproducibility a defined surface potential of the probe sheet is important. The method becomes problematic during the coating of dielectrics or reactive coating processes. The probe must shield against an external charge carrier, for example, created from a second electron beam evaporator. The ion probe is suitable in particular for exact regulation of high coating rates of metals in a non reactive atmosphere. The advantage is the fast response time and the small technical expenditure in comparison to other methods [9].

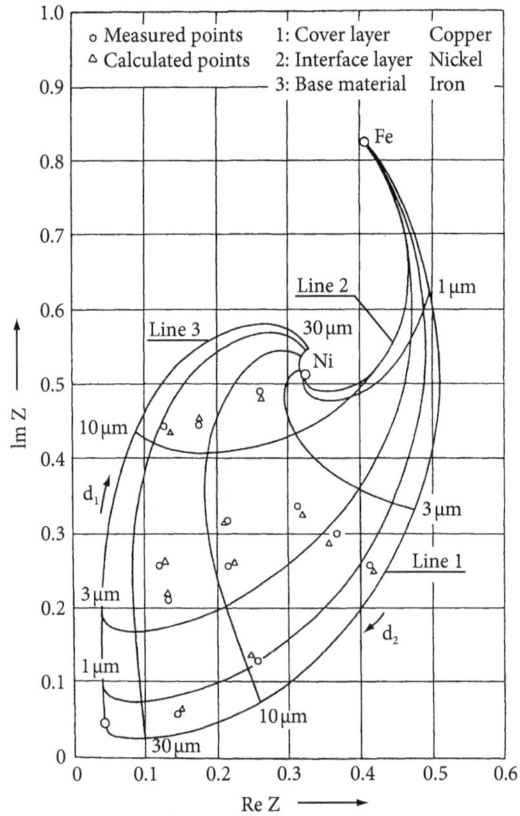

Fig. 11.2 Impedance diagram of an eddy current probe at the system Ni/Cu on Fe. Operating frequency: 1 MHz

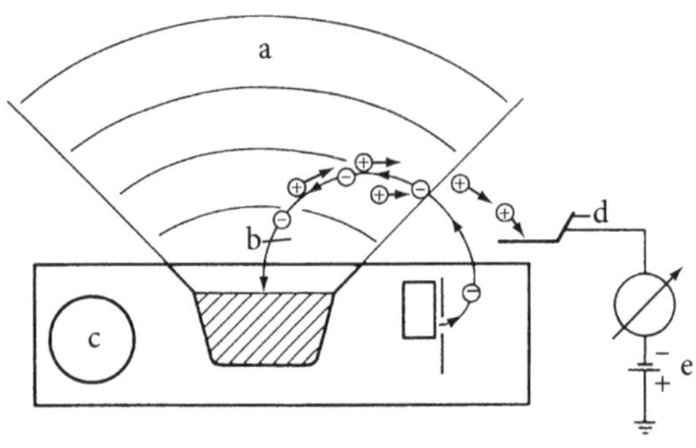

Fig. 11.3 Functional principle of an ion probe: integrated into an electron beam evaporator. *a* Vapor, *b* ionized vapor, *c* solenoid coil, *d* probe disk, *e* probe voltage

Table 11.1 Some comparative data for the use of ion probe, mass spectrometer, and EIES

Measurement principle	Ion probe	Mass spectrometer	EIES
Vapor source	ES evaporator	No restriction	No restriction
No pressure dependence	Below 10^{-5} mbar	Below 10^{-5} mbar	Below 10^{-5} mbar
Evaporate from alloys with 1 source	Integral measurement	Selective measurement	Selective measurement
Measuring accuracy, rate	±3 %	±5 %	±2 %
Rate zone	0.1–100 nm s^{-1}	10^{-3}–10 nm s^{-1}	0.05–1000 nm^{-1}
Application	High rates metals	Low rates in ultra high vacuum	Evaporation of alloys reactive method
Difficulty	Reproducibility	e/v overlap by residual gas	Poor photon emitters
	Probe surface	Disturbance through ES evaporator	
	Dielectrics	Isotopes	
Advantage	Simply	Sensitive	Selectively long term stability universal

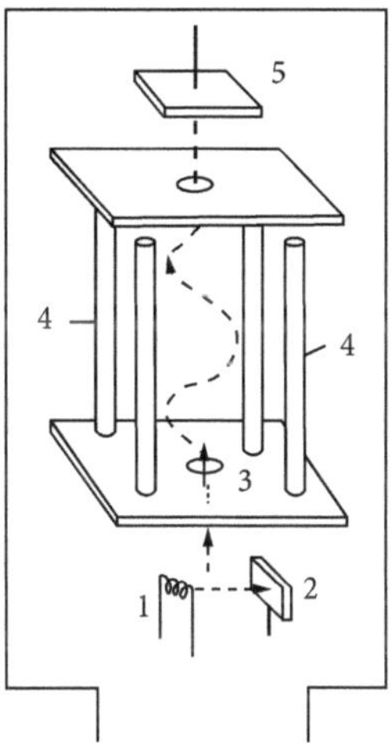

Fig. 11.4 Principle of a quadrupole mass spectrometer. *1* Filament, *2* electron collector, *3* entrance aperture, *4* quadrupole rods, ion collector

11.2.1.2 Mass Spectrometer

11.2.1.2.1 Quadrupole Mass Spectrometer

A mass spectrometer separates ions according to their charge/mass relationship. For rate control the quadrupole mass spectrometer [10], developed for the rest gas analysis, is usually used. The quadrupole mass spectrometer does not require a magnetic field, and in consequence, is much less bulky than magnetic types. The instrument (Fig. 11.4,) consists of four cylindrical rods to which a combination of DC and AC potentials are applied. For a given applied frequency only ions of a particular value of e/m pass through the spectrometer to the collector.

Ions of different e/m are collected by altering the AC frequency. The quadrupole is able to detect partial pressure of the order of 10^{-12} mbar. If the atoms or molecules flow through a window into the ionization measuring cell, a fraction is ionized, about 10^{-4} by electron bombardment, and measured after selection in the mass filter by a Faraday cage or a photomultiplier. The sensitivity of the quadrupole permits rate measurements up to 10^{-3} nm s^{-1}. The measuring cell must shield carefully against radiance, heat, and scattered electrons. The upper limit is given by space charge effects in the ionization chamber and lies approximately at 10 nm s^{-1}.

With the multiplex technique the periodic sampling of several mass values, and multicomponent and reactive processes can be controlled. Of special interest is the rate regulation with the mass spectrometer in the molecular beam epitaxy, because multicomponent films with extremely small rates are deposited.

11.2.1.2.2 Time-of-Flight Mass Spectrometer

The time-of-flight spectrometer consists of a pulsed ion source and an ion collector at the opposite end of an evacuated tube [11]. The ions are created by electron bombardment and accelerated out of source towards the collector by either one, or a series, of electric fields. The short pulse of ions traverses a tube of length L; the transit time t of ions of a particular m/e value, with velocity v is decided by

$$t = L/v\,. \tag{11.1}$$

If the potential difference through which the ions are accelerated is initially V, then

$$v = (2Ve/m)^{0.5} \tag{11.2}$$

and

$$t = \left[L/(2V)^{0.5}\right](m/e)^{0.5}\,. \tag{11.3}$$

At the collector at the far end of the tube, ions will, therefore, arrive at times decided by m/e. The heavier ions require the longer travel times. Thus each source pulse results in a mass spectrum that can be analyzed on a laptop. Time-of-flight instruments are usually able to detect partial pressures of the order of 10^{-9} mbar.

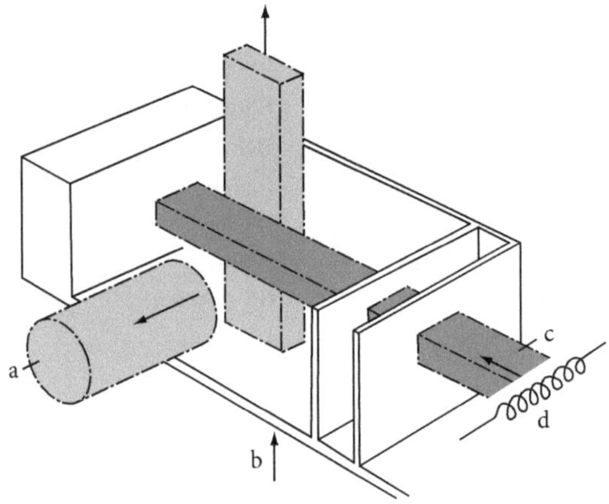

Fig. 11.5 Functional principle of an EIES measuring head. *a* Photons, *b* evaporated material, *c* electrons, *d* heating spiral

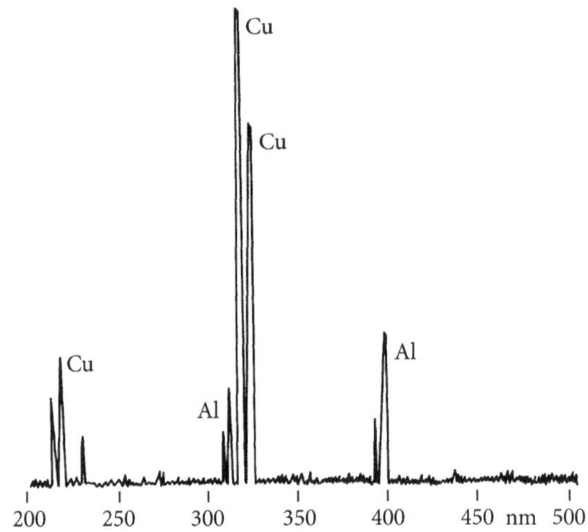

Fig. 11.6 EIES spectrum of an Al/Cu alloy

11.2.1.3 Electron-Induced Emission Spectrometry (EIES)

Electron-induced emission spectrometry offers interesting possibilities for rate control. The vapor flux, which passes through a window in the measuring head, becomes excited by means of electron collision with the emission of light (Fig. 11.5).

The photon flux of the excited atoms or molecules is proportional to the particle flux. The coating rate can be measured with high sensitivity (10^{-2} nm s^{-1}) and small response time (ms) [7]. A presumption is the selection of intensity and perfectly separate spectral lines, which for most materials are in the UV (unltraviolet) to VIS (visible) range, for gases also in the IR (infrared) range (Fig. 11.6).

With unknown emission lines a monochromator is used to determine the high intensity emission lines. For well-known emission lines edge filter are used. Several components of the coating material can be detected; with reactive processes a material component with the reactive gas can also be detected. The EIES rate measurement is relatively insensitive to outside influences.

11.3 Film Thickness and Rate of Deposition Measurements with Quartz Crystal Oscillators

11.3.1 Introduction

Quartz crystal oscillators are used in vacuum coating equipments in the optical and the semiconductor industry for regulation of the coating rate and monitoring of the film thickness. These devices operate on the principle that the resonant frequency of a piezoelectric crystal is changed by added weight. The frequency change is used to measure the coating rate.

11.3.1.1 Frequency Measuring Method

A quartz crystal plate is accurately ground to a thickness l, and metal electrodes evaporate on each side. When the electrodes are attached to an oscillating circuit, resonance occurs at a frequency f, given by

$$f = v_T/2l = A/d \quad \text{(Hz)}, \tag{11.4}$$

where v_T is the velocity of sound across the quartz wafer and A is the frequency constant. For the usual orientation along which quartz crystals are cut (called the AT cut) $A = 1661\,\text{kHz/mm}$. If an amount of material Δm is deposited on the crystal area F, the effective increase in film thickness is given by

$$\Delta l = \Delta m/\varrho F, \tag{11.5}$$

where ϱ is the quartz density. Using Eqs. (11.1) and (11.2) we obtain

$$\begin{aligned} \Delta f &= -f^2 \Delta m/A\varrho F \quad \text{or} \\ \Delta m/F &= -\Delta f/f^2 A\varrho \quad [\text{g/cm}^2]. \end{aligned} \tag{11.6}$$

Since the resonance frequency can be monitored continuously, the instrument can be used to measure the rate of deposition.

11.3.1.2 *Z*-Match®Method

If the system quartz oscillation and coating are regarded as coupled resonators from two layers, the coupling between the resonator layers of the sound identifier impedance relationship $z = z_F/z_Q$ depends on the deposited film and the quartz.

The definition equations for the sound identifier impedances are given by:

$$z_F = \varrho_F v_F = \sqrt{\varrho_F c_F}, \quad z_Q = \varrho_Q v_Q = \sqrt{\varrho_Q c_Q}, \tag{11.7}$$

where v_Q and c_Q are the phase velocity of the shear wave and the shearing stiffness constant regarding the considered piezoelectrically excited standing wave in the thickness direction, and v_F and c_F the appropriate parameters of the extrinsic film. The theory of the coupled resonator from two films gives the following analytic form:

$$\begin{aligned} z_Q \tan Q + z_F \tan F &= 0, \\ F = \frac{f}{f_F}\pi, \quad Q &= \frac{f}{f_Q}\pi, \end{aligned} \tag{11.8}$$

By using Eq. (11.7) and the relationship

$$f_F = \frac{1}{2l_F}\sqrt{\frac{c_F}{\varrho_F}}, \quad f_Q = \frac{1}{2l_Q}\sqrt{\frac{c_Q}{\varrho_Q}} \tag{11.9}$$

inserted into Eq. (11.18), we obtain the relationship of the explicit mass loading frequency

$$\frac{m_F}{m_Q} = -\frac{z_F f_Q}{z_Q \pi f} \arctan\left(\frac{z_Q}{z_F} \tan \frac{\pi f}{f_Q}\right). \tag{11.10}$$

For the most usual coating materials, in Table 11.2 the z_F values (bulk values), valid for the respective solid material, are registered.

A homogeneous mechanical resonator possesses a trivial resonant frequency spectrum. With each harmonic wave frequency an integral multiple of the primary wave frequency is an integer multiple of the primary wave frequency. A resonator from two layers usually possesses $z_F \neq z_Q$, a nontrivial spectrum, with which in each case the nearby resonant frequencies exhibit unequal distances. For this reason, the resonant frequencies of a two or multilevel piezoelectrical resonator are called quasi-harmonic.

When we introduce two resonant frequencies of the spectrum one after another in Eq. (11.9), then we get two linear independent equations from each, which can be solved mathematically. Figure 11.7 shows the block diagram of a two-frequency film thickness measuring instrument.

Table 11.2 Density ϱ_F, specific acoustic impedance z_F of the coating material and usable coating range maximum, l_F for a) frequency measuring method, b) Z-match procedure, with acceptance of a tolerated systematic error of 2 %. The used AT cut parameters are: frequency constant of $A = 1661\,\text{kHz/mm}$, density $\varrho_Q = 2.649 \times 10^3\,\text{kg}\,\text{m}^{-3}$, sound identifier impedance $z_Q = 8.801 \times 10^6\,\text{kg}\,\text{m}^{-2}\,\text{s}^{-1}$ [12]

Symbol	Material	Density ϱ_F $[10^3\,\text{kg}\,\text{m}^{-2}\,\text{s}]$	Impedance z_F $[10^6\,\text{kg}\,\text{m}^{-2}\,\text{s}]$	a) Max. l_F	b) Max. l_F
Al	Aluminum	2.68	8.42	2.7	45.0
Ag	Silver	10.50	18.5	0.5	3.5
AgBr	Silver bromide	6.47	7.46	1.1	8.2
AgCl	Silver chloride	5.56	6.67	1.3	6.9
As	Arsenic	5.73	9.07	1.3	27.5
Au	Gold	19.29	23.9	0.3	1.8
B	Boron	2.54	22.6	2.2	13.7
Be	Beryllium	1.85	16.3	3.0	20.8
C	Carbon	2.25	2.70	9.8	4.1
CaF_2	Calcium fluoride	3.18	11.3	1.7	16.7
Cd	Cadmium	8.64	15.6	0.6	4.7
CdS	Cadmium sulfide	4.83	8.63	1.5	43.3
CdTe	Cadmium telluride	5.85	8.98	1.3	35.8
Co	Cobalt	8.80	26.8	0.6	4.0
Cr	Chrome	7.18	28.6	0.8	4.9
Cu	Cooper	8.93	20.5	0.6	4.1
Fe	Iron	7.86	25.4	0.7	4.4
Ga	Gallium	5.85	4.41	1.3	3.1
GaAs	Gallium arsenide	5.31	5.54	1.4	4.8
Ge	Germanium	5.35	16.9	1.0	7.2
In	Indium	7.30	10.5	0.8	8.5
InSb	Indium antimonide	5.76	11.4	1.0	9.2
Ir	Iridium	22.46	68.6	0.2	1.5
KCl	Potassium chloride	1.98	4.29	3.9	9.3
LiF	Lithium fluoride	2.64	11.9	2.1	18.8
Mg	Magnesium	1.74	5.50	4.2	14.8
MgO	Magnesium oxide	3.58	21.5	1.5	10.3
Mn	Manganese	7.40	24.3	0.7	4.7
Mo	Molybdenum	10.20	34.5	0.5	3.2
NaCl	Sodium chloride	2.17	5.61	3.4	12.7
Nb	Niobium	8.75	18.0	0.6	4.4
Ni	Nickel	8.80	28.3	0.6	4.0
Pb	Lead	11.22	7.86	0.7	6.1
PbS	Lead sulphide	7.50	15.4	0.7	5.4
Pd	Palladium	11.50	22.9	0.5	3.0
Pt	Platinum	21.40	37.0	0.3	1.5
Sb	Antimony	6.62	11.4	0.8	8.0
Se	Selenium	4.82	10.2	1.1	14.1

Table 11.2 Continued

Symbol	Material	Density ϱ_F [10^3 kg m^{-2} s]	Impedance z_F [10^6 kg m^{-2} s]	a) Max. I_F	b) Max. I_F
Si	Silicon	2.32	12.3	2.4	20.6
SiO_2	Silicon dioxide	2.20	8.28	3.3	45.0
Sn	Tin	7.30	11.2	0.8	7.5
Ta	Tantalum	16.60	33.9	0.3	2.1
Te	Tellurium	6.25	9.80	0.9	12.6
Ti	Titanium	4.50	14.0	1.2	9.4
U	Uranium	18.70	39.5	0.3	1.8
V	Vanadium	5.80	16.6	0.9	6.6
W	Tungsten	19.30	55.2	0.3	1.7
WC	Tungsten carbide	13.80	54.9	0.4	2.4
Y	Yttrium	4.50	10.8	1.2	13.1
Zn	Zinc	7.10	16.7	0.8	5.4
ZnO	Zinc oxide	5.61	15.7	1.0	7.2
ZnS	Zinc sulphide	40.90	11.3	0.1	1.3
ZnSe	Zinc selenide	5.26	12.2	1.0	9.1

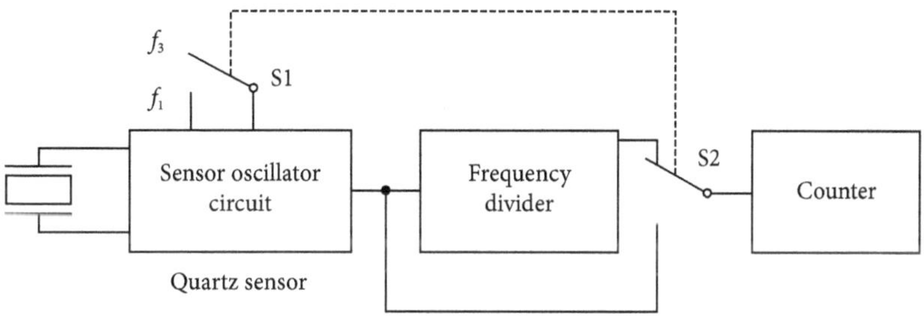

Fig. 11.7 Block diagram of the two-frequency oscillator

The sensor oscillator circuit contains a band-pass filter in the feedback circuit, which transmissions ranging from switch S1 can be switched from the primary wave to the third quasi-harmonic wave. At the same time, switching switch S2 to a 1 : 3 frequency divider, the following frequency counter for the basic and harmonic wave excitation can exhibit practically the same frequency range.

Equation (11.4) describes the piezoelectrically active quartz resonator part by a homogenous expression for a passive foreign film. A more exact analysis leads to an accurate description for the resonant frequencies of the two-film resonator for a linear model with short circuit electrodes. An AT cut quartz is used. We then obtain:

$$K(Q) z_Q \tan Q + z_F \tan F = 0 \tag{11.11}$$

$$K(Q) = \frac{1 - \dfrac{2k^2}{Q} \tan \dfrac{Q}{2}}{1 - \dfrac{k^2}{Q} \tan Q}, \tag{11.12}$$

$$\frac{m_F}{m_Q} = -\frac{z_F f_Q}{z_Q \pi f} \arctan\left(K(Q) \frac{z_Q}{z_F} \tan \frac{\pi f}{f_Q}\right), \tag{11.13}$$

where k is the natural oscillation of the effective electromagnetic coupling coefficient. In contrast to Eq. (11.8), Eqs. (11.10) and (11.11), and (11.13) contain a function $K(Q)$, which takes a value below 1 within the interesting range. $K(Q)$ z_Q can be regarded as the effective specific acoustic impedance of the quartz material with short circuit electrodes.

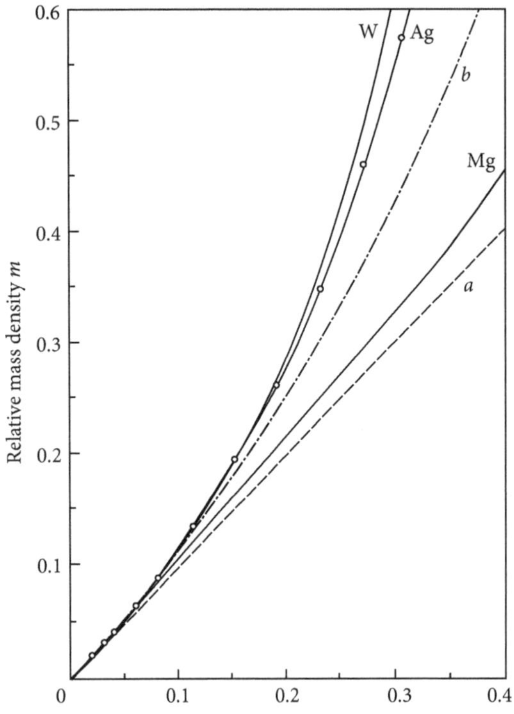

Fig. 11.8 Dependence of the sensor characteristic of film thickness measuring quartz on coating material (in the example W, Ag, Mg) according to the two-layer resonator model Eq. (11.13) and/or in good approximation after Eq. (11.10). For the comparison, the *a* frequency measuring method is also shown

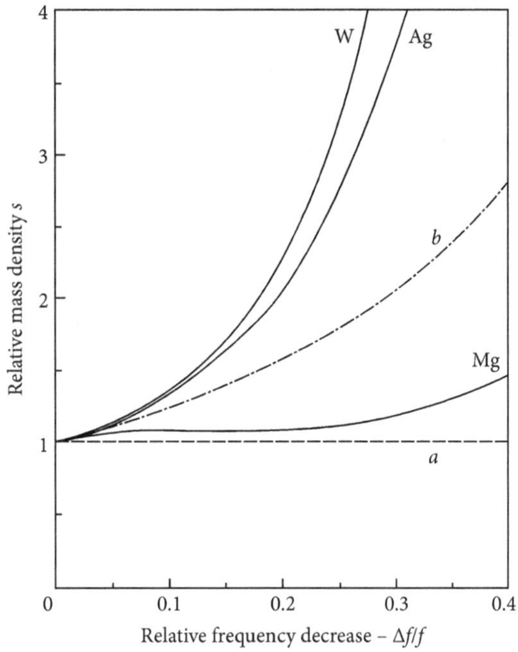

Fig. 11.9 Dependence of the relative mass sensitivity of a layer thickness measuring quartz (related to the sensitivity of the unloaded quartz) of the preloading and of the coating material according to the two-layer resonator model. *a* frequency measuring method, *b* period time measurement method

11.3.1.3 Characteristic Sensor Curves

The values (Eqs. (11.13) and (11.10)) are calculated with the help of the sensor functions for f and z calculated m values, which typically differ by only 0.3 % from each other, so that the diagram of the sensor characteristic and/or mass sensitivity in the Fig. 11.8 shows both equations within the drawing accuracy and the material dependence of the sensor characteristic. For practice, the derivation of the sensor characteristic represented in Fig. 11.9 is important, because this shows the reciprocal S of the respective sensitivity of a measuring quartz as a function of the coating material and of the preloading. From the curves it can be recognized that the rate values supplied by unloaded and loaded measuring quartz can differ with measuring instruments that do not consider the z value of the coating material above far more than 100 %.

11.3.1.4 Film Relaxation

For the permanent coating of a measuring quartz, after completion of the coating process, a pronounced, exponentially decayed frequency drift of the quartz at a high loading toward to higher frequency values can be observed [13]. This frequency drift to higher frequency values leads to lower film thickness measurement values (Fig. 11.10).

Very high preloading of the sensor quartz and very high coating rates and hard coating material can lead to an error in the thickness indication up to 10 %, caused by the film relaxation.

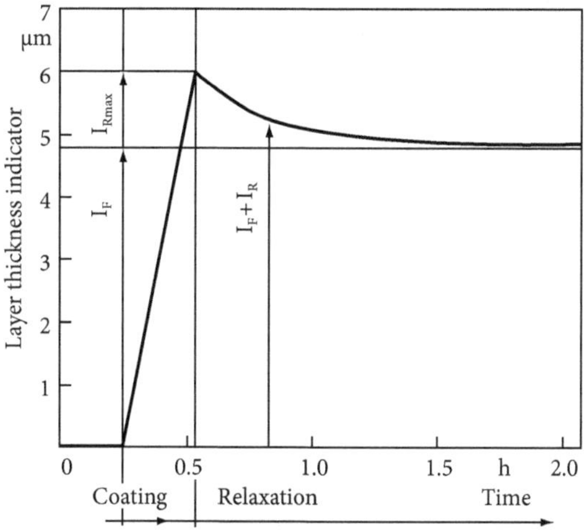

Fig. 11.10 Schematic representation of film relaxation. l_F... correct the measured value of the film thickness after relaxation of the film, l_g... indicate errors in the film thickness, $l_{R_{max}}$... max errors due to the relaxation (relaxation hub, $l_F + l_{R_{max}}$... indication of the film thickness measuring instrument at the end of the coating process)

11.4 Optical Measuring Procedures

11.4.1 Introduction

Optical films are measured about a fractional amount from a reference wavelength λ_0 (construction wavelength) up to multiples of this wavelength. The number of films in a film package is between 1 for a single film coating for antireflection and over 100 for an infrared block filter system. The reflection curves and the transmission curves of the finished film systems must usually lie within given ranges of tolerance. This is possible only if the individual films of the film system are prepared with appropriate accuracy.

The thickness of the films must be measured during the coating procedure and be switched off when the prescribed film thickness is reached. In the case of optical measuring of the film thickness the reflection or transmission of a test glass or a substrate is pursued during the coating. Reflection or transmission shows a nonlinear dependence on the film thickness. Automation of the coating process with optical measurement requires a complex control system [14].

Advantages of the optical measuring of film thickness are:

- Possibility of direct measurement of the desired properties
- Utilization of error compensation effects to increase the reproducibility of the production process
- No limitation on film thickness (important for IR films)
- Large flexibility
- Possibility of error diagnostics at the coating process.

In some cases, the optical measurement is even the only possibility to determine the thicknesses of applied films. This is, for instance, the case with coil coating equipment and continuous tunnel equipment.

11.4.2 Systematics of Optical Measuring of Film Thickness

Figure 11.11 shows an overview of the optical measuring methods, distinguished according to the nature of the measurement (monochromatic or spectral) and the position of the measurement (indirectly on test glasses or directly on the substrates).

During monochromatic measurement of the coating process only the changes of the intensity from the measuring wavelength are used. The measuring wavelength λ_0 is filtered out of the wideband measuring light by filters or monochromators and can be changed between two layers to achieve optimal measuring conditions.

With spectral measurement, with the help of a spectrometer reflection or transmission will periodically pass through the spectrum within an imaginary broad spectral range. The minimum frequency of these periodic spectral measurements depends on the velocity of the coating process. Figure 11.12 shows the different methods of indirect and direct measurements and illustrates some typical applications.

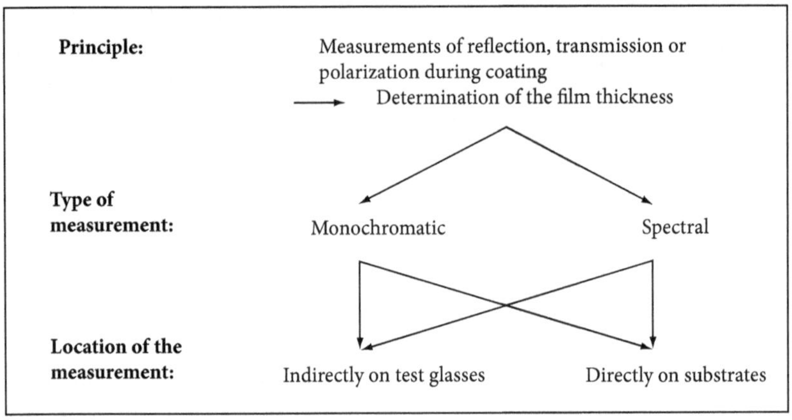

Fig. 11.11 Variation options of optical measurement

Indirect measurement

- Only 1 test glass
- 2 test glasses alternated (A/B method)
- Test glass exchange for each 2 layers
- Any combination

Direct measurement

- Continuously
- Intermittently

Application in:

Indirect: Batch equipments; Flow path equipments

Direct: Batch equipments; Flow path equipments; Belt coating equipments

Fig. 11.12 Methods of the indirect and direct measurement

11.4.2.1 Monochromatic Optical Measuring of Film Thickness

Figure 11.13 shows the spectrum of the reflection curve of a high breaking germanium film with the refractive index 4.1 on glass measured with monochromatic light as a function of the film thickness applied, in each case relative to the measuring wavelength nd/λ_m (curve a). The reflection or transmission curve (monitor curve) is periodic and passes through at optical film thicknesses that correspond to multiples of $\lambda_m/4$ extreme values.

Depending on the desired film thickness the coating process must be terminated when a certain intensity value is reached or when passing through one of the extreme values of the monitor curve. In the first case we speak of level monitoring and in the second case of turning value monitoring [15].

The attainable switch-off accuracy shows a complicated dependence on the film thickness because of the nonlinear process of the monitor curve. The resulting film thickness or switch-off errors depend on the quality of the photometer

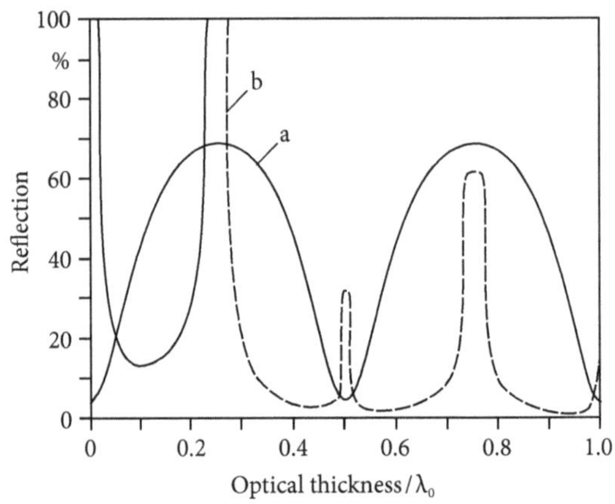

Fig. 11.13 Monitor curve of a germanium film on glass (*curve a*) and associated curve of the switching off errors for noise of the reflection measured value of 0.5 % (*curve b*)

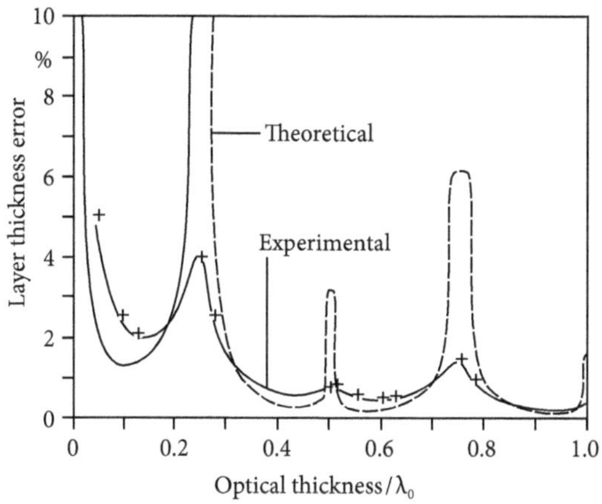

Fig. 11.14 Comparison of the experimentally determined layer thickness errors with the theoretical error curve

used, its calibration, and the measuring environment. A deciding influence is due to is due to the linearity of the photometer and hence the internal noise of the photometer, as well as the overlaid mechanical oscillations of the coating equipment to the measured value resolution.

In case of the deposition of a germanium film with a measuring wavelength of 2000 nm, the reflection can be calculated a measured value dissolution of about 0.5 %, shown in Fig. 11.13. With this value the relative film thickness error, represented by curve b, is $\Delta d/d$. In each case, its identification in the reflection interval of the 0.5 % part of the monitor curve on the film thickness axle was projected and the appropriate film thickness interval Δd was read off.

The switch-off errors are thus reversely proportional to the respective gradient of the monitor curve. They are the largest in the extreme values of the monitor curve. Curve b in Fig. 11.13 considers the possible existing nonlinearity of the photometer. The theoretical error curve is shown in Fig. 11.14.

The measurements essentially confirm the theoretical considerations; however, the extreme values of the practically identified thickness errors are compared in relation to the errors with other film thicknesses that were not as highly reinforced as was theoretically expected. This fact can be attributed to the nature of the method used in this case for the detection of the extreme values.

For the measurements (Fig. 11.14), the extreme values were determined by differentiation of the measuring signal [16], which ensures a substantially more exact detection of the extreme values as for the calculation of switch-off errors the presupposed comparison of intensity values.

11.4.2.1.1 Problems During the Transformation of Monochromatic Measuring of Film Thickness in Practice

When planning a coating process we can choose between different monitor strategies. The chosen monitor strategy has an important influence on the switch-off errors and thus on the reproducibility of the entire coating process [15]. Due to the geometry of the workpiece to be coated it is meaningful to use a deposited test glass to determine the film thickness.

For exact film thickness measurements a positive geometrical displacement between test glass films and substrate films is favorable. Such a displacement exists, if the film thickness on the test glass is larger than that on the substrate. Frequently, therefore, the geometrical displacement between test glass and substrate is adjusted. The geometrical displacement usually depends on the coating material. The reasons for this are the following:

- Different evaporation lobe of the different materials, e.g., for TiO_2 and SiO_2
- Separate sources of coating for each material with different positions in the coating recipient
- Different condensation conditions on test glass and substrate.

A further aspect, which must be considered during the planning of coating equipment is the physical displacement. A physical displacement between test glass and substrate occurs with temperature differences and different condensation conditions, which both concern different precoating and coating angles. Consequences of the physical displacement are refractive index differences between test glass films and substrate films.

11.4.3 Calculation of Optical Film Systems

The propagation of a planar, monochromatic and linear polarized light wave in a medium can be described by the following complex description of the electrical field intensity:

$$E = E_0 \exp\{\mathrm{i}[(2\pi/T)t - (2\pi N/\lambda)z]\} , \quad (11.14)$$

where E is the momentary electrical field intensity, E_0 the amplitude of the electrical field, i the imaginary unit, T the period duration, λ the wavelength of the light wave, z the local coordinate, N the complex refractive index of the medium ($N = n - \mathrm{i}k$), n the refractive index of the medium and k the extinction coefficient of the medium. The electrical field intensity E and the magnetic field intensity H are coupled over the so-called optical admittance y with one another.

$$H = yE . \quad (11.15)$$

The optical admittance of a medium can be calculated out the complex refractive index N of the medium and the admittance of the vacuum $y_0 (y_0 = \sqrt{\varepsilon_0/\mu_0})$,

$$y = N\, y_0 . \quad (11.16)$$

Except for the proportionality factor and dimension factor y_0, the optical admittance is equal to the refractive index of the medium concerned.

If a planar light wave under an angle α strikes any boundary surface between two media with different refractive indices, then the wave is partly reflected and partly passed. Reflection and transmission can be calculated from the condition that the components of the electrical and magnetic field intensity tangential to the boundary surface at the passage through the boundary surface must be constant. These two tangential components are connected with one another over a modified optical admittance η,

$$H = \eta\, E . \quad (11.17)$$

With the angular incidence of light on the boundary surface it must be differentiated between the optical admittance for parallel $\eta_{\parallel}$ and perpendicularly $\eta_{\perp}$ polarized light,

$$\begin{aligned} \eta_{\parallel} &= y/\cos(\alpha) \\ \eta_{\perp} &= y/\cos(\alpha) . \end{aligned} \quad (11.18)$$

For the reflectance at this boundary surface we obtain:

$$R = \left| \frac{\eta_0 - \eta_1}{\eta_0 + \eta_1} \right|^2 . \quad (11.19)$$

Here, the indices denote the incidence medium 0 and the emergence medium 1. Inserting in Eq. (11.19) the relations of Eq. (11.18), we obtain the well-known Fresnel equations for the reflection of a plane wave at the boundary surface of two media. If such a polarized, planar, monochromatic light wave transits a coated medium, then it is reflected several times back and forth at the boundary surfaces.

The overlaying partial waves interfere with each other and generate electric and magnetic fieldes. The field components tangential to the boundary surfaces of the coated media E and H are connected with one another at a position with the tangential field components E' and H' at another position by the following relationship matrices:

$$\begin{bmatrix} E \\ H \end{bmatrix} = \begin{bmatrix} M_{11} & M_{12} \\ M_{21} & M_{22} \end{bmatrix} \begin{bmatrix} E' \\ H' \end{bmatrix} . \quad (11.20)$$

In particular, this is valid also for the connection of the electromagnetic fields of two adjacent boundary surfaces a and b, since the tangential components of the field force at these boundary surfaces are continuous.

For the matrix we then obtain:

$$\begin{bmatrix} E_\mathrm{a} \\ H_\mathrm{a} \end{bmatrix} = \begin{bmatrix} \cos(\delta) & (\mathrm{i}/\eta)\sin(\delta) \\ \mathrm{i}\,\eta\sin(\delta) & \cos(\delta) \end{bmatrix} \begin{bmatrix} E_\mathrm{b} \\ H_\mathrm{b} \end{bmatrix} . \quad (11.21)$$

The components of the matrix depend only on the properties of the film at both boundary surfaces. This is, therefore also called the characteristic matrix of a film.

The optical admittance η of the coating material can be calculated from the relations (11.18).

Thereby it must be differentiated between parallel and perpendicularly to the plane of incidence of the polarized light.

The phase factor δ calculates itself after Eq. (11.22):

$$\delta = 2\pi N d \cos(\alpha)/\lambda\,, \qquad (11.22)$$

where
N is the complex refractive index of the film,
d the geometrical thickness of the film and
α the propagation angle of the light in the film relative to the axis of propagation.

In the case of a film system with several layers, the field forces at the entrance of the layer system with the help of E_0, H_0, and at the exit of the layer system at the entrance in the substrate E_S, H_S, can be connected by multiplication of the characteristic matrices of all layers with one another. Thus, it is valid that

$$E_0 \begin{bmatrix} 1 \\ Y \end{bmatrix} = \prod_{r=1}^{m} \begin{bmatrix} \cos(\delta_r) & (\mathrm{i}/\eta_r)\sin(\delta_r) \\ \mathrm{i}\,\eta_r \sin(\delta_r) & \cos(\delta_r) \end{bmatrix} \times \begin{bmatrix} 1 \\ \eta_S \end{bmatrix} E_S\,. \qquad (11.23)$$

Similarly to Eq. (11.13), an entrance admittance Y of the film system is defined. η_S is the optical admittance of the substrate for the tangential field components. Alternatively to Eq. (11.23) we can also write:

$$\begin{bmatrix} B \\ C \end{bmatrix} = \boldsymbol{M} \begin{bmatrix} 1 \\ \eta_S \end{bmatrix}, \qquad (11.24)$$

with

$$\boldsymbol{M} = M_1 M_2 M_3 \ldots M_m \qquad (11.25)$$

and

$$Y = C/B\,. \qquad (11.26)$$

The characteristic matrix $\boldsymbol{M}$ of the total film system is equal to the product of all characteristic matrices of the single films.

The reflection of the film system can be calculated in analogy to the reflection of a boundary surface between two homogeneous media after Eq. (11.19). For η_1 must in this case be the admittance of the complete film system inserted in Eq. (11.26). We obtain

$$R = \left|\frac{\eta_0(M_{11} + \eta_S M_{12}) - (M_{21} + \eta_S M_{22})}{\eta_0(M_{11} + \eta_S M_{12}) + (M_{21} + \eta_S M_{22})}\right|^2, \qquad (11.27)$$

where the matrix elements M_{ij} of the complete system are similar to the matrices elements of Eq. (11.20). In the matrix of a single layer according to Eq. (11.21), the imaginary unit i is also included in the definition of the array elements M_{12} and M_{21}.

The phase factor δ of a layer in Eq. (11.22) is proportional to the reciprocal wavelengths $1/\lambda$. For theoretical analysis it is convenient to represent the characteristics of the film systems as a function of λ_0/λ (Fig. 11.15). λ_0 is thereby a reference wavelength or construction wavelength. Introducing λ_0 into the equation for δ, we obtain in the case of absorption free films

$$\delta = 2\pi(nd/\lambda_0)\cos(\alpha)(\lambda_0/\lambda)\,, \qquad (11.28)$$

where n is the real part of the complex refractive index N. With adsorption freedom we have $N = n$. The product of the refractive index n and the geometrical thickness d is called the optical layer thickness nd. The optical film thicknesses are frequently indicated relative to the reference wavelength.

11.4.3.1 Characteristics of a Single Layer

The array elements of the main diagonals of the characteristic matrix of a single film are as in Eq. (11.23) (where $M_{12} = M_{21}$). With a given optical film thickness of a film nd the phase δ in Eq. (11.28) depends on the respective wavelength. In practice, those wavelengths are of special importance, for which the optical thickness of the film corresponds to a multiple of $\lambda_0/4$ and/or $\lambda_0/2$. With these wavelengths the reflection and/or transmission curves of the layers exhibit extreme values. We speak then of $\lambda/4$ and/or $\lambda/2$ layers.

By perpendicular incidence of light the optical admittance of the materials can be replaced by the refractive index. The characteristic matrices for $\lambda/4$ and $\lambda/2$ layers in this case have

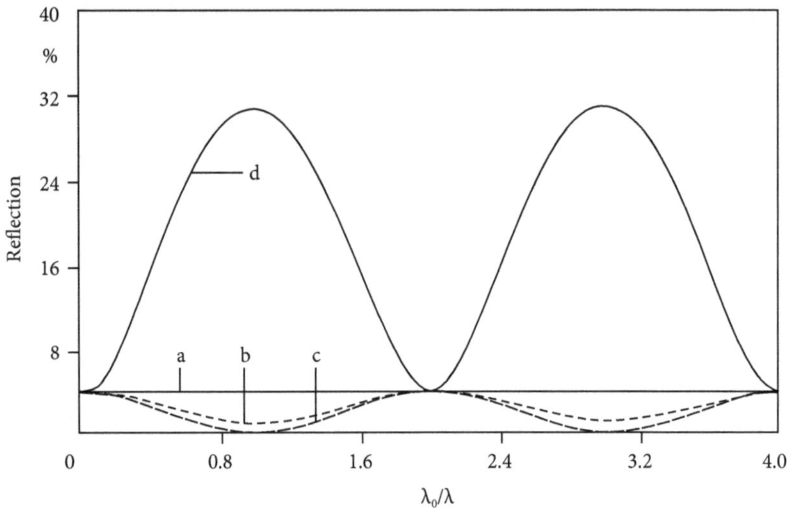

Fig. 11.15 Spectral reflection curves of some $\lambda_0/4$ layers; $n_S = 1.52$. *a* Reflection of the uncoated substrate, *b* $n_1 = 1.38$, *c* $n_1 = 1.233$, *d* $n_1 = 2.30$

the following form (n_1 = refractive index of the layer): $\lambda/4$ layer

$$M_{\lambda/4} = \begin{bmatrix} 0 & \frac{i/n_1}{0} \\ i\,n_1 & \end{bmatrix} \quad (11.29a)$$

λ/2-layer

$$M_{\lambda/2} = \begin{bmatrix} -1 & 0 \\ 0 & -1 \end{bmatrix}. \quad (11.29b)$$

If these layers are on a substrate with the refractive index n_s, then with Eqs. (11.29a,b) and (11.23) we obtain for the reflection

$$R_{\lambda/4} = \left| \frac{n_0 - n_1^2/n_s}{n_0 + n_1^2/n_s} \right|^2 \quad (11.30a)$$

$$R_{\lambda/2} = \left| \frac{n_0 - n_s}{n_0 + n_s} \right|^2. \quad (11.30b)$$

The reflection of the $\lambda/2$ layer is independent of the film refractive index equal to the reflection of the uncoated substrate. This is also valid if the $\lambda/2$ layers are determined within a multiple layer system. In place of the substrate refractive index in Eq. (11.30b) the admittance of the layer system under the $\lambda/2$ layer is used.

With a $\lambda/4$ layer, in accordance with Eq. (11.30a), depending on the respective refractive index of the layer one can generate reflection increases or reflection reductions. Figure 11.15 shows some calculated spectral reflection layer curves of different refractive indexes, whose thickness in each case directly corresponds to a quarter of the wavelength λ_0. The reflection reduction with a monolayer is ideally ($R = 0$), if in Eq. (11.30a) the numerator becomes 0.

For the layer refractive index from the condition it follows that

$$n_1 = \sqrt{n_0 n_s}. \quad (11.31)$$

This must be equal to the geometric mean of the refractive indices of the two adjacent media, in the case of antireflection coating on an air/glass boundary surface ($n_s = 1.52$), $n_1 = 1.233$.

A comparison with Eq. (11.26a) shows that the substrate coated with a $\lambda/4$ layer operates as if it would have an actual refractive index of n_1^2/n_s. This value is called the effective refractive index of the $\lambda/4$ single layer

$$n_{\text{eff}} = n_1^2/n_s. \quad (11.32)$$

In practice, there usually are not any coating materials available with the calculated refractive index to fulfill the amplitude conditions accurately and to thus reach the zero of reflection.

In order to achieve with a given application as good an approach of the actual reflecting property to the desired value as possible, the layer thicknesses are often chosen as deviating from $\lambda/4$. Thus the possibility exists to change both the height of the reflection minimum and the width of the range with low reflecting property.

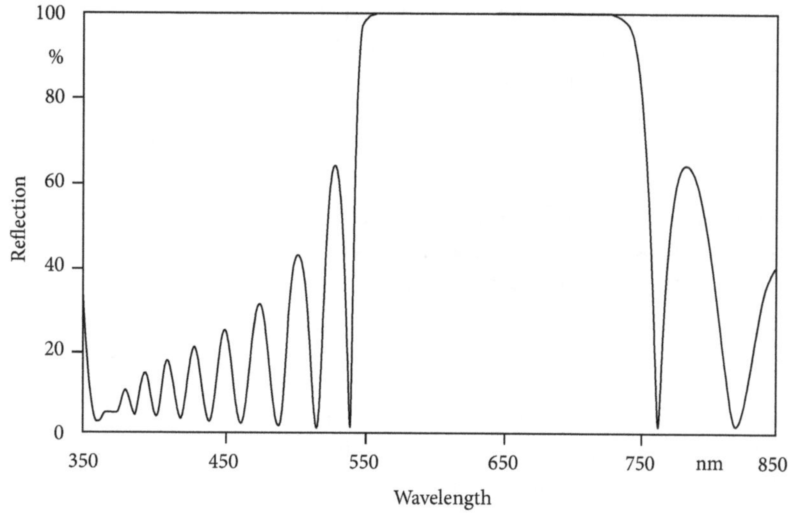

Fig. 11.16 He/Ne laser mirror with the layer structure glass (HL) 10 H/air, $\lambda_0 = 632.8\,\text{nm}$. Glass: BK 7. H: $\lambda_0/4$ TiO_2. L: $\lambda_0/4$ SiO_2

11.4.3.2 Laser Mirrors

In the simplest case, laser mirrors consist of a change of high and low refracting layers, in each case with an optical layer thickness of $\lambda/4$. A typical layer system consists of

$$\text{glass}/(\text{HL})^p\ \text{H}/\text{air}\,, \qquad (11.33)$$

with

$$\text{H}\lambda/4\,\text{TiO}_2 \quad \text{or} \quad \lambda/4\,\text{Ta}_2\text{O}_5$$

and

$$\text{L}\lambda/4\,\text{SiO}_2\,,$$

where p is an integer and indicates how often the pair of layers (HL) must be applied; p usually is between 10 and 15. As a result of repetitive use of the concept of the effective refractive index of λ_0 layers from Sect. 11.4.3.1, the reflection of the layer system (Eq. (11.33)) arises also for the construction wavelength

$$R = \left| \frac{n_0 - (n_H/n_L)^{2p}\, n_H^2/n_S}{n_0 + (n_H/n_L)^{2p}\, n_H^2/n_S} \right|^2 . \qquad (11.34)$$

Table 11.3 shows $n_0 = 1.00$ for the refractive indices; $n_H = 2,3$; $n_L = 1.46$; and $n_S = 1.52$ are the reflection values according to Eq. (11.29) for different exponents p and the layer numbers following from p.

Table 11.3 Calculation of the He/Ne laser mirror using the dispersion data and the absorption data

p	Layer number	Reflection [%]
2	5	82.9650
4	9	97.0154
6	13	99.5092
8	17	99.9202
10	21	99.9870
12	25	99.9979
14	29	99.9997

Table 11.3 shows that with increasing layer number the reflection increases at first very fast. With 13 layers, 99.5 % reflection is reached. In practice, maximum attainable reflection is limited by dispersion and adsorption in the layers and thereby crucially depends also on the quality of the coating process.

For application in laser gyroscopes, mirrors with the smallest dispersions and absorptions possible are needed. Mirrors manufactured by ion beam sputtering operate satisfactorily. With this coating process, a degree of reflection of up to 99.985 % is achieved. With the application in high power lasers the destruction threshold can be improved by detuning the top layers. With laser mirrors, according to Eq. (11.33) the electrical field forces at the boundary surfaces between the layers are maximum and by existing boundary surface absorptions during excess of a power limit always lead to the destruction of the mirror.

Figure 11.16 shows the calculated reflection curve of a laser mirror with as high a reflection as for a He/Ne laser with a layer structure according to Eq. (11.34).

11.5 Determination of the Film Thickness by X-Ray Emission and X-Ray Fluorescence

Microprobe and X-ray fluorescence analysis are used routinely for the determination of film composition and film thicknesses ex situ, see Chap. 10. The same principles can also be used for in-situ analysis of thin films during their generation. Figure 11.17 shows the structure of an arrangement, with the film thickness and film composition by sputtering to be controlled intermittently.

For the analysis the sample is moved from the sputtering position to the measuring position. The excitation can take place, for example, via electron bombardment or via with X-ray analysis. The electron beam thereby leaves a small screen, which the pressure difference between source (10^{-5} mbar) and sputtering chamber ($\sim 10^{-2}$ mbar) sustained. The screen can be covered alternatively by metal foil, for example, a 25 µm thick Ni foil. Under bombardment of the electrons the Ni foil sends X-rays, which can be used for X-ray fluorescence excitation of the sample. The characteristic X-ray emitted by the sample is checked and analyzed by a Si(Li) detector. The very sensitive Si(Li) crystal is protected by a Be window. Contamination by sputtered material is prevented with an exchangeable polymer foil.

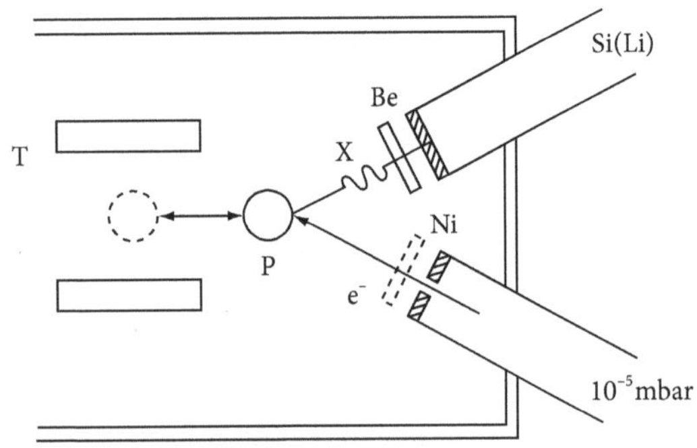

Fig. 11.17 Principle of an arrangement for in-situ analysis of films with X-ray emission; e: electron beam; X: emitted X-ray; P: sample; T: target

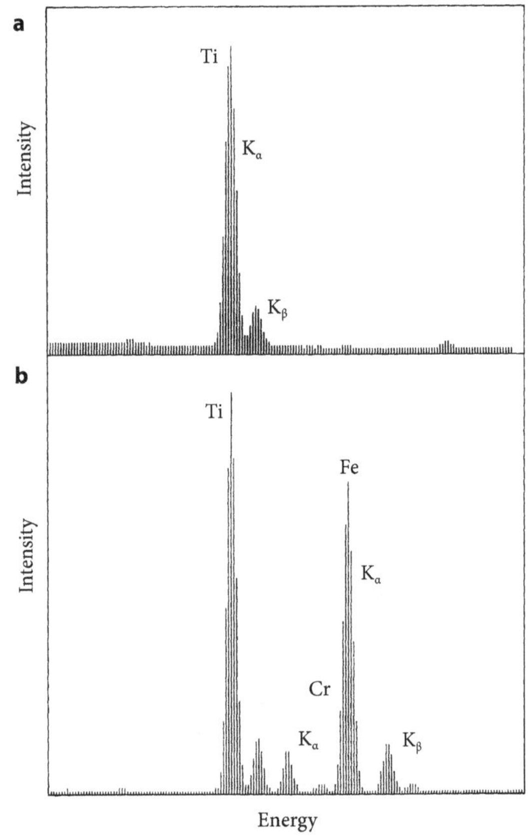

Fig. 11.18 X-ray spectrum of a TiN layer on a steel substrate. Energy K_α X-ray lines: Ti: 4.6 keV; Cr: 5.5 keV; Fe: 6.4 keV. **a** Excitation with electron beams, **b** excitation with X-rays

Since the Be window is penetrable only with X-rays with energies of more than 1 keV, the analysis can be only executed at elements with ordinal numbers $Z \geq 12$ (Mg).

Figure 11.18 shows the X-ray spectrum of a 5.3 µm thick TiN film on a steel substrate. With excitation with an electron beam (Fig. 11.18a) the penetration depths of the electrons do not reach the boundary surface film substrate. Electrons do not penetrate the substrate, in this case steel. From there, one receives only signals from the TiN film. In the case of excitation with X-rays, the penetration depth is so large that also atoms

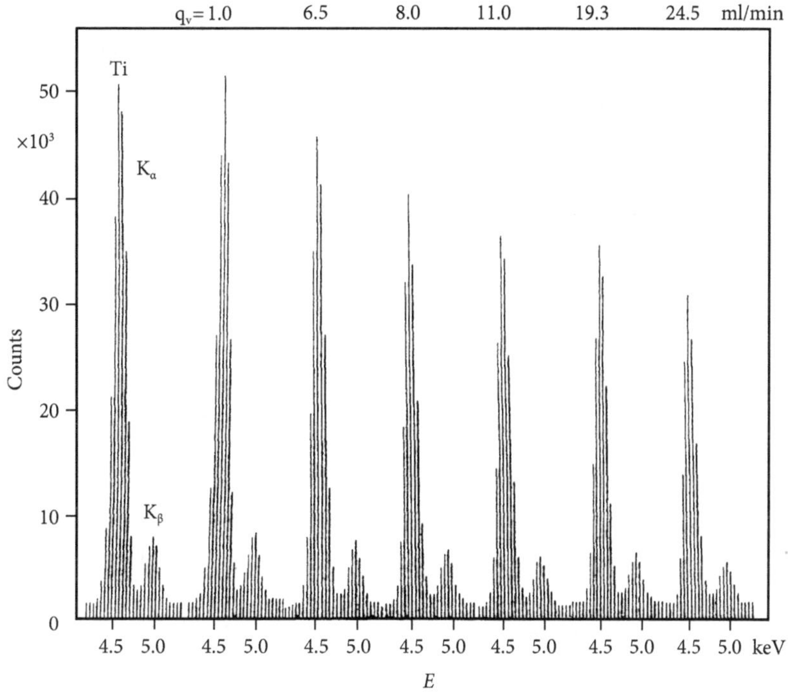

Fig. 11.19 X-ray spectra of $Ti_{1-x}N_x$ layers with reactive sputtering with different nitrogen fluxes

of the steel substrate are excited to emit X-rays (Fig. 11.18b).

In addition to the Ti lines from TiN, Fe and Cr lines of steel are also observed. In both cases we do not see lines of nitrogen, because the X-ray energy of nitrogen is (0.392 keV) and the nitrogen line is absorbed by the Be window. By comparison with well-known calibrate samples the concentration relationship N/Ti can be calculated from the intensity of the Ti lines obtained by electron beam excitation [17]. Additionally if we consider the relationship of the Ti and Fe lines won with X-ray excitation, then we can determine, with well-known film thickness, for example, the density of TiN as a function of the composition.

Through alternative excitation with electrons or X-rays we can obtain complementary statements. Figure 11.19 shows spectra of $Ti_{1-x}N_x$ layers, which resulted in the case of reactive sputtering with different nitrogen fluxes. The intensity of the K_α and K_β lines decreases with increasing nitrogen flux and within the range between 11.0 and 19.3 ml/min reaches the characteristic value of stoichiometric TiN.

The quantitative analysis is affected considerably by different factors [18], e.g., through

- the penetration depth of the electron or X-ray used for the excitation;
- the backscattering of high energy electrons;
- the absorption of the emitted X-ray on the way through the material; and
- the increase of the intensity through fluorescence.

These different effects must considered by means the so-called ZAK correction (ordinal number absorption fluorescence), which depends very strongly on the material composition (matrix effect).

With thin films (Fig. 11.20) the range of the exciting radiation is often of the order of magnitude of the film thickness and, therefore, two different materials, i.e., film and substrate are involved in the analysis.

Electrons of a 20 keV electron beam, e.g., in gold, have a penetration depth of approximately 0.3 μm. The excitation of the X-ray, however, only takes place via the electrons, whose energy still lies above the excitation energy. In gold, for

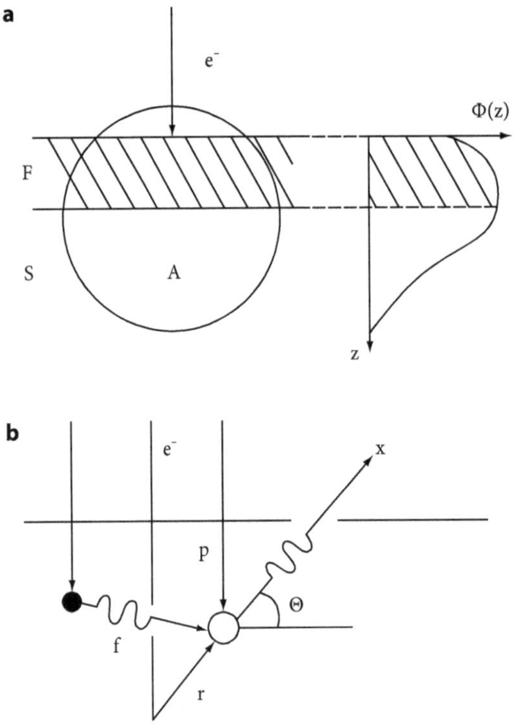

Fig. 11.20 Excitation of X-ray (X) by electrons (e^-) in a thin layer (F) on a substrate (S). **a** Excitation range (A) and excitation function $\Phi(z)$, **b** mechanisms of the excitation and the absorption; p: direct excitation, r: excitation through backscattered electrons, f: excitation by fluorescence; θ: exit angle

instance, this corresponds to a range starting at 0.1 μm.

Backscattered electrons with sufficiently high energy can contribute additionally to ionization. If the energy distribution of the electrons (particle flux density in the interval $\mathrm{d}E$) as a function of the depth z with $N(E, z)\,\mathrm{d}E$ is designated, then the excitation function $\Phi_\mathrm{A}(z)$ for an element A results through [19]

$$\Phi_\mathrm{A}(Z) = \int_0^{E_\mathrm{max}} q_\mathrm{A}(E) \times N(E, Z)\,\mathrm{d}E\,, \quad (11.35)$$

where $q_\mathrm{A}(E)$ is the ionization cross-section for the shell applied from the atom A. After the ionization X-rays develop from λ_A with the yield Y (λ_A). On the way to the surface absorption takes place, which depends on the absorption coefficient μ and the exit angle θ.

At a layer with thickness t and the atomic concentration $C_\mathrm{A}(z)$ the intensity per surface is given by

$$I(\lambda_\mathrm{A}) = \int_0^t C_\mathrm{A}(z) \times \exp\{-\mu z / \sin\Theta\} \times Y(\lambda_\mathrm{A}) \times \Phi_\mathrm{A}(z)\,\mathrm{d}z\,. \quad (11.36)$$

Apart from the correction of the absorption, which represents an effect of first order, a fluorescence correction, which is of second order, is necessary in many cases. Moreover, all X-ray quanta with sufficiently high energy can also contribute to the ionization. Since usually the films or film packages as well as the substrates contribute to the measuring signal, the different parts are to be considered with the integrations of Eqs. (11.35) and (11.36).

11.6 Atomic Emission Spectroscopy

This measuring technique uses the fact that in each plasma there are many excited atoms. Upon return into the initial state these atoms send photons of certain energy. With plasma-supported coating processes, spectroscopic analysis of these light emissions (see also Sect. 10.4) can serve as the basis of the control process [20].

Atomic emission spectroscopy is important as an automatic control measuring system with reactive sputtering. The reactive gas part of the gas flow in the discharge chamber determines whether the cathode is in the metallic or reactive operating condition.

Plasma emission spectroscopy determines this part directly in the discharge chamber near the cathode and thus within the range in which many of the film properties develop. Moreover, with this technology the active reactions of the target surface can be analyzed. Changes of the chemical compounds with otherwise constant conditions are reflected in the distribution of the intensity of the plasma emission.

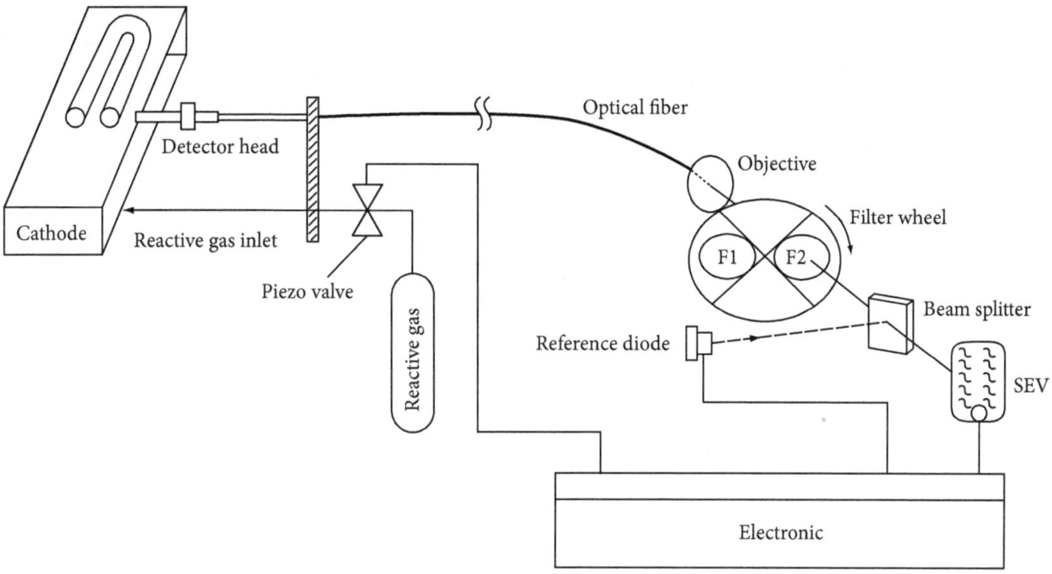

Fig. 11.21 Principal structure of a plasma emission monitor for two wavelengths

The intensity of an emission line characterizing the target material depends on the degree of the coverage of the target surface with reactive products. For a certain layer the optimal cathode condition is coupled with a certain line intensity. This condition can be kept by intensity measurement of the emission line and the adjustment of the reactive gas flow steered from it.

Figure 11.21 shows the principle of a plasma emission monitor. The detector head, i.e., the end of a fiber optic light guide consisting of several single fibers is approximately 20 mm above the target surface. A special construction protects the light guide against coating. The desired emission lines can be selected with a line filter, the equipment is laid out for the simultaneous observation of two lines.

Two measuring signals before the registration over the photoelectric multiplier are pulsed by means of a chopper wheel. For elimination of the equipment drift, additionally to the measuring signals a dark signal as well as a reference signal is measured. The latter consists of a light emitting diode that emits blue, which is held by means of regulated heating at constant temperature. The topped electronics always assigns the same output signal as the intensity of the reference light source. This output signal is the control signal for the flux control of the reactive gas.

Variation of the current for the reference source of light and thus its intensity and the sensitivity of the measuring system is used to regulate the brightness of the plasma.

For effective regulation a good alignment between emission automatic controllers and the closed loop control system is important. Large gas volumes after the inlet valve or a large distance between gas inlet and cathode decrease the effectiveness with fast process fluctuations. The effect of such a control is shown in Fig. 11.22.

For reactive coating of AlN using a metallic Al-target, cathode voltage, and total pressure in the discharge chamber as well as the emission signal of the Al are plotted as function of the coating duration.

Fig. 11.22 Dependence of important coating parameters on the coating time by insertion of a plasma emission automatic controller (**b**) and without regulation (**a**) with the example of reactive AIN coating

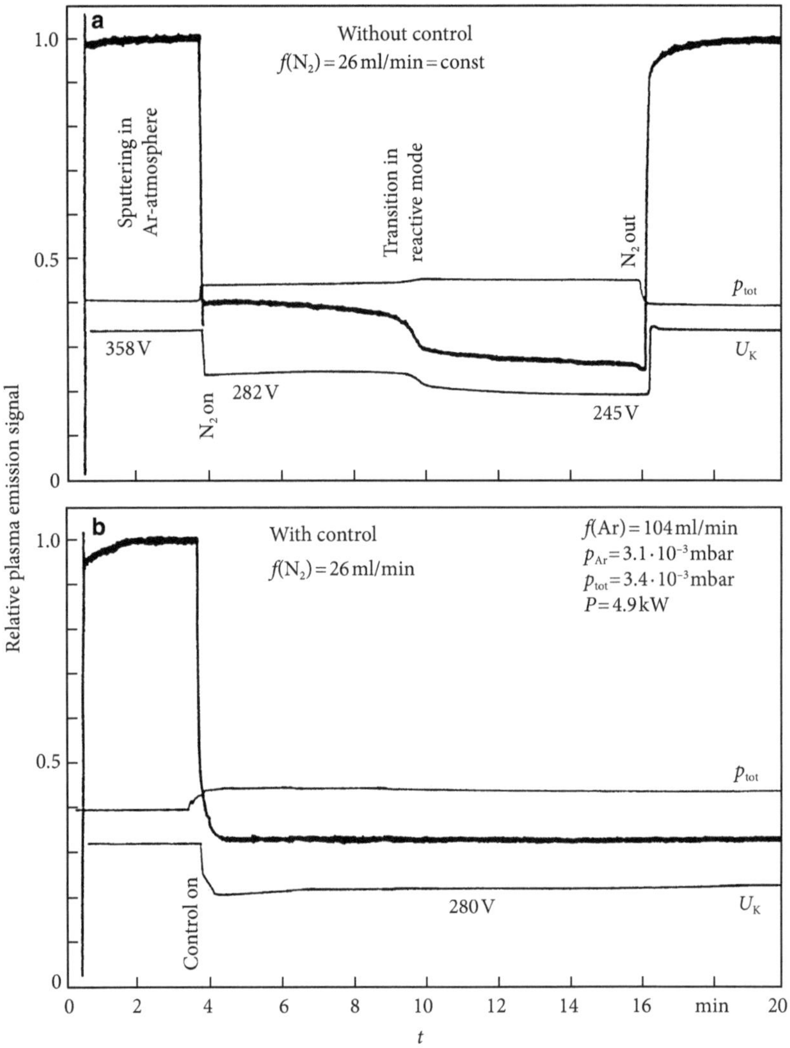

References

1. Stoll RL (1974) The analysis of eddy currents. Oxford University Press
2. Krawczyk A, Tegopoulos JA (1993) Numerical modelling of eddy currents. Clarendon Press
3. Malik RJ (1987) Electron beam source molecular-beam-epitaxy of III-V compounds. J Vac Sci Technol V 5(3):722–724
4. Jaffey AH, Flynn KF, Glendenin LE, Bentley WC, Essling AM (1989) Chemistry Division, Argonne National Laboratory, Argonne, Illinois 60439
5. Hegner F, Feuerstein A (1978) Aluminum-Silicon Metallization by Rate Controlled Dual EB-Gun. Solid State Technol Nov. 1978:49–54
6. Hudner J, Ohlen H, Stolt L, Östling M (1990) Coevaporation thin films of Y-Ba-Cu-O utilizing Quadrupole Mass Spectrometer Rate Control. J of the Less Common Metals 164 & 165:422
7. Lu C, Lightner MJ, Gogol CA (1977) Rate Contolling and Composition nalysis of Alloy Deposition Processes by Electron Impact Emission Spectroscopy (EIES). J Vac Sci Technol 14(1):103–107
8. Zinner E, Crozaz G- (1986) Secondary Ion Mass Spectrometry SIMS V. Springer
9. Seliger RL, Ward JW, Wang V, Kubena RL (1970) A high-intensity scanning ion probe with submicrometer spot size. Applied Physics Letters 34(5):310
10. Livonen A, Saintola R, Valli K (1990) Ion-guide quadrupole mass spectrometer. Phys Scr 42:133

11. Lipinsky D, Jede R, Tümpner H, Ganschow O, Benninghoven A (1985) High Sensitivity Quasisimultaneous Secondary Neutral, Secondary Ion and Residual Gas Mass Spectrometry by a new Electron Impact Postionizer. J Vac Sci Technol A 3:2035
12. Benes E (1984) Improved quartz crystal microbalance technique. J Appl Phys 56:608–626
13. Hertl S, Benes E, Wimmer L, Schmid M (1985) Investigation of quartz crystal thickness shear and twist modes using a new noninterferometric laser speckle measuring method 39th Annual Frequency Control Symposium. IEEE Xplore, Philadelphia, USA, pp 535–543
14. Zöller A, Boos M, Herrmann R, Klug W, Lehnert W (1989) Optical Tickness Monitoring of Dielectric Optical Filters Using a New In-Situ Photometer with High Signal Resolution and Excellent Long-Term-Stability. SPIE 1019:106–112
15. Zöller A, Herrmann R, Klug E, Zültzke W (1986) Optical monitoring: comparison of different monitoring strategies with respect to the resulting reproducibility to the completed laser systems. SPI 652:10–14
16. Herrmann R, Zöller A (1986) Automation of coating processes for the production of optical layer systems Annual Conference of China Optics Society in Guilin, China.
17. Stobiecki F, Stobiecki T, Kellner F, Thomas K, Röll K, Gärtner H (1991) Analysis of Ti-N Films by calibration of Ti X-ray spectra. Fresenius J Anal Chem 341:365–368
18. Scott VD, Love G (1983) Quantitative electron-probe microanalysis. Ellis Horwood Ltd., Chichester
19. Karduck P, Ammann N, Esser HG, Winter J (1991) Quantitative electron probe microanalysis for the characterisation of thin carbon-boron layers in fusion devices. Fresenius J Anal Chem 341:315–319
20. Schiller S et al (1987) The Use of a Plasma Emission Monitor for the Control of the Reactive Sputter Process. Proc. 6th Int. Conf. IPAT, Brighton, p 23

12 Measurements of Thin Layers After the Coating Process

H. Frey and T. Helmut

12.1 Measurements of Thermal Conductivity

12.1.1 Introduction

The thermal conductivity of thin films is important for the nucleation process and the film growth. The heat conductivity λ of a material is defined through:

$$\mathrm{d}\dot{Q} = -\lambda\,(\partial\delta/\partial s)\,\mathrm{d}A\,. \tag{12.1}$$

The heat flow $\mathrm{d}\dot{Q}$ (quantity of heat for each time unit) flows through, under the effect toward the surface-normal existing temperature gradient $(\partial\delta/\partial s)$ the area $\mathrm{d}A$.

12.1.2 Experimental Determination

The thermal conductivity of a thin metallic film can be indirectly determined from the surface resistance R using the Wiedemann–Franz law [1, 2]:

$$\lambda/(\sigma T) = L\,, \tag{12.2}$$

where

L is the Lorentz number ($L = 2.45 \times 10^{-8}$ $\mathrm{W}/\Omega\,\mathrm{K}$ at ambient temperature),
σ the electrical conductivity, and
T the absolute temperature.

12.2 Electrical Conductivity

12.2.1 Definition

The electrical conductivity of thin films can be calculated from the measurement of the surface resistance R. The surface resistance is the resistance measured between two electrodes that are attached to opposite sides of a film with square surface. The surface resistance $R_\square$ is independent of the size of the square surface, because for the resistance of a cuboid of the thickness d, width b, and length l, Fig. 12.1a is valid:

$$R = \varrho l/db\,. \tag{12.3}$$

With the square surface ($l = b$),

$$R = R_\square = \varrho/d \tag{12.4}$$

independently of the size of the surface of the square. The surface resistance has the dimension of a resistance.

12.2.2 Methods of Determination

The surface resistance is measured with the four-point-method [3, 4]. In addition, four punctual electrodes are set on the surface layer. The electrodes can be arranged linearly with equispaced distance from each other (Fig. 12.1b), or in four corners of a square (Fig. 12.1c).

During the linear arrangement a current flow I is generated by the outside two electrodes and

© Springer-Verlag Berlin Heidelberg 2015

H. Frey, H. R. Khan, *Handbook of Thin-Film Technology*, DOI 10.1007/978-3-642-05430-3_12

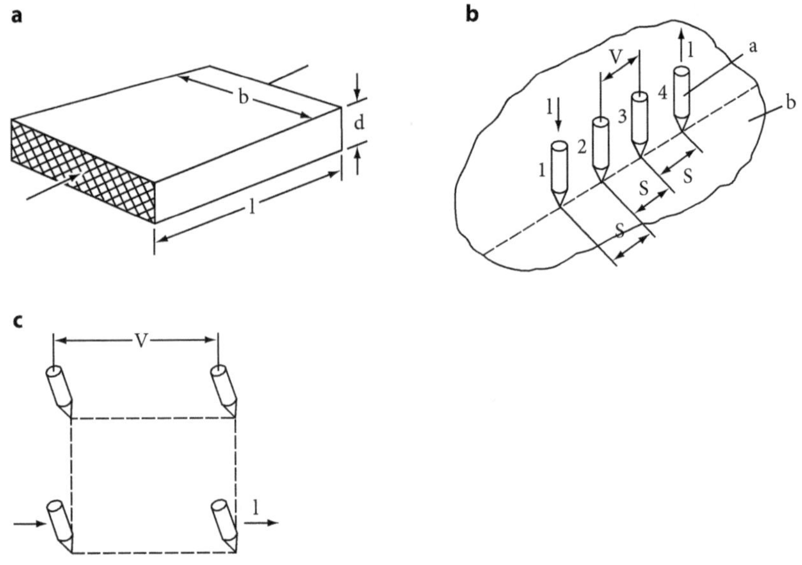

Fig. 12.1 **a** Definition of the surface resistance, d thickness of the layer, b width, l length, **b** linear four-point probe, S probe distance, a probes, b sample, **c** square four-point probe, I impressed current, V measuring voltage

the voltage drop U between the internal two electrodes is measured.

For the surface resistance it is valid that:

$$R_{\square} = 4.532\, U/I\,. \tag{12.5}$$

In the case of the square electrode arrangement, the current is sent by two neighboring electrodes and the voltage drop is measured between the two opposite electrodes [4]:

$$R_{\square} = 9.06\, U/I\,. \tag{12.6}$$

The relations (Eqs. (12.5) and (12.6)) are only valid if the diameter of the film surface against the electrode gap is large, otherwise geometrical correction factors for the measured values must be considered, which are calculated in Eqs. (12.5) and (12.6).

12.3 Magnetic Properties

12.3.1 Properties of Magnetic Layers

Magnetic thin films are predominantly used in data storage and in sensor technology. As material alloys of 3D-transition metals, iron, cobalt, and nickel or compounds with noble earth such as gadolinium, terbium, and dysprosium are applicable [5]. For magnetic recording alloys such as $Co_{80}Ni_{20}$, $Co_{75}Ni_{20}Cr_5$, or $Co_{75}Cr_{25}$ are used. Magneto optic memory films consist of the base material $Fe_{75}Tb_{25}$, which can be partly substituted by other transition metals or noble earth compounds.

Thin film magnetic heads or thin magnetic film sensors usually contain soft-magnetic materials, e.g., $Ni_{80}Fe_{20}$ (Permalloy) or FeAlSi (Sendust). Amorphous films of metal such as FeB, $Co_{91}Zr_9$ or $Fe_{75}Tb_{25}$ are being used increasingly. The magnetization curve $M(H)$, which represents the magnetization M as function of the magnetic field intensity H, supplies the most important parameter (Fig. 12.2).

The external field H is created by coils, electromagnets, or permanent magnets.

In air and/or vacuum the permeability μ_0 is coupled with the flux density (induction) $B_0 = \mu_0 H$, which is increased in a magnetic material by adjustment of the atomic magnetic moments:

$$B = \mu_0(H + M) \tag{12.7}$$

with

$$\mu_0 = 4\pi \times 10^{-7}\,\mathrm{Vs/Am}\,.$$

The magnetization M is given by the magnetic moment m of the sample per volume V: $M = m/V$. Instead of the description $M(H)$, also the quantity $B(H)$ is frequently used. The mag-

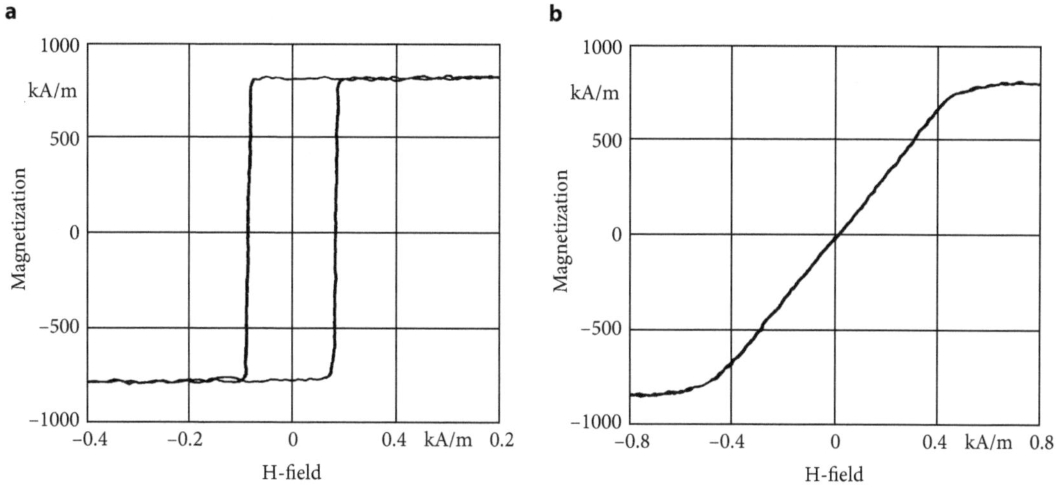

Fig. 12.2 Magnetization curves $M(H)$ of a NiFe layer with uniaxial anisotropy for different directions of the sample **a** parallel to the light axis, **b** perpendicularly to light axis

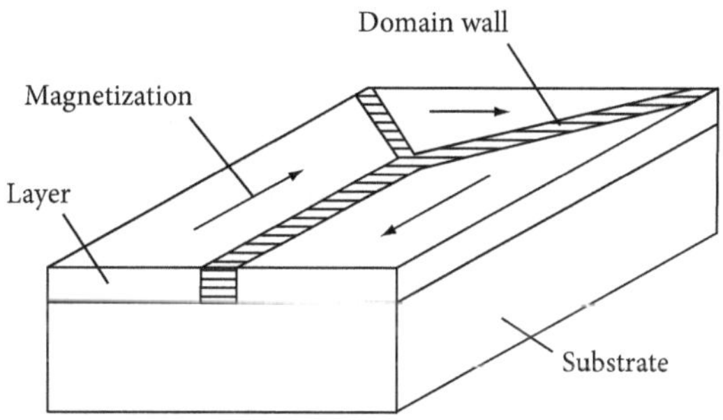

Fig. 12.3 Magnetic domains and domain walls (schematic)

netization curve $M(H)$ comes off by different superposition effects [6].

In principle, at ferromagnetic materials the atomic magnetic moments are oriented parallel to each other by quantum-mechanical coupling. Such an orientation cannot be reached by technically produced fields. In iron the effective field caused by atomic effects caused approximately 500 T, whereas with electromagnets approximately 2.5 T is attainable. Despite the magnetic orientation ferromagnetic materials normally appear outwardly almost non-magnetic, because the magnetization divides itself into magnetic ranges (domains, Weiss districts), Fig. 12.3.

Within a magnetic range the atomic magnetic moments are parallel oriented; the directions of magnetization of the domains, however, are different from each other. Depending upon the orientation of the domains, the entire magnetization of the sample can lie between zero and the saturation magnetization M_s. Differently oriented domains are separated through transition zones (domain walls, Bloch, or Néel walls); their width amounts to 100 nm. When is applying an outside magnetic field, ranges whose magnetization is under a small angle to the field, are preferred energetically in relation to other ranges.

12.3.2 Magnetic Anisotropy

Magnetic preferred directions (anisotropies) play a particular role with magnetic reversal thin films [6]. Since the dimensions in the film plane are substantially larger than the film thickness, there exists in principle a form anisotropy, which prefers energetically a magnetization parallel to the film plane. If the magnetization is taken out from the film plane, then magnetic poles on the surface films develop, which generate a magnetization that determines an opposite, demagnetizing field. Thus an energy increase is

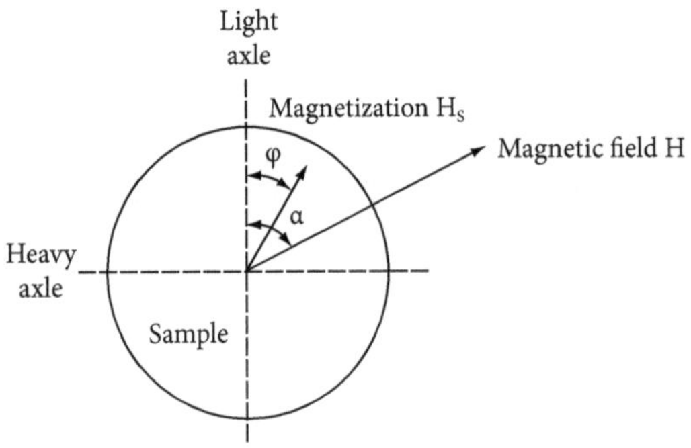

Fig. 12.4 Directions of the magnetization M_s and the field H in a circular layer with a uniaxial anisotropy

connected, which depends on the angle between the direction of magnetization and the film normal.

If the film is perpendicularly saturated to the film plane, then the power density has its maximum value $(1/2)\mu_0 M_s^2$; it becomes zero if the magnetization lies in the layer plane.

The form anisotropy represents a so-called uniaxial anisotropy, with that the unfavorable film (heavy axis) is perpendicular to the film plane and on each axis in the film plane a favorable direction (light axis) can form. Because of the form anisotropy the magnetization will usually remain in the film plane.

Additionally, uniaxial or multiaxial anisotropies can occur in the film plane [7, 8]. Also, films with perpendicular anisotropy can be generated, if the form anisotropy is partly or completely compensated by preparative procedures.

The power density of a uniaxial anisotropy is generally indicated by an anisotropy constant of K_n and described by the equation

$$E_k = K_u \times \sin^2 \varphi \,, \tag{12.8}$$

where

φ is the angle between the magnetization and the light axis (Fig. 12.4).

By uniaxial anisotropy perpendicularly to the layer plane the anisotropy constant must be corrected with the contribution of the form anisotropy. Thus the effective anisotropy constant is $K_{eff} = K_u - (1/2)\mu_0 M_s^2$.

12.3.3 Characteristics of Magnetization Reversal

With the existence of a uniaxial anisotropy the position of the magnetization can be calculated as a function of the field and therewith also the magnetization curve [6]. One accepts that the sample is homogeneously magnetized (saturation magnetization M_s) and that no domain fragmentation is present (in the domain model). Additionally to the anisotropic energy, the energy in the outside field (Zeemann energy) is considered. If the field H with the light axis includes the angle α (Fig. 12.4), the power density is

$$E = -\mu_0 H M_S \cos(\alpha - \varphi) + K_u \times \sin^2 \varphi \,. \tag{12.9}$$

The direction of the magnetization adjusts itself under the influence of the anisotropy and the outside field in such a way that the power density becomes minimal. Stability conditions supply criteria for magnetization reversal by homogeneous rotation. The magnetization curve can be calculated either from these formulas or be determined through a descriptive geometrical procedure. The limiting cases are important at which the magnetic field lies perpendicularly or parallel to the easy direction.

In the first case, the magnetization is turned continuously into the direction of the outside field by the outside field (Fig. 12.2b). The component of the magnetization in field direction, which is determined with the usual magnetization measurements, rises linear with the field strength and reaches saturation with the so-called anisotropy field strength

$$H_k = 2\, K_u / \mu_0 M_S \,. \tag{12.10}$$

This magnitude represents a characteristic measure of the strength of the anisotropy.

It leaves itself to approach from the magnetization curve in heavy axis (i.e., perpendicularly to the light axis) to determine. The anisotropy field strength during the magnetization reversal parallel to the light direction also plays a role. If the film is at first saturated in the positive direction, the magnetization remains stable also with application of a contrafield until the magnetic field reaches the value – H_k. There the magnetization jumps by homogeneous rotation into the opposite light direction. During the entire cycle, according to the domain model, one should receive a rectangular hysteresis with the values $M = \pm M_s$ and $H = \pm H_k$. Although the model represents an idealization, it is frequently used for discussion and analysis of thin films.

With real films further effects are added, which lead to more or less large deviations in the heavy direction or to ideal rectangle behavior in the light direction. By magnetization reversal domains are usually formed in the light direction, before the anisotropy field force H_k is reached. The coercive field force is not determined from the anisotropy, it is determined from the wall movements (Fig. 12.2a).

In the heavy direction the magnetization can split up when switching off the field into parallel ranges with opposite magnetization. Magnetoelastic energy by internal stresses and magnetostriction exchange energy by deviations of the magnetic moments from the ideal parallel position, crystal anisotropy for the individual crystallites, power density of the walls (wall energy), etc., which with real films can lead to further deviations from the ideal behavior.

12.3.4 Magnetic Measuring Methods (Overview)

To measure magnetic film properties basically three principles are used:

- A stray magnetic field is connected with the magnetic moment of the sample, $m = M \times V$, which is proportional to m; changes of the stray field can be proven inductively (induction magnetometer, vibrating sample magnetometer, SQUIDS) [9].
- In an outside homogeneous field the sample experiences a moment of torque; an inhomogeneous field causes a force on the sample. Both factors are proportional to m and can be measured mechanically (torsion magnetometer and gradient field magnetometers) [8].
- Optical measuring methods use the Kerr or Faraday effect [10]; with reflection at magnetic surfaces or with the passage through magnetic materials around a small angle θ, the oscillation plane from polarized light turns, so that it is proportional to the magnetization of the sample. This turn can be optically determined and supplies $M\ (H)$; the effects can also be used for the optical illustration of magnetic ranges (Kerr magnetometer and Kerr microscope).

12.3.5 Inductive Methods

With the induction magnetometer the sample periodically undergoes magnetization reversal by an exterior alternating field. The stray field of the sample produces a periodic magnetic change of the flux Φ_{St} in an induction coil and supplies an induction signal $U(t)$, which is proportional to the change of the magnetic moment:

$$U(t) \sim (\mathrm{d}\Phi_{St}/\mathrm{d}t) \sim (\mathrm{d}m/\mathrm{d}t)\,. \qquad (12.11)$$

By electronic integration of the voltage $U(t)$ we receive the magnetic moment $m(t)$. Since a relatively fast magnetization reversal (e.g., 50 Hz) is necessary air core coils must be used for the production of the outside field. The attainable field force is limited to approximately 50 kA/m. This method is only suitable for soft magnetic films for samples with a low coercive field force. Disturbance fields from the environment (e.g., from electric lines) must be eliminated by compensation circuits.

With the vibrating sample magnetometer (vibrating sample magnetometer, VSM) the sample is moved by means of a mechanical drive periodically in the outside field H transverse to the field direction (Fig. 12.5) [7].

The stray field of the sample, which is detected by induction coils, produces a periodic induction signal $U(t)$, which is proportional to the mechanical movement. With constant oscillation amplitude y_0 and constant frequency ω the amplitude U_0 of the induction signal is directly proportional to the magnetic moment: $U_0 \sim$

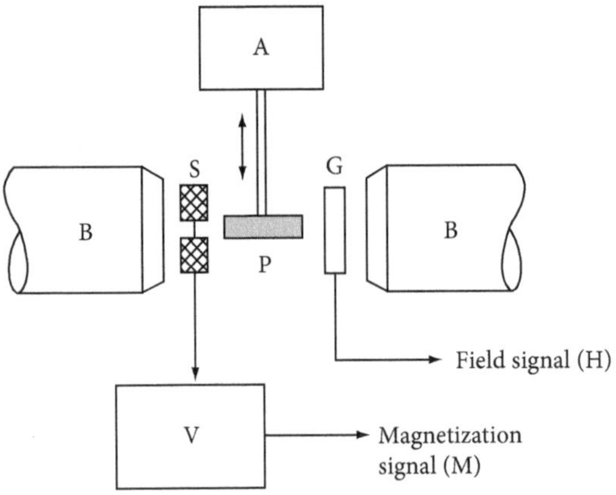

Fig. 12.5 Principle of a vibration sample magnetometer. *A* drive, *B* sample, *S* induction coils, *V* amplifier, *B* magnet, *G* hall probe

$\omega y_0 m$. For the low-noise amplification of the signal lock-in technology is used. Since the magnetic moment m as a function of the outside field strength is to be measured, the change of the field may take place only relatively slowly [measure time $t \gg (1/\omega)$], so that also electromagnets are suitable for the generation of the magnetic field.

The method can also be used for soft magnetic films (field generation preferentially by air core coils) and for hard magnetic films (field generation, e.g., by electromagnets).

The induction principle is also used by the so-called SQUID device (superconducting quantity interferometer device) [9]. Here, the stray field of a moved sample is collected by a superconducting ring, which contains a Josephson contact. Changes of the magnetic current, which is proportional to the magnetic moment of the sample, are proven by jumps of the Josephson current, which arise whenever a flux quantum leaves or enters the ring. With SQUID it is possible to extend into atomic dimensions of the magnetization.

12.3.6 Mechanical Measuring Methods

Mechanical measuring methods for the magnetic properties of thin films are based on measurement of the forces in an inhomogeneous field or of the torsion moment. The gradient field magnetometer (alternating gradient magnetometer, AGM) uses the principle of mechanical resonance [8].

The sample, which is fastened on a flexible bar, is exposed with a periodically changing gradient field, whose frequency lies near of the mechanical resonant frequency of the bar. The force on the magnetic sample stimulates the bar to resonant vibrations, whose amplitude is proportional to the magnetic moment m.

The amplitude can be measured, e.g., by the voltage induced in a piezo-crystal. The torsion moment, which experiences a homogeneous magnetized film in a magnetic field, is used for the measurement of the anisotropy constants. If the magnetization M_s of the sample lies outside the light axis, a torsion moment L_{magn} is generated, which tries to turn back the magnetization into the light axis (Fig. 12.4).

We obtain the torsion moment by the derivation of Eq. (12.8) after the angle

$$L_{magn} = (dE_k/d\varphi) = K_u \times \sin 2\varphi \,. \quad (12.12)$$

Since the position of the magnetization is fixed by the outside field, an opposite mechanical torsion moment is performed on the entire sample $L_{ext} = -L_{magn}$, which can be measured externally, e.g., over a torsion wire. By variation of size and direction of the attached field and by measurement of the external torque L_{ext}, the magnetic moment and anisotropy energy of the sample can be determined separately.

12.3.7 Optical Measuring Methods

The optical methods for investigation of magnetic thin films are based on the turn of the plane of polarization of light. Metallic films are usually not transparent; therefore the magneto-optic Kerr effect is predominantly used, which appears with reflection at magnetic surfaces [10]. If the magnetization lies in the film plane, then one differentiates between the longitudinal and transversal Kerr effect, depending on whether the magnetization is parallel or perpendicular to the plane of incidence of the light. Figure 12.6 shows

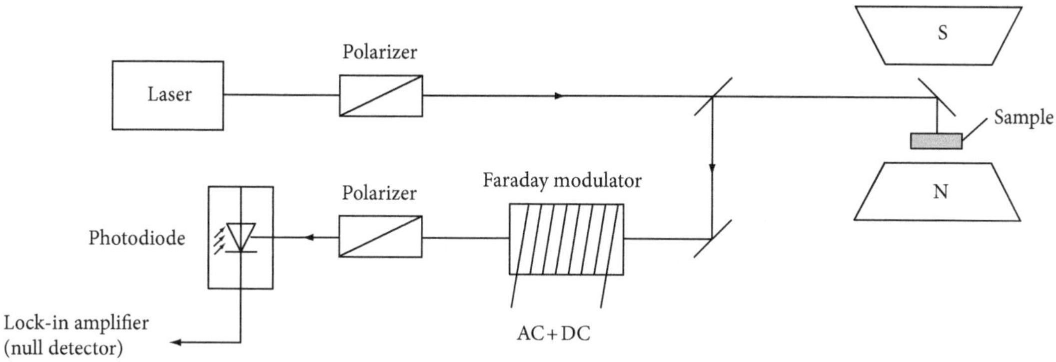

Fig. 12.6 Principle of a Kerr magnetometer

an arrangement for measure the polar Kerr effect. The light emitted by a laser strikes the sample surface over a polarizer and a beam splitter. After the reflection, the light in the beam splitter is turned back.

After reflection, the light in the beam splitter is deflected back and focused on a detector (photodiode) via a further polarizer. If the forward directions of the light to the two polarizers are accurately crossed, the intensity on the detector is zero, as long as no Kerr rotation exists. The occurrence of a Kerr rotation can, in principle, be measured by the increase of the intensity. Alternatively, the analyzer can be turned so far until the intensity at the detector again becomes zero. Both possibilities have technical problems because the electro-optical rotation lies mostly within the range of 0.5 to 1°. Frequently, a Faraday modulator is used, which turns back the polarization plane by a current applied from the outside into the original plane. The Faraday DC current is a measuring signal for the electro-optical rotation and so for the magnetization M of the sample.

An explicit increase can be reached if the Faraday DC current is overlaid with an additional AC current [11]. Longitudinal and transverse Kerr effects can be examined with an appropriate arrangement when the light beam strikes at a diagonal angle, e.g., at 45° to the film plane. Figure 12.7 shows comparison measurements for the magneto-optic double films TbFe/TbFeCo.

Additional SiN layers are suitable for an enhancement of the magnetic contrasts. By means of the Kerr effect the magnetization curves of the two magnetic layers can be determined separately. At the TbFeCo layer, which was measured through the substrate, an additional Faraday effect in the glass appears, which leads to a distortion of the hysteresis curve; however the substantial information (coercive field force and magnitude of the Kerr rotation) is not falsified. The sign of the hysteresis curve is inverted. The Kerr rotation is determined almost exclusively by the FeCo part of the transition metal.

The inversion of the hysteresis curve indicates that the magnetization of the TbFeCo layer is dominated by the rare earth part. The Kerr effect supplies detailed information about the local behavior of the individual layers; however no statement about the magnetic moments. These values result from the hysteresis curve, which was measured with the vibration magnetometer. The contributions can be determined by analysis of the curve shape. The magneto-optic Kerr effect can be used for the optical illustration of magnetic domains. Figure 12.8 shows the structure of a high resolution Kerr microscope [11, 12].

Similarly as with the Kerr magnetometer, the components are laser, polarizer, beam splitter (Smith reflector), and analyzer. The contrast is obtained with changes of intensity with crossed position of analyzer and polarizer. Figure 12.9 shows typical domain structures found in Tb_xFe_{1-x} films with different concentrations and different field forces.

Apart from the Kerr microscopy there are numerous other procedures for the illustration of

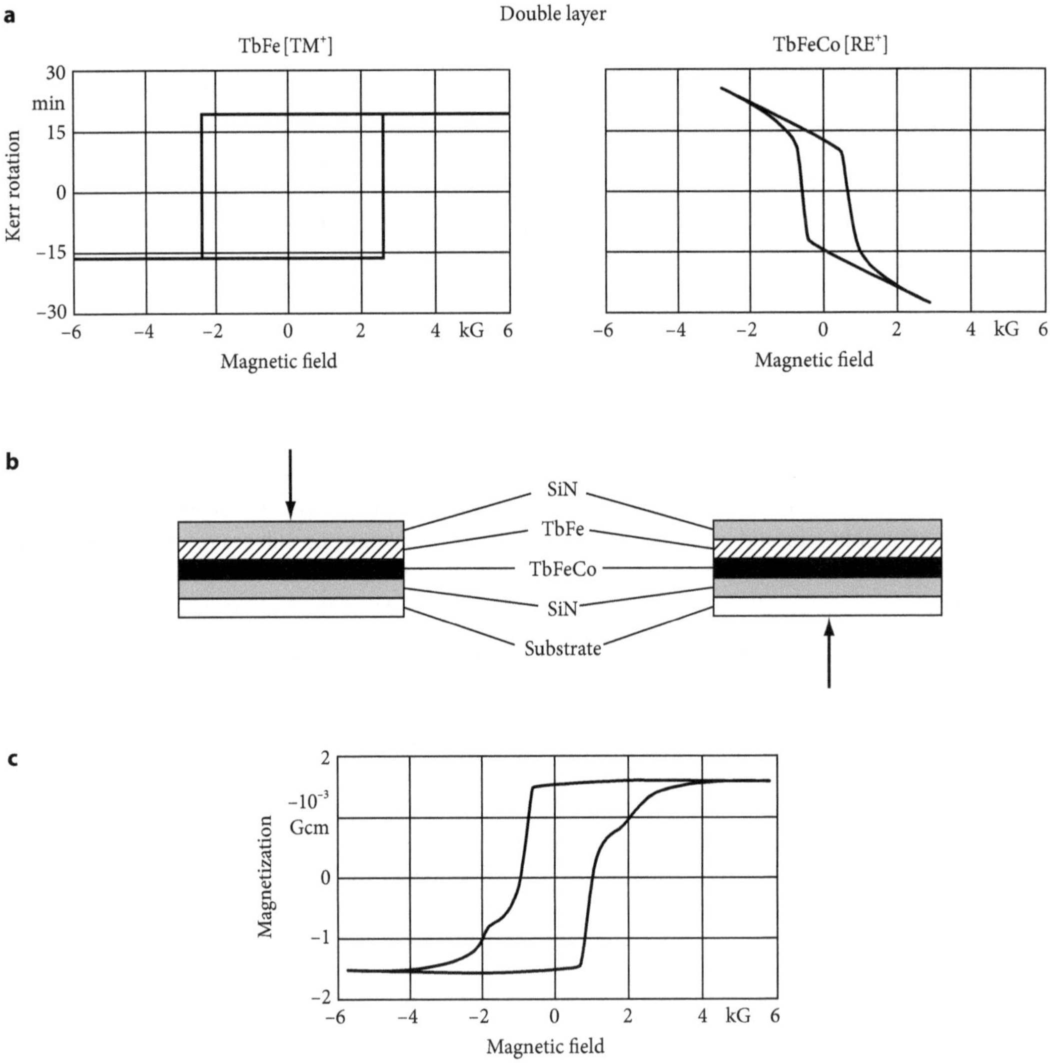

Fig. 12.7 Hysteresis curves of magneto-optic FeTb/FeTbCo double layers. **a** Measurements with the Kerr magnetometer, **b** layer structure, **c** measurements with the vibration magnetometer

magnetic surfaces. Particularly high resolutions can be reached with the magnetic force microscope (MFM).

12.4 Measurement of Color Characteristics

12.4.1 Introduction

Radiations that can cause color perceptions through stimulation of the retina are called color stimulus. A color stimulus is evaluated according to the three sensitivity functions of the eye (red, green, and blue sensitivity) and results in a color stimulus specification, which corresponds to a color perception. A color stimulus specification is described by three chromaticities and can be represented as a radius vector in the three-dimensional color space. The determination of the three chromaticities for an observer is called colorimetry or color measurement [13].

By self-luminous samples the color stimulus specification is given by the emitted radiation.

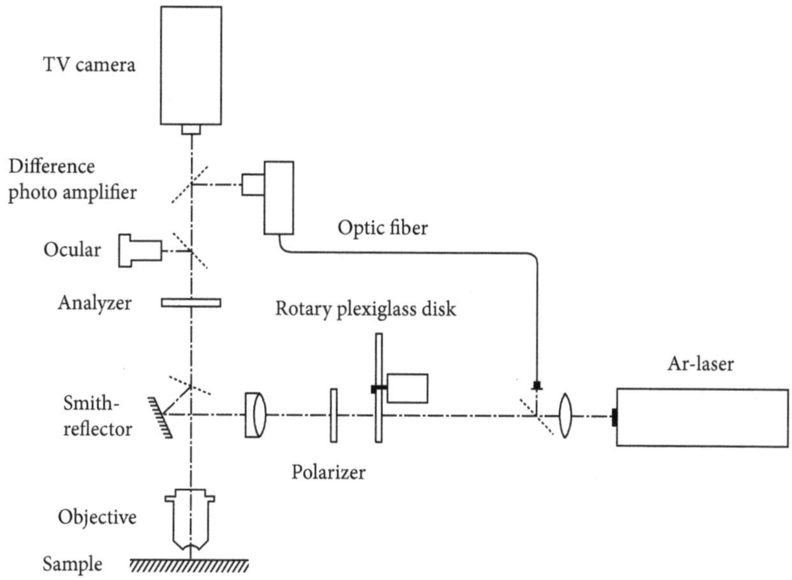

Fig. 12.8 Structure of a high resolution Kerr microscope

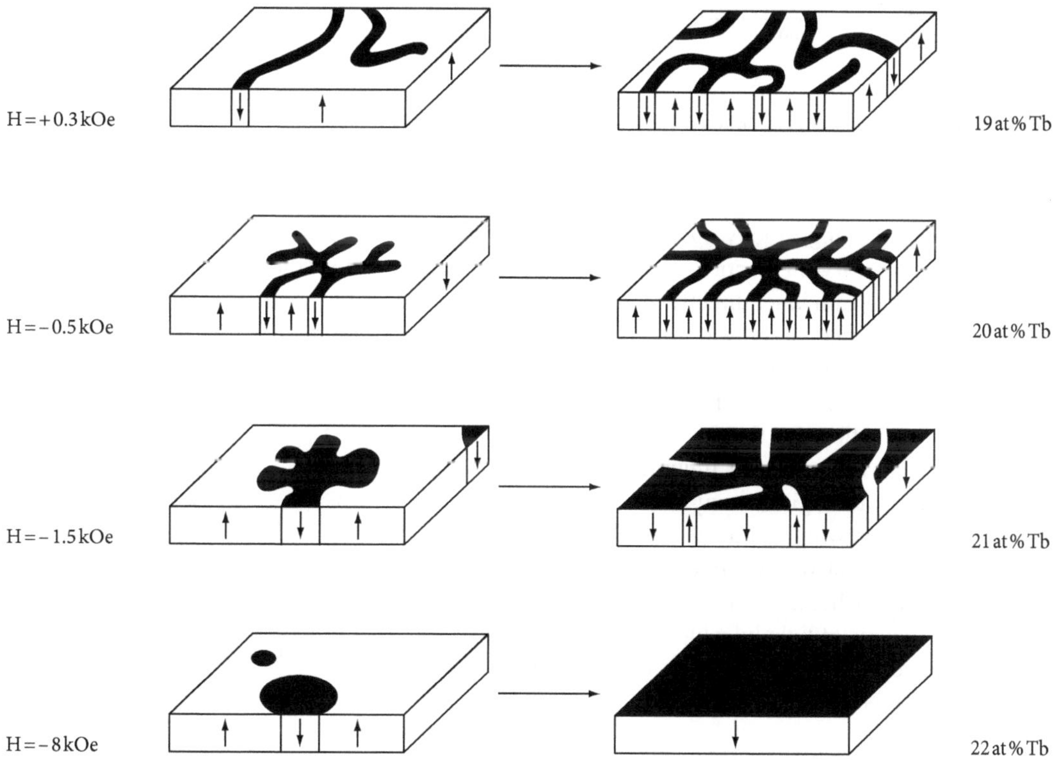

Fig. 12.9 Magnetic domains in $Fe_{1-x}Tb_x$ layers depending on different Tb (Terbium) concentrations and field forces

The color stimulus specification by not luminous samples results from the fact that the light striking the not luminous samples from a secondary light source passes through certain parts and is absorbed, scattered, and reflected. The respective parts are wavelength-dependent. For colorimetry

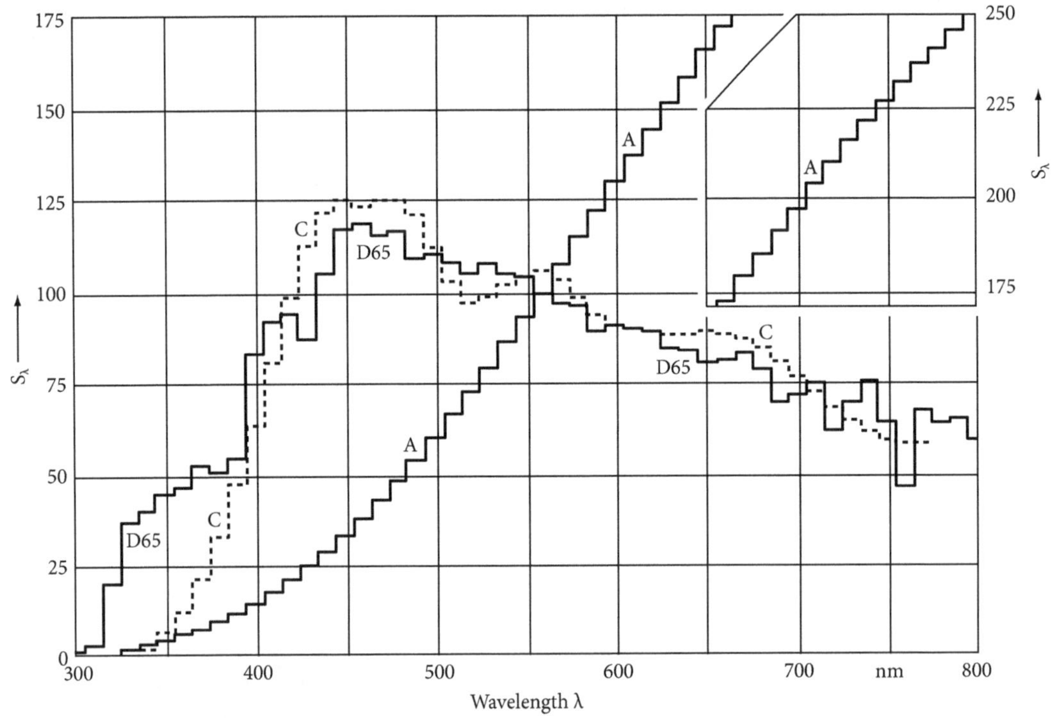

Fig. 12.10 Spectral power distribution S for the standard types of light A, C, and D65

with artificial lighting exactly specified sources of light are necessary. In practice, relative values are used; $\lambda = 560$ nm is set directly to 100.

The relative spectral power distribution of each light characteristic and is described or plotted in tables. Figure 12.10 shows the spectral radiation distribution for the standard types of light A, C, and D65. For an illuminated body, the spectral radiation gives many different color valences and therewith also a calorimetric measure on the illuminated color. In practice, we use only few types of light. At present there is still no artificial source of light by which these values can be realized. Xenon lamps with upstreamed UV filter are used.

12.4.2 Colorimetry

To generate a certain color by a mixture of basic colors, the respective parts of the basic colors are regarded as an index for the mixed color. These standard color values, which correspond to the coordinates in the color space and are represented by a space vector, are designated by X, Y, and Z, where the Y component is a degree of brightness.

For simplification we use the standard color content values x, y, and z instead of the standard color values X, Y, and Z, where it is valid that:

$$x = \frac{X}{X + Y + Z}, \quad (12.13)$$

$$y = \frac{Y}{X + Y + Z}, \quad (12.14)$$

$$z = \frac{Z}{X + Y + Z}. \quad (12.15)$$

There

$$x + y + z = 1, \quad (12.16)$$

then for the identification of a chromaticity the indication of two standard value of color contents, for example, x and y, is sufficient, while the third size is redundant and only has a control function.

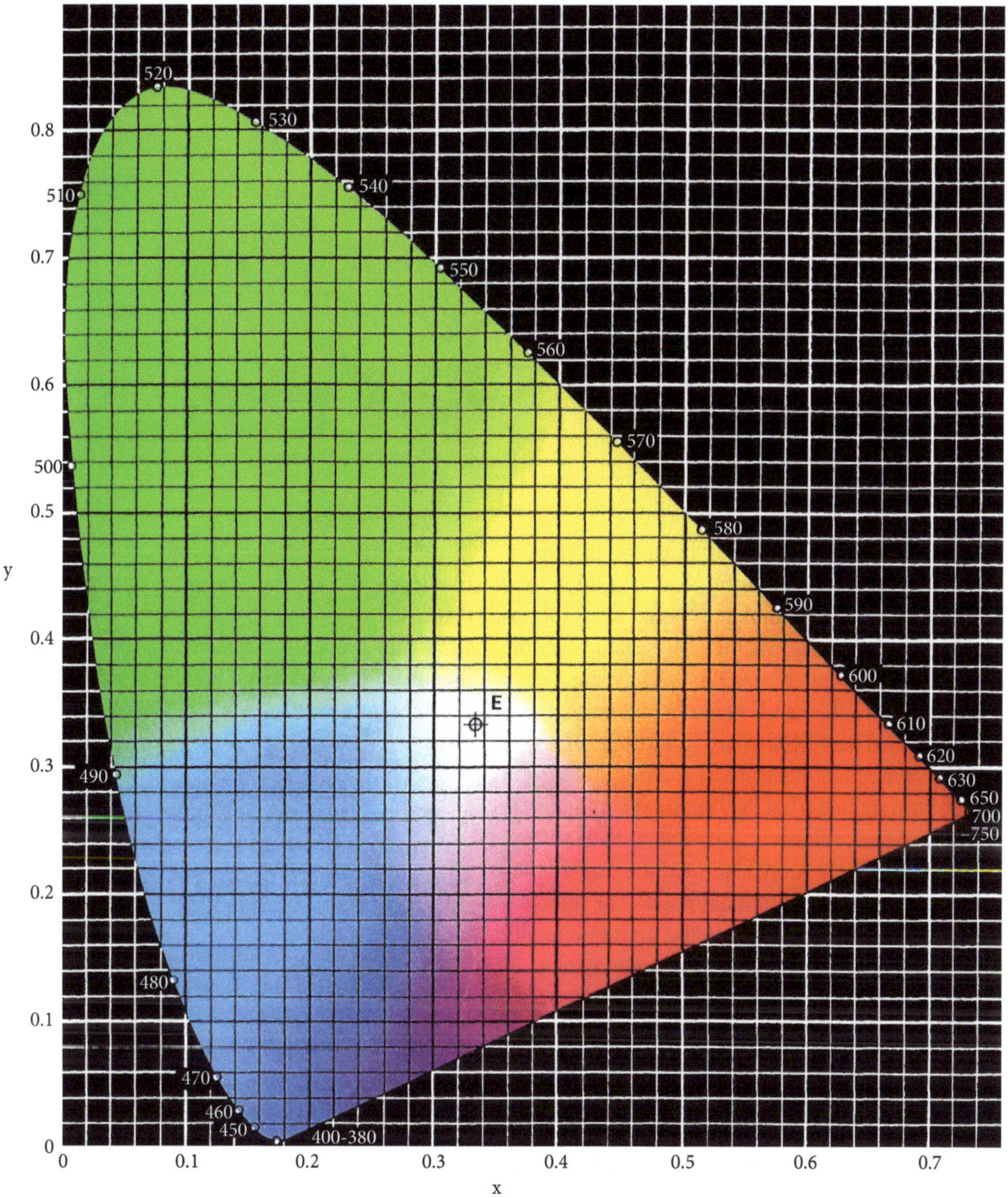

Fig. 12.11 Standard color chart according to DIN 5033

For an open identification of a color stimulus specification a further independent variable is needed as luminosity value. The values of color contents x and y can be used for a diagram of the standard value of color content in one plane. One uses a perpendicular coordinate system with the standard value of color content of x as abscissa and y as ordinate with same scales in both coordinate directions. Each chroma is assigned a chromaticity coordinate. A not closed curve ring develops; the connecting straight line of the two corner points is called the magenta straight line. All producible colors lie within this surface (Fig. 12.11).

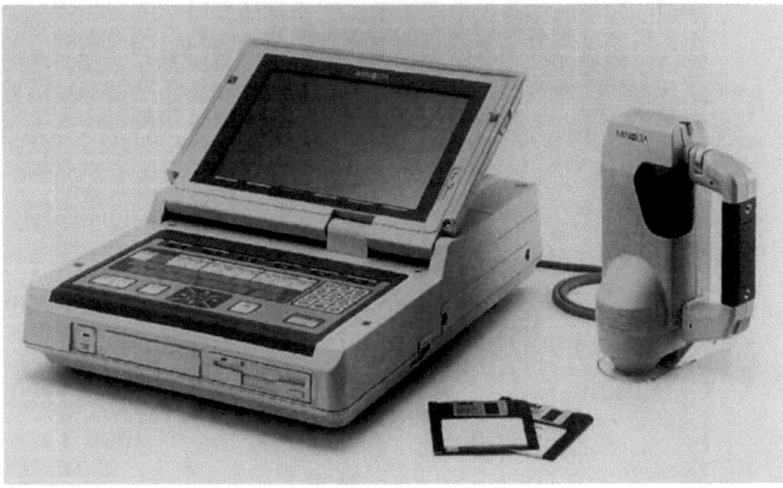

Fig. 12.12 Spectral photometers for colorimetry

With color sensibility it must be differentiated between color, colorfulness, saturation, and brightness. The illuminant point lies, for instance, in the center of the color chart with the coordinates $x = y = 0.33$. Gray and black bodies are in the same place of the color triangle because they differ only in the size of the Y value, which is a measure of the brightness. On each straight line from the illuminant point, to the side there are colors of same color shade, where the colors appear more saturated the further they are away from the illuminant point. The intersection of the straight lines with the limit line finally indicates the wavelength belonging to a color.

12.4.3 Color Measuring Instruments

As light source, modern color measuring devices use a xenon lamp with an upstream UV filter; a switch from D65 to other standard illuminants is, in general, possible. During the examination the deviation of the xenon light from the standard illuminants D65 is corrected electronically. Figure 12.12 shows a mobile spectral photometer for accurate colorimetry [14]. The light produced by the xenon lamp is diffuse scattered in the spherical mixing chamber and lights up the sample material constant over the measuring field of 8 mm from all space angles. Only the light rectangular reflected at the sample surface is collected by the measuring probe opening and led by means of a fiber glass optics to the sensor S_1 (Fig. 12.13).

A trap prevents the falsification of measurement results by gloss light. The sensor S_2 serves for lighting control.

12.5 Optical Properties of Thin Films

12.5.1 Reflectivity and Transmission

The reflectivity of a substrate covered with a thin film depends on the wavelength according to Fresnel's laws. As a consequence, also the transmission is wavelength dependent. The thin film parameters determining the reflectivity are the film thickness and the complex index of refraction $\tilde{n}$, which is determined by the index of refraction n and by the absorption coefficient α_B according to Beer's absorption law:

$$\Phi_{out} = \Phi_{in} e^{-\alpha_B z} .$$

This is the incident light power Φ_{in}, the transmitted light power Φ_{out}, and the sample thickness z.

The E field of a plane wave with wavelength λ_0 penetrating this material is described by

$$E(z,t) = E_0 e^{i(nk_0 z - \omega t)} e^{-\frac{\alpha_B}{2} z} , \qquad (12.17)$$

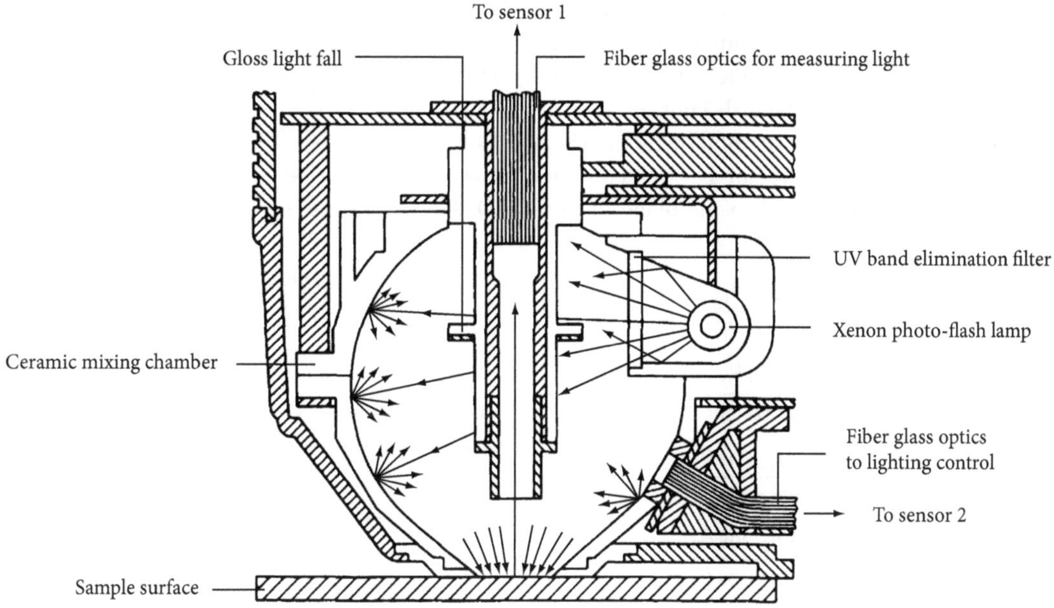

Fig. 12.13 Mixing chamber

with $k_0 = 2\pi/\lambda_0$. Setting

$$\kappa = \frac{\alpha_B}{2k_0},$$

one obtains

$$E(z,t) = E_0\,e^{i(\tilde{n}k_0 z - \omega t)},$$

with

$$\tilde{n} = n + i\kappa$$

as the complex index of refraction.

As shown in Fig. 12.14 a plane wave with amplitude 1 represented by the incident ray hits a thin film of thickness d on a substrate. The ray is partially reflected and transmitted at the interfaces air–film and film–substrate. The various partial waves are superimposed to generate the reflected wave. The complex amplitude of the reflected wave is then

$$\tilde{r} = \tilde{r}_1 + \tilde{r}_2(1-\tilde{r}_1^2)e^{-i\varphi} - \tilde{r}_1\tilde{r}_2^2(1-\tilde{r}_1^2)e^{-i2\varphi} + \tilde{r}_1^2\tilde{r}_2^3(1-\tilde{r}_1^2)e^{-i3\varphi}\ldots \tag{12.18}$$

$$= \tilde{r}_1 + \tilde{r}_2(1-\tilde{r}_1^2)e^{-i\varphi}\sum_{q=0}^{\infty}\left(-\tilde{r}_1\tilde{r}_2e^{-i\varphi}\right)^q \tag{12.19}$$

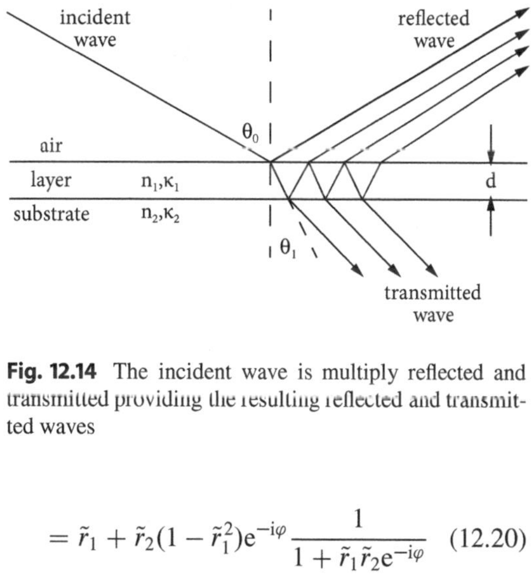

Fig. 12.14 The incident wave is multiply reflected and transmitted providing the resulting reflected and transmitted waves

$$= \tilde{r}_1 + \tilde{r}_2(1-\tilde{r}_1^2)e^{-i\varphi}\frac{1}{1+\tilde{r}_1\tilde{r}_2e^{-i\varphi}} \tag{12.20}$$

$$= \frac{\tilde{r}_1 + \tilde{r}_2e^{-i\varphi}}{1+\tilde{r}_1\tilde{r}_2e^{-i\varphi}}. \tag{12.21}$$

The sum expression in Eq. (12.19) is equal to the geometric series which is substituted by its sum formula in Eq. (12.20). A similar expression can be derived for the amplitude transmission $\tilde{t}$:

$$\tilde{t} = \frac{\sqrt{1-\tilde{r}_1^2}\sqrt{1-\tilde{r}_1^2}e^{-i\varphi/2}}{1+\tilde{r}_1\tilde{r}_2e^{-i\varphi}}. \tag{12.22}$$

Here $\tilde{r}_1$ is the complex amplitude reflection according to Fresnel's equations at the interface air–film and $\tilde{r}_2$ at the interface film–substrate [17]. We distinguish between perpendicular (s) and parallel (p) polarization, which means that the E field vector is oriented either perpendicularly or parallel to the plane of incidence:

$$\tilde{r}_{1s} = \frac{\cos\theta_0 - \tilde{n}_1\cos\theta_1}{\cos\theta_0 + \tilde{n}_1\cos\theta_1} \tag{12.23}$$

$$\tilde{r}_{1p} = \frac{\cos\theta_1 - \tilde{n}_1\cos\theta_0}{\cos\theta_1 + \tilde{n}_1\cos\theta_0} \tag{12.24}$$

$$\tilde{r}_{2s} = \frac{\tilde{n}_1\cos\theta_0 - \tilde{n}_2\cos\theta_1}{\tilde{n}_1\cos\theta_0 + \tilde{n}_2\cos\theta_1} \tag{12.25}$$

$$\tilde{r}_{2p} = \frac{\tilde{n}_1\cos\theta_1 - \tilde{n}_2\cos\theta_0}{\tilde{n}_1\cos\theta_1 + \tilde{n}_2\cos\theta_0}\,, \tag{12.26}$$

with $\tilde{n}_1$ as the complex index of the film and $\tilde{n}_2$ as the complex index of the substrate; θ_0 is the angle of incidence and θ_1 the angle of the ray refracted at the air–film interface. The phase shift after passing the film with thickness d twice is:

$$\varphi = k_0 \cdot \tilde{n}_1 \cdot 2d \cdot \cos\theta_1\,.$$

Note that both φ and $\cos\theta_1$ are complex, because $\tilde{n}_1$ is a complex number and because of the relation

$$\sin\theta_0 = \tilde{n}_1 \sin\theta_1\,.$$

$\tilde{n}_1$ is real only for dielectric films and substrates and thus θ_1. The reflectivity for p polarized light is then

$$R_p = |\tilde{r}_p|^2 = \left|\frac{\tilde{r}_{1p} + \tilde{r}_{2p}e^{-i\varphi}}{1 + \tilde{r}_{1p}\tilde{r}_{2p}e^{-i\varphi}}\right|^2$$

and for s polarized light

$$R_s = |\tilde{r}_s|^2 = \left|\frac{\tilde{r}_{1s} + \tilde{r}_{2s}e^{-i\varphi}}{1 + \tilde{r}_{1s}\tilde{r}_{2s}e^{-i\varphi}}\right|^2\,.$$

In a similar way, for the transmission for p polarized light we have

$$T_p = |\tilde{t}_p|^2 = \left|\frac{\sqrt{1-\tilde{r}_{1p}^2}\sqrt{1-\tilde{r}_{2p}^2}\,e^{-i\varphi/2}}{1 + \tilde{r}_{1p}\tilde{r}_{2p}e^{-i\varphi}}\right|^2$$

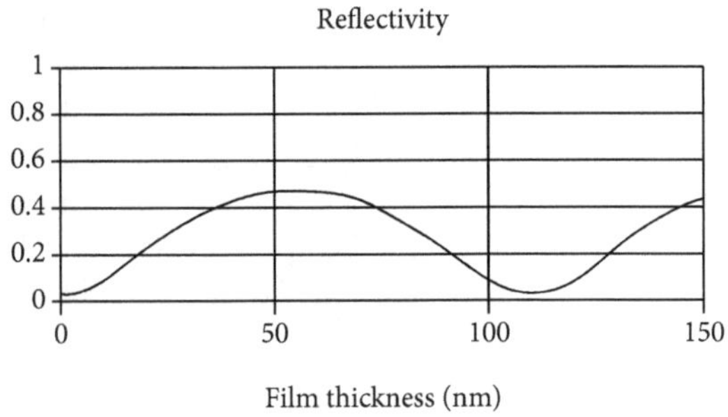

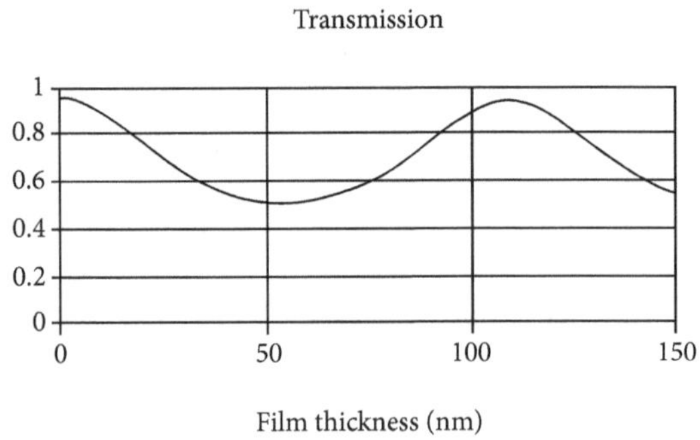

Fig. 12.15 Reflectivity R and transmission T as functions of the film thickness d. The material parameters for the thin film are $n = 2.9$, $k = 0.005$ and for the substrate $n = 1.52$, $k = 0$

and for s polarized light

$$T_s = |\tilde{t}_s|^2 = \left|\frac{\sqrt{1-\tilde{r}_{1s}^2}\sqrt{1-\tilde{r}_{2s}^2}\,e^{-i\varphi/2}}{1 + \tilde{r}_{1s}\tilde{r}_{2s}e^{-i\varphi}}\right|^2\,.$$

As an example, Fig. 12.15 shows the reflectivity $R(d)$ and transmission $T(d)$ as functions of the film thickness d for $\lambda = 632,8\,\text{nm}$, $n_1 = 2.9$, $\kappa_1 = 0.005$, $n_2 = 1.52$, and $\kappa_2 = 0.0$ for normal incidence. Thus, no distinction is made between s and p polarization.

12.5.2 Ellipsometry

When polarized light is reflected at a surface, the polarization state changes depending on the thickness, the index of refraction and the absorption coefficient of the thin film and the substrate. Ellipsometry is a technique that can measure these parameters with high accuracy [20]. The principle is rather simple. The change of the polarization state of a light beam incident onto the sample is measured after the reflection. The adequate formalism to describe the phenomena associated with ellipsometry is the Jones and Stokes formalism of polarization.

12.5.2.1 Jones and Stokes Vector Formalism and Polarization

The E field vector of a plane wave propagating in the z direction is

$$\vec{E} = \begin{pmatrix} |E_x| e^{i(-\omega t + kz)} \\ |E_y| e^{i(-\omega t + kz + \varphi)} \end{pmatrix}$$

$$= \begin{pmatrix} |E_x| \\ |E_y| e^{i\varphi} \end{pmatrix} e^{i(-\omega t + kz)} .$$

This wave is elliptically polarized because the E field vector describes an ellipse as a function of time in a plane normal to the propagation direction of the electromagnetic wave. Special cases are described below.

Linear Polarized Light.

$$|E_x| = 0 \quad \vec{E} = \binom{0}{1} \quad \text{p polarized}$$
$$|E_y| = 0 \quad \vec{E} = \binom{1}{0} \quad \text{s polarized.}$$

Circular Polarized Light.

$$|E_x| = |E_y| \quad \varphi = \pi/2 \quad \vec{E} = \frac{1}{\sqrt{2}}\binom{1}{e^{i\pi/2}} = \frac{1}{\sqrt{2}}\binom{1}{i} \quad +\sigma \text{polarized}$$
$$|E_x| = |E_y| \quad \varphi = -\pi/2 \quad \vec{E} = \frac{1}{\sqrt{2}}\binom{1}{e^{-i\pi/2}} = \frac{1}{\sqrt{2}}\binom{1}{-i} \quad -\sigma \text{polarized.}$$

The vectors $\binom{1}{0}$, $\binom{0}{1}$, $\frac{1}{\sqrt{2}}\binom{1}{i}$, and $\frac{1}{\sqrt{2}}\binom{1}{-i}$ are the basic Jones vectors [19]. The vectors $\binom{1}{0}$, $\binom{0}{1}$ and $\frac{1}{\sqrt{2}}\binom{1}{i}$, $\frac{1}{\sqrt{2}}\binom{1}{-i}$ are orthogonal vector pairs. Therefore, in general, any polarization state can be described as a superposition of the two circular polarization Jones vectors or the two linear polarization Jones vectors:

$$\begin{pmatrix} a \\ b e^{i\varphi} \end{pmatrix} = a \begin{pmatrix} 1 \\ 0 \end{pmatrix} + b e^{i\varphi} \begin{pmatrix} 0 \\ 1 \end{pmatrix} \qquad (12.27)$$

with the Jones parameters a, b, φ. The general case of elliptically polarized light means that the E field vector describes an ellipse in the plane orthogonal to the z axis (propagation axis) (see Fig. 12.16).

This ellipse can be described with the angle α between the principal axis of the ellipse and the y axis of the coordinate system and the ratio between the length of the two principal axes (ellipticity) ϵ.

These two quantities can be associated with the angular coordinates on a sphere. The angle 2ϵ plays the role of the declination angle and 2α the role of the azimuth angle. Thus, each point of this Poincaré-sphere (Fig. 12.17) corresponds to a unique elliptical polarization state. Of course, one can also use the Cartesian coor-

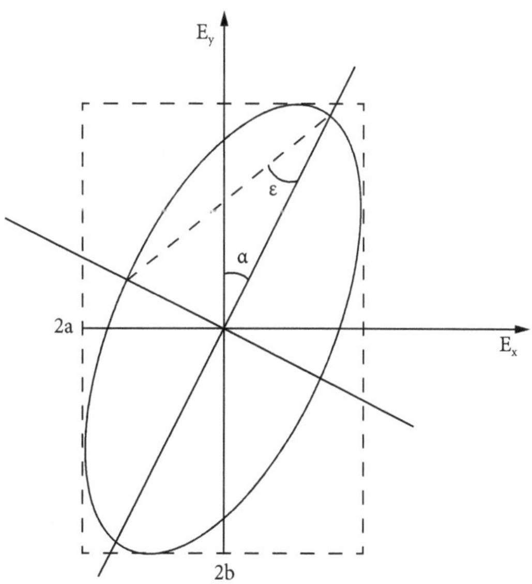

Fig. 12.16 Parameters of the elliptical polarization state. The E field describes an ellipse with tilt angle α and ellipticity angle ϵ. The ellipse inscribes a rectangle with side lengths $2a$ and $2b$

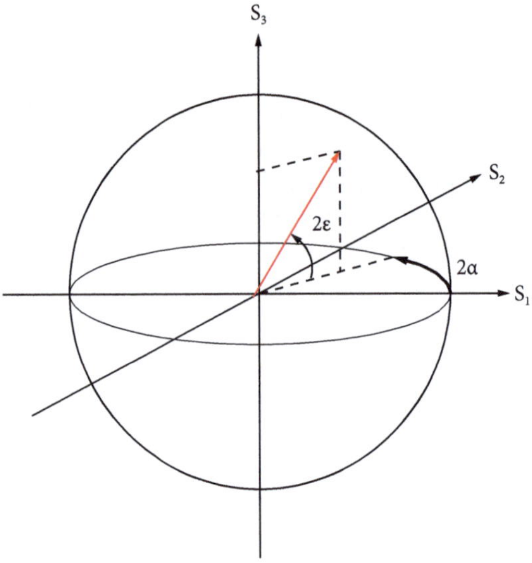

Fig. 12.17 Poincaré sphere

dinates (Stokes vector)

$$S_1 = S_0 \cos 2\epsilon \cos 2\alpha \quad (12.28)$$

$$S_2 = S_0 \cos 2\epsilon \sin 2\alpha \quad (12.29)$$

$$S_3 = S_0 \sin 2\epsilon \quad (12.30)$$

with

$$S_1^2 + S_2^2 + S_3^2 = S_0^2 \quad (12.31)$$

and the intensity S_0. The complete Stokes vector is

$$\vec{S} = \begin{pmatrix} S_0 \\ S_1 \\ S_2 \\ S_3 \end{pmatrix} = \begin{pmatrix} S_0 \\ S_0 \cos 2\epsilon \cos 2\alpha \\ S_0 \cos 2\epsilon \sin 2\alpha \\ S_0 \sin 2\epsilon \end{pmatrix}. \quad (12.32)$$

The following relations to the Jones parameters a, b, φ exist:

$$a^2 = (S_0 + S_1)/2 \quad (12.33)$$

$$b^2 = (S_0 - S_1)/2 \quad (12.34)$$

$$\varphi = \arctan(S_3/S_2) \quad (12.35)$$

or

$$\tan(2\alpha) = \frac{2ab \cos\varphi}{a^2 - b^2} \quad (12.36)$$

$$\sin(2\epsilon) = \frac{2ab}{a^2 + b^2} \sin\varphi\,. \quad (12.37)$$

Measurement of the Stokes Parameters. The components of the Stokes vector can be measured. The positive S_1 component indicated in the following as S_1^+ corresponds to p polarized light and the negative S_1 component indicated as S_1^- corresponds to s polarized light. Thus, if the elliptically polarized light beam hits a polarizing beam splitter, the p polarized component is transmitted (S_1^+) and the s polarized component is reflected (S_1^-).

In a similar way, S_2^+ and S_2^- can be measured with a polarizing beam splitter that is oriented 45° (45° -analyzer) relative to the orientation of the polarizer in the previous case (0° analyzer). The transmitted light corresponds to S_2^+ and the reflected light to S_2^-.

For the measurement of the circular polarized components S_3^+ and S_3^-, a $\lambda/4$ -plate in combination with a 45° analyzer is necessary. Only positive circular polarized light (S_3^+) is transmitted. The negative part S_3^- of S_3 is reflected by the 45° analyzing beam splitter.

The Stokes vector components are then:

$$S_1 = S_1^+ - S_1^- \quad (12.38)$$

$$S_2 = S_2^+ - S_2^- \quad (12.39)$$

$$S_3 = S_3^+ - S_3^-\,. \quad (12.40)$$

Because of energy conservation also the following relation is fulfilled:

$$S_0 = S_1^+ + S_1^- = S_2^+ + S_2^- = S_3^+ + S_3^-\,. \quad (12.41)$$

With Eqs. (12.38)–(12.41) we obtain

$$S_1 = 2S_1^+ - S_0 \quad (12.42)$$

$$S_2 = 2S_2^+ - S_0 \quad (12.43)$$

$$S_3 = 2S_3^+ - S_0\,. \quad (12.44)$$

With the additional condition

$$S_1^2 + S_2^2 + S_3^2 = S_0^2 \quad (12.45)$$

it is sufficient to determine S_3^+, S_2^+ and S_2^-. The Stokes parameter S_2 can be determined with Eq. (12.39), S_0 with Eq. (12.41), S_3 with Eq. (12.44), and S_1 with Eq. (12.45).

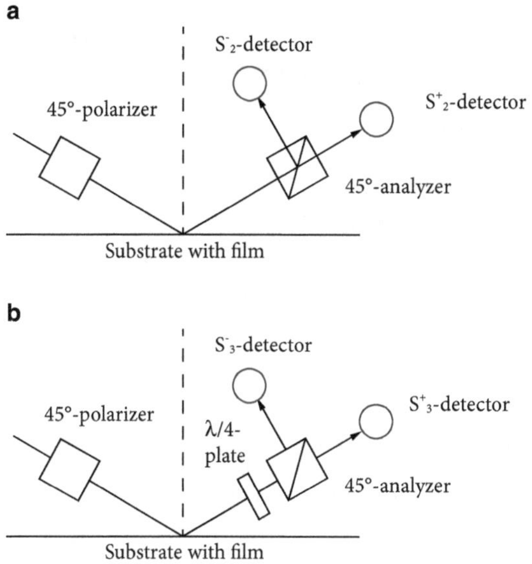

Fig. 12.18 **a** State of the ellipsometer with the $\lambda/4$ plate flipped out for the measurement of the S_1^+ and the S_1^- component of the Stokes vector. **b** State of the ellipsometer with $\lambda/4$-plate flipped in for the measurement of the S_3^+ and S_3^-component of the Stokes vector

A device that measures these components is an ellipsometer. It can be used to measure in particular the polarization state of the light beam reflected from a thin film layer (or multilayer) on a substrate. The principle setup of an ellipsometer is shown in Fig. 12.18.

The incident light beam is linearly polarized 45° relative to the plane of incidence by a polarizer. The reflected beam hits the 45° oriented polarizing beam splitter (analyzer). The two detectors at the two outputs of the analyzer measure S_2^+ and S_2^-.

In a second measurement a $\lambda/4$ plate is flipped in front of the analyzer. The optical axis of the $\lambda/4$ plate is oriented parallel to the plane of incidence. Therefore, positive and negative circular polarized light is detected by the two detectors at the output of the 45° analyzer, respectively, providing the signals S_3^+ and S_3^-. The light source can be a laser. If the change of the polarization state due to the reflection is measured across a broad spectral range a white light source is used (e.g., halogen lamp). In that case, a monochromator is used to filter out a tunable wavelength.

With Eqs. (12.33)–(12.35) the Jones vector components can be calculated from the Stokes vector components where a corresponds to parallel component E_p of the electrical field vector of the reflected beam and b is the orthogonal component E_s. The angle φ corresponds to the phase difference between the parallel and the orthogonal component.

By convention the angles Ψ and Δ are used to describe the polarization state of the reflected light beam; they are related to the Jones vector components by the following relations:

$$\Psi = \arctan \frac{E_\mathrm{p}}{E_\mathrm{s}} = \arctan \frac{a}{b} \tag{12.46}$$

$$\Delta = \varphi\,. \tag{12.47}$$

Instead of inserting a quarter-wave plate an alternative approach is to rotate a half-wave plate in front of the analyzer or to rotate the analyzer itself with a certain frequency (rotating element ellipsometry). The detected intensity behind the analyzer is

$$I(t) = I_0[1 + A\cos(2\omega t) + B\sin(2\omega t)]\,. \tag{12.48}$$

The signal is Fourier analyzed. The Fourier coefficient A is related to the Jones coefficients a, b by

$$a^2 = 1 + A \tag{12.49}$$

$$b^2 = 1 - A\,. \tag{12.50}$$

With Eq. (12.46) it follows that

$$\tan\Psi = \sqrt{\frac{1+A}{1-A}}\,. \tag{12.51}$$

The phase shift ϕ of the oscillating function $I(t)$ in Eq. (12.48) is

$$\tan(\phi) = B/A\,. \tag{12.52}$$

As can be easily seen from Fig. 12.16 that the phase ϕ is determined by the tilt angle α when the analyzer is rotated. It is $\phi = 2\alpha$. Thus

$$\tan(2\alpha) = B/A\,. \tag{12.53}$$

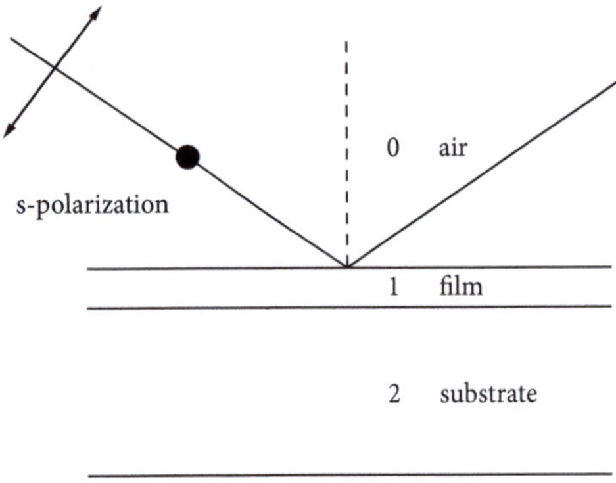

Fig. 12.19 Reflection at the film layer

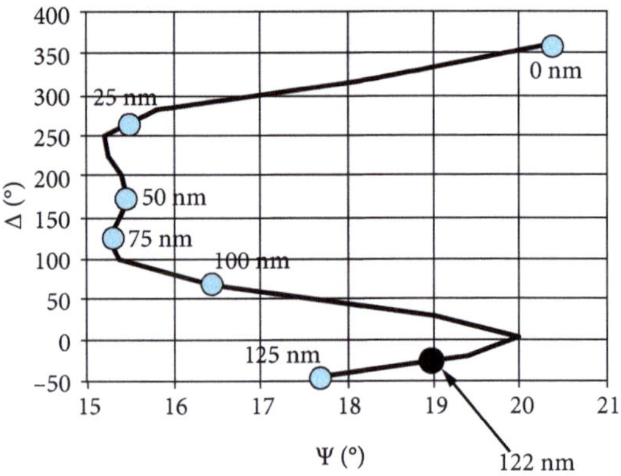

Fig. 12.20 Parameterized Ψ-Δ plot. The parameter is the film thickness d. The material parameters in this example are $n_1 = 2.9$, $\kappa = 0.005$ for the thin film and $n_2 = 1.52$, $\kappa = 0$ for the substrate

With Eqs. (12.53), (12.49), (12.50), (12.36), and (12.47) it follows that

$$\cos\Delta = \frac{B}{\sqrt{1-A^2}}. \quad (12.54)$$

From Eqs. 12.54 and (12.51) the ellipsometric angles Δ and Ψ can be determined.

In another type of ellipsometer the polarization state of the incoming beam is adjusted so that the polarization of the reflected beam is linear and blocked by the analyzer. The adjustment of the elliptical polarization state can be achieved, for example, by rotating the polarizer and a quarter-wave plate appropriately. This nulling method is sensitive but needs several motors for the rotations.

The angles Ψ and Δ are related to the Fresnel reflection coefficients $r_p = E_p^{out}/E_p^{in}$ and $r_s = E_s^{out}/E_s^{in}$ of the p and s polarized component by:

$$\frac{r_p}{r_s} = \tan\Psi e^{i\Delta}, \quad (12.55)$$

with E_p^{out} as the amplitude of the reflected p polarized beam, E_s^{out} as the amplitude of the reflected s polarized beam, E_p^{in} as the amplitude of the incident p polarized beam and E_s^{in} as the amplitude of the incident s polarized beam.

The parameters Ψ and Δ are related to the film thickness d, the complex index $\tilde{n}_1 = n_1 + i\kappa_1$ of the film and the complex index $\tilde{n}_2 = n_2 + i\kappa_2$ of the substrate. With Eqs. (12.55) and (12.21) we obtain:

$$\tan\Psi e^{i\Delta} = \frac{\tilde{r}_{1p} + \tilde{r}_{2p}e^{-i\varphi}}{1 + \tilde{r}_{1p}\tilde{r}_{2p}e^{-i\varphi}} \cdot \frac{1 + \tilde{r}_{1s}\tilde{r}_{2s}e^{-i\varphi}}{\tilde{r}_{1s} + \tilde{r}_{2s}e^{-i\varphi}}, \quad (12.56)$$

with the Fresnel reflection coefficient $\tilde{r}_{1p}$ for the p polarization component, $\tilde{r}_{1s}$ for the s polarization component at the interface between air and the film, and $\tilde{r}_{2p}$ for the p polarization component and $\tilde{r}_{2s}$ for the s-polarization component at the interface between film and substrate (Fig. 12.19).

The parameters n_1, κ_1, and d are the unknown film parameters that will be determined by the ellipsometer measurement. However, Eq. (12.56) cannot be solved for all three unknowns because there are only two equations relating the real and imaginary part of the left and right sides of Eq. (12.56), respectively. Thus at least one parameter needs to be known in advance. In practice, either the film material is known and the thickness is determined or the thickness is measured with an alternative technique and the material constants n_1 and κ_1 are determined. Of course, ellipsometry can also be used to measure n and k of the pure bulk material. The identical formalism can be applied with $d = 0$.

As an illustration, Fig. 12.20 shows the solution of Eq. (12.56) as a curve in the Ψ-Δ plane with the thickness d as the curve parameter and $n_1 = 2.9, \kappa_1 = 0.005$ as the parameters of a ZnO

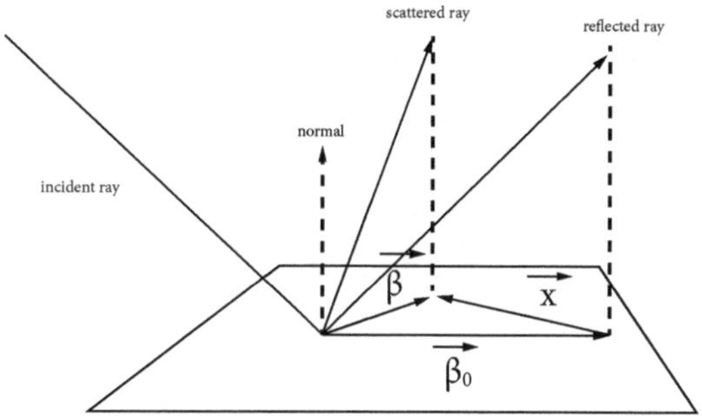

Fig. 12.21 The incoming ray is scattered at the interception point with the surface. The specular beam fulfills the reflection law

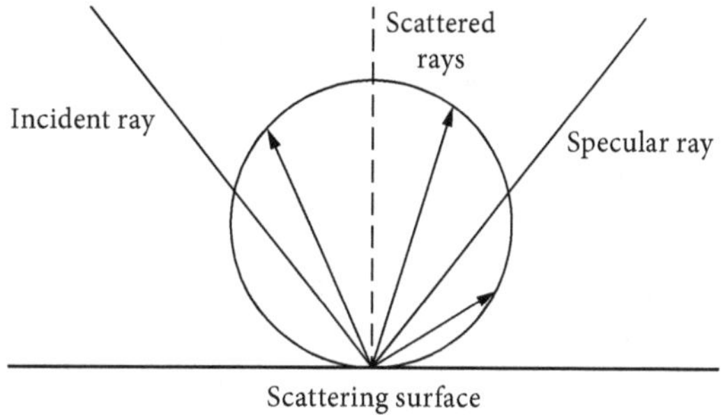

Fig. 12.22 Angular intensity distribution of a Lambertian scattering surface

film on a BK7 glass substrate with $n_2 = 1.52$ and $\kappa_2 = 0$. The probe wavelength is $\lambda = 632.8$ nm and the incidence angle $\theta_0 = 70°$. The ellipsometer provides the result $\Delta = -31°$ and $\Psi = 19°$. The thickness parameter d corresponding to these values can be determined from Fig. 12.20 as 122 nm.

12.5.3 Scattering

Besides reflectivity, absorption and transmission thin films can also be characterized by their scattering properties. In terms of ray optics scattering can be described as the stochastic change of the propagation direction of a light ray at the surface interface. Even if the surface is not actually reflective there is always the specular ray, which changes its direction according to the reflection law.

Scatter Models. Scatter models are defined in terms of an intensity probability distribution function. When a ray is scattered, a new direction of propagation is chosen. Figure 12.21 shows the vectors used to describe the various scattering models.

The normal vector $\vec{N}$ defines the orientation of the surface at the ray–surface interception point. The incident ray is described by $\vec{I}$, the specular ray vector is $\vec{R}$, and the scattered ray vector is $\vec{S}$. The vectors $\vec{N}$, $\vec{I}$, $\vec{R}$, and $\vec{S}$ are all unit vectors. The projection of the specular and scattered ray vectors onto the surface are denoted by $\vec{\beta}_0$ and $\vec{\beta}$, respectively. The vector $\vec{\beta}_0$ has a length equal to $\sin\theta_r$, while $\vec{\beta}$ has a length of $\sin\theta_0$, where θ_r and θ_s denote the angles between the normal vector and the specular and scattered ray angles, respectively. The scatter vector is denoted $\vec{x}$ depending on the scatter model chosen. The most important models are the Lambertian, Gaussian, and the ABg models.

Lambertian Scattering. Lambertian scattering means the angular intensity distribution function is

$$I(\theta_s) = I_0 \cos\theta_s \tag{12.57}$$

(see Fig. 12.22). Here I_0 is the intensity at $\theta_s = 0$. Note that Lambertian scattering is independent of the angle of incidence θ_0. Most diffuse surfaces are very nearly Lambertian.

Gaussian Scattering. The Gaussian scattering distribution function is of the form:

$$I(|\vec{x}|) = I_0 \mathrm{e}^{-|\vec{x}|^2/\sigma^2}, \tag{12.58}$$

where I_0 is again a normalizing constant and the surface type-dependent constant σ characterizes the width of the distribution.

ABg Scattering Model. The ABg scattering model is a widely used method for defining the bidirectional scattering distribution function

(BSDF). BSDF is defined as the scattered radiance per unit incident irradiance, or

$$\mathrm{BSDF}(\theta_\mathrm{i},\phi_\mathrm{i},\theta_\mathrm{s},\phi_\mathrm{s}) = \frac{\mathrm{d}L_\mathrm{s}(\theta_\mathrm{s},\phi_\mathrm{s})}{\mathrm{d}E_\mathrm{i}(\theta_\mathrm{i},\phi_\mathrm{i})}, \quad (12.59)$$

where θ is the declination angle measured from the normal, and ϕ is the azimuthal angle, and the subscripts i and s refer to incident and scattered directions, respectively. The general term BSDF can refer to two separate functions, BRDF and BTDF, for reflective and transmitted distributions, respectively.

BSDF Properties. For many optical surfaces, BSDF is independent of the incident direction if it is plotted as a function of $|\vec{x}| = |\vec{\beta} - \vec{\beta}_0|$ instead of angles of incidence. This quantity is the distance between the scattered and unscattered ray vectors when projected down to the scattering surface. It is generally adequate to use this scattering model when the scattering is mainly due to random isotropic surface roughness, and the scale of the roughness is small compared to wavelength of the light being scattered. These assumptions are generally valid for polished optical surfaces.

The ABg BSDF Model. When BSDF is plotted as a function of $|\vec{x}|$, the resulting data may usually be fit to a function of the form

$$\mathrm{BSDF} = \frac{A}{B + |\vec{x}|^g}. \quad (12.60)$$

The ABg BSDF model is named such because of the three fit parameters A, B, and g. If g is 0 (note that g may be any value, positive or negative, but typically it is between 0 and 3), then BSDF is constant in the direction cosine space:

$$\mathrm{BSDF} = \frac{A}{B + 1}, \quad (12.61)$$

and the resulting scattering is Lambertian.

12.5.4 Color

The color of thin films is determined by the spectrum of the reflected or transmitted light. Color is strictly speaking not a pure physical quantity but related both to physics and subjective perception. However, it is possible to attribute to each color a set of values which can be related to the measurable light spectrum. The procedure how to calculate these color coordinates from a given light spectrum is regulated in various international standards. The most common standard is the CIE (1932) standard [18].

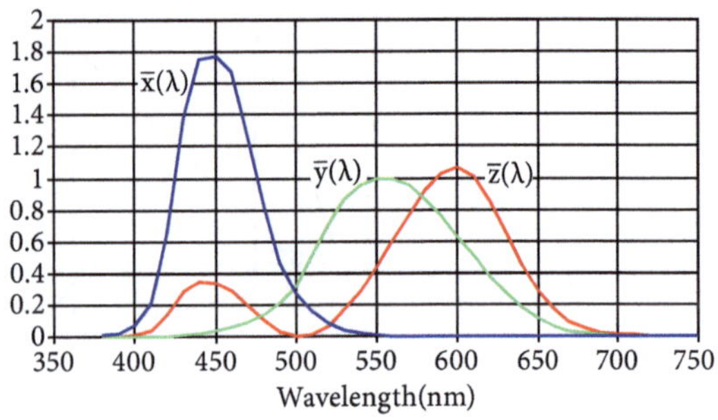

Fig. 12.23 Standardized normal color matching functions

Given the spectral radiant flux $\Phi_\mathrm{e}(\lambda)$ of a light source, we first calculate the normalized spectrum

$$S(\lambda) = \frac{\Phi_\mathrm{e}(\lambda)}{\Phi_\mathrm{e}(\lambda_0)},$$

with λ_0 as the wave length of the maximum of the spectrum. If the spectral transmission or reflectivity of the sample is described by the normalized spectrum $\varphi(\lambda)$ the so-called tristimulus coordinates are

$$X = k \int_{380\,\mathrm{nm}}^{780\,\mathrm{nm}} \varphi(\lambda) S(\lambda) \cdot \bar{x}(\lambda) \cdot \mathrm{d}\lambda$$

$$Y = k \int_{380\,\mathrm{nm}}^{780\,\mathrm{nm}} \varphi(\lambda) S(\lambda) \cdot \bar{y}(\lambda) \cdot \mathrm{d}\lambda\,,$$

$$Z = k \int_{380\,\mathrm{nm}}^{780\,\mathrm{nm}} \varphi(\lambda) S(\lambda) \cdot \bar{z}(\lambda) \cdot \mathrm{d}\lambda\,.$$

The standardized normal color matching functions $\bar{x}(\lambda)$, $\bar{y}(\lambda)$, and $\bar{z}(\lambda)$ are shown in Fig. 12.23.

The constant k is determined by the condition for a "white standard" ($\varphi(\lambda) = 1$)

$$Y = k \int_{380\,\mathrm{nm}}^{780\,\mathrm{nm}} S(\lambda) \cdot \bar{y}(\lambda) \cdot \mathrm{d}\lambda = 100\,. \quad (12.62)$$

The xy chromaticity coordinates are defined as:

$$x = \frac{X}{X+Y+Z} \quad (12.63)$$

$$y = \frac{Y}{X+Y+Z}\,. \quad (12.64)$$

With x and y the chromaticity can be visualized in a two-dimensional color map. The neutral point is at $x = y = 0.33$ (black, gray, white). Light characterized by a single wavelength is called a spectral pure color. The chromaticity coordinates of pure colors comprise a triangle within the color map. Note that two different spectra may have the same chromaticity coordinates causing the same color impression.

The radial distance within the color triangle to the neutral point is a measure for the color saturation. The azimuthal position represents the color tone. The spectrally pure colors (determined by one wave length) are located on a line limiting the color triangle which comprises all possible colors.

12.6 Spectrometers

The spectral properties of thin films can be measured with spectrometers. Although there is a large variety of different types and principles of spectrometers, all spectrometers have in common a light source, some wavelength selecting component, and a light detector. The different wavelengths of a spectrum can be separated into an angular spectrum (grating and prism spectrometers) or transformed into a time-dependent signal which is finally Fourier transformed (Fourier transform spectrometers). Another important class are variable narrow-band optical filters like Fabry–Perot spectrometers. In thin film technology mainly grating spectrometers and Fourier transform spectrometers are used.

12.6.1 Grating Monochromators and Spectrometers

The dispersive element of a grating spectrometer is an optical grating. Although prisms have been the prototypes of dispersive components since Isaac Newton, today, they play only a major role in UV spectrometry. The main reason is that optical gratings achieve a higher dispersion in the visible and infrared regime. Another advantage is that the dispersion of gratings is linear, whereas the dispersion of prisms depends on the nonlinear dispersion characteristics of the applied glass.

12.6.1.1 Optical Grating

Optical gratings normally used in spectrometers are reflection gratings. Their substrates may be either flat or curved. Also the line structure may be parallel or curved depending on whether the grating is not only a dispersive but also an imaging device. The simplest type of grating is a periodic line structure with a certain line separation (grating constant). The periodicity of the grating is also given in lines per millimeter. The structure may be generated mechanically with a ruling machine that engraves the line structures into the substrate with a diamond tip. The second common method is to structure a photoresist by illuminating it with interfering laser light and transfer the generated structure by anisotropic etching techniques into the substrate. If a light beam hits the grating the beam is spread into the various diffraction orders according to the grating equation [3]

$$m\lambda = -g(\sin\alpha_{\mathrm{in}} + \sin\alpha_m)\,, \quad (12.65)$$

with m as the number of the diffraction order, λ the wavelength, g the grating constant, α_{in} the incidence angle, and α_m the diffraction angle of the diffraction order m. According to Eq. (12.65) the diffraction angle of a certain order depends on the wavelength λ. The angular dispersion of the grating is given by:

$$\frac{\mathrm{d}\alpha}{\mathrm{d}\lambda} = \frac{m}{g\cos\alpha_m}\,. \quad (12.66)$$

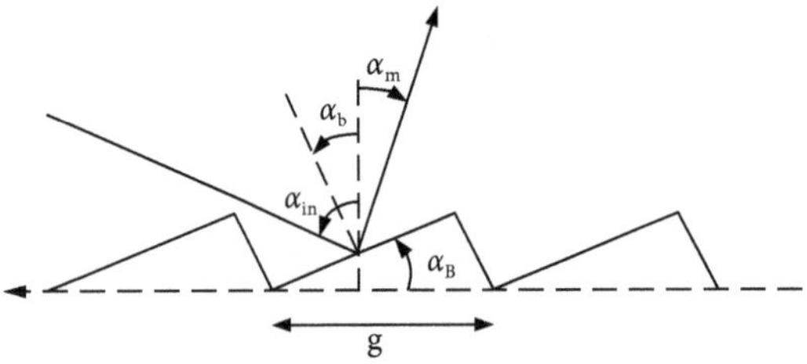

Fig. 12.24 Blazed grating. The blaze angle is chosen so that the respective diffraction order angle is equal to the reflection angle

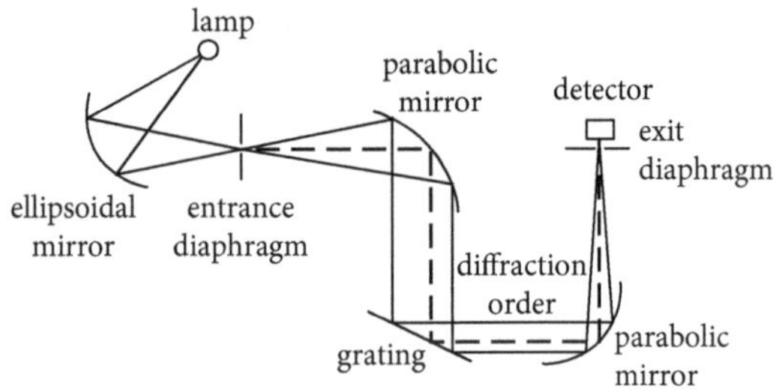

Fig. 12.25 Monochromator setup

The smaller the grating constant g the better the angular separation of the different wavelengths. The purpose of a grating in a spectrometer is to separate different wavelength components into different angles. Thus, a high dispersion is one aspect. The angular resolution of the grating is also limited by the diffraction-induced divergence of the light beam. It is necessary to illuminate the grating with a beam bundle with a large diameter to minimize diffraction. The wavelength resolution $\delta\lambda$ of a grating is

$$\delta\lambda = \frac{\lambda}{mN}, \qquad (12.67)$$

with N the number of lines illuminated by the light beam. The beam must be broad to illuminate a high number of lines.

Besides the resolution of the spectrometer, also a broad spectral range should be covered by the spectrometer. This is given by

$$\Delta\lambda = \frac{\lambda_{\min}}{m}, \qquad (12.68)$$

with the minimum wavelength $\lambda_{\min}$ of the spectrum.

Equation (12.65) determines the direction of the diffraction orders but it gives no information about the strength of the different orders. The grating would be ideal if all the light went into a single diffraction order. In fact, this can be achieved with a blazed grating. The surface of such a grating has a sawtooth shape as shown in Fig. 12.24. The surface sections between the lines are tilted by the blaze angle α_B. The angle of incidence α_{in} must fulfill the blaze condition

$$\alpha_{in} = 2\alpha_B - \alpha_m . \qquad (12.69)$$

The incoming beam and the outgoing diffraction order m are, of course, related by the grating condition Eq. (12.65). The blaze condition says that the incoming beam and the outgoing beam of the diffraction order fulfill the reflection law with respect to the tilted surface segments. With the grating equation Eq. (12.65) and the blaze condition Eq. (12.69) one obtains

$$m\lambda = -g[\sin\alpha_{in} + \sin(2\alpha_B - \alpha_{in})] . \qquad (12.70)$$

The blaze condition can be fulfilled for a broad range of wavelengths by choosing the appropriate angle of incidence. The diffraction efficiency of a blazed grating also depends on the orientation of the polarization of the light. A high efficiency can only be achieved if the polarization of the light is parallel to the line orientation of the grating.

Figure 12.25 shows the principal setup of a monochromator. A bright light source such as, for example, a discharge lamp is imaged with an ellipsoidal mirror onto the entrance diaphragm of the monochromator. The divergent light emerging from the diaphragm is collimated with a parabola mirror. The collimated beam hits a reflection grating. If the grating is blazed, in the ideal case all light is diffracted into a specific diffraction order. The diffracted light bundles of the different wavelengths are collimated and focused onto the plane of the exit diaphragm of the monochromator.

Only the bundle of one specific wavelength is focused onto the exit diaphragm. The wavelength can be tuned by rotating the optical grating. The resolution of the monochromator is determined by Eq. (12.67). The number of grating lines illuminated by the incoming collimated light beam determines the resolution. However, this relation is only valid if the radius of the diaphragm $r_{\text{diaphragm}}$ is so small, that the divergence angle φ of the light beam emanating from the entrance diaphragm is determined by the diffraction related Airy formula:

$$r_{\text{diaphragm}} = 0.61 \frac{\lambda}{\text{NA}}, \tag{12.71}$$

with the numerical aperture $\text{NA} = \sin\varphi$. In order to make optimal use of the light from the light source, the angle of convergence of the light focused onto the entrance diaphragm of the monochromator must also be equal to φ.

An alternative to the application of elliptical and parabolic mirrors in the system described above is to collimate the incoming light by a spherical mirror. Spherical mirrors are easier to manufacture but generate astigmatism if they are tilted. After the collimated beam passes the grating a second tilted spherical mirror focuses the light onto the exit diaphragm. Additional astigmatism is contributed. The astigmatism causes a line shaped image of the entrance diaphragm. Fortunately, the line image is oriented normal to the angular dispersion of the grating. In this case, the entrance and exit diaphragms are slits.

The monochromator can be used as a wavelength tunable light source, which can be applied, for example, to measure the wavelength dependence of transmission or reflectivity of thin films. Monochromators are also used as tunable light sources in spectral ellipsometry. In contrast to monochromators, grating spectrometers have a fixed grating. The spectrum is detected simultaneously with a detector array (Fellgett advantage). The disadvantages are that the resolution is limited by the size and the number of detectors on the detector array. Also optical aberrations limit the resolution and efficiency.

12.6.2 Fourier Transform Spectrometers

Grating spectrometers are very common because of their simple and robust design. CCD (charge-coupled device) line sensors make data acquisition easy. In the infrared regime, however, CCD lines are expensive or not available with the necessary pixel number. In this field, fast Fourier transform (FFT) spectrometers are preferred. The concept of these spectrometers is based on the Wiener–Khinchin theorem, which relates the Fourier transform of the temporal coherence function of a light source to its optical spectrum.

12.6.2.1 Temporal Coherence Function

The temporal coherence function of a light beam can be measured with a Michelson interferometer, as shown in Fig. 12.26. The interference signal at the detector is measured as a function of the arm length difference Δz. For this purpose, the arm length is changed by moving one mirror at a constant speed.

The electrical field of the light beam is

$$E(t) = E_0(t) \cdot \mathrm{e}^{-\mathrm{i}\omega t}.$$

The amplitude $E_0(t)$ is assumed to be complex and stochastic. The power density is

$$I = \frac{1}{2\eta_0} \langle E^*(t) E(t) \rangle \tag{12.72}$$

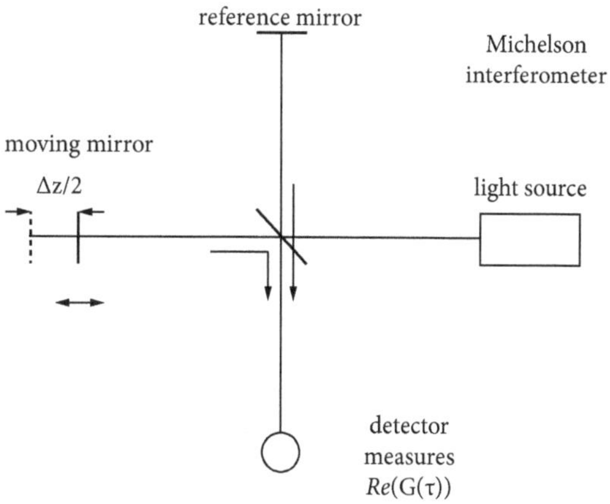

Fig. 12.26 Michelson interferometer setup

$$= \frac{1}{2\eta_0} \lim_{T\to\infty} \frac{1}{2T} \int_{-T}^{T} E^*(t)E(t)\mathrm{d}t\,, \quad (12.73)$$

with the impedance of the vacuum

$$\eta_0 = \sqrt{\frac{\mu_0}{\epsilon_0}} = 377\,\Omega\,.$$

In favor of simplification one defines $u(t) = \frac{E(t)}{\sqrt{2\eta_0}}$ such that

$$I = \langle u^*(t)u(t)\rangle\,.$$

At the detector of the Michelson interferometer the amplitudes of the partial beams of the interferometer are superimposed. The amplitude function of one partial beam at the detector is

$$u_1(t) = \frac{1}{2}u(t)\,, \quad (12.74)$$

the other one is

$$u_2(t) = \frac{1}{2}u(t+\tau)\,, \quad (12.75)$$

where $\tau = 2\cdot\Delta z/c$ is the difference in the time of flight caused by the arm length difference Δz. The intensity at the detector is then

$$I_{\mathrm{Det}}(\tau) = \langle (u_1^*(t) + u_2^*(t))(u_1(t) + u_2(t))\rangle \quad (12.76)$$

$$= \frac{1}{2}I + (\langle u_1^*(t)u_2(t)\rangle + \langle u_2^*(t)u_1(t)\rangle)\,. \quad (12.77)$$

The temporal coherence function is defined as

$$G(\tau) = \langle u^*(t)u(t+\tau)\rangle \quad (12.78)$$

$$= \lim_{T\to\infty} \frac{1}{2T} \int_{-T}^{T} u^*(t)u(t+\tau)\mathrm{d}t\,. \quad (12.79)$$

Therefore,

$$I_{\mathrm{Det}}(\tau) = \frac{1}{2}I + \frac{1}{4}(G^*(\tau) + G(\tau))\,. \quad (12.80)$$

With

$$G(\tau) = \int_{-\infty}^{\infty} u^*(t)u(t+\tau)\mathrm{d}t$$

we obtain with the correlation theorem of Fourier transform theory

$$\tilde{G}(\nu) = \mathrm{FT}\{G(\tau)\} = \tilde{u}^*(\nu)\tilde{u}(\nu) = S(\nu)\,,$$

with the Fourier transformation FT and the power spectrum $S(\nu)$ of the light source. In other words, the Fourier transform of the coherence function is equal to the power spectrum $S(\nu)$ of the light source. This is the Wiener–Khinchin theorem, which allows to calculate the power spectrum by Fourier transforming the coherence function.

In practice, the various Fourier transform spectrometers comprise a Michelson interferometer with variable arm lengths. The arm length difference is changed with constant speed. The light from the two arms is superimposed at the detector. The detector signal is amplified and recorded as a function of time. The signal is Fourier transformed by a microprocessor in order to obtain the complex signal spectrum. The absolute square of the complex spectrum is the power spectrum corresponding to the optical spectrum of the light source.

12.7 Microscopy

12.7.1 Optical Microscopes

A microscope typically consists of an objective, a tube lens, and an eye piece, as shown in Fig. 12.27. The object is in the front focal plane of the objective. The tube lens focuses the image into the back focal plane of the tube lens.

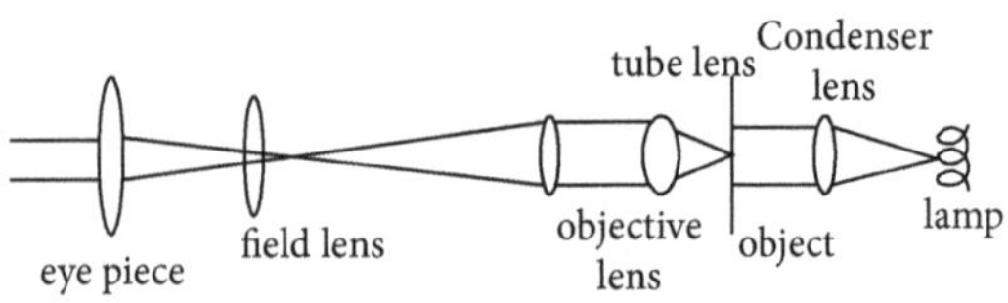

Fig. 12.27 Microscope schematic

The advantage of this geometry is that the rays from a object point are parallel behind the objective, which is necessary when polarizers are inserted, e.g., in polarization microscopy. The intermediate image is either transferred to an image capturing device like a CCD or a film, or it is looked at by an observer through an eye piece. The eye piece plays the role of a magnifier lens magnifying the intermediate image. The total magnification is the product of the magnification of the objective in conjunction with the tube lens.

Besides the magnification, the main parameters characterizing an objective are its numerical aperture and the working distance (distance between object plane and lens vertex). The numerical aperture is defined by

$$\mathrm{NA} = n \sin \alpha \,. \qquad (12.81)$$

The angle α characterizes the divergence of those rays emanating from an object point that are accepted by the objective and focused to the intermediate image plane. n is the index of refraction of the medium between object and objective. Normally this will be air with $n = 1$. To increase resolution immersion oil with $n > 1$ can be used to fill the space between the object and the objective to increase the numerical aperture NA.

The aperture diaphragm of the objective is normally located near the back focal plane of the objective. It determines the aperture angle α and defines the exit pupil of the objective. If the diaphragm is the back focal plane all chief rays from object points going through the center of the aperture diaphragm are parallel in the object space. This telecentric optics guarantees that the image size does not change when the object is defocused. This is useful in particular when the microscope is used to measure distances within the object.

The eye piece consists of a field lens and the ocular lens. The field lens is necessary to avoid the key hole effect. Without the field lens the observer's eye pupil would accept only ray bundles with small angles relative to the optical axis. Thus, the observer would see only a small portion of the object field. If the field lens is in place (it is normally located near the intermediate image plane), the ray bundles emanating from peripheral object points still enter the eye's pupil. The refractive power of the field lens is chosen such that the field lens together with the ocular lens (and optionally the tube lens) image the exit pupil (aperture stop) of the objective into the exit pupil of the microscope, which is situated tens of millimeters behind the eye piece at the position of the observers's eye pupil.

Illumination. The object is illuminated by the condenser optics. The light may transmit the object if it is transparent (transmitted light microscopy) or the object is illuminated through the objective (incident light microscopy). The light cone emanating from a point of the light source is collimated by the condenser lens. The parallel beam bundle hits the object. According to Ernst Abbe any object can be thought of as being a superposition of sinusoidal gratings. This model is justified by the Fourier transform theory stating that any (integrable) function can be represented as a superposition of harmonic functions. This is certainly true in particular for the function describing the spatial distribution of the transmission of the object. For this reason it is sufficient for the further discussion to focus on grating objects. The parallel beam bundle hitting the object is then split into the various diffraction orders. The objective focuses the diffraction orders into the plane of the aperture stop, which is located in the back focal plane of the objective. The tube lens collimates the diffraction orders hitting the intermediate image plane as plane waves with different angles of incidence. As shown in Fig. 12.28 the first orders and the zeroth order interfere. We may ignore the interference between the first orders when we assume that the grating is weak such that the zeroth order is strong compared to the first orders.

The interference pattern can be interpreted as the image of the optical grating. The distance of the maxima of the interference pattern is determined by the diffraction angle θ, given by the

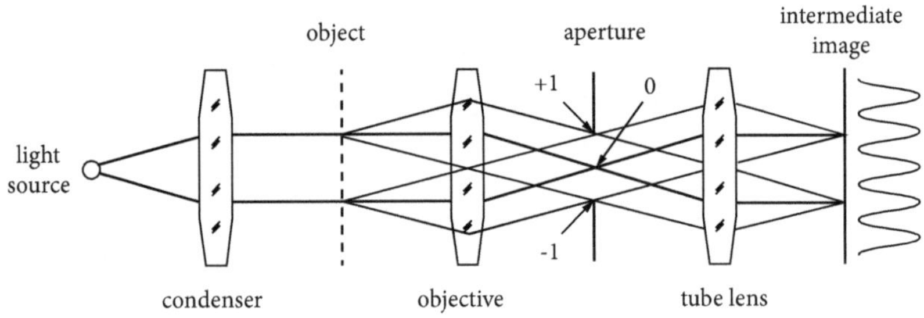

Fig. 12.28 Coherent illumination

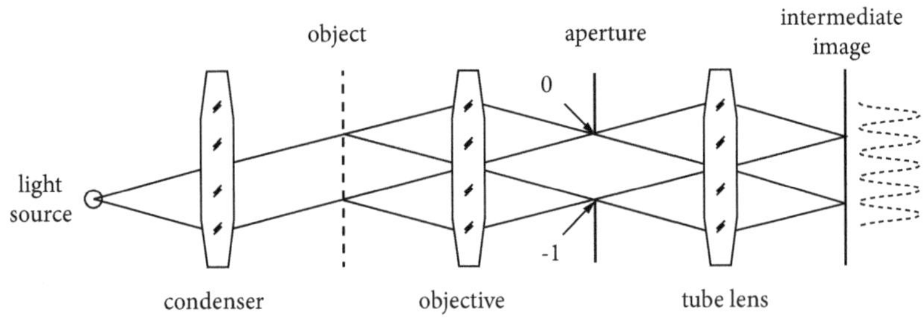

Fig. 12.29 Oblique illumination

diffraction equation of the optical grating

$$g \sin\theta = \lambda\,. \tag{12.82}$$

Figure 12.28 also shows that interference can only occur when the first diffraction order is fitted into the aperture stop. The radius a of the aperture stop determines the numerical aperture NA of the objective. It is

$$\mathrm{NA} = \sin\theta_{\max}\,,$$

with $\theta_{\max}$ as the diffraction angle, where the first diffraction order still passes the aperture stop. With paraxial approximation we have $\mathrm{NA} = a/f$. The resolution limit is achieved when the grating constant g is so small that

$$g_{\mathrm{grenz}} = \frac{\lambda}{\mathrm{NA}}\,.$$

In Fig. 12.28 the illuminating wave hits the grating perpendicularly. The first diffraction grating still passes the aperture stop.

The resolution limit can be improved by a factor of 2 by choosing oblique illumination, as shown in Fig. 12.29.

The minimum grating constant $g_{\min}$, which can still be resolved is then

$$g_{\min} = \frac{\lambda}{2\cdot\mathrm{NA}}\,. \tag{12.83}$$

So far we have discussed point-like light sources. Normally extended light sources such as, for example, halogen lamps are used. Figure 12.29 shows the zeroth and first order for the case of an off-axis light source point. The interference patterns generated by the individual points of the light source in the intermediate image plane are identical. Because of the symmetric setup the exit angles θ_0, θ_{-1} of the respective diffraction orders at the grating and the angles of incidence at the intermediate image plane are equal. In a real microscope the magnification is higher than 1. Nevertheless, the magnified interference patterns from each point of the light source are still identical. In other words, the contrast is not diminished when an extended light source is used. In addition, the advantage of Koehler's illumination is that an extended light source provides more light than a point-like source. The highest efficiency is achieved when the image of the light source fills the aperture of the objective. Therefore, the numerical aperture of the condenser should be equal to the numerical aperture of the objective. Another advantage is that the maximum resolution is twice the resolution achievable with coherent illumination because also the oblique zeroth order waves and their higher diffraction orders contribute to the imaging process.

Useful Magnification. The magnification of the microscope objective is noted there. This means the ratio between the size of the image of the object in the intermediate image plane and the size of the object. If an eye piece is used, or a camera, the image size can be magnified further. In principle there are no limits, in particular when the image is recorded electronically. The distance of the above-mentioned interference fringes in the intermediate image plane corresponding to the resolution limit is, according to Eq. (12.83):

$$\Delta = \frac{\lambda}{2 \cdot \mathrm{NA}} \cdot M_{\mathrm{obj}} , \qquad (12.84)$$

with the magnification of the objective M_{obj}. With an eye piece the magnification M_{ep} is given by the magnification of a magnifier lens:

$$M_{\mathrm{ep}} = \frac{s}{f} , \qquad (12.85)$$

with the normal visual distance $s = 0.25\,\mathrm{m}$ and the focal length f of the eye piece. The resolution of the human eye is $\delta_{\mathrm{eye}} = 1'$. Therefore, the magnification of the eye piece is sufficient as long as the eye can resolve the magnified structures of the intermediate image of the microscope. The focal length f of the eye piece is then:

$$f = \frac{\Delta}{\delta} .$$

With $\lambda = 550\,\mathrm{nm}$ and Eqs. (12.84), (12.85) follows

$$M_{\mathrm{ges}} = M_{\mathrm{ep}} M_{\mathrm{obj}} = 300\,\mathrm{NA} .$$

Because the resolution of the eye is normally worse than $1'$ the recommended useful magnification is between 500 NA and 1000 NA.

Polarization Microscopy. Because thin films are normally homogeneous, the polarization of polarized light is not changed in transmitted light microscopy. However, dust particles or mechanical stress may depolarize or change the polarization state. In a polarization microscope the specimen is illuminated with linearly polarized light. For this purpose, a linear polarizer is normally located underneath the condenser optics, which can be flipped in and out. Behind the objective another polarizer (analyzer) is placed with an orientation perpendicular to the orientation of the polarizer of the condenser. If the specimen does not change the state of polarization, no light reaches the image plane. In the case of optically anisotropic materials, the reflected light (incident light microscopy) or the transmitted light (transmitted light microscopy) is elliptically polarized. In a polycrystalline specimen, the effect may vary strongly from one crystallite to another, so that the crystallite structure appears with high contrast in the intermediate image. Scattering object details often display a depolarizing effect, which means that they appear brighter in the image than smooth objects. Polarization microscopes require specially selected objectives and condenser lenses that exhibit as little strain-induced birefringence as possible. The analyzer and polarizer should be arranged in a parallel ray path.

Dark Field Microscopy. As discussed above polarization microscopy is one tool to image scattering disturbances on thin films. An even stronger method is dark field microscopy. In a dark field microscope only the light diffracted by the object reaches the image plane. The ray path of a dark-field microscope is shown in Fig. 12.30. The light source is imaged into the plane of the ring-shaped aperture stop of the condenser. Figure 12.30 shows the light pencil emanating from a point of the intermediate image of the light source in the ring aperture.

The condenser lens collimates the light. The parallel light bundle passes the object plane but does not enter the objective lens. However, if there is a scattering particle in the object plane part of the illuminating light is scattered into the objective and is focussed into the image plane. This makes it possible to make visible very weak objects that would be swamped by the

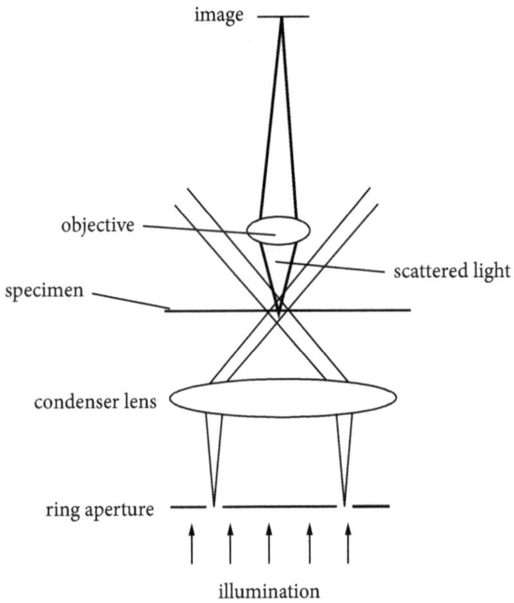

Fig. 12.30 Dark field illumination

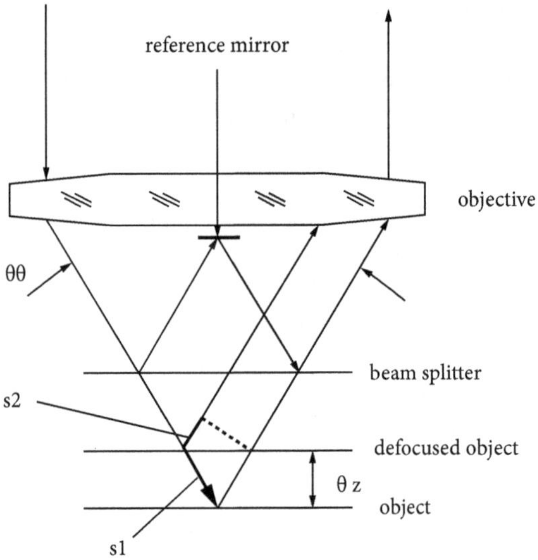

Fig. 12.31 Mireau interference microscope

undiffracted light in normal bright field illumination.

Interference Microscopy. Thin film surfaces are often reflecting and plane structures. The topography is very flat. Deviations from the plane reference are in the nanometer range. A plane wave reflected at this surface carries the height deviations of the sample in their wave fronts. In interference contrast microscopy the phase distribution impressed on the illuminating wave field by a phase object is imaged. The microscope interferometers include the Mireau and Linnik interferometer, the differential interference contrast (Nomarski DIC), and the Interphako method. Figure 12.31 shows the schematic setup of a Mireau interferometer objective.

The Mireau interferometer consists of an objective lens and a beam splitter plate between lens and object. Part of the incident light is reflected by the beam splitter and is focused on a small mirror in front of the objective lens. The other part is reflected by the object. If the object surface height changes or the object is defocused by Δz, the optical path length of the reference path changes by Δs. It is $\Delta s = 2s_1 - s_2$ with $s_1 = \Delta z/\cos\theta$ and $s_2 = 2\tan\theta\sin\theta\Delta z$. Therefore,

$$\Delta s = 2\Delta z\cos\theta\,, \tag{12.86}$$

with the angle of incidence θ of the reference beam. The resulting phase shift of the probe beam is

$$\Delta\varphi = k\cdot\Delta s = 2k\Delta z\cos\theta\,, \tag{12.87}$$

with $k = 2\pi/\lambda$. A phase shift of 2π and, therefore, a shift of the interference pattern by one period is achieved when the sample surface height changes by

$$\Delta z_{2\pi} = \frac{\lambda}{2\cos\theta}\,. \tag{12.88}$$

Figure 12.32 shows an example.

Differential Interference Contrast (Nomarsky DIC). The illuminating wave shown in Fig. 12.33 is linearly polarized. It strikes a Wollaston prism, splitting the principal ray of the illuminating wave into two rays.

The Wollaston prism is located in the back focal plane of the objective such that the object is illuminated by two foci with the spacing d. The

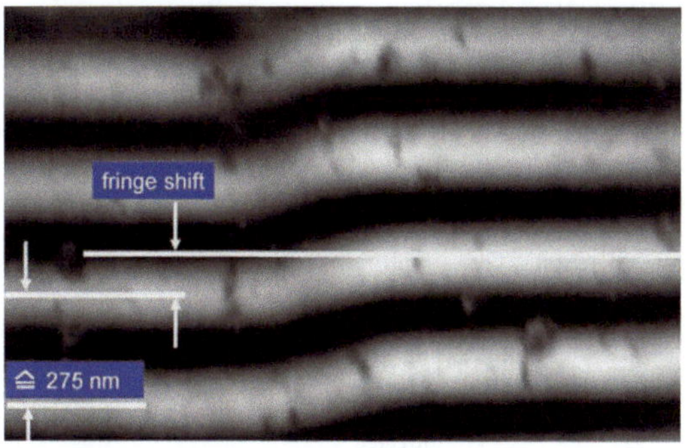

Fig. 12.32 Interferogram of the edge of a ZnO film on a glass plate. The fringe separation corresponds to $\lambda/2 = 275\,\text{nm}$. From the fringe shift a step height of 105 nm can be derived

reflected waves experience a phase shift difference

$$\delta = 4\pi\Delta z/\lambda \tag{12.89}$$

due to the step structure of the object shown in Fig. 12.33. Both waves are recombined in the Wollaston prism and superimposed in the intermediate image, where they interfere. With this method, phase differences between adjacent object points separated by the distance d are represented as brightness contrast in the intermediate image. Thus, the contrast of this technique depends on the change of the object topography. It is a very sensitive method to image, for example, crystallite boundaries on thin films or very small step heights.

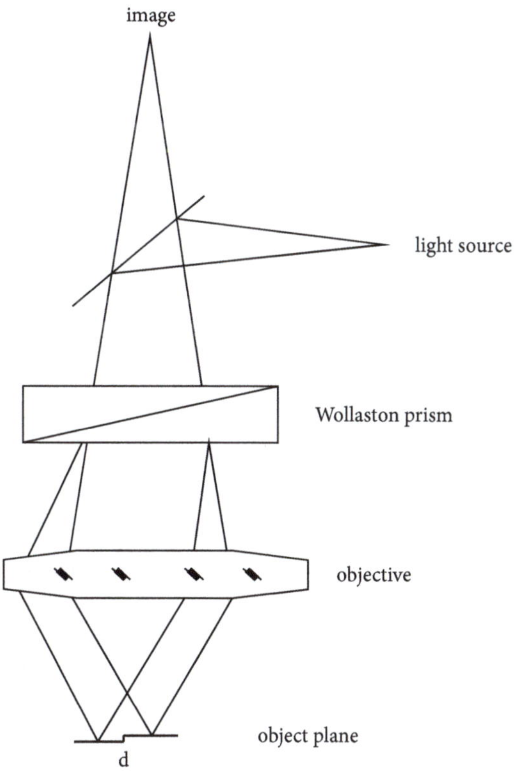

Fig. 12.33 Differential interference contrast microscopy (DIC)

12.7.2 Electron Microscopy

Electron microscopes are important tools in thin film technology, first of all because of their high resolution. Scanning electron microscopes (SEM) can show surface structures without complicated preparation techniques. The electron beam can be used to excite atoms to emit X-rays. The typical spectra can be used to identify the element distribution on the surface and within the bulk material with nanometer resolution. Transmission electron microscopy (TEM) provides even resolution, which allows resolution of individual atoms. More important information is the crystallite structure which can be monitored by electron beam diffraction. A very important part of electron microscopy is sample preparation. For example, the resolution of atoms with TEM can only be achieved if the sample is very thin. Different layers of a thin film probe can only be distinguished with SEM if a proper cross-section of the film can be prepared. Particularly in the last few years, ion beam etching has been established as a very elegant and universal preparation technique in practical electron microscopy.

Electron microscopes use electron beams to generate images with very high resolution. The beam is focused by magnetic and electrostatic fields guiding the electrons with the Lorentz and Coulomb forces. The setup of electron microscopes can be divided into electron source, condenser, objective, and projective system.

Although the particle character of electrons determines most technical characteristics of electron microscopy, the ultimate resolution limit of electron microscopes is determined by the wave character of the electrons. The minimum spatial distance d_{min} that can be resolved corresponds to the resolution limit in light microscopy:

$$d_{min} = \frac{\lambda_{de\,Broglie}}{2\alpha},$$

with the de Broglie wavelength $\lambda_{de\,Broglie}$ and 2α as the angle of divergence of the focused electron beam. The de Broglie wavelength depends on the electron mass m_e and the electron velocity v:

$$\lambda_{de\,Broglie} = \frac{h}{m_e \cdot v}.$$

In contrast to photons, which propagate with the constant speed of light, electrons can be accelerated or reduced by appropriate electrostatic potentials. As a consequence, the wavelength and thus the resolution of the microscope can be easily changed by changing the electron speed. On the other hand, in electron microscopes the numerical aperture angle α is much lower compared to numerical apertures NA used in light microscopy because the aberrations of magnetic lenses increase dramatically with α.

Electron Source. The electrons for the beam of an electron microscope are generated by an electron gun. It emits electrons (e.g., thermally) into the vacuum and accelerates them between a cathode and an anode. The gun brightness is

$$\beta = \frac{j}{\pi\alpha^2}, \tag{12.90}$$

with the beam current density j and the beam divergence angle 2α. The gun brightness β corresponds to the radiance L_e of an optical light source. The current density of the cathode emission can be estimated from Richardson's law

$$j = AT^2 \exp(-\phi/kT), \tag{12.91}$$

where $k = 1.38 \cdot 10^{-23}$ J/K is Boltzmann's constant, T is the cathode temperature, ϕ is the work function of the cathode, and A is a constant depending on the cathode material. Therefore,

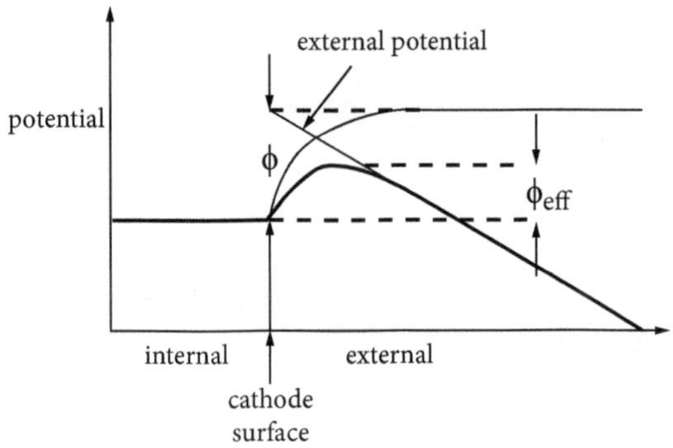

Fig. 12.34 The work function ϕ is reduced by the Schottky effect. The *thick line* shows the effective potential barrier at the cathode with the reduced work function ϕ_{eff}

a high current density j can be achieved with a high cathode temperature T and a cathode material with a low work function ϕ. Applying a high field strength E to the cathode the work function ϕ can be further reduced by the Schottky effect, leading to an effective work function

$$\phi_{eff} = \phi - \sqrt{\frac{e^3|E|}{4\pi\epsilon_0}}. \tag{12.92}$$

The Schottky effect is caused by the mirror charges induced by the emitted electrons in the metallic cathode. An electron leaving the cathode still experiences an attractive force. Thus, the potential barrier is not a step function. Together with the superimposed external electrical field between cathode and anode the potential barrier shown in Fig. 12.34 is received.

Most metals melt before they reach a sufficiently high temperature for thermionic emission. With tungsten working temperatures T between 2500 and 3000 K can be achieved. Its melting point is at 3650 K. Lanthanum hexaboride (LaB_6) is used as cathode material at lower temperatures T = 1400–2000 K because its work function is only 2.7 eV (4.5 eV for tungsten).

The work function of the field emission gun can be further reduced with a ZrO_2 coating applied to the surface of the gun tip. These Schottky field emission guns show a better stability compared to cold FEGs (field emission gun). However, the source size is slightly larger.

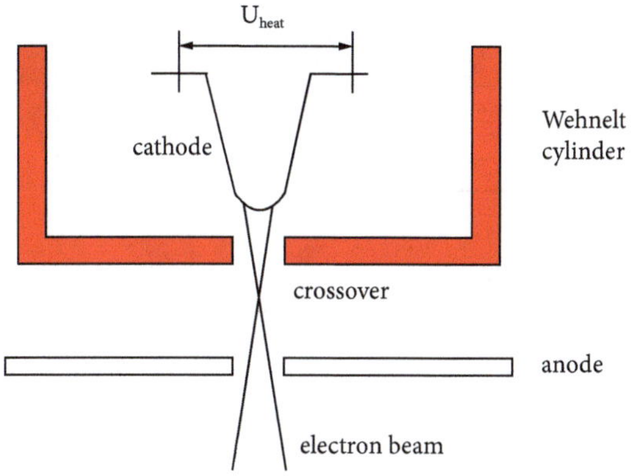

Fig. 12.35 Thermionic gun

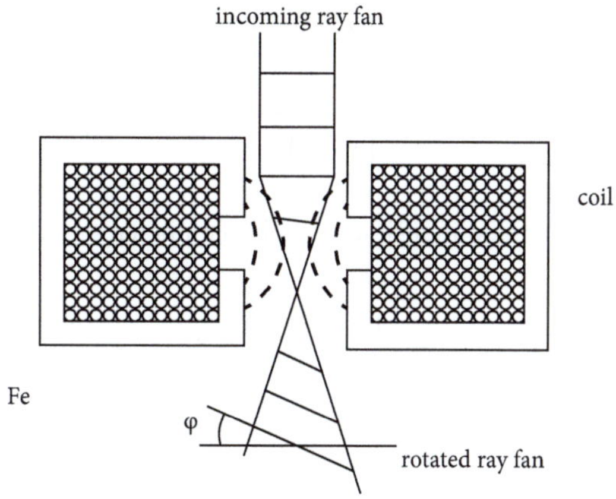

Fig. 12.36 Electron lens. The electrons follow spiral-shaped trajectories through the lens. A parallel incoming ray fan is focused and rotated by the magnetic lens

Tungsten sources in thermionic guns are most widely used because of their high current stability and relatively low cost. The thermionic gun is a triode consisting of a negatively charged tungsten wire cathode heated resistively, a Wehnelt cylinder, and an anode, as shown schematically in Fig. 12.35. These components are maintained at different potentials, giving a beam energy of 0.1–40 keV for SEMs and 100–400 keV for TEMs. The Wehnelt cylinder acts as an electrostatic lens generating an intermediate image of the cathode with a diameter d_0 (crossover). The distribution of the electron energy $f(E)$ is determined by the Maxwell-Boltzmann distribution and, therefore, by the cathode temperature T

$$f(E) = \frac{E}{(kT)^2} \exp(-E/(kT)) , \qquad (12.93)$$

with the Boltzmann constant k. The full-width half maximum of this distribution is $\Delta E = 2.45kT$. For a cathode temperature of $T = 2500$ K the half-width is $\Delta E = 0.5$ eV. The typically observed energy spread is, however, 1–2 eV. This can be explained by the repulsive Coulomb interaction of the electrons in the crossover zone (Boersch effect).

It is desirable to have a low energy spread ΔE to minimize chromatic aberration. Therefore, LaB_6 cathodes are preferable because they can be operated at lower temperatures. FEGs provide a high gun brightness. Because they are operated at ambient temperature their energy spread ΔE is low. This is important for high resolution SEM and TEM imaging and microanalysis. The small source size of FEGs also guarantees a high spatial coherence for special TEM imaging techniques such as atomic structure imaging and electron holography.

Electron Lens. Electron beams can be focused with magnetic and electrostatic fields. In magnetic fields the trajectory of electrons is determined by the Lorentz force:

$$\boldsymbol{F}_{\text{Lorentz}} = -e(\boldsymbol{v} \times \boldsymbol{B}) ,$$

with electron velocity $\boldsymbol{v}$, electron charge e, and the magnetic field $\boldsymbol{B}$ of the electron lens. An electron beam can be focused by a magnetic dipole field generated by a magnetic coil as shown in Fig. 12.36.

The electrons follow spiral-shaped trajectories through the lens. A parallel incoming ray fan is focused and rotated by the magnetic lens. The B field along the axis is

$$B_z = \frac{B_0}{1 + (z/a)^2} , \qquad (12.94)$$

where a is the half-width at half maximum of $B_z(z)$ and B_0 is the maximum field at the lens center. The focal length of this lens is then

$$f = \frac{a}{\sin\left(\pi/\sqrt{1+k^2}\right)}, \quad (12.95)$$

with the dimensionless field parameter

$$k^2 = \frac{e^2 B_0^2 a^2}{8 m_e E}, \quad (12.96)$$

where m_e is the electron mass, e the electron charge, and E the electron energy. The angle of rotation is

$$\varphi = k \frac{\pi}{\sqrt{1+k^2}}. \quad (12.97)$$

Objective lenses normally operate with k^2 between 2 and 3 providing a focal length in the order of a. Note that the focal length depends on the electron energy E.

Magnetic lenses are used in all types of condenser, projector, and objective systems (Fig. 12.37). The advantage of magnetic lenses compared to electrostatic lenses is that no high voltages are needed but instead high currents, which are easier to handle.

The Lorentz force depends on electron speed but the Coulomb force does not:

$$\boldsymbol{F}_{\text{Coulomb}} = -e\boldsymbol{E},$$

with the electric field $\boldsymbol{E}$ of the electrostatic lens. Therefore, magnetic lenses can be combined with electrostatic lenses to reduce chromatic aberrations.

Electron Lens Aberrations and Their Correction. Like spherical glass lenses electron optical lenses show aberrations. In contrast to glass lenses magnetic field lenses not only focus the electron beam but also introduce a rotation of the ray fan. As a result electron optical lenses display also third-order anisotropic astigmatism, distortion, and coma.

The aberrations depend on the angle of divergence 2α of the beam pencil emanating from the object point, the distance of the object point from the optical axis, and the symmetry and strength of the lenses. Spherical aberration, for example, is caused by the fact that the refractive power of the lens is different for marginal rays and paraxial rays. The marginal rays constitute a blurred image spot at the paraxial image plane. The corresponding spot diameter in the object plane is

$$d_s = 2C_s\alpha^3, \quad (12.98)$$

with the spherical aberration coefficient C_s, which decreases with increasing field parameter k and thus with increasing lens strength.

In addition to monochromatic aberrations, chromatic aberration also plays an important role. Because of the energy spread ΔE of electrons emitted by electron guns and the velocity dependence of the Lorentz force electrons with different energy are focused to different image planes. Spherical aberration can be reduced by locating the specimen within the magnetic field (immersion lens). Astigmatism, which is primarily caused by symmetry distortions of the magnetic lenses, can be corrected by quadrupole fields (stigmators). A reduction of chromatic aberrations is achieved by using FEGs with reduced ΔE.

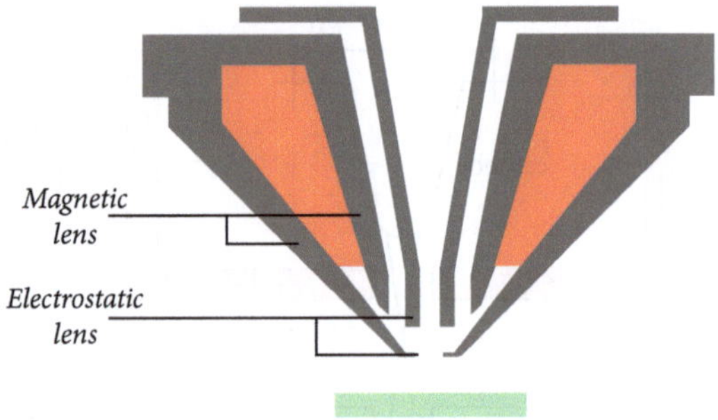

Fig. 12.37 Low voltage scanning electron microscope lens (Zeiss Gemini)

Scanning Electron Microscopy. In a scanning electron microscope (SEM) (the schematic setup is shown in Fig. 12.38) a fine convergent beam of electrons is scanned across the surface of a bulk specimen such as, for example, a thin film sample on a substrate. The electrons are extracted from

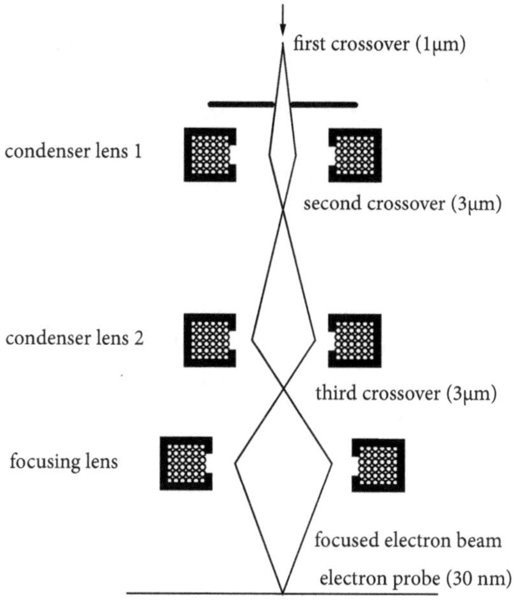

Fig. 12.38 Scanning electron microscope setup

the cathode's thermionic source and focused with a Wehnelt cylinder into the first crossover with a diameter of about (30 μm). A mechanical aperture stop allows us to change the size and divergence angle of the beam determining the probe aperture α_p at the specimen. The probe aperture α_p, the probe diameter d_p, and the electron probe current I_p are not independent because of the constancy of the gun brightness in the electron optical system (Lagrange theorem). Therefore, they cannot be changed independently.

The first crossover is further demagnified by two condenser lenses until the third crossover reaches a diameter of about 0.3 μm. Finally, the beam is focused onto the specimen by the objective lens to a probe diameter in the nanometer range. The probe is scanned across the sample with scanning coils periodically deflecting the beam in x and y directions. The electrons emitted from the probe–specimen interaction zone are gathered by detectors. The detector current is recorded and displayed. The SEM has a high depth of field because the probe aperture α_p is rather low. Another advantage is that only minimal specimen preparation is necessary. Normally, the specimens must be coated with a conducting film (e.g., gold) to avoid charging of the specimen through the high energy electron beam. However, also isolators can be imaged if the beam energy is low enough so that emitted electrons are replaced by electrons from the probe beam.

The electrons of the focused probe in the SEM interact elastically and inelastically with the specimen. The elastic scattering processes are Rutherford and Mott scattering. Inelastic scattering is caused by inner and intra-band transitions in the valence and conduction band, by inner shell excitation and plasmon excitation. The scattered electrons remitted from the specimen can be detected. Back scattered electrons (BSE) are generated in a zone that is much larger and deeper than the focus spot. They are multiply scattered electrons of the probe beam with a very broad energy distribution. In addition to BSE, secondary electrons (SE) are also important for image generation. They are generated directly at the probe beam focus.

Conventionally, the boundary energy between SE and BSE electrons is 50 eV. The SE electrons are further classified into SE1 and SE2 electrons. SE1 electrons are secondary electrons that are directly excited by the primary electrons to an energy level high enough to overcome the work function. Their contribution to the SE signal is about 20–50 % and their energy is typically 2–5 eV. SE1 electrons emanate from a film that is only a few nanometers thick below the specimen surface and are scattered back into a very small angle. Therefore, SE1 electrons are capable of forming an image of the surface (topographic contrast). SE2 electrons are secondary electrons generated by BSE electrons. Their energy is higher and they emanate from a volume within the specimen which is much broader and deeper than the probe focus. They are also scattered into a much broader angle than the SE1 electrons.

X-Ray Microanalysis. The incident electron beam may excite electrons in an inner shell of specimen atoms causing its ejection and resulting in the formation of an electron hole within the atom's electronic structure. An electron from an outer shell soon fills the hole, and the excess energy of that electron is released in the form of an X-ray photon. The energy spectrum of these X-ray photons is characteristic for the specific element.

X-ray spectrometers can be interfaced to a SEM to obtain elemental data from the beam–specimen interaction where X-rays with a characteristic spectrum are emitted from specimen atoms that have been excited by the probe beam.

An energy dispersive X-ray spectrometer uses a cooled, reverse-bias p-i-n Si(Li) crystal or an intrinsic Ge crystal. Absorption of each X-ray photon leads to the formation of electron–hole pairs, which are swept away by the applied bias to form a charge pulse. Each charge pulse is proportional to the energy of the incoming X-ray photon. This type of spectrometer is very efficient for X-ray photon energies in the range of 2.5–15 keV, but inefficient with low resolution at low energies (< 1 keV). In this case, a wavelength dispersive spectrometer (WDM) is preferable; it uses an analyzing crystal to diffract X-rays of specific wavelength λ according to Bragg's law

$$m\lambda = 2d \sin\theta \,, \tag{12.99}$$

where d is the diffracting plane spacing, m is the diffraction order, and θ the angle of incidence of the X-ray beam on the crystal. The diffracted X-rays are detected by a proportional counter. A slit aperture in front of the detector selects a certain diffraction angle. The angle of incidence θ, which is related to λ by Eq. (12.99), is changed by rotating the crystal. The detector signal is registered as a function of θ and, therefore, according to Eq. (12.99) as a function of λ, providing a high resolution X-ray spectrum.

The focused electron beam generates X-rays from a small interaction volume within the specimen, generally 1 μm in depth and width. With this method, it is possible to detect a very small quantity of an analyte ($< 10^{-16}$ g).

Low Voltage SEM. High energy electron beams are used to avoid chromatic aberration because the thermal energy spread of the electrons is small compared to the electron energy. Another reason is to achieve high beam brightness as discussed above. As a result, high resolution and small probe diameters are attainable. Low density specimens with no conductive coating require very low electron energies in order to achieve a charge equilibrium on the specimen surface. Besides the increase of the chromatic aberration the electron interactions at crossover points in the electron optical system further enhance the energy spread (Boersch effect). Also the beam brightness is limited.

The SEM optics shown in Fig. 12.39 avoids the disadvantages of a low voltage electron microscope by avoiding crossovers between electron source and specimen. The main improvement is the combination of a booster system, constant high voltage (10 kV) between source and objective, and a chromatic electrostatic/magnetic lens, decelerating the electron beam. Both spherical and chromatic aberration are reduced when the electron energy is reduced. This is achieved by the electrostatic field of the electrostatic lens, which decelerates the electrons to low energies when they reach the magnetic field lens.

In addition the energy spread of electrons is reduced by using a field emission tip (cathode). The Schottky emitter has a virtual source diameter of 15 nm. No real crossover (only virtual crossover) exists, thus, reducing the Boersch effect.

The secondary electrons (SE) generated directly at the specimen surface are emitted into the electrostatic lens and focused by the objective such that the electron trajectories optimally reach an annular SE detector (in-lens detector) located within the objective lens. The BSE electrons emanating from deeper zones of the specimen are successfully suppressed by this arrangement. Thus, the SE image describes the specimen surface much more precisely than an image acquired with a classical detector located in the specimen chamber which collects both SE and BSE electrons.

The left-hand part of Fig. 12.40 shows the image of an underetched silicon carbide sample acquired with an Everhart Thornley detector. The image shows deep etching grooves, covered by a very thin film. The right-hand side picture shows the same detail acquired with an in-lens detector selecting primarily SE electrons generated at the surface of the thin film.

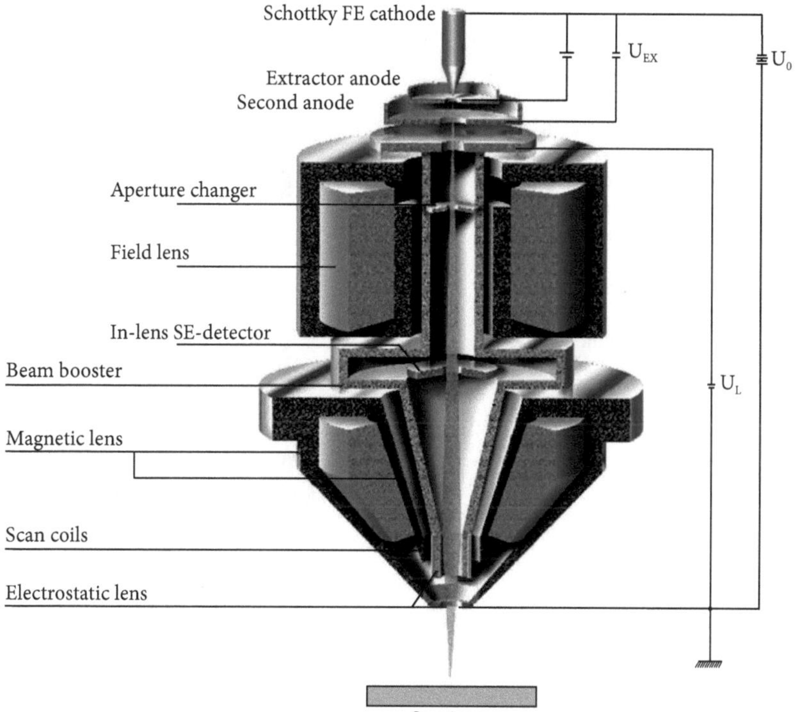

Fig. 12.39 Gemini scanning electron microscope setup

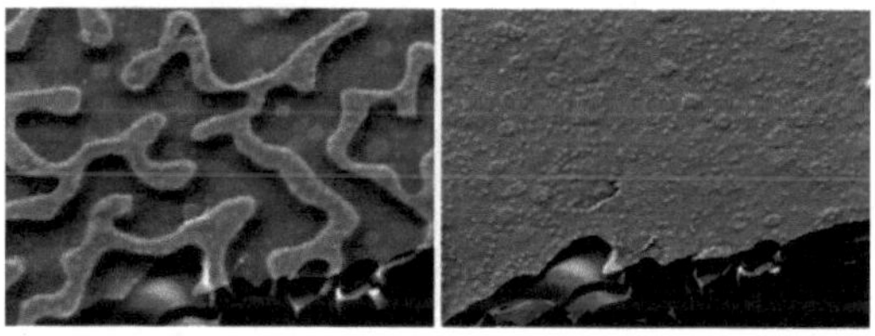

Fig. 12.40 In-lens detection

Transmission Electron Microscopy. The left-hand side diagram of Fig. 12.41 shows the schematic setup of a transmission electron microscope. An area with a diameter of 1–10 μm of a thinned specimen is irradiated with a nearly parallel beam of high energy electrons (100–400 keV) emitted from the electron gun and collimated by the condenser lens. The objective lens forms an intermediate image by collecting electrons that have either been transmitted with little or no change in angle, as in bright field imaging (BF), or scattered by the specimen, as in dark field imaging (DF). The intermediate image is further magnified by the projector lens system generating the final image onto the viewing screen or CCD. The image generation can be understood in terms of Abbe's theory of the microscope because of the wave nature of the electrons (de Broglie waves). The plane electron waves from the condenser are diffracted by the specimen. This consists of atoms comprising periodic structures or at least structures that can be thought of as being composed of periodic structures according to Fourier theory. The diffraction orders are superimposed in the image plane by the projection system generating an interference pattern that represents the image of the sample.

The image of the diffraction orders (diffraction image) can be directly viewed by choosing an appropriate coil current of the projector lens system to focus the diffraction image onto the image plane (Fig. 12.41).

The basic difference of image formation in light microscopy and TEM is that specimens for light microscopy can be prepared with a thickness as small as the light wavelength. Specimens for TEM are to be regarded as three-dimensional bulk structures because the de Broglie wave-

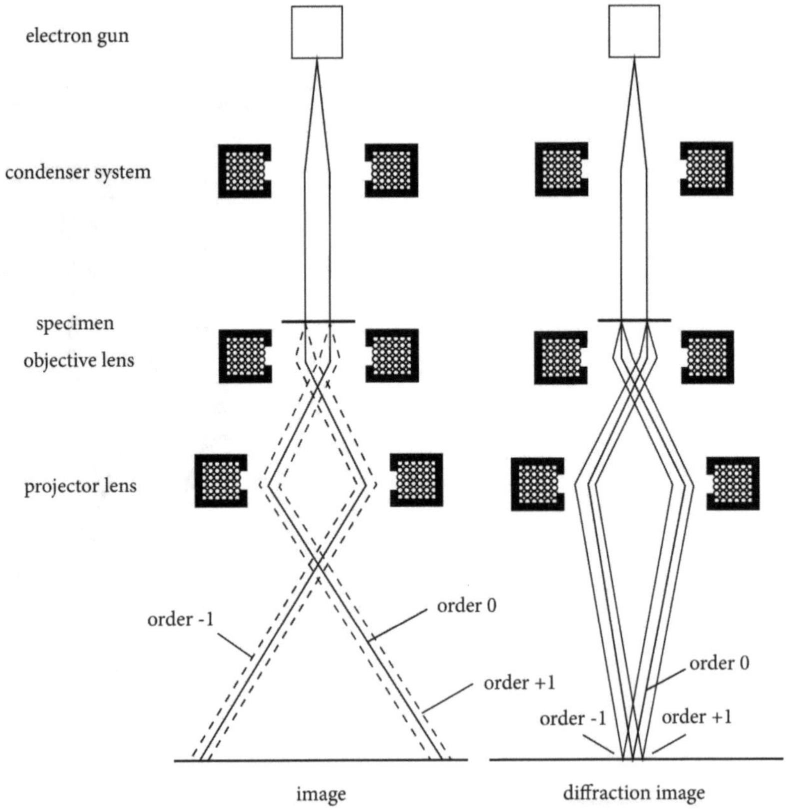

Fig. 12.41 *Left*: image formation of the transmission electron microscope as a superposition of interfering diffraction orders in the image plane. *Right*: by reducing the power of the projector lens the diffraction orders are focused on the "image" plane. This diffraction diagram can be used to investigate the crystal structure of the sample

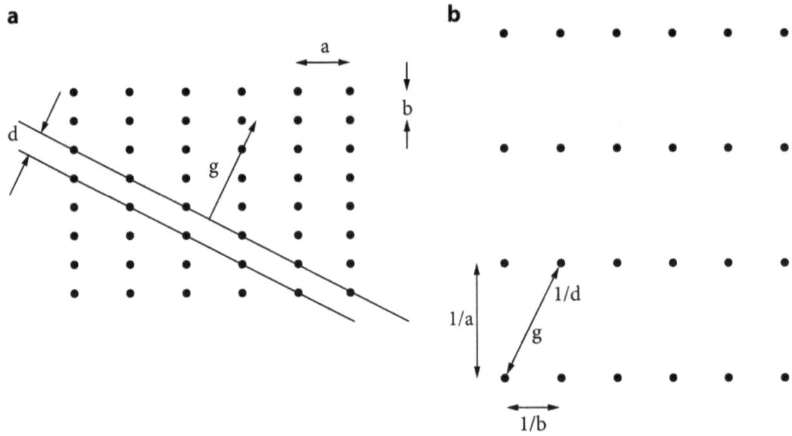

Fig. 12.42 **a** Crystal lattice, **b** reciprocal lattice

length is smaller than the sample thickness. Therefore, a parallel electron beam passing even a thin single crystal specimen is diffracted according to Bragg's law, which describes diffraction by a three-dimensional periodic lattice structure. The Bragg condition for diffraction is

$$\boldsymbol{k} - \boldsymbol{k}_0 = \boldsymbol{g} \,. \tag{12.100}$$

Here $\boldsymbol{k}_0$ is the k-vector of the incoming electron wave with length $|\boldsymbol{k}_0| = 2\pi/\lambda$, $\boldsymbol{k}$ the k-vector of the outgoing diffracted electron wave with the same length $|\boldsymbol{k}| = |\boldsymbol{k}_0|$ (but different directions), and $\boldsymbol{g}$ a reciprocal lattice vector with length $2\pi/d$ normal to the lattice planes with spacing d; λ is the de Broglie wavelength of the electrons.

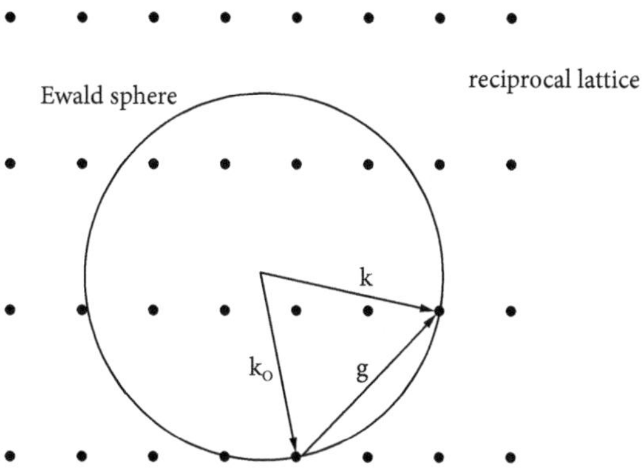

Fig. 12.43 Ewald sphere. Condition for Bragg reflection

Figure 12.42 shows the geometrical situation. The reciprocal lattice is given by the normal vectors $\boldsymbol{g}$ of the lattice planes of a crystal. The length of $\boldsymbol{g}$ is $2\pi/d$ with d as the corresponding lattice plane distance. The sphere with radius k in Fig. 12.43 is the Ewald sphere. Due to the momentum conservation law we have

$$\boldsymbol{g} = \boldsymbol{k}_0 - \boldsymbol{k} . \qquad (12.101)$$

Because $\boldsymbol{k}_0$ and $\boldsymbol{k}$ have the same length, Bragg's law can be written with Eq. (12.101) as

$$\frac{|\boldsymbol{g}|}{2} = |\boldsymbol{k}| \sin\theta \qquad (12.102)$$

or

$$\lambda = 2d \sin\theta , \qquad (12.103)$$

with 2θ the angle between $\boldsymbol{k}_0$ and $\boldsymbol{k}$. The angle θ is known as the Bragg angle. Thus, the directions of the various diffraction orders characterized by the directions of $\boldsymbol{k}$ represent the various lattice plane orientations of the crystal specimen.

12.8 Permeation

For thin film technology permeation is of fundamental importance. The metallic, glass, or rubber walls of vacuum chambers or pipes are more or less permeable to gases. The permeation mechanism can be atomic or molecular. For monatomic gases the permeation gas flow is calculated as:

$$I_P = A\,P\,\frac{p_2 - p_1}{d} , \qquad (12.104)$$

where
I_P is the permeation gas flow,
A is the surface,
P is the permeation constant,
p_1, p_2 are the gas pressures at the two surfaces of the solid body, and
d is the thickness of the solid body.

With diatomic gases dissociation usually arises. We then obtain the following equation:

$$I_P = A\,P\,\frac{\sqrt{p_2} - \sqrt{p_1}}{d} . \qquad (12.105)$$

The permeation constant is gas and material-dependent and increases exponentially with rising temperature.

At ambient temperature and higher temperatures up to some hundreds of degrees for glass and metals – exceptions are silver for oxygen and palladium for hydrogen – the permeation is extremely low and does not have any considerable influence on the pressure ratios in a vacuum recipient, but it is not negligible for elastic seals. The permeation of elastic seals determines the attainable final vacuum pressure beside the pumping speed of the pumps [15]. Permeation is also important for food packaging [16].

The permeation of atmospheric gases through metal walls does not include rare gases (He, Ar, Ne, Kr, Xe), since no noble gas diffuses through metals at any temperature under purely thermal activation. There can be penetration of noble gas ions under a potential gradient, or noble gases can be formed in situ in the metal interior by a nuclear disintegration process.

The permeability of aluminum for hydrogen is very small. It is negligible in all cases except in ultrahigh vacuum chambers, or with thin-walled chambers heated at high temperatures. Copper is a metal with low permeability for all gases, including hydrogen. Nickel has a higher permeability for hydrogen. The permeation of hydrogen through steels increases with increasing carbon content; low carbon steels are thus preferred as vacuum containers. Organic polymers are permeated by all gases, including noble gases.

12.8.1 Measurement Principles

With all measuring instruments, the foil or film to be examined is clamped gas-tightly between two chambers. A chamber contains the measuring gas, and the part of the gas that permeates through the sample is quantitatively measured in the second chamber [25].

In principle, three measuring methods are used:

- Absolute pressure measuring method
- Isostatic measuring method
- Quasi-isostatic measuring method.

12.8.1.1 Absolute Pressure Measuring Method

This measuring method is only used with test gas in two chambers. The permeation process is based on the absolute pressure gradient between the two chambers. Figure 12.44 [25] shows the different absolute pressure measuring possibilities.

Figure 12.45 shows a calibration device with measurable variable pressure and volumes. The permeation constant P can increase the curve when a steady state of the gas passage prevails over the linear part of the pressure.

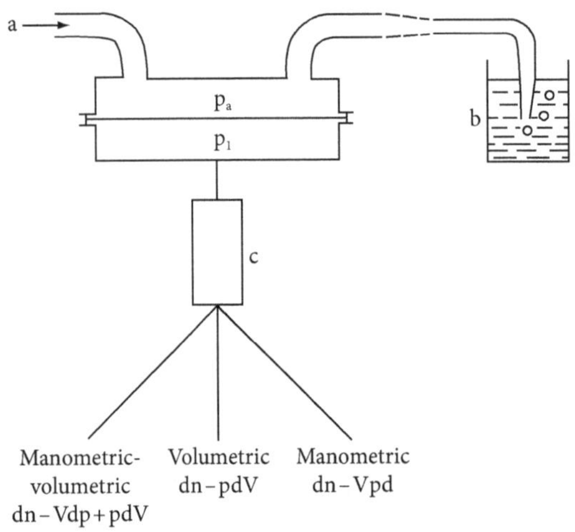

Fig. 12.44 Absolute pressure measuring methods. p_B atmospheric pressure, p_1 pressure in the measuring chamber a test gas, b bubble counter, c detection system

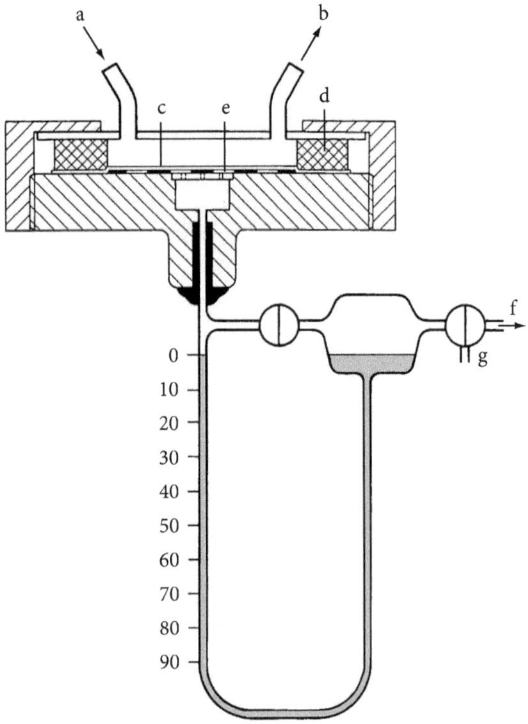

Fig. 12.45 Calibration device with measurable variable pressure and measurable variable volume. *a* gas inlet, *b* gas outlet, *c* sample (foil), *d* seal, *e* filter paper, *f* pump, *g* ventilation

The permeation constant P is calculated with constant volume as:

$$P = \frac{T_0 V_M (p_2 - p_1) d}{T\, p_0 A (p_G - p_1) t}, \qquad (12.106)$$

where

p_0 1013 mbar,
T_0 273 K,
T is the temperature of the measuring chamber,
V_M is the volume of the measuring chamber,
A is the surface of the samples (foils) to be examined,
d is the thickness of the foil to be examined,
p_G is the pressure in the gas supply chamber,
p_1 is the pressure in the measuring chamber at the beginning of the measurement,
p_2 is the pressure in the measuring chamber at the end of the measurement, and
t is the duration of the measurement.

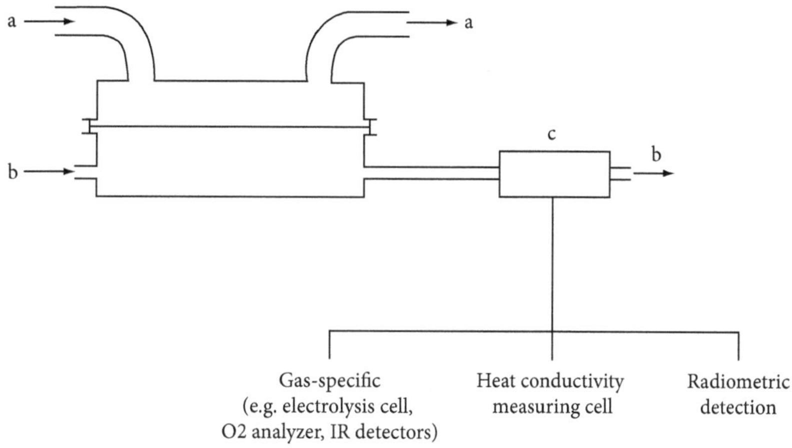

Fig. 12.46 Pattern of a measuring instrument working according to the isostatic principle. *a* test gas, *b* feed gas, *c* sensor

12.8.1.2 Isostatic Measuring Method

With the isostatic measuring method different gases can be in both chambers in addition to the measuring gas. These gases may not react with the sample, and the total pressure must be equal on both sides of the sample. Moreover, a pressure gradient for the measuring gas must exist. Both chambers are in pressure balance with the external atmosphere. Figure 12.46 schematically shows a measuring instrument functioning according to this principle.

12.8.1.3 Quasi-Isostatic Measuring Method

With this method one of the two chambers is locked, clamped between the sample. Pressure equalization with atmospheric air or with the other chamber does not take place. In one of the two chambers a little over- or low pressure develops, which is connected with an easy deflection of the sample. The permeated mass of the gas is determined by means of gas chromatography [25].

12.9 Mechanical Stresses in Thin Films

12.9.1 Introduction

Mechanical stresses in thin films influence the optical, electrical, magnetic, mechanical and chemical properties, and the adhesion.

The stresses σ in thin films consists of the two parts σ_T and σ_i:

$$\sigma = \sigma_T + \sigma_i \,. \qquad (12.107)$$

σ_T are thermal stresses, which develop if the coefficients of dilatation of film and substrate are different and if the measurement takes place at another temperature than with the coating procedure.

We have:

$$\sigma_T = E_S\,(\alpha_S - \alpha_U)\,(T_B - T_M)\,, \qquad (12.108)$$

where

E_S is the modulus of elasticity of the film,
α_S is the middle thermal coefficient of expansion of the film,
α_U is the middle thermal coefficient of expansion of the substrate,
T_B is the substrate temperature during the coating, and
T_M is the substrate temperature during the measurement.

Factors are residual gas insertion into the developing film during the coating, condensation velocity, substrate temperature, film thickness, film structure, and bias [26, 27]. By insertion of residual gas atoms into the film, compressive stress can be converted into tensile stress [28], Fig. 12.47. The stresses are measured as a thin substrate from quartz or glass is coated and the stress due to the deflection of the substrate is measured. The separation of the two parts σ_T and

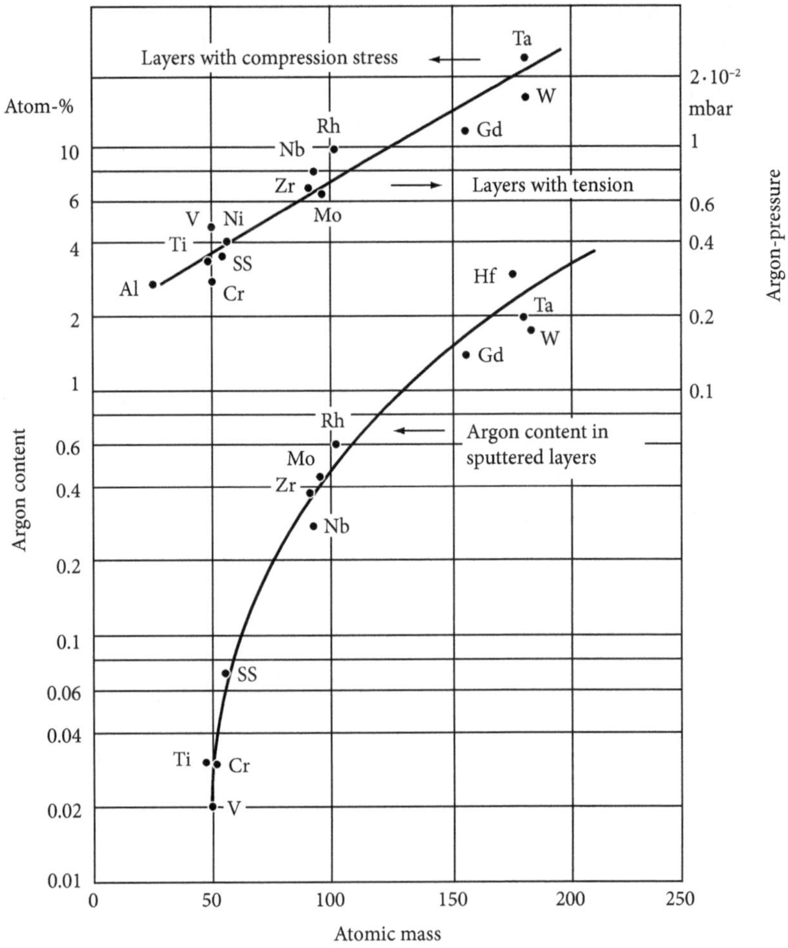

Fig. 12.47 Argon pressure at the transition of compression stresses to tensile stresses. A cylindric magnetron cathode was used for the film generation. The argon pressure with the recording of the lower curve amounted to 1.3×10^{-3} mbar

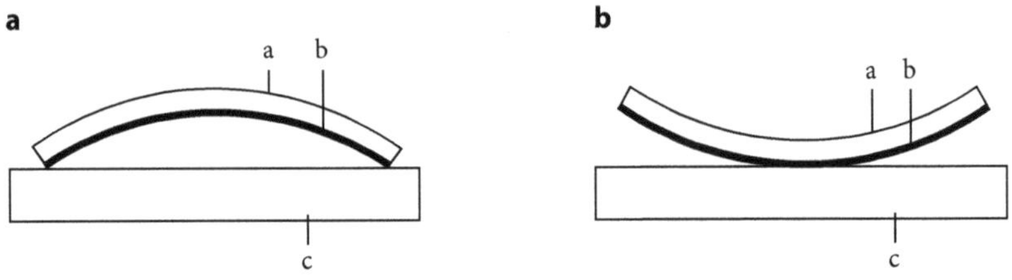

Fig. 12.48 Substrate deformation. **a** Compression stress, **b** tensile stress. *a* Substrate, *b* layer, *c* reference plate

σ_i is possible as follows: σ is measured at the ambient temperature and at the coating temperature.

At coating temperature $T_B = T_M$, is $\sigma_T = 0$ and according to Eq. (12.107) $\sigma = \sigma_i$. For ambient temperature then $\sigma_T = \sigma - \sigma_i$, if σ is the measured stress at ambient temperature. The deformation of the coated substrates can be determined either optically [29], electrically [30], or by weighing [31].

12.9.2 Interference Optical Measurement

The coated and sagged substrate is put with the film side on a flat glass plate. With tensile stress it lays upon the ends, during compression stress in the center of the substrate (Fig. 12.48).

With the arrangement the curve radius R_S of the substrate can be determined according to

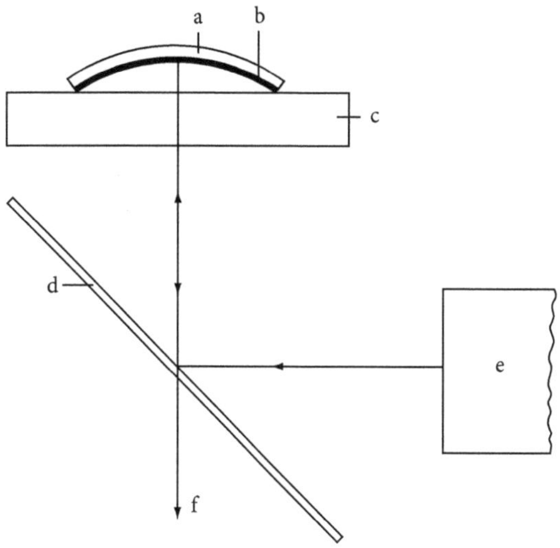

Fig. 12.49 Schema for measurement of film stresses by interference. *a* Substrate, *b* layer, *c* planar glass plate, *d* semitransparent layer, *e* lamp, *f* to the camera

Fig. 12.49; it is

$$R_S = \frac{D_m^2 - D_n^2}{4\lambda(m-n)}, \qquad (12.109)$$

with D_m diameters of the m-th, D_n diameter of the n-th Newton's ring, and λ the wavelength of the irradiated light.

For the total stress of a thin film we obtain:

$$\sigma = \frac{E_S d_S^2}{6(1-\nu_S)\,d_F}\left(\frac{1}{R_{S1}} - \frac{1}{R_{S2}}\right), \qquad (12.110)$$

with

E_S the modulus of elasticity of the substrate,
ν_S Poisson's constant of the substrate material,
d_S the thickness of the substrate,
d_F the thickness of the film,
R_{S1} and R_{S2} the curve radius of the substrate before and after coating.

12.9.3 Electrical Measurements

With this measuring method [30], the substrate is deposited on the averted side of the coating source before application of the analyzing film with an electrically conductive layer, for example, silver.

The substrate and a metallic plate that is opposite it at a distance of approximately 1 mm form a capacitor, which is part of a high frequency oscillation circle. Changes in the capacity develop if the substrate bends because of the internal stresses. These changes of capacity can be detected on the exit of the resonant circuit.

12.9.4 Laser Scanning Method

The measurement principle of the laser scanning method is shown in Fig. 12.50.

A laser beam is turned back by a rotary mirror and a lens in such a way that it always strikes the substrate perpendicularly. The reflected beams converge at an ideally planar substrate surface and strike on one point of the photodiode. If the sample is bent by a coating, then the laser beams do not converge. The deflection, which depends on the curve radius and the respective mirror position, is measured by a photodiode with a position-dependent display [31].

Since the actual surfaces deviate more or less greatly from the ideally flat surface, it is, in principle, necessary to measure the deflection of the laser on the same distance by each substrate before and after coating. The laser scanning method can be used also in situ.

12.10 Hardness Measurements

12.10.1 Introduction

For the hardness measurement of thin films we usually use the microhardness methods after Vickers and Knoop. The test weights lie between 0.02 and 2 N. At ultra micro hardening measurements Saleh and Teich [32] used test weights between 2×10^{-2} and 10^{-5} N.

During the hardness measurement a diamond pyramid indentation is generated and measured optically. The surface area of the Vickers pyramid is quadratic, the angle between the opposite pyramid surfaces amounts to 136°. The Knoop pyramid has a rhombic surface area, whereby the longitudinal diagonal and the transverse diagonal

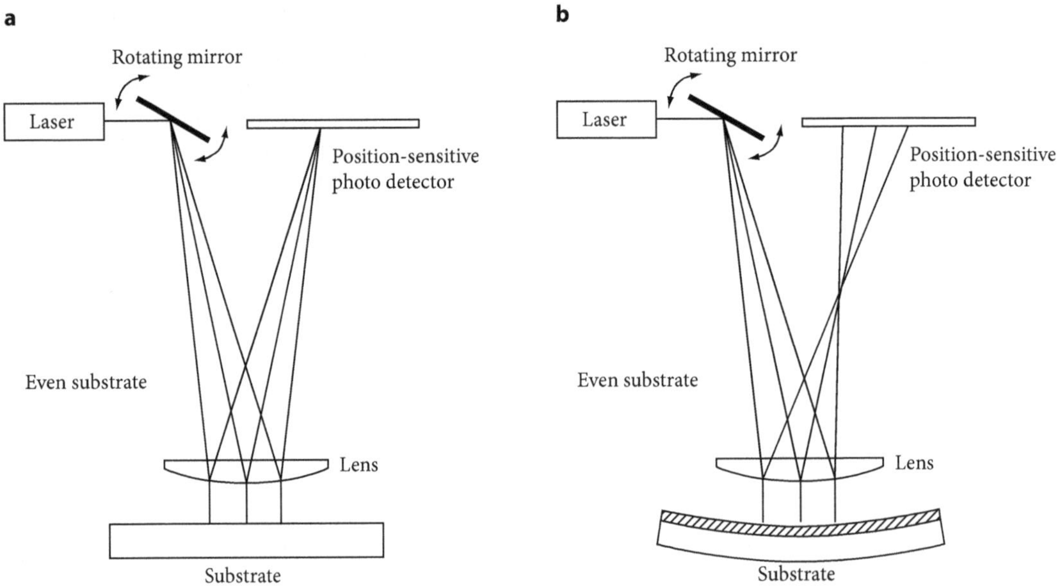

Fig. 12.50 Principle of the laser scanning method. **a** Uncoated, planar substrate, **b** coated substrate with layer stress

of this surface are 7 : 1, and the angles in paired opposed edges 172° 30′ and 130°.

The microhardness (N/mm^2) can be removed from tables [33] or calculated from the following equations:

$$HV = \frac{18{,}192 \times p}{d^2} \tag{12.111}$$

$$HK = \frac{139{,}596 \times p}{d^2}\,. \tag{12.112}$$

By the Knoop method the length of the larger indentation diagonals is to be used for d.

12.10.2 Influence of Measurement Conditions on the Hardness Values

Microhardness measurements of thin films are uncertain and multiple measurements for the calculation of average values are necessary. For critical analysis the test conditions should examine the following.

- Control of the test load.
  The precision with which the load is defined depends on the type of mechanical device used. The absolute error during the microhardness measurement is larger, the smaller the test load.
- Movement.
  With calculation of hardness one accepts that the diamond penetrates perpendicularly to the sample surface; otherwise for a lateral component the impression is larger, and thus the calculated hardness value is larger than the actual one.
- Penetration velocity.
  In the case of too small penetration velocity too small harnesses arise. Usually the penetration velocities lie between 15 and 70 μm/s.
- The size of the impressions.
  With each microhardness measurement the test load and loading period are to be indicated so that the measured values can be verified. According to ISO 4516-1980 (E), the instruction 67-70 HK 0.025/20 means that the Knoop hardness of 67 to 70 with a test load of 0,025 was measured with a test load of $0.025 \times 9.81\,N = 0.245\,N$ and a loading period of 20 s [34].
- Surface finish.
  A microhardness measurement of a heterogeneous material requires as large an impres-

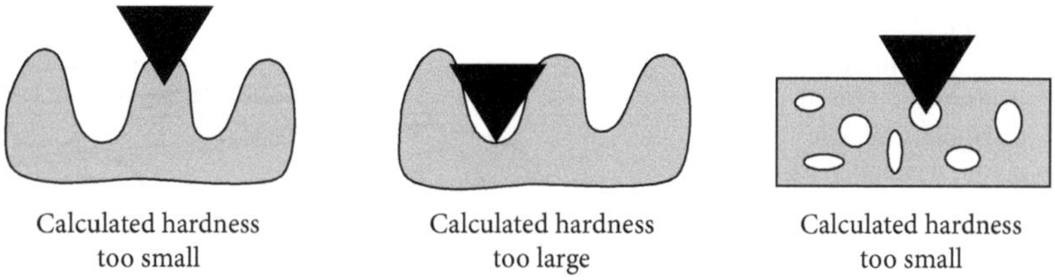

Fig. 12.51 Geometrical parameters during the determination of the hardness of thin films

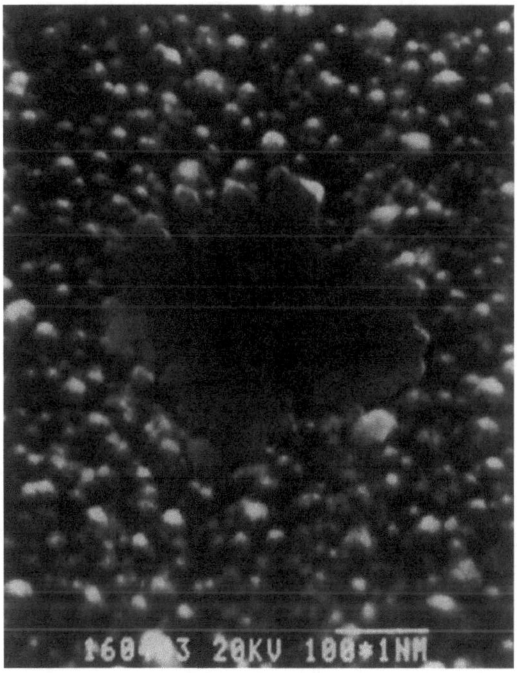

Fig. 12.52 Test impression of a Vickers diamond on a rough TiN layer (REM photo)

sion as possible, since with too small impressions the hardness of individual structural constituents and not those of the entire material are measured. At rough surfaces the measurement of the impression diagonals is inaccurate. If the diamond tip strikes a cuspid of the film, then the diamond tip penetrates less, the calculated hardness value is larger. If the diamond tip strikes a cavity the calculated hardness value is too small (Fig. 12.51).

With porous material the impression resistance also degrades if the diamond encounters a pore. A too small hardness value is measured (Fig. 12.52).

12.10.3 Measuring Sets and Their Application

The measurement of the hardness of films $< 1\,\mu\text{m}$ is not possible with commercial layer thickness measuring instruments. Because of the small film thickness low test loads must be used. For the interpretation the dissolution with optical microscopes is not sufficient. Ultra microhardness measuring instruments in combination with a SEM (scanning electron microscope) are used [35]. The Vickers pyramid (Fig. 12.53), is fixed on a single flat spring. The test load is applied by the electromagnetically deformed double spring arm. The test load is measured by means of resistance strain gauges. Figure 12.54 shows a hardness testing set integrated into a SEM.

The measuring methods described so far have a fundamental error: they measure only the plastic deformation, as it is present after the unloading by the test load. The part of the plastic deformation is test load-dependent and material-dependent.

Figure 12.55 shows the test load dependence by the load-dependent and material-dependent elastic portion, which is small for metals and very large for plastics. In order to measure both the elastic and the plastic part, a hardness measurement under loading is necessary.

The schema of commercial equipment is shown in Fig. 12.56.

The test load is generated over an electrodynamic transformer. Figure 12.57 shows the impression depth as a function of the test load F. If we apply the impression depth as a function of the root from the test load, the result is a straight line, whose slope is a measure for the hardness.

For the slope we obtain:

$$m_{\mathrm{F}} = \sqrt{\frac{1}{H \times 26.43}} \tag{12.113}$$

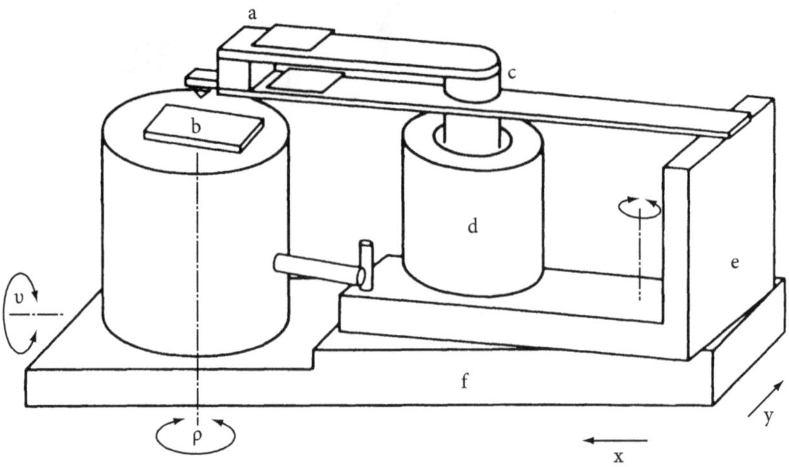

Fig. 12.53 Pattern of an ultra microhardness testing instrument. *a* Strain gauge, *b* sample, *c* double spring, *d* coil, *e* clutch, *f* baseplate

or

$$H = \sqrt{\frac{1}{26.43\, m_F^2}} . \qquad (12.114)$$

By analyzing the impression depth as a function of the image of the test load the parts of the flexible and plastic deformation can be determined (Fig. 12.58).

12.11 Adhesion

12.11.1 Introduction

The adhesion force between a substrate and a film is the work, which is necessary for the complete separation the film from the substrate. The bounding can come by electrostatic, chemical or van-de-Waals forces.

Fig. 12.54 Ultra microhardness testing instrument in a REM chamber

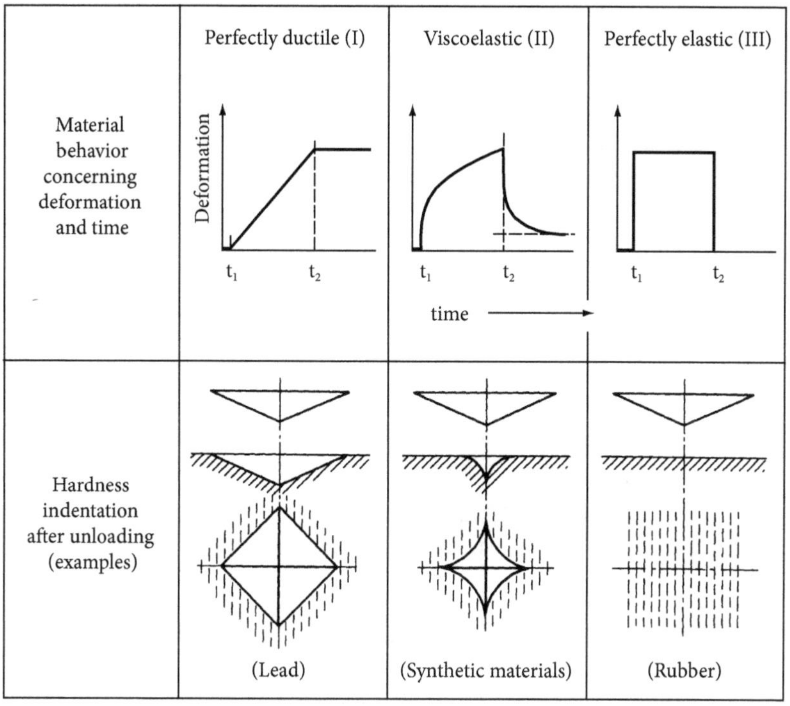

Fig. 12.55 Deformation behavior of different materials during hardness measurement and the shape of the hardness impressions derived from it

The adhesion based on the surface properties, is named basis adhesion. It is independent of the thickness of the together-adjoined bodies, geometry, the temperature, the measuring technique and the effect of outside forces. During the measurement of the adhesion still further factors a role is playing, for example the stresses within the film. The measured adhesion is always smaller than the basis adhesion.

If the separation takes place inside the film or the substrate, then the adhesion force are more largely than the cohesion force, one speak then of cohesion break.

Basic knowledge of the adhesion is only partly present. Only empirical results essentially exist. Reason is when analyzing the adhesion many influencing parameters are not detectable.

Numerous investigations analyze the mechanism of the adhesion, the adhesion by different material combination and the measuring instruments [36].

12.11.2 Measuring Instruments and Measuring Methods

12.11.2.1 Adhesive Tape Test

The tape test is subjective. It depends on the type of tape, the contact pressure on the surface, the impact time of the glue up to the offtake, the roughness of the test surface, and the quickness and the angle at the offtake. The adhesive tape test is suitable only for first orienting analyses.

12.11.2.2 Grid Cut Test

Somewhat more exact results can be obtained by the grid cut test as compared to the tape test. For the grid test a grid with 25 squares is used. The grid cut test is depends on the proportional part of the replaced film. The adhesion is divided into six stages. The estimation criteria are arranged in Table 12.1.

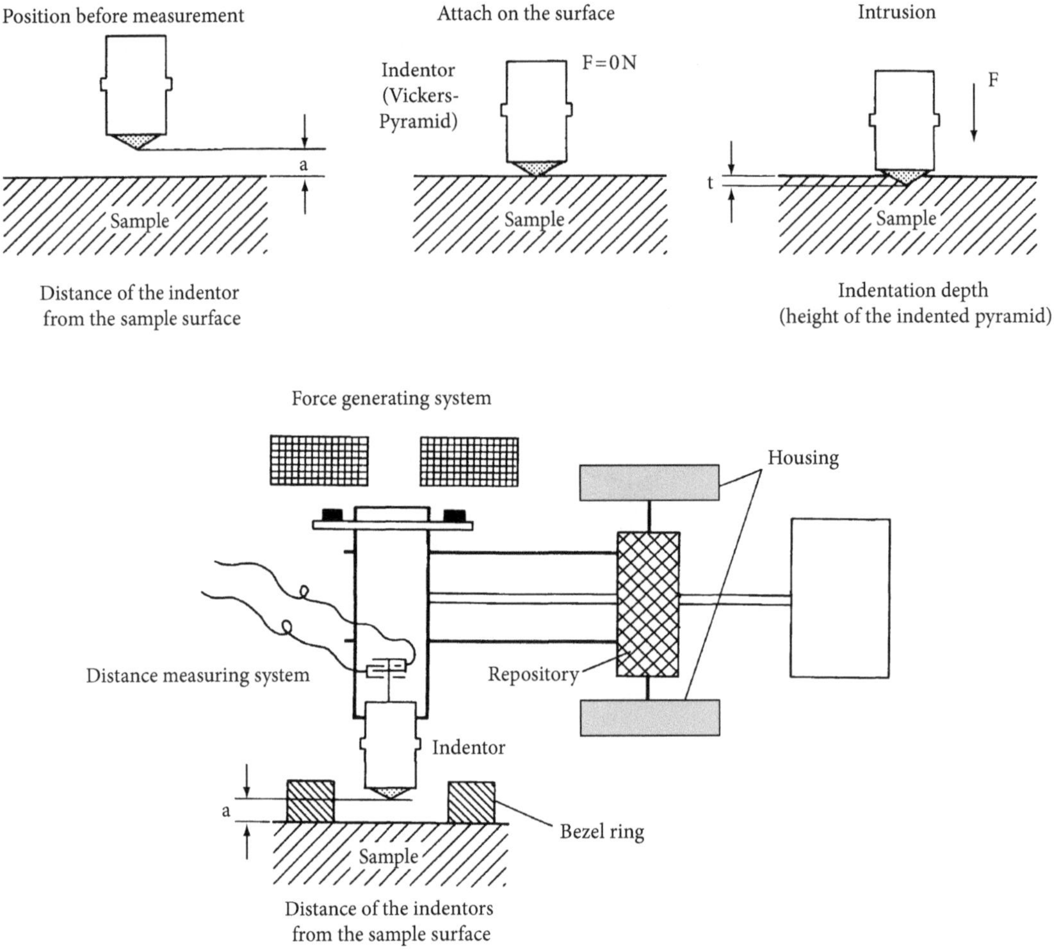

Fig. 12.56 Schema of a hardness testing instrument under loading

12.11.2.3 Quantitative Stripping Test

With the stripping test [37] it is possible to make a quantitative statement about the adhesion of thin films on substrates. A cylinder is fixed on a face surface by means of a contact layer with very good adhesion characteristics on the film present at the substrate. With a measuring device one determines which force is necessary in order to loosen the cylindrical body from the support plate, whereby the separation between the individual layers, within these layers, or also within the substrate can take place. The result is falsified, since with application of the contact layer a change in the internal stress of the film takes place and thereby the adhesion between substrate and film changes.

12.11.2.4 Ultra Centrifuge Test

With the ultra centrifuge test no influence between results of measurement and measuring method occurred. The film is on a rotor, which turns so fast that the film is replaced by the centrifugal force. For the calculation of the adhesion the thickness and density of the film deposited on the rotor are necessary.

Under the condition that the film thickness is negligibly small in relation to the rotor diameter, we obtain:

$$A = \frac{4\pi N^2 R d \varrho}{g}, \tag{12.115}$$

with

A the adhesion,

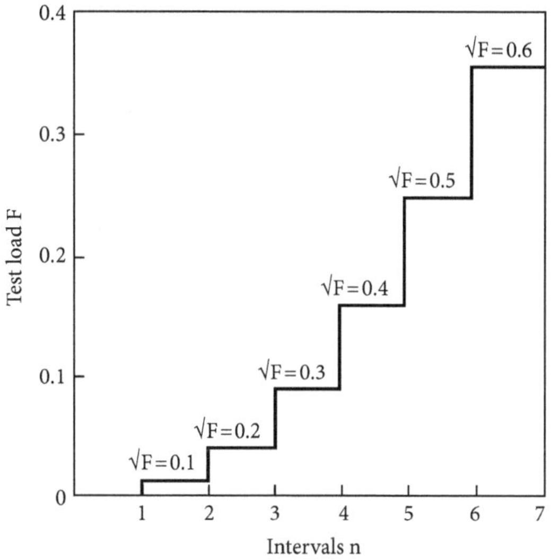

Fig. 12.57 Selection of the test load

R the radius of the rotor,
d the thickness of the layer,
N the rotations for each time unit,
ϱ the density of the film, and
g the gravitation acceleration.

For chromium films on steel substrates with the ultra centrifuge method adhesive strengths were determined with 11,500 N/cm^2.

With the ultra centrifuge test the selection of the rotor material plays an important role. The investigation of very small film masses must be large – due to the centrifugal forces necessary for the replacement of the films – according to numbers of revolutions. Therefore, with this measuring method the limitation is given with the rotor material.

12.11.2.5 Film Stripping in the Magnetic Field

With this method a current flows through the layer, where a magnetic field is overlaid in such a manner that a force turned away by the substrate develops. This method functions only with electrically nonconductive substrates and electrically conducting layers [37].

12.11.2.6 Ultrasonic Test

With an ultrasonic microscope it is possible to analyze films that are not optically transparent. The image contrast is based, among other things, on the elastic properties and the density of the body to be examined.

The lateral dissolution (maximum 0.3 μm) and the impression depth (maximally 5 mm) depend

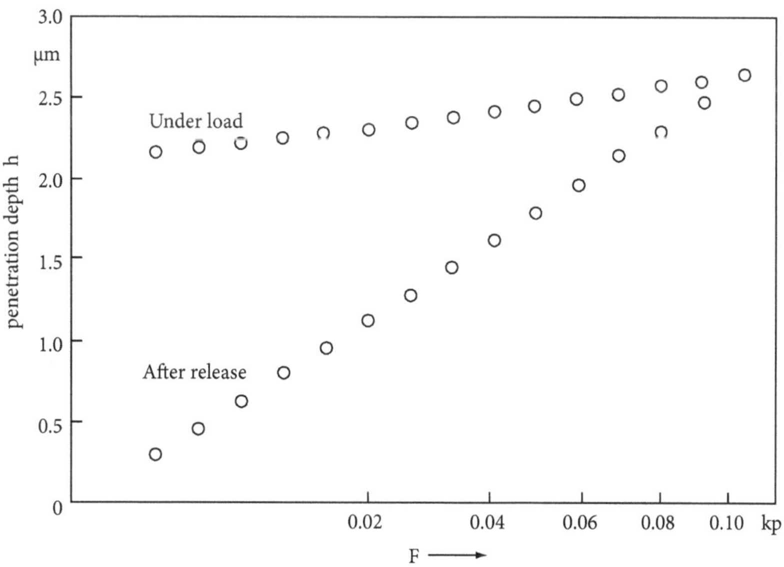

Fig. 12.58 Comparison of hardness under loading and after unloading

Table 12.1 Evaluation criteria of the adhesive strength after the lattice cut test

Grid cut characteristic value	Description	Picture
Gt 0	The edges of cutting are perfectly smooth, no section of the layer chipped off	–
Gt 1	At the intersection points of the grid lines small fragments of the layer delaminated; chipped off surface about 5 %.	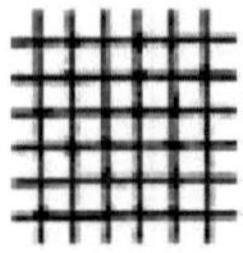
Gt 2	The layer delaminated off edges of cutting and/or at the intersections of the grid lines; delaminated surface about 15 % of the sections	
Gt 3	The layer partly or completely delaminated along the edges of cutting in broad strips; and/or the layer delaminated from individual sections totally or partly; delaminated surface about 35 % of the sections	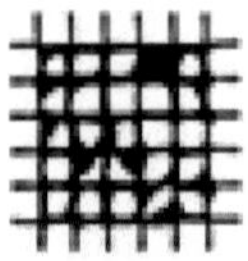
Gt 4	The layer delaminated along the edges of cutting in broad strips and/or of individual sections totally or partly; delaminated surface about 65 % of the sections	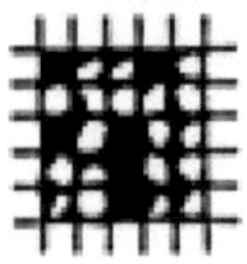
Gt 5	Delaminated surface more than 65 % of the sections	–

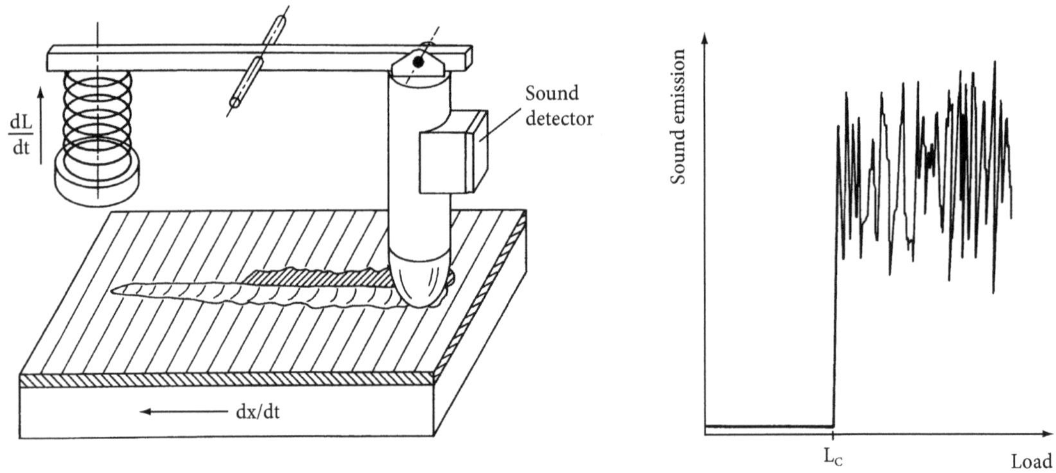

Fig. 12.59 Schema of scratch test equipment and a typical ultrasonic curve as a function of the load; L_c critical load

on the ultrasonic frequency used, the material, and the sound coupling. Usual frequencies are between 10 MHz and 2 GHz [38]. At 2 GHz the maximum impression depth amounts to some micrometers, at 0.1 MHz to approximately 1 mm.

By using the entire frequency range ultrasonic microscopy offers the possibility to sustain a large amount of information without destruction of the inside of the layer, e.g., about different phases, grain boundaries, diffusion zones, chem-

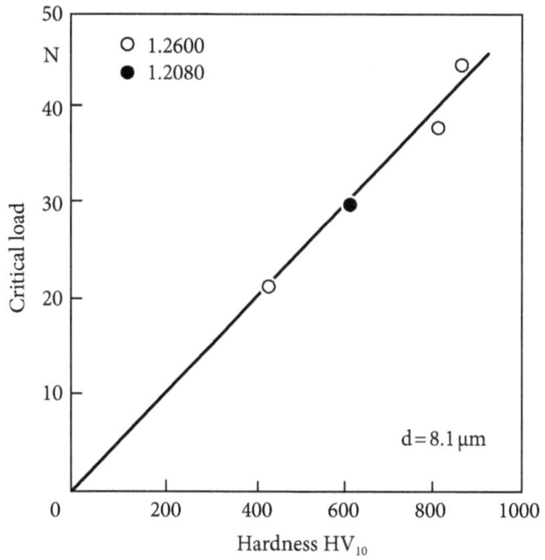

Fig. 12.60 Critical load as function of the base material for two different steel grades; thickness the TiC layer 8.1 µm

ical inhomogeneities, and internal stresses. Apart from layer thickness measurements also qualitative statements about the adhesive strength of a thin film can be made [39].

12.11.2.7 Laser Method

With the laser method a high energy pulsed laser beam is directed toward the not coated substrate side. The generated shock waves arrive the film–substrate interface. Only point-shaped delaminations [37] develop. The threshold energy is a measure for the adhesion.

12.11.2.8 Scratch Test

With the scratch test is a conus-shaped steel shaft, tungsten carbide, or diamond is led in a perpendicular direction across the test sample loaded with a certain weight. The sample is fixed on a cross table and can be shifted. With small loads an elastic deformation develops, which decreases with unloading. With increasing load a plastic deformation is added, which induces a scratch trace. The load is increased until during a critical load the deformation of the film on the substrate appears. The film displacement can be determined by reflection or transmission measurement, by interferometric measurement, by REM, or by dielectric layers on the metallic substrate by measuring the electrical resistance [37]. By means of a sound emission detector the noise of the film rupture can be diagnosed.

Figure 12.59 shows the schematic structure of scratch test equipment and an ultrasonic curve as a function of the load.

With increasing substrate hardness the critical load becomes larger. A typical dependence is represented in Fig. 12.60. The critical load also increases with the film thickness (Fig. 12.61).

The saturation value at the two upper curves is probably due to deformation of the steel substrates.

12.12 Roughness of Solid Surfaces

12.12.1 Introduction

The roughness of a substrate surface is of essential importance for thin film technology. It influences nearly all properties of deposited films. Spikes and holes increase the scattered light and thereby decrease, e.g., with highly reflecting films, the directed reflection. They also affect the stability of monofilms or film systems, because the roughness of substrates is the starting point for film infiltrations with impurities and thus degrade film adhesion and corrosion behavior [40]. Unevenness is also a cause for degradation of electrical film properties, because dielectric films are no longer cover the surface, lose their isolation ability between two metallic layers, and thus short-circuits develop [41]. Finally, also the frictional behavior of thin films is affected by unevenness.

12.12.2 Definitions

The most important parameters for measuring roughness are (Fig. 12.62) the following:

- Reference distance l, the test length used for the interpretation.

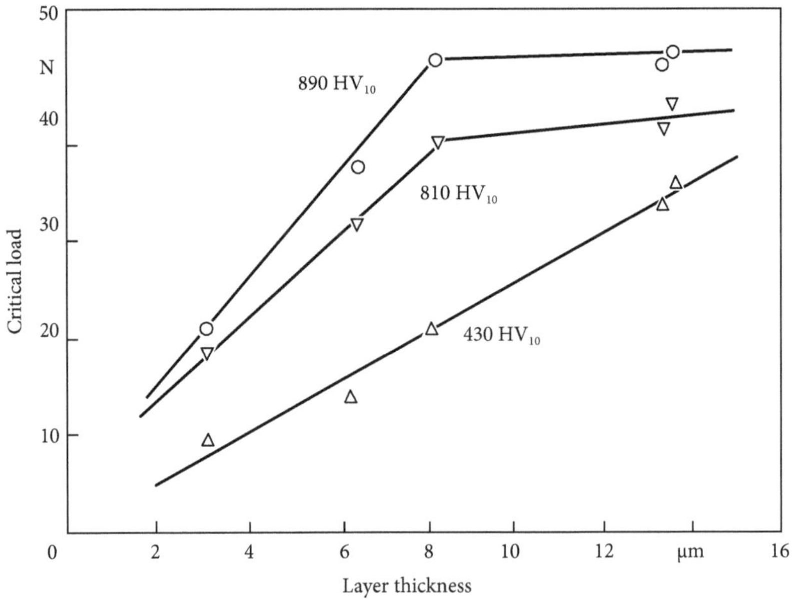

Fig. 12.61 Critical load as a function of the TiC film thickness for steel substrates of different hardness

- Reference profile, the profile, at which the shape deviations are referred.
- Mean profile, reference profile, which is shifted in such a way within the reference distance that the sum of the enclosed of surface part A_{oi} above and A_{ui} below the mean profile is equal.
- Ground profile, reference profile, which is shifted so far perpendicularly to the geometrically ideal profile that it affects the furthest point of the actual profile T away from the reference profile.
- Roughness R_t, distance of the ground profile from the reference profile.
- Smoothing depth R_p, middle distance of the reference profile from the actual profile.
- Average rough value R_a, arithmetic average value of the absolute values of the distances h_i of the actual profile from the mean profile:

$$R_a = \frac{1}{l} \int_{x=0}^{x=1} |h_i| \times dx \,. \tag{12.116}$$

12.12.3 Measuring Instruments and Measuring Methods

12.12.3.1 Mechanical Measurement (Stylus Method)

A method for roughness measurement is the mechanical scanning of the surface. The vertical

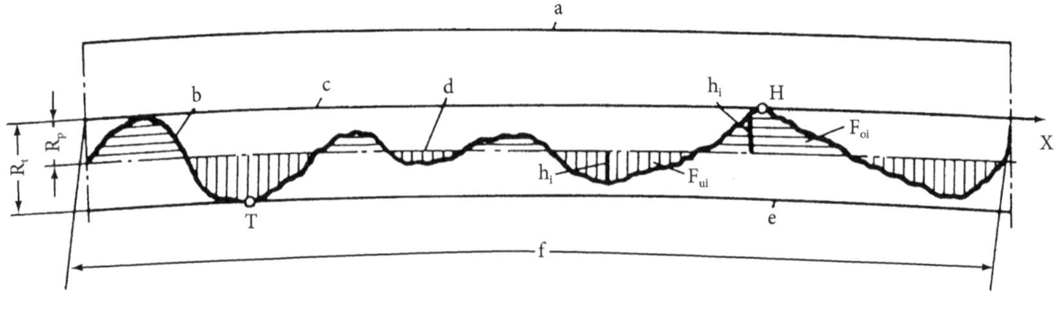

Fig. 12.62 Situation of the reference profile to the actual profile. *a* Geometrically ideal profile, *b* actual profile, *c* point of reference, *d* middle profile, *e* ground profile, *f* roughness profile distance *l*

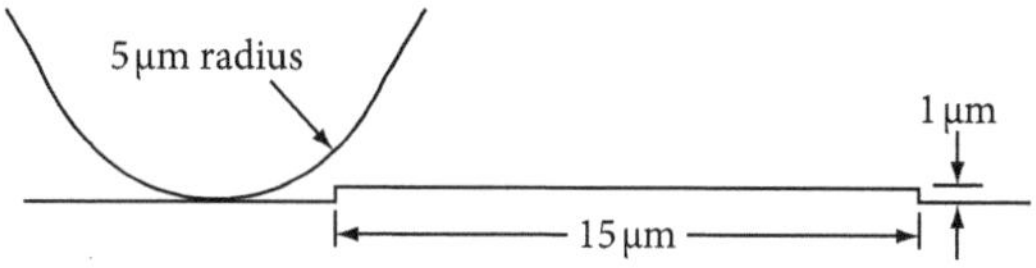

Fig. 12.63 Schema of a tip measurement with exemplary statements of size

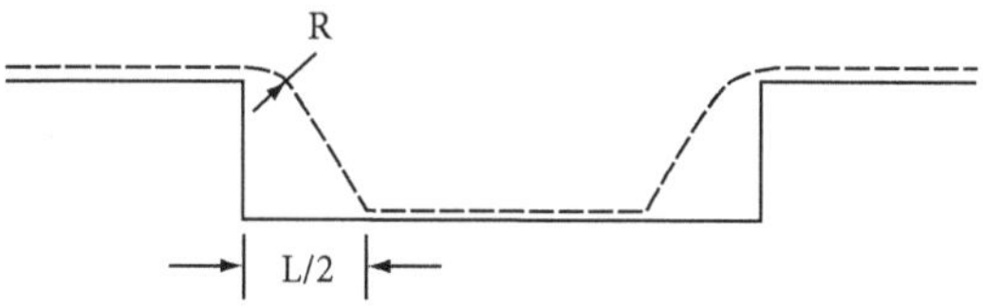

Fig. 12.64 Example of the deviation of a profile taken up with the stylus method from actual geometry, *L* reduction of the profile distance

movement of a stylus is amplified electronically. Deviations from the ideal surface of the substrate or the applied film are recorded. Exact roughness measurements require particularly pointed tips with very small bearing pressures. The function of the tip radius is shown in Fig. 12.63.

Figure 12.64 shows that at an edge profile the measured width of a valley fails more narrowly, where it is larger than the curve radius of the tip.

The advantage of the stylus method is that with accordingly small curve radius of the tip of the scanning pin and with small scanning speed real surface profiles are recorded. For exact measurement a REM must be used [42].

The main disadvantage is the surface contact. A slow measurement is the result. With soft materials this can lead to a destruction of the surface. A further disadvantage is the linear scanning, which does not lead to good reproducibility of the measured values.

12.12.3.2 Roughness Measurements by Microscope Interferometers

The microscope interferometer is based on the Michelson interferometer (contribution optics, temporal coherence function). With a shift of the sample, changes from the sample beam travel some distance and from there the phase opposite the reference beam also changes (Fig. 12.65). At each point of the interferogram the brightness of the wave and thus the level of the surface on the appropriate place of the sample surface can be calculated.

The phase difference of all positions may not distinguish more than one wavelength. Three wavelengths must be used to contain all necessary information, in order for us to be able to determine for each point on the examined surface the phase difference between reference and sample beam. From the calculated phase differences we can analyze the topography of the sample. Unevenness of less than 1 nm can still be dissolved.

12.13 Measuring of Film Thickness

The methods for the measuring of optical thickness are treated in detail in the chapter on optics. In the following, only methods that are not based on optical systems are described.

12.13.1 Film Thickness Determination by Particle Emission

The methods presented in this chapter are usually designated by users and equipment manufacturers as methods for measuring film thickness. However, with these not the film thickness is measured, but the mass allocation of a detected sample. The transformation of a dimension into another one is usually possible without difficulties. With procedural errors, like strong oxygen insertion into a metal film, the mass allocation remains nearly unchanged, the geometrical film thickness can, however, be substantially larger.

12.13.1.1 Beta Backscattering Method

The principle of the beta backscattering method for measuring the film thickness is shown in Fig. 12.66.

A screen separates the measuring surface from the material to be measured. From the screen center a beta sends emitters (electrons), which come from a nuclear disintegration, beta particles (electrons). The electrons are partly elasti-

Fig. 12.65 Principle of the microscope interferometer

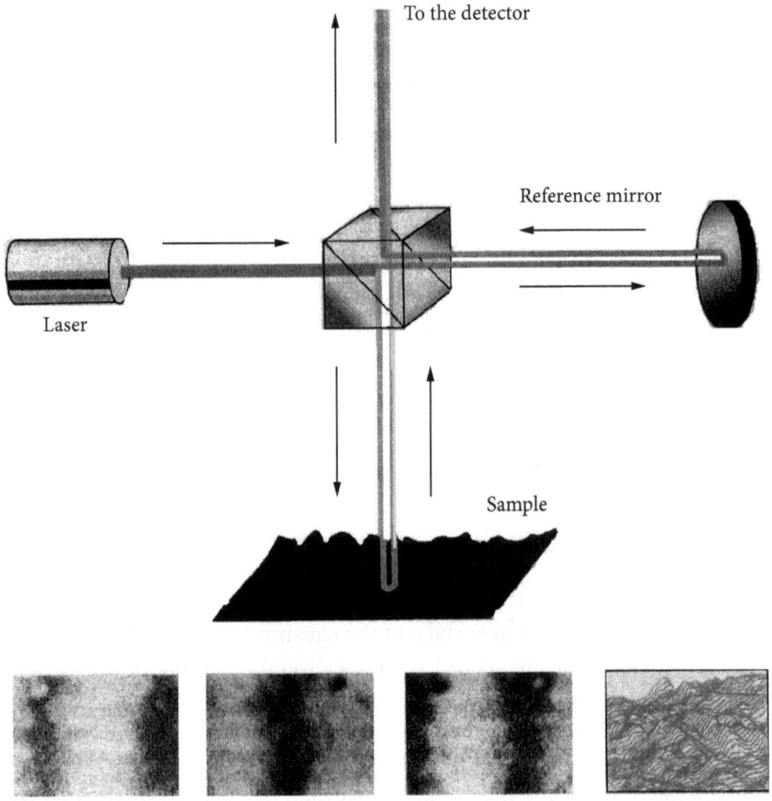

Fig. 12.66 Principle schema of β backscattering thin film measuring equipment

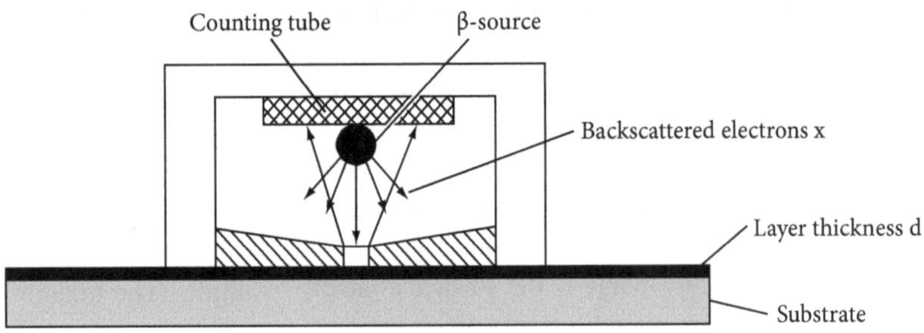

cally backscattered on the sample. Some of the backscattered electrons are collected by a detector. With homogenous samples, the backscatter coefficient depends practically only on the ordinal number Z of the material. The backscatter coefficient is a measure for the film thickness. For conditions with constantly held backscatter coefficient, the counting rate is directly indicated from the reflected electrons. Figure 12.67 shows the connection between measured standardized counting rate and film thickness for gold on nickel.

The standardized counting rate is $x_n = (x - x_0)/(x_s - x_0)$, where x_0 is the counting rate of the uncoated substrate, x_s the counting rate of the substrate with saturation film thickness, and x the counting rate of the sample with interesting film thickness.

An important condition for the applicability of the beta backscattering method is that the backscatter coefficients of the substrate material and the film material are sufficiently far apart. Only then does film thickness variation cause a clearly detectable change of the counting

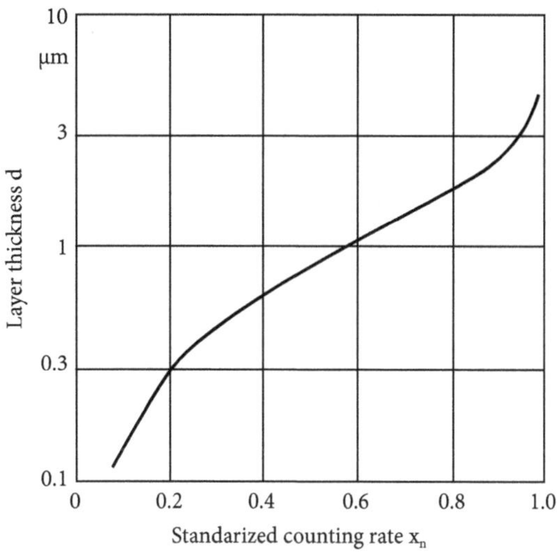

Fig. 12.67 Dependence of the layer thickness on the standardized counting rate with the example of gold on nickel

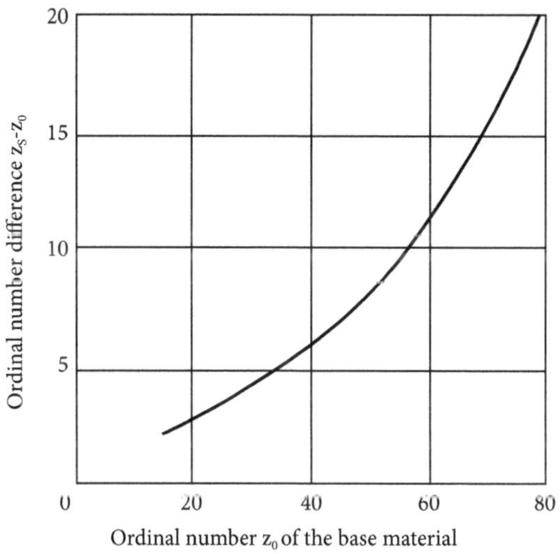

Fig. 12.68 Necessary ordinal number difference as function of the ordinal number of the basic material

rate of backscattered electrons. The necessary ordinal number difference as function of the ordinal number of the basic material is shown in Fig. 12.68.

The beta backscattering method is problem-free with layers of metal on metal substrates. Complications can occur if substrate and/or film materials from alloys, oxides, or composite materials are present. Here it must be proven in individual cases whether this measuring method also gives usable results for the available application [43].

12.13.1.2 Other Film Thickness Measuring Methods

Further methods to measure film thickness are:

- X-ray fluorescence method [44]
- Measuring of film thickness with EDX
- Measuring of film thickness by weighing.

12.14 Determination of Pinhole Densities

Films generated by PVD (Physical Vapor Deposition) processes usually various defects. Some of these are due to the so-called pinholes, holes that go through the film up to the substrate.

Pinholes can result from impurity particles that were on the substrate and were not removed by cleaning before coating. On the other hand, evacuation can whirl up impurity particles, which are deposited on the substrate. A further source of pinholes are coatings on screens or cover plates, which peel off due to high internal stresses or thermal expansion.

A further cause for pinholes is the coating process itself. Thus, with incorrect processing, vaporization can lead to splashes of metal burning holes into the foil. In reactive sputtering processes arc generation can be a problem; temporary discharges between carriers and substrate can lead to punctual defects. Flaking of screens can be minimized, as the screens are held at constant temperature during all phases of the coating process. Another way is to adapt the screen material regarding its thermal expansion behavior on the growing layers.

Pinhole densities on somewhat optically reflecting samples can be determined quickly with laser scanning techniques [45]. A laser beam scans over the sample; its reflected light is measured directly by a detector. If the laser beam encounters a punctual defect, the light is scattered, and the detector registers a decrease of the directly reflecting intensity.

The pinholes in the insulator foils are determined by measuring the capacitance by putting the foil between two capacitor plates. The capacitance is a measure of the pinhole density.

References

1. Franz R, Wiedemann G (1853) Über die Wärme-Leitungsfähigkeit der Metalle. Annalen der Physik und Chemie 8(Bd. LXXXXIX):1853
2. Dua AK, Agarwala RP (1971) Thermal conductivity of thin films of alkali metals. Thin Solid Films 10:137–139
3. Patrick WJ (1966) Measurement of resistivity and mobility in silicon epitaxial layers on a control wafer. Solid-State Electronics 9(3):203–211
4. Maissel LI, Glang R (1970) Chapter 13. In: Maissel LI, Glang R (eds) Handbook of Thin Film Technology. McGraw-Hill, New York
5. Ishikawa Y, Miura N (1991) Physics and Engineering Application of Magnetism. Springer, Berlin
6. Stoner EC, Wohlfahrth EP (1991) A mechanism of magnetic hysteresis in heterogeneous alloys. IEEE Trans on Magnetic 27:4475–3518
7. Foner S (1975) Further improvements in vibrating sample magnetometer sensitivity. Rev Sci Inst 46:1425–1435
8. Flanders PJ (1988) An alternating-gradient magnetometer. J Appl Phys 63:3940–3945
9. Solymar L (1972) Superconductive Tunneling and Application. Chapman and Hall, London
10. Hornauer H, Atommono TM, Röll K (1990) A Kerr magnetometer using Faraday modulation technique. J Magn Magnetic Materials 83:551–552
11. Rohrmann H, Hoffmann H (1989) High-resolution Kerr observation of magnetic domains. Thin Solid Films 175:273–279
12. Hamrle J et al (2007) Huge quadratic magneto-optical Kerr effect and magnetization reversal in the Co_2FeSi compound. J Phys D: Appl Phys 40:1563
13. Wyszecki G (1978) Colometry. In: Driscoll WG, Vaughan W (eds) Handbook of Optics. McGraw-Hill, New York
14. Schrift Nr. 537 316/8.88 der Minolta GmbH, Ahrensburg, Germany
15. Durrant BG (1984) Barrier & Sealing – The prospects for metalized PP, Paper, Film & Foil Converter, pp 61[12-11]
16. Krauss WM (1984) Objective/subjective views of metalized food packaging. Paper, Films & Converter 55:58–60
17. Born M, Wolf E (2003) Principles of optics, 7th edn. Cambridge University Press
18. MacAdam DL (1985) Color measurement, 2nd edn. Springer
19. Pedrotti FL, Pedrotti LS (1993) Introduction to Optics, 2nd edn. Prentice Hall
20. Azzam RMA, Bashara NM (1992) Ellipsometry and polarized light. North-Holland
21. Lawes G (1987) Scanning electron microscopy and x-ray microanalysis. Wiley
22. Reimer L (1985) Scanning electron microscopy: physics of image formation and microanalysis. Springer
23. “Handbook of microscopy” VCH, 1997
24. Saleh BEA, Teich MC (2007) Fundamentals of Photonics. Wiley
25. Tu KN, Rosenberg R (1988) Analytical techniques for thin films Treatise on Materials Science and Technology, vol. 27. Academic Press, Inc., Boston
26. Windischmann H (1991) Intrinsic stress in sputtered thin films. J Vac Sci Technol A 9(4):2431–2436
27. Klokholm E (1968) Intrinsic stress in evaporated. J Vac Sci Technol 6:138–140
28. Orent T (1991) Stress modification in sputtered zinc sulfide and zinc oxysulfide thin films. J Vac Sci Technol A 9(4):2447–2452
29. Rossnagel SM, Glistrap P, Rujkorakarn R (1982) Stress measurement in thin films by geometrical optics. J Vac Sci Technol 21(4):1045–1046
30. Klokholm E (1969) An apparatus for measuring stress in thin films. The. Review of Scientific Instruments 40:1054–1058
31. Nix WD, Mehl RF (1989) Mechanical Properties of Thin Films. Metallurgical Transaction A 20:2217–2244
32. Bangert H, Kamintschek A et al (1986) Ultramicrohardness Measurement on Aluminum Films Evaporated under Various Conditions. Thin Solid Films 137:193–198
33. ISO 409/1: Metallic materials – Hardness test – Tables of Vickers hardness values for use in tests made on flat surfaces – Part 1: HV 5 to HV 100. Ed.: International Organization for Standardization, 409/1-1982 (E).
34. ISO 4516: Metallic and related coatings – Vickers and Knoop microhardness tests – Ed.: International Organization for Standardization, 4516-1980 (E).
35. Fischer-Cripps AC (2004) Nanoindentation. Springer
36. Mittal KL (1975) Surface chemical citeria for maximum adhesion and their verification against the experimentally measured adhesive strength values. In: Lee LH (ed) Adhesion Science and Technology, 9th edn. Plenum Publishing Corporation, New York, pp 129–168

37. Valli J (1986) A review of adhesion test methods for thin hard coatings. J Vac Sci Technol A 4(6):3007–3014
38. Ganz S, Krämer KF (1990) Zerstörungsfreie Untersuchung mit hochauflösender Akustomikroskopie vin 10 MHz bis 2 GHz. Kunststoffe 80:602–606
39. Hoppe M, Bereiter-Hahn J (1985) Applications of Scanning Acoustic Microscopy – Survey and New Aspects. IEEE Transactions on Sonics and Ultrasonics SU-32(2):289–301
40. Kienel G (1992) Plasma-assisted vacuum coating processes. Int J of Modern Phys B 6:1–24
41. Marton D, Fine J (1990) Sputtering Induced Surface Roughness of Metallic Thin Films. Thin Solid Films 185:79–90
42. Bennett JM (1985) Comparison of instruments for measuring step heights and surface profiles. Reprint from Applied Optics 24:3766–3772
43. Ott A (1974) Neue Erkenntnisse bei der radiometrisch-elektronischen Messung von Schichtdicken. Chemie Ingenieur Technik 46(19):805–811
44. Rößiger V, Thomas HJ (1990) Quantitative XRF Analyisis of Surface Layers: Procedure of the Determination of Thickness and Composition. X-Ray Spectrometry 19:211
45. Licari JJ, Enlow LR (1988) Hybrid Microciruit Technology Handbook. Noyes Publications, Park Ridge

Nanoparticle Films

13

H. R. Khan

This chapter gives a short introduction to nanomaterials and a description of the synthesis of nanoparticle films. It further describes in detail two nanoparticle film systems, i.e., magnetic nanocomposites in which magnetic nanoparticles are randomly distributed and nanostructured magnetic nanocomposites in which magnetic nanoparticles are embedded in a structured insulating matrix, and nanoparticle ZnO-based films.

13.1 Introduction to Nanomaterials

In 1959 at the annual meeting of the American Physical Society, [1] in his historical lecture said "There is plenty of room at the bottom" and opened a new field of solid state physics and the discussion of synthesizing materials on a smaller scale. This was the beginning of the fields of nanotechnology and nanoscience. In the 1980s two major developments, the birth of cluster science and the invention of scanning tunnel microscopy (STM by Rohrer and Binning) took place. This technique makes it possible to observe very small material structures down to individual atoms. The atomic force microscopy (ATM) was invented about 5 years after the invention of STM. These developments contributed to the discoveries of many nanomaterials such as fullerenes, carbon nanotubes, nanocrystalline semiconductors, and magnetic nanocrystals and composites.

Particles with sizes range between 1 and 100 nm are defined as as nanomaterials. These nanomaterials may have different geometrical shapes such as globular, plate-like, rod-like, or more complex geometries. Spherical particles smaller than 10 nm are defined as clusters, and the number of atoms in a cluster increases with the diameter of the cluster. As the number of atoms in a cluster increases, there is a critical size of the cluster above which a particular bond geometry that is characteristic of the bulk solid is energetically preferred, such that structure switches to that of bulk. Materials below a critical dimension of 100 nm exhibit very different physical and chemical properties compared to their bulk counterparts. Moreover bulk materials have constant physical properties regardless of their size, but with nanodimensions this is often not the case. Just to mention a few, the size-dependent properties are quantum confinement in semiconductors particles, surface plasmon resonance in some metal particles, and superparamagnetism in magnetic particles.

Nanoparticles have a very high surface area to volume ratio. This provides a tremendous driving force for diffusion at elevated temperatures. The dramatic changes in physical properties of the materials such as melting point, optical properties (such as color due to changes of the optical band gap and wavelength of the optical transition), mechanical hardness, magnetic properties (such as coercivity, permeability, saturation magnetization, magnetic ordering) and chemical properties (such as ionization potential, catalytic activity, or selectivity) take place after

© Springer-Verlag Berlin Heidelberg 2015

H. Frey, H. R. Khan, *Handbook of Thin-Film Technology*, DOI 10.1007/978-3-642-05430-3_13

the transition from the bulk to nanoparticle state of a material.

Nanoparticle films are thermodynamically unstable compared to macroscopic films of larger grain sizes. The direct characterization of the nanoparticle size and shape as well the strain in a nanoparticle film is important after the synthesis of nanoparticle films. This can be done by using high resolution electron microscopy (HTM) or scanning electron microscopy (SEM). These methods are limited because they provide measurements on a very small localized surface area and only provide data of a relatively small number of average grain sizes and shapes. In addition to microscopic techniques, a more conventional method that can be used is the x-ray diffraction technique. This technique is based on the analysis of the line widths of the x-ray Bragg peaks of the (Θ–2Θ) x-ray diffractograms. P. Scherrer [2, 3] correlated the particle size in a film or in a powder to the full width at half of the maximum of the x-ray diffraction line:

$$\Delta\Theta = k\lambda/L\,\cos\Theta\,.$$

Here λ is the x-ray wavelength, k the Scherrer constant (which is 0.9 for the spherical particles), 2Θ is the scattering angle in the (Θ–2Θ) x-ray diffractogram, and L is the volume-weighted average column length and is equal to $0.75D$, where D is the grain diameter of the spherical particle. This way of determining the particle size is applicable to smaller nanograin sizes (larger broadening of the diffraction peaks) and in the absence of microstrain.

The x-ray diffraction technique also allows the determination of the microstrain in the nanoparticles (4). Microstrain in nanoparticles describes the relative mean square deviation of the lattice spacing from its mean value and depends on the grain size. This shows that there is a strain gradient in a nanoparticle, but x-ray diffraction gives a homogenous volume averaged value of the microstrain. The nanoparticle size D and the microstrain ε are related by the equation:

$$\Delta\Theta\,\cos\Theta = \lambda/D + 4\varepsilon\,\sin\Theta\,,$$

where ε and D are determined from the $\Delta\Theta\,\cos\Theta$ versus $\sin\Theta$ graphic plot. It should be noted that not only the x-ray diffraction line width but also the shape of the line changes with the nanoparticle size. Smaller size nanoparticles have higher microstrain. The lattice contraction of the smaller particles is a result of the coordination of the surface atoms and extra pressure is exerted due to the surface stress.

13.2 Synthesis of Nanoparticle Films

The most commonly used methods to deposit nanoparticle films are physical methods such as electron beam deposition (EBD), pulsed laser ablation or deposition (PLD), and ion beam deposition (IBD) or ion-assisted beam deposition (IABD). Chemical methods such as chemical vapor deposition (CVD), sol–gel, and electrochemical deposition techniques are also used to deposit nanoparticle films. These deposition techniques are described in detail in various chapters of the book (Chapters 6, 7, 8, 9).

First to compare the merits and demerits of a physical and a chemical method e.g., the CVD and PLD techniques, the synthesis of nanoparticle diamond films is described in short as an example. The electron beam evaporation technique to deposit the films of magnetic nanocomposites, ordered films of ferromagnetic nanoparticle films using electrochemical technique, and the ion beam deposition technique to deposit the nanoparticles films of ZnO are described below.

13.2.1 Diamond Films Deposited by Chemical Vapor Deposition (CVD) and Pulsed Laser Ablation (PLD) Techniques

CVD is a chemical process to deposit films of materials having polycrystalline, nanocrystalline, or amorphous microstructures. There exist variations of the CVD technique that differ in the chemical reactions and are classified depending on the operating pressure, operating plasma, and the precursor used. Operating pressure-based CVDs are subclassified as atmospheric pressure (APCVD), low pressure (LPCVD) and ultra high vacuum (UHCVD) CVDs. LPCVD works at reduced pressures (mTorr) to reduce the unwanted gas impurities in the deposited film. UHCVD works at low pressures, typically below 10^{-8} Torr. MOCVD is based on using metalorganic precursors. Microwave plasma enhanced (MEPCVD) is a technique in which the chemical reaction on the substrate is enhanced by the plasma and the deposition temperature can be reduced substantially. The CVD technique is widely used to deposit the nanoparticle films of those materials that are available as gases or volatile liquids.

The morphology of the nanoparticle film depends on the reactant gases, their mixing ratio, and the substrate temperature.

A typical example is the nanocrystalline diamond film deposited using CVD and acetylene and hydrogen as chemical precursors. At low partial pressures, highly crystalline primary sp_3-bonded diamond film is obtained. As the partial pressure increases, a more disordered film is formed, which contains the disordered structure graphite in which nanocrystalline diamond clusters are embedded. By adjusting the partial pressure, a high quality nanocrystalline diamond film mostly sp_3 bonding is obtained, which has a smoother surface and improved electronic properties. Nanocrystalline diamond films can be grown from methane or fullerene precursors with or without the addition of argon or molecular hydrogen as carrier gases. The films contain mostly sp_3 bonding with 1 to 2 % sp_2 bonding residing at the grain boundaries.

On the other hand, in the PLD technique, a solid target is irradiated with a laser beam and the material ablated from the target is deposited as a film on a substrate. In this method, no special gaseous precursors as in the case of the CVD technique are required. Diamond films can be deposited using a graphite target. The laser plasma discharge containing C^{3+} and C^{4+} is accelerated and deposited onto a cold substrate or at room temperature. Hard dense nodules of diamond of 10–50 nm in size are formed. Coatings of 2 to 5 μm thickness can be deposited on a variety of substrates, such as steel, whereas poor adhesion and chemical reactions occur in CVD deposition.

Diamond is chemically inert and is transparent from UV to the IR spectrum. This property can be used to deposit protective coatings on transparent substrates. The main problem is the surface roughness of diamond films, which causes attenuation and scattering of the incident light. Nanocrystalline diamond films are much smoother and show enormous potential in diamond optics. Nanocrystalline films have a surface roughness of about 8–40 nm in a 1 um thick film. With this surface roughness, there is a significant reduction of visible light transmission, but there is no reduction in the transmission of the IR spectrum. Moreover, the optical transparency is about 85 % for radiation of wavelengths above 700 nm.

13.2.2 Magnetic Nanocomposites

Magnetic nanoparticles (particles smaller than 100 nm in size) that are randomly or orderly distributed in a nonmagnetic conducting or insulating matrix are defined as magnetic nanocomposites. The matrix may be metallic, semimetallic, semiconducting, or insulating. These magnetic nanocomposites exhibit magnetic properties that are very different from the bulk. In the last years, much effort has been made to synthesize and understand the magnetic and magnetoresistive properties of these composites for practical applications. A multilayer system based on the alternate layers of magnetic and nonmagnetic (metallic or insulating) films of nanometer thickness shows enhanced magnetoresistance and has been called giant magnetoresistance (GMR) after its discovery by Grünberg et al. or tunnel magnetoresistance (TMR). Magnetic nanocomposites may be pictured as three-dimensional multilayer systems. There is a similarity between the magnetic multilayer and magnetic nanocomposite systems. Magnetic multilayers and magnetic nanocomposites are useful in applications such as magnetic information storage media, magnetic sensors and in high frequency devices. Present computers are already equipped with reading heads based on the nanodimensional magnetic multilayers. These magnetic multilayers may be substituted by magnetic nanocomposites, which are much easier to prepare in an economical way.

Much effort has been made to understand the influence of particle size on magnetic parameters such as remanence M_r, coercivity H_c, and saturation magnetization M_s. These are important magnetic parameters from the application point of view. They change drastically with the dimension of the magnetic particles. A typical example is the assembly of iron nanoparticles. A 20 nm Fe particle produced by evaporation exhibits coercivity up to two orders of magnitude higher than bulk Fe, and the saturation magnetization varies between 20–90 % of the bulk, depending on the iron particle size. The maximum coercivity of a ferromagnetic material is observed when the particle size reaches the critical size of a single domain structure. The coercivities of nanoparticles of Fe, Co, and Ni agree very well with the theoretically calculated values based on the single domain structure. Frenkel and Dorfman predicted that magnetic particles consisting of single domains can exist, and Kittel estimated that a transformation from multidomain to a single domain occurs at a particle size of about 15 nm in a common ferromagnetic element. The remanence M_r of a magnetic material arises because the magnetic moments of some parts of the sample have energy barriers that they cannot reverse without the assistance of an external magnetic field or thermal energy. Remanence is related to the intrinsic properties of the material. If the concentration of magnetic particles in a composite varies, then magnetic interaction is affected due to the variation of the interparticle spacings. Therefore, the remanence curves are a guide to the semi-quantitative assessments of the magnetic interactions. The particle size of a magnetic component in a composite may be varied down to a single domain structure, and the composite may show a behavior ranging from ferromagnetic to superparamagnetic and various degrees of coercivities. The magnetic component in a composite may be a single element such as Fe, Co, or Ni or alloys such as FeNi, CoFe, etc. ([5–9]).

After the discovery of GMR [10] in magnetic and nonmagnetic multilayers, as mentioned earlier, nanocomposites based on the disordered distribution of the magnetic nanoparticles distributed in a metallic matrix such as Ag, Cu, etc., also showed a GMR effect [11]. The origin of GMR in these multilayers and composites is from the spin-dependent scattering of the conduction electrons on the magnetic nanoparticles and is size dependent.

If the matrix is insulating, such as Al_2O_3 or SiO_2, then after a threshold concentration of the magnetic nanoparticles, the composite transforms from the insulating to the conducting state. This transformation is caused by the tunneling of the electrons through the insulating barrier separating the two magnetic nanoparticles. The magnetoresistance of these nanocomposites is called tunnel magnetoresistance (TMR) [12, 13] TMR is very sensitive to the magnetic field applied and may be used to make magnetic field sensing devices. Magnetic nanocomposites may be synthesized by PVD techniques, as mentioned earlier. Each technique has its own merits and demerits.

A series of magnetic nanocomposite films such as Al–Co, AlCo–O, Co–Cu, Co–Ag, Fe–Ag, FeNi–Ag, FeRu, and CoFe–Al_2O_3 have been investigated [14–23]. Two magnetic nanocomposite systems are described here: one disordered (magnetic nanoparticles randomly distributed) and the other one ordered (magnetic nanoparticles are arranged orderly) in an insulating matrix such as Al_2O_3.

13.2.3 Disordered Magnetic Nanocomposites $(Co_{50}Fe_{50})_x(Al_2O_3)_{100-x}$

Nanocomposite films composed of nanometer-sized magnetic particles distributed in an insulating matrix and exhibiting the TMR effect have received much attention, not only for their potential applications but also for their value in fundamental understanding. Composite films are traditional model systems for critical percolation behavior studies and the investigation of basic magnetic and transport properties of granular ferromagnets. There is a wealth of information in the literature on this topic [24–32].

In most cases, the films are prepared by magnetron sputtering using composite targets [24–26, 30], tandem deposition [31], or reactive sputtering [32]; other techniques are rarely used. Also the preparation and characterization of SiO_2-based films are often reported [24–26, 30, 31], while data on Al_2O_3-based films are lacking. Recently, a number of publications on the magnetic and transport properties of Al-oxide-based films prepared by reactive sputtering [32] or ion beam deposition [33, 34] appeared. Although the microstructure, transport, and optical properties of electron beam evaporated films have been studied [29], there is a lack of work on magnetotransport for this type of film.

The analysis of microstructure, concentration dependence of magnetoresistance, and correlation between transport and magnetic properties for electron beam evaporated composite films consisting of magnetic nanoparticles of $Co_{50}Fe_{50}$ in insulating matrix of Al_2O_3 are described here.

Thin $(Co_{50}Fe_{50})_x(Al_2O_3)_{100-x}$ composite films with a magnetic volume fraction of $0.07 < x < 0.52$ and a thickness of about 400 nm are deposited on glass substrates at room temperature in vacuum of 10^{-4} Pa using the dual electron beam deposition technique [35]. The thickness and composition of the films is determined using optical measurements of the thickness profiles of the reference films deposited from individual targets and using energy dispersive analysis of x-rays (EDAX). The bulk densities of $Co_{50}Fe_{50}$ and Al_2O_3 are used in the calculations. The crystalline structure of the films is investigated by the (ϑ–2ϑ) Bragg–Brentano x-ray diffraction (XRD) technique using CuK_α radiation and transmission electron microscopy (TEM). The samples for TEM studies are prepared by polishing off the original samples substrate with further ion milling of the film in Ar plasma. The electrical resistance (ρ) and magnetoresistance (MR) are measured by a four-point technique in the temperature range of 5–300 K using the quantum design physical property measurement system. MR is measured in magnetic fields (H) up to 5T in current-in-plane configuration and different geometries:

1. with H in the film plane and parallel current (I; L – longitudinal geometry).
2. H in the film plane but perpendicular to I (T – transverse geometry).

The MR values are determined as $\mathrm{MR} = [\rho(H) - \rho(0)]/\rho(0)$, where $\rho(0)$ and $\rho(H)$ are the resistance of the film in zero and in an applied magnetic field, respectively. Room temperature (RT) magnetic hysteresis loops are measured in the fields up to 5T using the quantum design SQUID magnetometer with H parallel to the film plane.

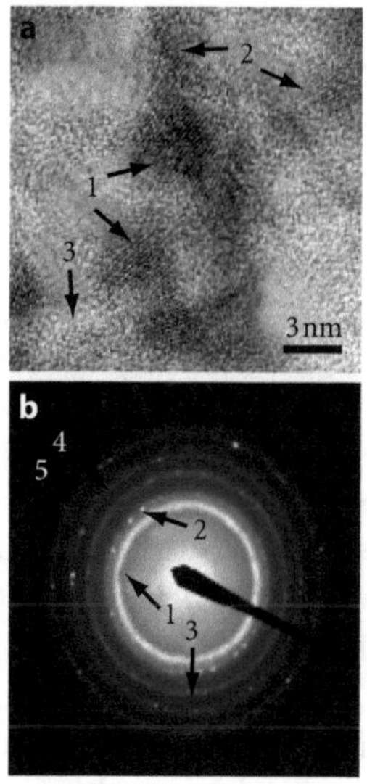

Fig. 13.1 High resolution transmission electron micrograph (**a**) and an electron diffraction pattern (**b**) of a deposited film $(Co_{50}Fe_{50})_{0.16}$–$(Al_2O_3)_{0.84}$

The lattice structure as analyzed by XRD demonstrated that the films are x-ray amorphous. X-ray diffractograms showed only a broad peak that corresponded to ⟨110⟩ body centered cubic (bcc) $Co_{50}Fe_{50}$. A high resolution transmission electron micrograph and the associated electron diffraction pattern of a deposited film with $x = 0.16$ are shown in Fig. 13.1.

Two types of crystalline particles embedded in the amorphous matrix can be observed. The first is spherical crystalline particles with an average size of about 2 nm (Fig. 13.1a, marked as 1). The second is small crystalline regions with the size of a few lattice periods (Fig. 13.1a, marked as 2 and 3). The size of the latter crystalline regions is about 1 nm. The formation of such a structure is typical for composite films and is caused by the low migration rate of oxide clusters that prevent coalescence and recrystallization of the metal phase [26, 27]. Analysis of the diffraction rings allows us to conclude that the film consists of hexagonal close packed (hcp) Al_2O_3 and bcc $Co_{50}Fe_{50}$ particles. The first line can be considered as $\langle 110 \rangle$ hcp Al_2O_3. The second one can be attributed to superposition of $\langle 110 \rangle$ bcc $Co_{50}Fe_{50}$ and $\langle 113 \rangle$ hcp Al_2O_3. The third line is $\langle 211 \rangle$ hcp Al_2O_3. The fourth and fifth lines are the result of the superposition of $\langle 200 \rangle$ bcc $Co_{50}Fe_{50}$ + $\langle 300 \rangle$ hcp Al_2O_3 and $\langle 211 \rangle$ bcc $Co_{50}Fe_{50}$ + $\langle 2010 \rangle$ hcp Al_2O_3, respectively.

Additional structural information can be obtained and the evolution of the magnetic structure with concentration x can be conveniently studied by measuring the magnetic properties. Magnetization measurements demonstrate that the magnetic phase of the films changes from superparamagnetic to ferromagnetic with increasing x. The RT magnetization curves are shown in Fig. 13.2 for films with $x = 0.09, 0.12, 0.16, 0.24$, and 0.50.

No visible hysteresis is found for films with $x < 0.16$. The magnetic phase in the films of these compositions should be considered at RT as an ensemble of small noninteractive magnetic particles in superparamagnetic state [25]. With increasing x a gradual transition from the superparamagnetic to the ferromagnetic state occurs. For $x > x_c$, the shape of the hysteresis loop shows contributions of the ferromagnetic and the superparamagnetic phases. For $x = 0.50$, a narrow loop width is observed and the magnetization is almost saturated in fields higher than 1000 Oe. However, the residual magnetization (remanence) is lower than $0.15M_s$ (M_s is the saturation magnetization). This type of magnetic state transition and the shape of the magnetization loops are typical for composite films [25, 31] but seem to be different from

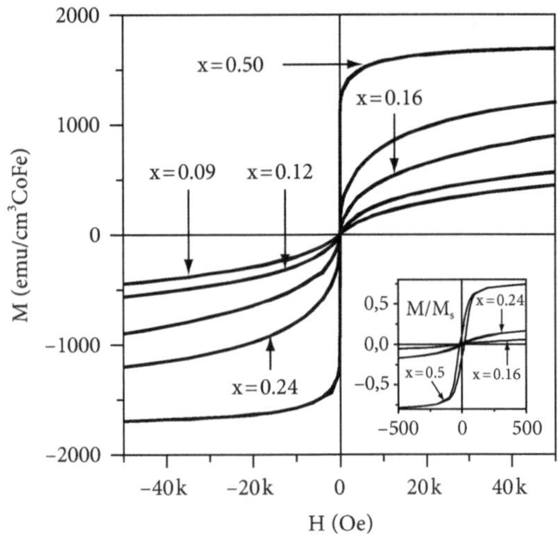

Fig. 13.2 Magnetic hysteresis loops with in plane magnetic field for $(Co_{50}Fe_{50})_x$-$(Al_2O_3)_{100-x}$ composite films of different compositions at T = 300K

those described recently for discontinuous metal-insulator multilayers [34], where a sharp onset of ferromagnetism was observed. Probably this difference is determined by the microstructure of the films and the dissimilarity in the topology of the ferromagnetic phase.

An estimation of the magnetic cluster sizes was made according to the model proposed [24] using a bimodal cluster size distribution and with the assumption of spherical magnetic clusters. In the model fitting, the following equation is used:

$$M = M_s\big(W_1 L(V_1 H/k_B T) + (1 - W_1) L(V_2 H/k_B T)\big), \quad (13.1)$$

where L is the Langevin function, V_1 and V_2 are the two volumes of the particles, k_B is the Boltzmann constant, T is the temperature, and W_1 is the volume fraction of the particles with the volume V_1.

The spontaneous magnetization $M_s \sim 1840$ emu/cm^3 was determined from measurements of the CoFe reference film. This value is within 5 % of the bulk value for $Co_{50}Fe_{49.7}$ permendur alloy. For film of volume fraction $x = 0.09$ the best fit of the experimental curves gave radii of

$r_1 \sim 1.2\,\text{nm}$ ($W_1 \sim 0.14$) and $r_2 \sim 0.4\,\text{nm}$ ($1 - W_1 \sim 0.86$). For films with $x = 0.16$ we had $r_1 \sim 1.8\,\text{nm}$ ($W_1 \sim 0.24$) and $r_2 \sim 0.5\,\text{nm}$ ($1 - W_1 \sim 0.76$). The estimated cluster sizes are consistent with the values of TEM observations. It is to be pointed out that the derived size of the particles does not change significantly with x, while the volume fraction, $1 - W_1$, of the smaller particles decreases. This should be attributed to the coalescence processes that take place with changing x. The progressive coalescence is also reflected in the concentration dependence of the resistance. The analysis of ρ vs. x dependence is used to determine the percolation threshold and is important for the study of the topology of the segregated conducting phase. The general rule of this dependence is the dramatic decrease of the resistance in the vicinity of the percolation threshold [24–30]. For the series of films studied, the percolation threshold was determined from the ρ vs. x curve using the general effective media equation [28]:

$$\frac{(1-x)\left(\rho_{\text{m}}^{1/t} - \rho_{\text{h}}^{1/t}\right)}{\left(\rho_{\text{m}}^{1/t} + A\rho_{\text{h}}^{1/t}\right)} + x\frac{\left(\rho_{\text{m}}^{1/t} - \rho_{\text{l}}^{1/t}\right)}{\left(\rho_{\text{m}}^{1/t} + A\rho_{\text{l}}^{1/t}\right)} = 0\,, \qquad (13.2)$$

where ρ_{h}, ρ_{l}, and ρ_{m} are the resistivities of the insulator phase, the metal phase, and the whole film, respectively; x is the volume fraction of the metal phase, and t is an exponent. Also, $A = (1 - x_{\text{c}})/x_{\text{c}}$, where x_{c} is the percolation volume fraction of the metal phase.

The percolation parameters are determined as $x_{\text{c}} \sim 0.17$ and $t \sim 2.3$; the parameters ρ_{l} and ρ_{h} are estimated to be $\sim 100\,\mu\Omega\,\text{cm}$ and $\sim 150\,\Omega\,\text{cm}$, respectively (Fig. 13.3a). The value of ρ_{l} is comparable to a pure CoFe film deposited under the same conditions for thickness calibration. The resistance of a reference film with polycrystalline structure is measured to be $50\,\mu\Omega\,\text{cm}$. The enhanced ρ_{l} value can be attributed to the nanoscale size and interface effects [26, 27]. The low value of ρ_{h} is caused by disorders in the composite film structure as well as partial alloying of Al_2O_3 with Co and Fe during film deposition [30].

The value of $x_{\text{c}} \sim 0.17$ agrees well with the theoretical prediction for a three-dimensional system consisting of uniformly spherical particles ($x_{\text{c}} = 0.16 \pm 0.02$) [28]. The derived value of t is somewhat higher than the universal value $t_{\text{un}} \sim 1.9$ but is within the range $1.5 < t < 4$ predicted by the tunneling percolation model [36]. It has been shown both experimentally and theoretically that values of the percolation parameters are sensitive to the topology of the conducting phase, i.e., dimensionality, particle size distribution, tendency to form specific (columnar or layered) structures. In continuous granular composites with variable particle size distribution these parameters become nonuniversal and can vary in a wide range. They should be determined for each given case of the percolation structure.

It is to be pointed out that the percolation threshold should not be mixed with the point where the transition from metallic to insulator conductance takes place. It is known [26, 29] that transport in composites can be described using three different regimes of conductance – metallic, transition, and insulator. The electrical transport in metallic regime is due to conduction along the metal phase and is determined by electron scattering against grain boundaries, impurities, and phonons. This regime is observed for the composition above x_{c} when the film consists of insulator inclusions in the metallic matrix. In the insulator regime the conductance is governed by thermally activated tunneling between separated metal particles. Such a conductance mechanism is typical for films of compositions below x_{c}. In the transition regime the conductance is due to a mixture of metallic conductance along the percolation cluster and tunneling along the chain of isolated particles. The transition from metal to the insulator regime of conductance is usually determined at the point where the temperature coefficient of resistance (TCR) is equal to zero. This condition should be fulfilled at x and T where the contributions from tunneling and percolation conduction are equal. In a disordered nanoscale cermet structure the composition at which transition from

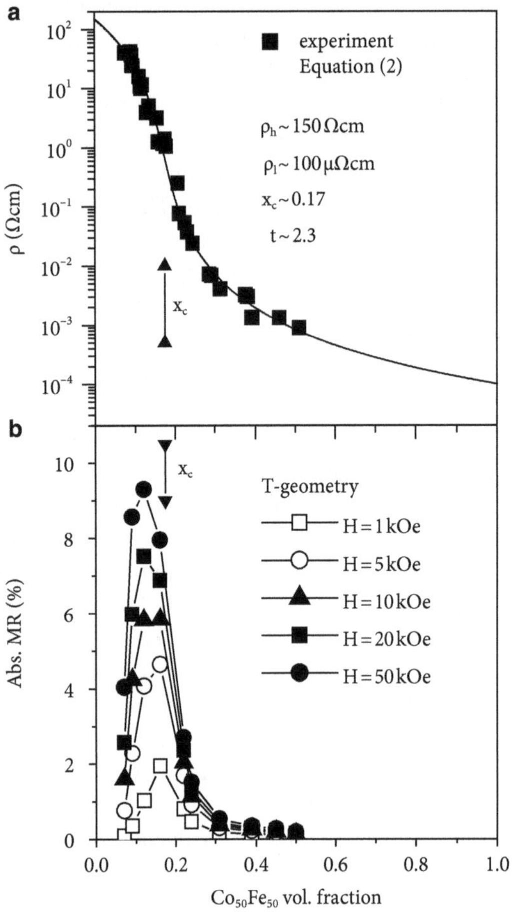

Fig. 13.3 The dependence of electrical resistance (**a**) and magnetoresistance (**b**) for $(Co_{50}Fe_{50})_x$–$(Al_2O_3)_{1-x}$ composite films as a function of composition

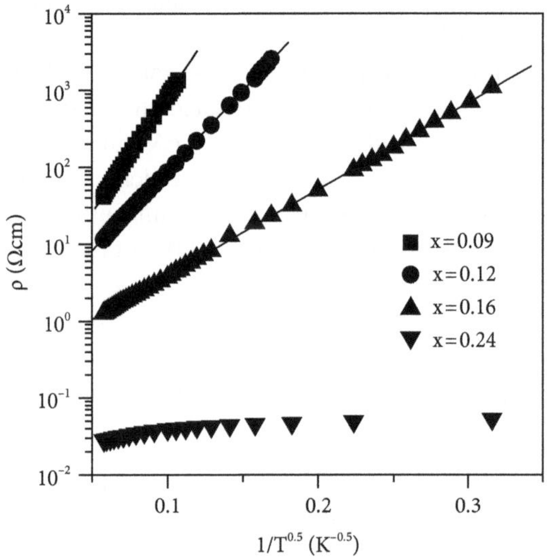

Fig. 13.4 Temperature dependence of resistance for $(Co_{50}Fe_{50})_x$–$(Al_2O_3)_{1-x}$ granular films. *Solid lines* represent results of the experimental data fitting using the $\ln(\rho) \sim (E_c/T)^{0.5}$ model with $E_c = 0.104$ eV, 0.05 eV and 0.015 eV for $x = 0.09$, 0.12, and 0.16, respectively

metal to insulator conductance takes place can be different from the percolation threshold, although for films with larger metallic particles the difference can be small.

The percolation threshold for electron beam deposited Co_x–$(Al_2O_3)_{100-x}$ films is determined from the ρ vs. x curve to be $\sim$ 0.25, while TCR at RT changes sign at $x \sim 0.7$. Contrarily for SiO_2 – based magnetron sputtered films [24–26, 30, 31] the sign change of the TCR and a dramatic increase of the film resistance both occur within a narrow range of $0.4 < x < 0.65$. This fact points to the important influence of microstructure, type of insulator matrix, and deposition technique on the properties of the films.

Using the Al_2O_3 matrix and the electron beam evaporation technique one can produce films with metallic particles of 2–3 nm in diameter [29], while the films with the SiO_2 matrix prepared by magnetron sputtering are characterized by particles of sizes 4–5 nm [25, 30, 31]. However, it is found that in all the cases the TCR at RT changes sign for $\rho \sim 10^{-2}$–$10^{-3}\,\Omega$ cm, which corresponds to the minimum conduction of the metal phase [26, 29]. For our films with $x < 0.16$ negative TCR and an exponential law $R \sim \exp(2(E_c/k_B T)^{0.5})$ indicating tunneling type of conductance is observed (Fig. 13.4).

Here E_c is an activation energy necessary to create a pair of charged particles during a thermally activated and/or bias assisted tunneling process in a system with distribution of particle sizes and interparticle distances [26]. The value of E_c decreases with x due to an increasing number and size of metallic particles and decreasing distances between them. For $x > 0.16$, the conductance still has activation characteristics but the temperature dependence cannot be described by using the above-mentioned law. Negative TCR suggests nonmetallic conductivity such as semiconductor-like or amorphous-like activity [24, 27]. The TCR at RT becomes positive for $x > 0.35$, while it is still negative for temperatures below 270 K. Metal conductance in a wide temperature range was observed only for $x > 45$. Thus in our case, the transition to metallic conductance takes place at similar ρ values (Fig. 13.3a). However, the value of x_c and the composition of the transition to the metallic regime of conductance are lower than reported earlier for Al_2O_3-based films.

The correlation between ρ and MR as a function of x can be seen in Figs. 13.3 and 13.5.

For $x < 0.16$, the TMR effect increases with x and the applied field. The dependencies of TMR vs. H coincide for transverse T and longitudinal L geometries in this composition range, indicating that the system is isotropic (e.g., Fig. 13.5, $x = 0.09$). Moreover, for compositions $x < x_c$ no hysteresis behavior in low fields and no saturation of TMR in fields up to 50 kOe were observed. These data are in good agreement with the magnetization measurements and confirm that the system at RT consists of small superparamagnetic granules.

A maximum isotropic TMR value of 10 % in a magnetic field of 50 kOe and at RT is observed. It is to be noted that the maximum of the TMR vs. H curve is shifted towards lower concentration with increasing magnetic field. At the same time, the saturation of TMR is more pronounced for films with a composition close to x_c. These effects are similar to the recently published data on TMR in $Co_{80}Fe_{20}/Al_2O_3$ discontinuous multilayers [34]. It is shown there that the highest value of TMR in granular systems should be observed in the case of superparamagnetic particles with totally uncorrelated moments. Such conditions will be fulfilled for values of x sufficiently lower than x_c. With an approaching percolation threshold a transition of the magnetic state of the films occurs, i.e., magnetic moments of the clusters become correlated [31, 34]. The differences for the TMR response in L and T geometries in the lower fields in the vicinity of x_c (e.g. $x = 0.12, 0.16$) can also be attributed to formation of long chains of coalesced clusters with strong shape anisotropy. The mag-

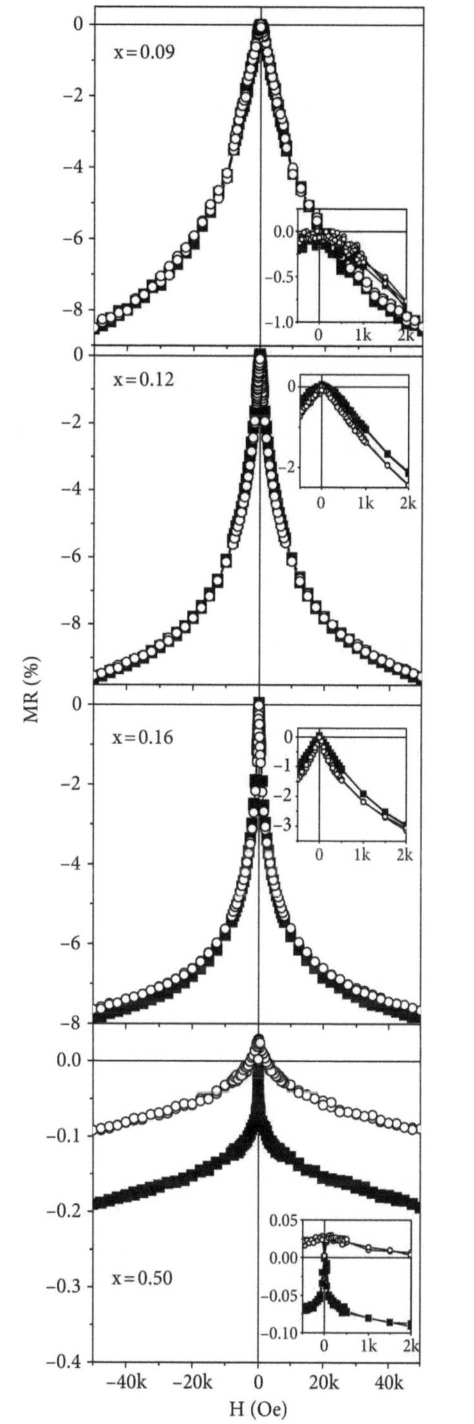

Fig. 13.5 Dependence of magnetoresistance vs. applied magnetic field for $(Co_{50}Fe_{50})_x$–$(Al_2O_3)_{1-x}$ composite films in L(○) and T(■) geometries

netic moments of these chains can be correlated due to dipole–dipole or in-plane inelastic interactions [31, 37, 38].

Such correlations can enhance low field TMR response as it occurs in discontinuous metal-insulator multilayers [33, 37]. As a result, TMR decreases faster in the lower fields, the TMR vs. H curve shows a tendency for saturation in high fields, while the TMR becomes lower compared to the system with uncorrelated magnetic moments. For $x > 0.17$, the percolation cluster already exists in the film, and the value of TMR decreases rapidly with x. For higher x, a component of anisotropic MR is observed at low fields in L geometry (e.g., $x = 0.50$). Such MR behavior in L geometry is typical for ferromagnetic metal films and gives evidence of a continuous ferromagnetic metal phase. In general, the MR vs. x dependence is typical for granular ferromagnets [24, 25, 33, 34], i.e., the maximal values of TMR are observed for compositions below the percolation threshold of the magnetic metal. Nevertheless, it should be pointed out that the value of the percolation threshold and the concentrations at which the maximal TMR effect is observed are much lower than reported earlier. This can be caused by different deposition procedure and by the properties of the Al_2O_3 matrix as well. It is also to be noted that the TMR for the films investigated is higher than reported earlier for magnetron sputtered films under the similar measurement conditions [24, 25, 31]. But it is of order of magnitude for current-perpendicular-to-plane measurements of Co–Al–O films prepared by reactive sputtering [32]. This indicates that the electron beam evaporation technique allows the preparation of better quality films.

To summarize, a general correlation between the microstructure and topology of the metal phase and magnetotransport is demonstrated. The maximum value of the TMR is found for compositions just below x_c. For those x values, films consist of superparamagnetic particles separated by tunneling barriers. Upon forming the percolation cluster, a network of connected magnetic particles with multidomain structure builds and the MR decreases. In certain regions above x_c, the MR value is determined both by the percolation cluster and the dispersed metal particles. With a further increase of x the contribution of the percolated clusters becomes dominant and the films show magnetic and transport properties of ferromagnetic type. With these compositions they should be considered as a magnetic metal matrix with insulator inclusions. A high value of TMR, the absence of hysteresis, and sizable anisotropy in higher fields for $x < x_c$ makes electron beam deposited films of these compositions very attractive candidates for intermediate magnetic fields sensor fabrication.

13.2.4 Nanostructured Nanocomposites Films of Co and CoFe

Nanostructured films offer a possibility of developing new kinds of electronic, optoelectronic, and magnetic devices. In the longitudinal continuous film recording media, the magnetic domains are not isolated and coupled. The present recording density is limited to about 100 Gbits/in^2. Recording density increases at a rate of about 60 % per year. To increase the recording density, the size of the magnetic domains (bits) has to be reduced. Decreasing bit size, while achieving a satisfactory signal-to-noise ratio, requires decreasing grain size. Decreasing the grain size further would make the magnetization of the grains thermally unstable or supermagnetic. To overcome this problem, an alternative solution is to structure the ferromagnetic film in the submicron or nanometer range and isolate the micro/nanoparticles such that they are not coupled electromagnetically. Such nanostructured films based on single domain ferromagnetic disks can be used as high-density storage media. A recording density of 65 Gbits/in^2, consisting of isolated arrays of 50 nm Ni nanowires on a disk obtained by electron beam lithography, has been reported [59, 60]. The lithographic technique consists of three steps: film deposition on a substrate, deposition of photosensitve layer, exposing the photosensitve layer coated film by electron beam or X-rays depending on the size of the required nanoparticles. This technique is expensive and time consuming. Another possibility of nanostructuring ferromagnetic films is the electrodeposition of ferromagnetic metals or alloys in the pores of anodic alumina or other porous templates.

The nanostructuring of ferromagnetic films by electrodeposition in the pores of anodic alumina is described here as an example. Pores of size ranging between 15 and 200 nm diameter can be prepared by anodic oxidation of aluminum or aluminum (Mg, Si) substrates [61]. Nanowires of metals or alloys of length in micrometer range can be fabricated by filling the pore by A.C. electrolysis [62, 63]. These arrays of ferromagnetic nanowires prepared by this technique show distinct magnetic properties, e. g., easy axis of magnetization perpendicular to the substrate plane, enhanced coercivity (H_c), and remanent magnetization (M_r) [64]. The fabrication of nanowire arrays of Co, FeNi, and Ni by electrochemical deposition in the pores of anodic alumina and their structural and magnetic properties [65–67] has been reported. These ferromagnetic nanodisks or wires have a potential of applications such as high-density recording media and sensors, e. g., reading heads in computers or high-sensitivity micromagnetic sensors. The advantage of electrochemical deposition is that wires of length in excess of 2 μm can be fabricated which is rather difficult by lithographic techniques. CoFe ferromagnetic alloys are used as soft magnetic materials in various applications. Arai et al. [68] reported electrodeposition of Co–Fe in the pores of anodic alumina and their magnetic properties. The fabrication of 18 nm diameter arrays of Co and $Co_{90}Fe_{10}$ nanowires by the electrochemical method and the investigation of their structural and magnetic properties are described here as a typical example in this area of research. The structural and magnetic data of nanowire arrays are compared with continuous films of Co and $Co_{90}Fe_{10}$.

An Al (1 % Mg) substrate was anodically oxidized using a H_2SO_4 electrolyte at 0 °C and the D.C. voltage of 25 V. Pores of diameter ~18 nm, the interpore distance of 42 nm, and a pore density of 10^{10} pores/in^2 were obtained. Figure 13.6 shows an SEM image of the anodically oxidized surface. The pore size distribution estimated from 50 pores was about 18 ± 3 nm. Co metal and CoFe alloys were deposited in the pores using a sulfate bath and A.C. electrolysis (14 V, 50 Hz), and arrays of Co and $Co_{90}Fe_{10}$ nanowires of 18 nm diameter and 3 μm height were obtained. The films of 1–2 μm thickness of Co and $Co_{90}Fe_{10}$ were also deposited on the copper substrates using the same electrolyte and a current density of 10 mA/cm^2. The thicknesses of the films and nanowires were determined from the cross-selections obtained by SEM micrographs. The chemical composition is determined by energy dispersive X-ray analysis. The crystallographic structure of Co and $Co_{90}Fe_{10}$was investigated by the X-ray diffraction technique.

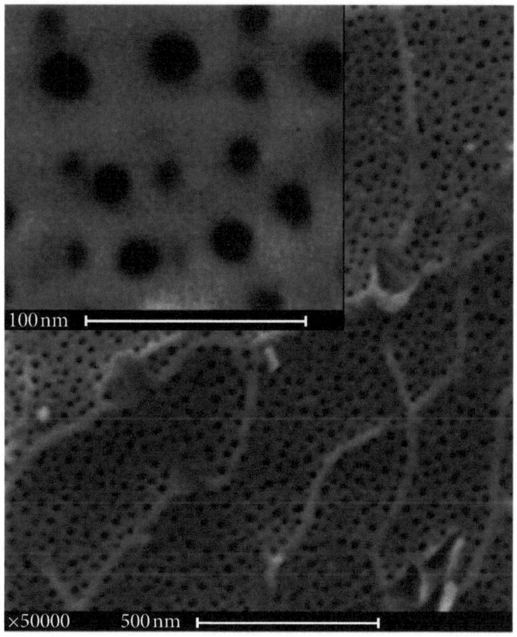

Fig. 13.6 SEM micrograph of anodic alumina surface with 18 nm diameter pores

The Co film on copper substrate consists of face-centered cubic (fcc) and hexagonal close packed (hcp) phases as shown by the X-ray reflections in Fig. 13.7. The $Co_{90}Fe_{10}$ film consists of only the fcc phase. The nanowires of cobalt (18 nm diameter and 3 μm height) consist of only the hcp phase with a strong preferred orientation of ⟨002⟩ reflection (c-axis oriented perpendicular

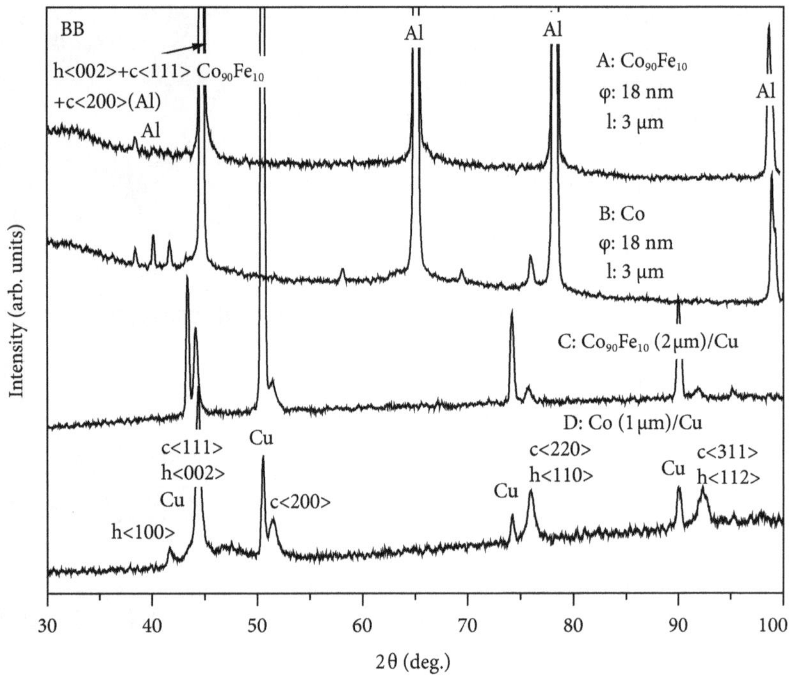

Fig. 13.7 (θ–2θ) Bragg-Brentano X-ray diffraction patterns of **a** $Co_{90}Fe_{10}$ and **b** Co nanowires of 18 nm diameter, **c** $Co_{90}Fe_{10}$, and **d** Co films

to the substrate). There is an overlap of the hcp $\langle 002\rangle$ Co, cubic $\langle 111\rangle$ Co, and cubic $\langle 200\rangle$ Al X-ray diffraction lines because they lie close to each other. The X-ray diffractogram of $Co_{90}Fe_{10}$ shows no cubic phase and overlapping of the preferred crystallographic orientation of the hcp $\langle 002\rangle$ line. In the metallurgical binary Co–Fe phase diagram, hcp CoFe solid solution exists up to 4.5 at% Fe, and the mixture of the hcp and fcc phases and an fcc phase up to 10 at% Fe.

The magnetic hysteresis loops of Co and $Co_{90}Fe_{10}$ films and nanowires are shown in Fig. 13.8 Co and $Co_{90}Fe_{10}$ films show an easy axis of magnetization parallel to the substrate plane. The saturation of the magnetization of $Co_{90}Fe_{10}$ films takes place slowly as compared to the cobalt film. For example, in the case of a cobalt film, saturation of magnetization occurs in a field of 2500 Oe, whereas for $Co_{90}Fe_{10}$ film, saturation of magnetization takes place in a field of 5000 Oe.

The coercivity H_c (⊥) (⊥ – field perpendicular to the substrate) values of the Co and $Co_{90}Fe_{10}$ films are smaller (109–288 Oe) as listed in Tab. 13.1. The hysteresis loop of 18 nm diameter and 3 μm length Co arrays shows that the easy axis of magnetization is perpendicular to the substrate and parallel to the axis of nanowires. The values of H_c (⊥) and M_r/M_s (⊥) of Co nanowire arrays are 1188 Oe and 81 %, respectively. The saturation of magnetization takes place in a field of 2500 Oe. The hard axis of magnetization is parallel to the substrate plane and the H_c (II) and M_r/M_s (II) (II – field parallel to the substrate) values are smaller as listed in Tab. 13.1. The hysteresis loop of $Co_{90}Fe_{10}$ nanowires is similar to Co and the easy axis of magnetization is perpendicular to the substrate but the H_c (⊥) and M_r/M_s (⊥) values increased to 2275 Oe and 96 %. The saturation of the magnetization also takes place at a higher field, e. g., 3500 Oe. These data show large change in the magnetic properties when Co is alloyed with Fe and the lateral dimension of the film is reduced to 18 nm.

13.2.5 Nanoparticle Transparent and Electrically Conducting Zinc Oxide Films

Transparent conducting oxides (TCO) are an essential part of technologies that require both large area electrical contact and optical transmissiv-

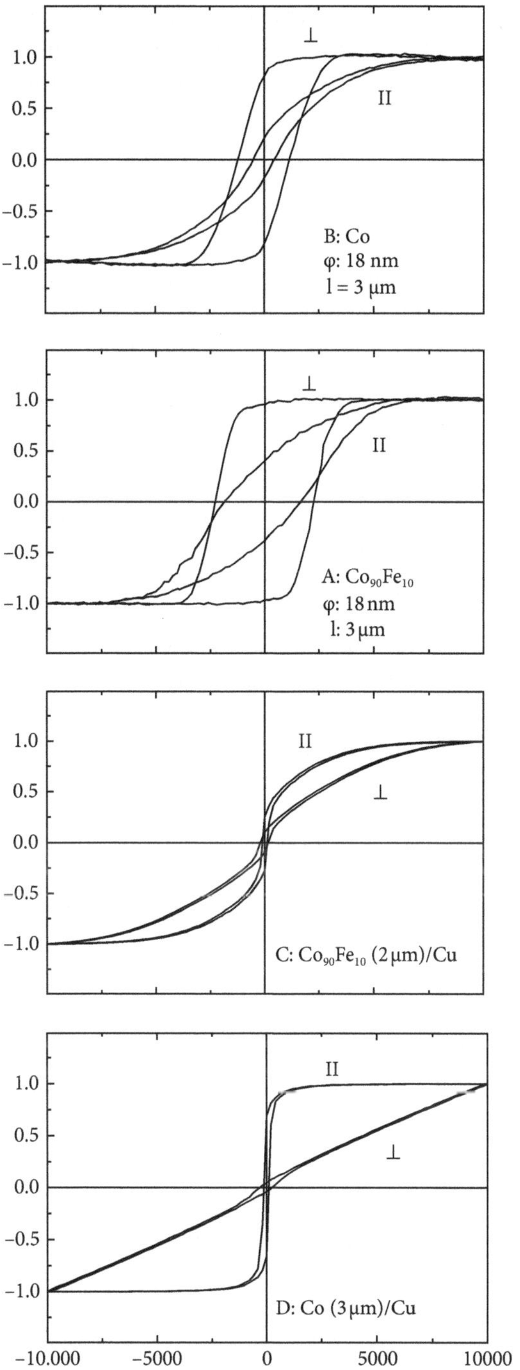

Fig. 13.8 Hysteresis loops of **a** $Co_{90}Fe_{10}$ and **b** Co nanowires of 18 nm diameter; **c** $Co_{90}Fe_{10}$ and **d** Co films

Table 13.1 Coercivity Hc and ratio of remanent and saturation magnetizations (Mr/Ms) of the films and nanowires of Co and Co90Fe10 with magnetic field parellel and perpendicular to the substrate

		H_c (Oe)	M_r/M_s
Co Film (3 μm)/Cu	II	120	0.67
	⊥	288	0.04
Co $\emptyset = 18$ nm; $l = 3\,\mu$m	II	480	0.19
	⊥	1188	0.81
$Co_{90}Fe_{10}$ Film (2 μm)/Cu	II	109	0.26
	⊥	187	0.10
$Co_{90}Fe_{10}$ $\emptyset = 18$ nm; $l = 3\,\mu$m	II	1769	0.39
	⊥	2275	0.96

ity in the visible part of the spectrum. High transparency combined with electrical conductivity larger than $10^3\ \Omega^{-1}\ cm^{-1}$ are required. The applications of the TCOs are in antireflection coatings, infrared reflectors, electrode in flat-panel displays, transparent heaters, antistatic coatings, electromagnetic transparent shielding, and light-emitting and detecting devices (LEDs) [69–72]. The most widely used TCO is indium-tin oxide (ITO). ITO has low resistivity about $10^{-3}\ \Omega$ cm and is 70 %–90 % transparent in the visible part of the spectrum. But the problems with the ITO are that it has an insufficient electrical conductivity for high-density display applications, it is unstable in hydrogen plasma, degrades in high-temperature applications, is corrosion sensitive, and forms hydrate [73–75]. The synthesis and properties of many In- and Sn-based oxide films deposited by different techniques [76–79] have been reported. Currently under development are light weight, flexible display technologies that will require deposition of higher performance (lower resistivity, higher transmissivity, defect free, nanocrystalline, epitaxial) TCOs on heat-sensitive substrates, such as polymers. For the optoelectronic devices, such as LEDs, lasers, etc., there is a need to deposit inexpensive, defect free n- and p-doped nanocrystalline, epitaxial films of transparent semiconductors. Zinc oxide (ZnO) offers a better alternative to ITO. ZnO is a direct band gap semiconductor with a band gap of 3.4 eV. The advantages of ZnO as TCO in many applications are low resistivity of the order of $10^{-4}\ \Omega$ cm obtained by doping with Al and Ga, etc. It is chemically and thermally stable above 500 °C. Transmissivity is larger than 90 % in the

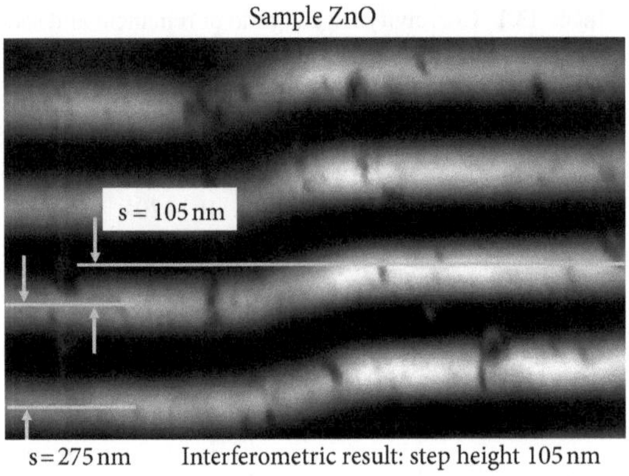

Fig. 13.9 Interference pattern from a ZnO film on glass substrate

visible portion of the spectrum. ZnO films are resistant against hydrogen plasma treatment and do not degrade when processed for solar cell applications [80–84]. Zinc (Zn) metal is cheaper and less toxic. The demonstration of the p-type doping of ZnO has intensified further interest for its applications in LEDs, detectors, and transparent electronics. ZnO has several advantages over GaN, for example, a larger exciton energy of about 60 meV compared to 25 meV for GaN. Therefore, it is possible to build brighter light sources than obtained by GaN. The ion-beam-deposition (IBD) technique offers many advantages over other techniques. For example, the microstructure, texture epitaxy [85–87], and the adhesivity of the deposited films can be controlled by varying the energy of the ions or by ion-assisted IBD from solid targets or gaseous sources. A continuous deposition at lower temperatures on flexible substrates, such as polymers as well as doping the films during deposition, is also possible. In this chapter, deposition of the c-axis-oriented nanocrystalline ZnO- and Al-doped ZnO films on glass substrates by the IBD technique (see Chap. 8) and the investigation of their structural and optical properties are described as an example.

A radio frequency (RF) ion beam source generating ions of 2 keV energy, mixtures of oxygen and argon as working gases, Zn (99.99 % pure), and Zn+Al composite targets are used for the sputter deposition of the films. Glass substrates placed on the water-cooled substrate holder are used for the films deposition.

The measurement of the film thickness d is made using interference microscopy and ellipsometry. The measurement of the optical transmission T in the wavelength range of 400–1000 nm is made by a photometer.

The parameters of the films, such as the refractive index n, extinction coefficient k, and the thickness d are determined using techniques, such as interferometry, ellipsometry, and reflectometry (see Sects. 12.5.1 and 12.5.2). Interferometry measurement provides a rough estimation of the film thickness d. Ellipsometry measurements provide the angles $\Delta(n,k,d)$ and $\Psi(n,k,d)$.

Reflectometry provides, according to Fresnel equations, reflectivity $R(n,k,d)$ and transmittivity $T(n,k,d)$. From these four relations, the three unknown parameters n, k, and d can be determined [88].

Fig. 13.10 SEM micrograph of Al-doped ZnO film – $Zn_{50.4}O_{48.1}Al_{1.5}$

Element	Mass %	Atom %
O	20.2	50.39
Zn	78.77	48.09
Al	1.03	1.52

$Zn_{50.4}O_{48.1}Al_{1.5}$

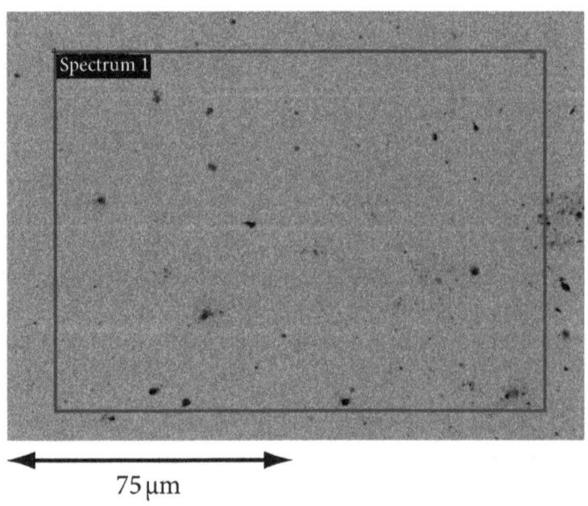

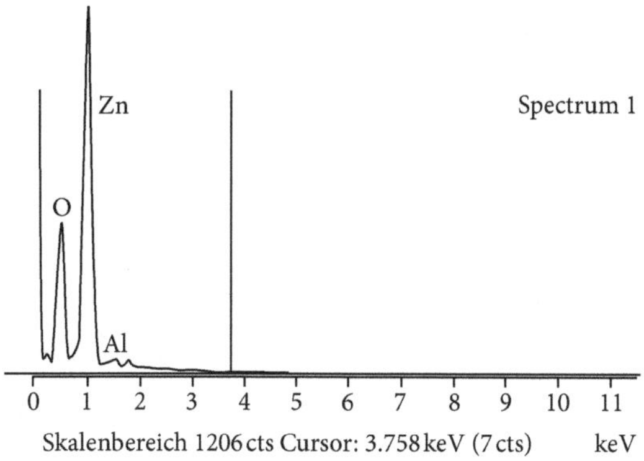

Fig. 13.11 EDXA spectrum of $Zn_{50.4}O_{48.1}Al_{1.5}$ film

Figure 13.9 shows the interferometric measurement of the film thickness

The morphology and chemical composition of the films are investigated using the SEM and EDXA techniques. The lattice structure and the crystallite size are determined by X-ray diffraction.

Figure 13.10 shows the SEM micrograph of Al-doped ZnO film-$Zn_{50.4}O_{48.1}Al_{1.5}$.

The chemical composition of the film as determined from the EDXA spectrum (Fig. 13.11) is $Zn_{50.4}O_{48.1}Al_{1.5}$.

Figure 13.12 shows the X-ray diffractogram of Al-doped ZnO film with ⟨002⟩ axis orientation and the nanocrystalline size of 10 nm. The broad diffraction peak at the $2\,\theta$ value of 23 degree originates from the glass substrate.

Figure 13.13 shows the transmission measurements of two ZnO films of different thicknesses (d) and crystallite sizes (D). A 5 % shift of the band edge is clearly observed due to the difference in crystallite sizes.

In summary, the thin nanocrystalline films of compositions ZnO and $Zn_{50.4}O_{48.1}Al_{1.5}$ are de-

Fig. 13.12 X-ray diffraction pattern of $Zn_{50.4}O_{48.1}Al_{1.5}$ film

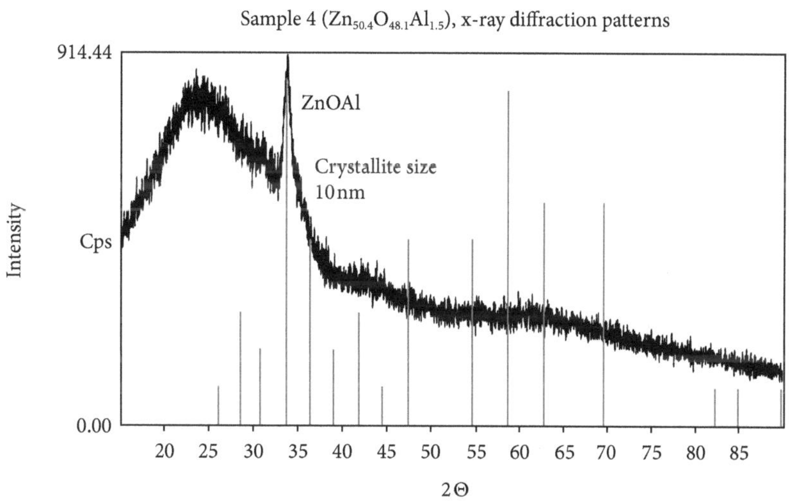

Fig. 13.13 Film 1 (*grey*): $D = 18$ nm, $d = 122$ nm; Film 2 (*black*): $D = 13$ nm, $d = 75$ nm

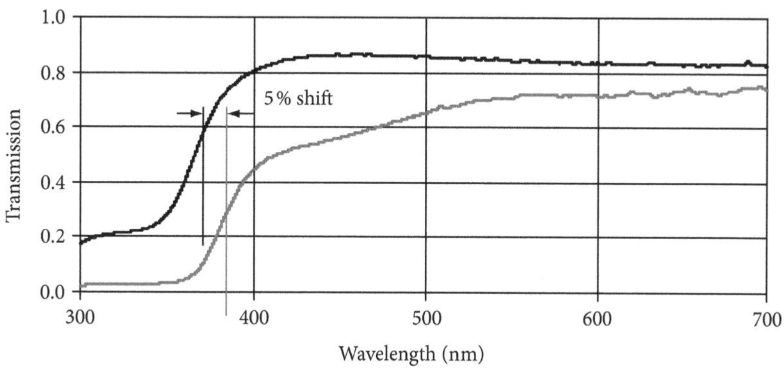

posited on glass substrates by the IBD technique. All films are the c-axis oriented and have an hcp lattice structure.

ZnO film of 122 nm thickness has a crystallite size of 18 nm. The derived n and k values are 2.9 and 0.005, respectively. The maximum transmission in the wavelength range of 580 and 620 nm is 80 %.

ZnO film of 75 nm thickness has a crystallite size of 13 nm. The derived n and k values are 2 and 0.05, respectively. The maximum transmission in the wavelength range of 580 and 1000 nm is 80 %. Both films show two different values of n and low values of k indicating the crystallite size dependence of n and low absorption coefficient ($4\pi k/\lambda$) showing the dielectric character of the films. The transmission versus wavelength curves show a 5 % band edge shift, indicating the band edge dependence on the crystallite size.

$Zn_{50.4}O_{48.1}Al_{1.5}$ film of 143 nm thickness has a crystallite size of 10 nm. The derived n and k values are 2.0 and 0.03, respectively. The maximum transmission in the wavelength range of 570 and 670 nm is 82 %. The resistivity is reduced by Al doping by a factor of 3 compared to pure ZnO films.

References

1. Feynmann RP (1960) Caltech: Engineering and Science 23:22–36
2. Scherrer P (1918) Göttinger Nachrichten 98(2):98
3. Natler H, Schmelzer M, Löffler MS, Krill CE, Fitch A, Hempelmann R (2000) J Phys Chem B 104:246
4. Cullity BD (1978) Elements of X-ray Diffraction. Addison-Wesley, Reading, MA, p 284
5. Tasaki IA, Takao M, Tokunaga H (1974) Jpn J Appl Phys 13:27
6. Kneller EF, Loborsky FE (1963) J Appl Phys 34:656
7. Gong W, Li H, Zhao Z (1991) J Chen, J Appl Phys 69:5119
8. Frenkel J, Dorfman J (1930) Nature 126:274
9. Kittel C (1946) Phys Rev 70:965
10. Grünberg P, Schneider R, Pang Y, Brodsky MB, Sowers H (1986) Phys Rev Lett 57:2242
11. Berkowitz AE, Mitchell JR, Carey MJ, Young AP, Zhang S, Spada FE, parker FT, Hutten A, Thomas G (1992) Phys Rev Lett 68:3745
12. Gittleman JL, Golstein Y, Bozowsky S (1972) Phys RevB 5:3609
13. Mitani S, Fujimori H, Ohnuma S (1997) J Magn Magn Mater 165:141
14. Khan HR (1997) Mat. Res. Soc. Symp. Proc. Vol.438, Page 531; Publisher: Materials Research Society, Pittsburg, Pennsylvania, U.S.A.
15. Khan HR (1997) J Magn Magn Mater 165:297–300
16. Granovsky A, Khan HR, Rudnikov V, Rudnikova M (1997) proc. 2nd International Workshop on Itinerant Electron Magnetism: Fluctuation Effects and Critical Phenomena Moscow, Russsia, Sept. 1997.
17. Prudnikov V, Granovsky A, Prudnikova M, Khan HR (1998) Proc. Itinerant Electron Magnetism Conference. Fluctuation Effects. Kluwer Academic Publishers, pp 353–362
18. Khan HR, Granovsky A, Brouers F, Ganshina E, Clerk JP, Kuzmichev M (1998) J Magn Magn Mater 186:53–56
19. Pohorilyi AN, Kravets AF, Shipil EV, Vovk AY, Kim CS, Khan HR (1998) J Magn Magn Mater 186:87–96
20. Pohorilyi AN, Kravets AF, Shipil EV, Vovk AY, Kim CS, Khan HR (1999) J Magn Magn Mater 196-197:43
21. Vovk AY, Wang J-q, Zhou W, Pgoriliy AN, Shipil EV, Kravetz AF, Khan HR (2002) J Appl Phys 91(12):10017
22. Pohoriliy AN, Kravets AF, Shipil EV, Poyalovsky DY, Vovk AY, Kim CS, Prudnikova MV, Khan HR (2003) Thin Solid films 423:218–223
23. Qadri SB, Keller TM, La Laskoski M, Little CA, Lubitz P, Osafsky MS, Khan HR (2007) Appl Phys A 86:391
24. Honda S, Okada T, Nawate M, Tokumoto M (1997) Phys Rev B 56:14566
25. Gerber A, Milner A, Groisman B, Karpovsky M, Gladkih A, Sulpice A (1997) Phys Rev B 55:6446
26. Abeles B, Sheng P, Coutts MD, Arie Y (1975) Adv Phys 24:401
27. Butylenko AK, Vovk AY (1998) Met Phys Adv Tech 17:593
28. McLachlan DS, Blaszkiewicz M, Newnham RE (1990) J Am Ceram Soc 73:2187
29. Niklasson GA, Tasaki CGGlIA, Takao M, Tokunaga H (1984) Jpn. J. Appl. Phys. 13, 27 (1974) ranqvist. J Appl Phys 55:3382
30. Barzilai S, Goldstein Y, Balberg I, Helman JS (1981) Phys Rev B 23:1809
31. Sankar S, Berkowitz AE, Smith D (2000) Phys Rev B 62:14273
32. Mitani S, Takanashi K, Yakushiji K, Fujimori H (1998) J Appl Phys 83:6524

33. Kakazei GN, Freitas PP, Cardoso S, Lopes AML, Pereira de Azevedo MM, Pogorelov YG, Sousa JB (1999) IEEE Trans Magn 35:2895

34. Kakazei GN, Pogorelov YG, Lopes AML, Sousa JB, Cardoso S, Freitas PP, Pereira de Azevedo MM, Snoeck E (2001) J Appl Phys 90:4044

35. Pohorilyi AN, Kravets AF, Shipil' EV, Vovk AY, Kim CS, Khan HR (1998) J Magn Magn Mater 186:87

36. Balberg I (1987) Phys Rev Lett 59:1305

37. Sankar S, Dieny B, Berkowitz AE (1997) J Appl Phys 81:5512

38. Kravets AF, Lesnik NA, Rokhlin MI, Wigen PE (2001) IEEE Trans Magn 37:2219

39. Dawar AL, Joshi JC (1984) J Mat Sci 19:1

40. Chopra KL, Major S, Pndya DK (1983) Thin Solid Films 102:1

41. Yamamoto S, Yamanaka T, Ueda Z (1987) J Vac Sci Technol A 5:1957

42. Minami T, Sato H, Nanto H, Tanaka S (1985) Jpn J Appl Phys 24:L781

43. Giulio MD, Manno D, Micocci G (1993) Solar Energy Mater Solarces 31:235

44. Morgan DV, Aliyu YH, Bunce RW (1981) Thin Solid Films 77:91

45. Kim H, Pique A, Horwitz JS, Mattoussi H, Murata H, Katafi ZH, Chrisey DB (1999) Appl Phys Lett 74:3444

46. Qadri SB, Skelton EF, Lubitz P, Nguyen NV, Khan HR (1996) Thin Solid Films 290-291:80–83

47. Qadri SB, Khan HR, Skelton EF, Lubitz P (1998) Surf and Coatings Techn 100-101:94–97

48. Qadri SB, Kim H, Khan HR, Pique A, Horwitz JS, Chrisey D, Skelton EF (2000) J Mat Res (Rapid Communications) 15(1):21

49. Qadri SB, Kim H, Khan HR, Pique A, Horwitz J, Chrisey D, Kim WJ, Skelton EF (2000) Proc. ICMC TF-2000, San Diego, Calif. U.S.A. und Thin Solid Films, pp 377–378, pp 750–75

50. Bogamadova AM, Djabrailov AM, Mamedov VV et al (1995) Thin Solid Films 260:19–29

51. Sorg Y, Kim ES, Kapila A (1995) J Elec Mat V24:83

52. Hu J, Gordon RG (1992) J Appl Phys 71(2):880

53. Major S, Banergee A, Chopra KL (1984) Thin Solid Films 122:31

54. Jin ZC, Hamberg I, Granquist CG (1988) J Appl Phys 64(10):5117

55. Khan HR, Frey H (1996) Surface and Coatings Technology 81:307–311

56. Khan HR, Frey H, Banhart F (1996) J of Nuclear Instruments and Methods in Physics research B 112:289–293

57. Khan HR, Frey H, Banhart F et al (1996) In: Hemment PLF (ed) Proc. EMRS Symposium 53 "Ion Beam Processing of Materials and deposition Processes of Protective Coatings". Elsevier, p 289

58. Born M, Wolf E (1975) Principles of Optics. Pergamon Press

59. Chou SY and Krauss PR (1996) Magn J. Magn. Mater. 155:151

60. Miramond C, Fermon C, Rousseaux F, Decanini D, and Carcenac F(1997) J Magn Magn Mater 165:500

61. J.P. O'Sullivan and Wood GC (1970), Proc Soc London A 317:511, Furnex RC, Trans Inst Metal Finish 61 (1983)

62. Goad DGW and Moskovits M (1978) Appl J Phys 49:2929

63. Presten CK and Moskovits M (1988) Phys J Chem 92:2957

64. Shiraki M, Wakui Y, Tokusima T and Tsuya N (1985) IEEE Trans. Magn., MAG-21:1465

65. Khan HR, Loebich O and Rauscher G (1996) Thin Solid Films, 275:207–209

66. Khan HR and Petrikowski K (2000) J Magn Mater 215–216:526–528

67. Khan HR and Petrikowski K (2000) Mater Res Symp Proc 581:245

68. Arai KI, Kang HW and Ishiyama K (1991) IEEE Trans Magn 27:4906

69. Dawar AL and Joshi JC (1984) Mater J Sci 19:1–23

70. Chopra KL, Major S and Pandya DK (1983) Thin Solid Films 102:1–46

71. Yamamoto S, Yamanaka T and Ueda Z (1987) Vac J Sci Technol A5:1957

72. Minami T, Sato H, Nanto H and Tanaka S (1985) Japan Appl J Phys 24:L781

73. Giulio MD, Manno D and Micocci G (1993) Solar Energy Mater Solarces 31:235

74. Morgan DV, Aliyu YH and Bunce RW (1981) Thin Solid Films 77:91

75. Kim H, Pique A, Horwitz JS, Mattoussi H, Murata H, Katafi ZH and Chrisey DB (1999) Appl Phys Lett 743:444

76. Qadri S B, Skelton E F, Lubitz P, Nguyen N V and Khan H R (1996) Thin Solid Films 290–291:80–83

77. Qadri SB, Khan HR, Skelton EF and Lubitz P (1998) Surf Coatings Tech 100–101:94–97

78. Qadri SB, Kim H, Khan HR, Pique A, Horwitz JS, Chrisey D, and Skelton EF (2000) Mater J Res 15:21

79. Qadri SB, Kim H, Khan HR, Pique A, Horwitz J, Chrisey D, Kim WJ and Skelton EF (2000) Proc ICMC TF–2000, San Diego, CA, Thin Solid Films 377–378:750–775

80. Bogamadova AM, Djabrailov AM, Mamedov VV et al. (1995) Thin Solid Films 260:19–29
81. Sorg Y, Kim ES and Kapila A (1995) Elec J Mat 24:83
82. Hu J and Gordon RG (1992) Appl J Phys 71:880
83. Major S, Banerjee A and Chopra KL (1984) Solid Films, 122:31
84. Jin ZC, Hamberg I and Granqvist CG (1988) Appl J Phys 64:5117
85. Khan HR and Frey H (1996) Surface Coatings Technol 81:307–311
86. Khan HR, Frey H and Banhart F (1996) Nucl J Instrum Methods Phys B 112:289–293
87. Khan HR, Frey H and Banhart F (1996) Proc EMRS Symposium 53, "Ion Beam Processing of Materials and deposition, Processes of Protective Coatings" p 289; Editors: Hemment PLF et al., Elsevier
88. Born M and Wolf E (1975) Principles of Optics, Pergamon Press

Index

© Springer-Verlag Berlin Heidelberg 2015
H. Frey, H. R. Khan, *Handbook of Thin-Film Technology*, DOI 10.1007/978-3-642-05430-3

MIX
Papier aus verantwortungsvollen Quellen
Paper from responsible sources
FSC® C105338

If you have any concerns about our products,
you can contact us on
ProductSafety@springernature.com

In case Publisher is established outside the EU,
the EU authorized representative is:
Springer Nature Customer Service Center GmbH
Europaplatz 3, 69115 Heidelberg, Germany

Printed by Libri Plureos GmbH
in Hamburg, Germany